Dora Godan

Pest Slugs and Snails

Biology and Control

Translated by
Sheila Gruber

47 Color Photographs in 7 Plates
125 Black and White Photographs and Drawings

Springer-Verlag
Berlin Heidelberg New York 1983

Dr. DORA GODAN
Federal Biological Research Centre for Agriculture and Forestry
Königin Luise-Straße 19
D-1000 Berlin 33

Translator:
SHEILA GRUBER
Nibelungenstraße 32
D-5300 Bonn 2

Title of the German Edition:
Dora Godan, Schadschnecken und ihre Bekämpfung
© 1979 by Eugen Ulmer GmbH & Co., Stuttgart, Germany
ISBN 3-8001-3044-0

ISBN 3-540-11894-2 Springer-Verlag Berlin Heidelberg New York
ISBN 0-387-11894-2 Springer-Verlag New York Heidelberg Berlin

Library of Congress Cataloging in Publication Data. Godan, Dora, Pest slugs and snails. Translation of: Schadschnecken und ihre Bekämpfung. Bibliography: p. . Includes indexes. 1. Gasteropoda-Control. 2. Gasteropoda. I. Title. SB998.G37G6313 1983 632'.643 83-424.

Typesetting, printing and bookbinding: Brühlsche Universitätsdruckerei, Giessen
2131/3130-543210

Preface

This monograph on pest slugs and snails reviews the problems they create as plant pests in horticulture, agriculture, and forestry, and also as intermediate hosts for parasitic trematodes, cestodes and nematodes which cause worm diseases in man and domestic animals. Here only those vector snails which inhabit flooded or constantly irrigated fields, water storage reservoirs or farm ponds are considered. Reference is also made to the role of pulmonates as carriers of the agents of disease – viruses, bacteria, fungi and worms' eggs, which may be transmitted to man with inadequately cleaned vegetables and fruit. The use of molluscs as indicators of chemical pollution of soil and water, methods of mass rearing of experimental animals and also marking methods are all discussed, as are threshold limit, critical number and the prognosis of damage in plant protection. A classification, an identification key and a systematic check-list of both pest slugs and snails and their predators are provided. The book reviews the biology, physiology, metabolism, reproduction and dispersal of freshwater and terrestrial gastropods, and also the ecological factors which allow a population explosion to occur, thus increasing the likelihood of damage to crops, or the spread of worm diseases of man and domestic animals. The interaction of parasitic worm larvae and their snail hosts is also discussed. These aspects all form an essential basis for the resolution of problems of control.

More than half the book is concerned with the problems of chemical and non-chemical control. This includes biological control, which makes use of parasites, predators and disease, or the manipulation of metabolism, sex transformation, pheromones and the so-called "density factor". Legislative control measures in the USA and Canada against "quarantine pest gastropods" are also described.

No such review of pest slugs and snails and their control has thus far been available. It is hoped that this monograph will fill the gap and furthermore, that questions and problems raised here will stimulate research in economic malacology.

The book will be of interest not only to applied biologists and those working in the field of plant protection, but also to plant pathologists, physiologists, doctors and veterinarians and to students of these disciplines.

The author gratefully acknowledges the great interest of Professor Dr. G. Schuhmann, President of the Federal Biological Research Centre for Agriculture and Forestry, and the financial support of the Deutsche Forschungsgemeinschaft.

The author wishes to thank Mrs. S. A. GRUBER, Bonn, for translating the book from German into English, and both R. ULMER, Ulmer Verlag, and D. CZESCHLIK, Springer-Verlag, whose collaboration made the production of the English edition possible.

Berlin, Spring 1983 DORA GODAN

Contents

Introduction

According to GÖTTING (1974) there are estimated to be 128,000 species of living molluscs and these include approximately 12,000 species of pulmonates, both Basommatophora and Stylommatophora (BOSS 1971). Many of them, however, are accorded little attention in relation to their importance as agricultural pests. International trade has been responsible for worldwide dispersal of many slugs and snails and continues to play this role today. Both pest species and others which, though normally of no special significance to man, can yet become harmful in particular circumstances, are characterized by a great variety of bodily form and behaviour.

Like all animals, as long as they are not considered by man to be either useful or destructive, gastropods play a role in the balance of nature: in the production of humus, in the disposal of plant and animal waste products, in the control of fungi, algae and lichens, and also as predators and parasites, while in their turn the molluscs are hunted and eaten by predators.

Slugs and snails are of importance to man because of the damage they do in agriculture, horticulture and forestry, so that in permanently moist regions or in rainy months and years control measures are necessary. Furthermore, they are of importance in medical and veterinary practice, since they serve as intermediate hosts for certain parasitic worms of man and his domestic animals.

Gastropods not only attract attention as pests, but like oysters, mussels and octopuses, they may be of commercial value (CADART 1955, BRANDT 1968). They are eaten as food in the mediterranean region and in earlier times in Europe were popular fare during lent. Today they are regarded as a delicacy mainly in France and Italy, and to an increasing extent in the Federal Republic of Germany.

Molluscs are also used in scientific research, as in the study of drug action, using molluscan hearts; of hormones, enzymes and antitoxins, especially in relation to immunological haematology, while their shell has served as a model for the study of the process of calcification of bones. Furthermore, molluscs are suitable for determining the degree of chemical pollution of coastal waters, lakes, rivers and land areas, and also as biological indicators of radioactive pollution. *Arion rufus* (L.), for example, is used to determine ^{54}Mn on land, *Physa acuta* Drap. and several Unionidae for estimating ^{137}Cs in water. Finally, freshwater and terrestrial snails are used as ecological indicator species. More recently, their shells have been used in palaeontology for establishing the temperature and salt content of prehistoric seas. Snails are also used in great numbers for the official testing of pesticides (molluscicides).

The following works can be recommended for general information about molluscs: EHRMANN 1956 (1933); FISCHER 1959; FRETTER and PEAKE 1975, 1979; FRÖMMING 1954; GÖTTING 1974; GRASSÉ 1968; HYMAN 1967; JANUS 1958, 1965;

LINDNER 1975; MALEK and CHENG 1974; MEAD 1979; MORTON 1967; MORTON and JONGE 1964; THIELE 1963; ZILCH and JAECKEL 1962.

In common with other animals, innumerable species of molluscs are today threatened with extinction as a result of changes in their environment (deforestation in Africa, sheep grazing in Australia), pollution, especially of seas, lakes and rivers, and increasingly as a consequence of excessive harvesting by commercially organized divers along the coastlines (ANT 1973). As a consequence, many molluscan species have recently been placed on the list of endangered animals (Rote Liste, eds. J. BLAB, E. NOWAK, W. TRAUTMANN, H. SUKOPP, Naturschutz Aktuell, Nr. 1. 1978). Several of the gastropods named in this monograph are described in the Rote Liste as being threatened with extinction in West Germany; yet they may constitute a pest if they occur in sufficient numbers in other countries. The following species are considered to be "in danger of extinction" or "seriously threatened": *Gyraulus laevis* (Alder), *Albida frumentum* (Draparnaud), *Candidula unifasciata* (Poiret), *Helicella itala* (Linnaeus); as "threatened": *Radix auricularia* (Linnaeus), *Viviparus viviparus* (Linnaeus); as "potentially threatened": *Ancylus fluviatilis* Müller, *Candidula caperata* (Montagu), *Candidula gigaxii* (Pfeiffer), *Ceciliodes acicula* (Müller), *Cepaea silvatica* (Draparnaud), *Cepaea vindobonensis* (Férussac), *Cernuella neglecta* (Draparnaud), *Deroceras rodnae* (Grossu and Lupu), *Helicella bolenensis* (Locard), *Helix aspersa* Müller, *Monacha cantiana* (Montagu).

1 The Biology of Gastropods

1.1 Morphology and Anatomy

Snails and slugs (Gastropoda), in common with mussels (Lamellibranchia) and octopuses, cuttlefish and squids (Cephalopoda), belong to the phylum Mollusca. They may be aquatic, terrestrial or amphibious. Marine molluscs have appendages of the body which serve as gills; these may lie within the mantle cavity, but in some species the cavity has ceased to function as a gill chamber. Aerial respiration is found in terrestrial and some freshwater gastropods, in which the vascular wall of the mantle cavity acts as a lung. These species belong to a group of molluscs, the Pulmonata, which originated in the sea and subsequently moved onto land. Later, in the course of evolution, several such species returned to the water, yet retained the lung as a respiratory organ; they live today on river banks, in ponds, lakes, streams and on agriculturally flooded fields. A survey of the general biology of molluscs is given by PURCHON (1977).

The **body** of gastropods can be divided into four sections: the *head* with mouth, sensory tentacles and simple eyes; the *foot* with its creeping sole; the *mantle*, a fold of skin over the back, which forms a respiratory cavity, envelops the visceral mass and which has a thin layer of tissue which, together with the epithelium of the adjacent upper surface of the body secretes the shell; and lastly the *visceral mass*, containing the main internal organs of the body, lies within the shell (Fig. 4) or, in the case of slugs, under the mantle shield (Figs. 1 and 5).

The **skin** is involved in regulation of body water content; it consists of a simple layer of epithelial cells which are ciliated. It contains sensory cells which are concentrated in particular regions, the tentacles; the lips of the mouth; the anterior dorsal and lateral surfaces of the foot; the edge of the respiratory aperture or pneumostome and the mantle (ZYLSTRA 1972, and others). In terrestrial pulmonates, the skin is formed by an epithelium which is as well supplied with enzymes and mucous cells as the epithelium of the vertebrate intestine (NEWELL 1973, 1977). In freshwater snails, like *Lymnaea stagnalis* (L.), respiration takes place through both the skin and the lung. The skin of terrestrial snails and slugs contains special glands which secrete a covering of slime, which serves as a protection against desiccation. In addition, mucus, on which the animal crawls, is produced by the pedal gland which opens at the front of the foot. The skin of terrestrial molluscs is of importance as the structure which is exposed to contact with pesticides; for this reason, experimental work on it, like the study of REYGROBELLET (1975), is desirable.

On the head, the large retractable tentacles are the most noticeable feature. The eyes appear as black dots; they are found at the tips on the tentacles in the Stylommatophora (which can retract their tentacles) and at the base of the tentacles in the Basommatophora (which have non-retractable tentacles) (Fig. 2). The eyes are ca-

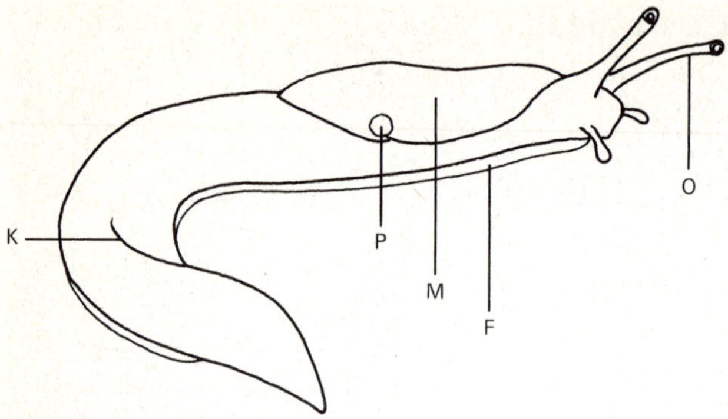

Fig. 1. Diagram of *Limax*: *K* keel on the back; *P* pneumostome; *M* mantle; *F* foot, foot fringe; *O* optic tentacle

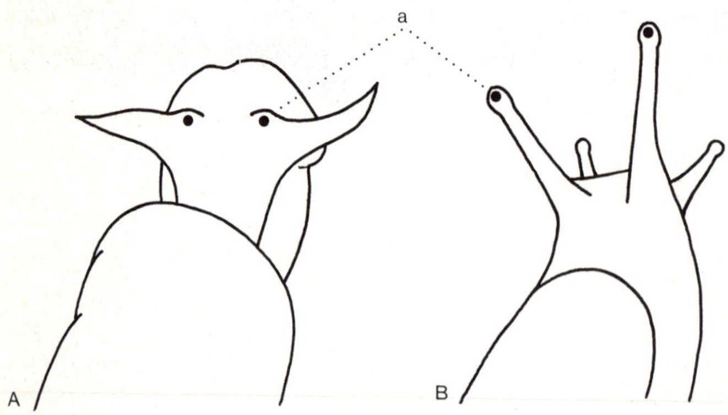

Fig. 2. Diagram of the head: *A* the eye at the basis of the great tentacle (Basommatophora); *B* the eye at the tip of the larger posterior or optic tentacle (Stylommatophora), *a* eyes

pable of regeneration if experimentally destroyed, as has been shown in *Helix aspersa* Müller (Eakin and Miller Ferlatte 1973). The structure of the molluscan eye, which is particularly well developed in the cephalopods with cornea, lens and retina, has been described by Mayes and Hermans (1973), Moczar and Moczar (1973) and by Röhlich and Töröck (1963). If the eyes are located at the tips of the tentacles, a second pair of tentacles occurs near the mouth. On the optic tentacles there are sensory cells which serve for distant chemical perception, or smell, while the oral tentacles have sensory cells for close range chemical perception, or taste (Kittel 1956). In the Basommatophora, which are all aquatic, there is only one pair of long or squat triangular tentacles, at the base of which the eyes are situated.

The **visceral mass** is bilaterally asymmetrical, usually spirally coiled to form a right-handed helix (dextral coiling), less commonly as a left-handed one (sinistral coiling). This asymmetry is the result of a twisting (torsion) which occurs in the course of embryonic development. In snails, the viscera lie within the whorls of the shell, and in slugs, under a thick mantle shield which often contains platelets or granules of calcium carbonate, and beneath which lies the *respiratory cavity*. The latter is enclosed by both the *mantle* and the dorsal body surface. In marine molluscs the respiratory cavity houses the gills, but in the secondarily aquatic fresh-

Fig. 3. Half-section of a gastropod shell (vineyard snail), columella (*left*), the last whorl is cut off (*right*)

water gastropods and terrestrial pulmonates it contains a lung, the walls of which are highly vascular. Air enters and leaves the respiratory cavity through an opening, the *pneumostome*, which is situated on the right of the body and opens and closes rhythmically. Its location, in relation to the middle of the mantle shield serves as a means of distinguishing between different families of slugs. In the **Arionidae** the pneumostome lies anterior to this mid-point; in the **Limacidae** and the **Milacidae** it lies posterior to it.

The mantle secretes the bivalve shell of mussels and the helical shell of snails. In slugs vestigial remains of a shell are found in the form of calcified platelets or granules in the Limacidae and Arionidae respectively, while a very small vestigial shell may be found at the posterior end of the foot in the Veronicellidae. In cuttlefish the shell, or cuttlebone, lies within the body. The oldest part of the shell is the apex, which is the start of the helix. The shell may be formed of only a few or many whorls, and their number is species-specific. The last whorl encloses an aperture, the mouth of the shell. The central axis around which the whorls coil is the columella. At the lower end of the shell the whorls may enclose a cavity, with the result that a broad, slit-like fissure, the umbilicus, is formed on the lower surface of the shell.

Figure 3 shows a half section of a snail shell with the columella and whorls, of which there are three or four as a rule in *Helix pomatia*. The direction of the spiral, which is genetically determined, can be established by viewing the shell from above: if the spiral is to the right, the coils run clockwise, as is the case in all recognized pest species. While the structure of the shell is a consequence of heredity, its thickness, colour, intensity of banding and growth rate are all influenced by environmental conditions.

Generally, contact between the body of the snail and the shell is easily loosened and the animal can be removed from the shell by a twisting movement. The only attachment is by the columella muscle, which arises from the foot musculature, runs upwards near the columella and is inserted on it. This muscle serves to draw

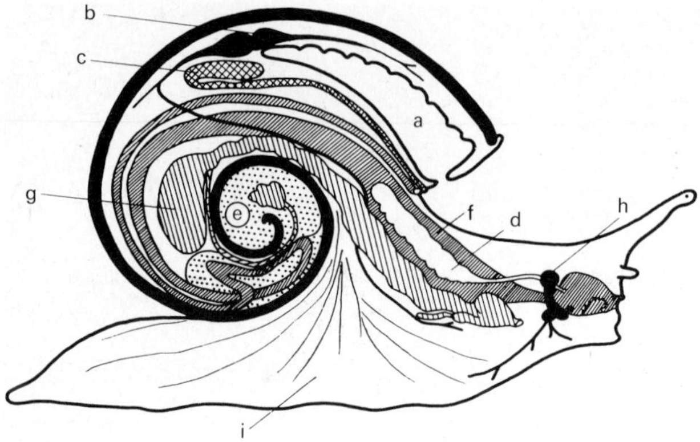

Fig. 4. Diagram of a generalized terrestrial gastropod (snail, Helicide), *a* mantle cavity with the lung; *b* heart; *c* kidney; *d* salivary gland; *e* digestive gland (hepatopancreas); *f* rectum; *g* reproductive system; *h* brain; *i* foot

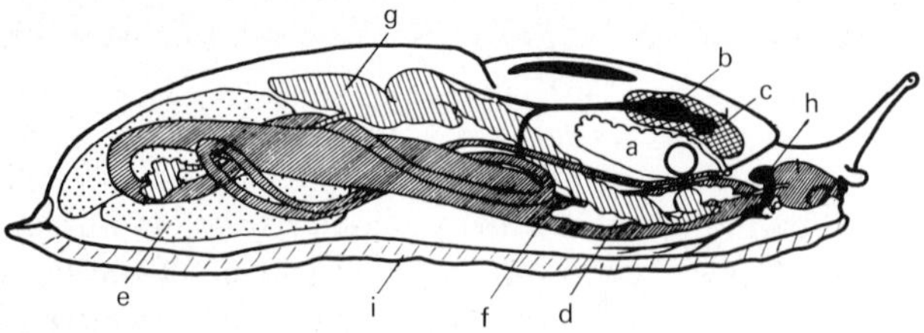

Fig. 5. Diagram of a generalized terrestrial gastropod (slug, Arionide), *a* mantle cavity with the lung; *b* heart; *c* kidney; *d* salivary gland; *e* digestive gland; *f* rectum; *g* reproductive system; *h* brain; *i* foot

the snail into the shell; the head is pulled in first and then the foot, with the sole turned inwards and the dorsal surface of the foot outwards. In the operculate snails the mouth of the shell is closed by a plate which is made either of an organic substance or of calcium carbonate; this operculum is borne on the upper surface of the foot (Plate IV/2). Most terrestrial snails have no operculum, but during hibernation the mouth of the shell is closed by a hard and chalky epiphragm which is cast off when activity is resumed in the spring.

The foot is essentially a powerful muscle, lying beneath the visceral hump. Locomotion is effected by waves of muscular contraction which start at the front and run posteriorly along the creeping sole; in some species, like *Pomatias elegans* (Müller), the sole consists of two parts. The appearance of the ventral surface or sole of the foot and also the colour of the slime are used to identify different species of slugs, as will be mentioned in the section on taxonomy. This applies also to the keel on the upper surface of the slug's body, that may be missing in some cases (Chap. 1.7.4).

The internal anatomy is dealt with comprehensively by RUNHAM and HUNTER (1970) and RIEDEL and WIKTOR (1974). The position of the various organs relative to one another is shown diagrammatically in Figs. 4 and 5 for snails and slugs respectively.

The **digestive system** comprises mouth, buccal cavity (buccal mass), oesophagus, salivary glands, crop, stomach, intestine, and liver (also called midgut gland or hepatopancreas). The hepatopancreas usually consists of two lobes, made up of a mass of ramifying tubules, and opens into the stomach. A characteristic feature of molluscs other than mussels is, apart from any special formation on the upper jaw, the radula, a tooth-covered rasp which serves to grate the food. It consists of

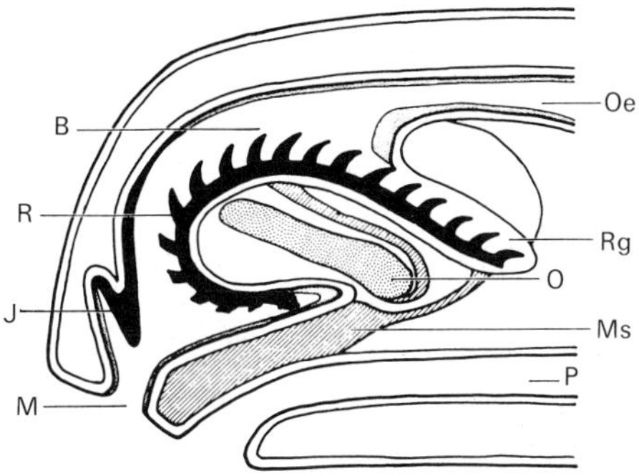

Fig. 6. Sagittal section through the head of a slug. *B* buccal cavity; *R* radula; *J* jaw; *M* mouth; *Oe* oesophagus; *Rg* radular gland; *O* odontophor cartilage; *Ms* muscles; *P* pedal mucus gland

many rows of chitinous teeth arranged on a horny membrane on the odontophore or "tongue" on the floor of the buccal cavity (Figs. 6, 7). New rows of teeth are constantly being produced from the floor of the buccal cavity, while older teeth are probably reabsorbed. The teeth in the centre of the radula differ in shape from those on the sides; their appearance and shape are taxonomically useful. MÄRKEL (1957) has given an account of the origin, structure and function as well as the toothmarks left by pulmonate radulae. Electronmicrograph studies of the teeth have also been made (RUNHAM 1969; THOMPSON and BEBBINGTON 1973; and others). The radula of carnivorous snails, such as *Conus* spp. (KOHN et al. 1972), is adapted to the specialized function of prey catching. In carnivorous pulmonates, furthermore, if the radula or proboscis are lost, they are immediately regenerated (CARRIKER et al. 1972; CARRIKER and VAN ZANDT 1972).

The **nervous systems** of all molluscs have certain basic features in common. That of gastropods comprises several pairs of ganglia, each pair being associated with important sense organs (Fig. 8). The cerebral ganglia lie above the buccal cavity and supply the tentacles and eyes. Below the alimentary canal and above the anterior end of the foot are the pedal ganglia; statocysts are located either on or near these ganglia, and serve as organs of balance (WOLFF 1977). The unpaired visceral ganglion also lies ventral to the digestive tract and near to it is a sense organ commonly found in molluscs, viz. the osphradium, a ciliated thickening of the epithelium of the mantle cavity, thought to be a chemoreceptor. The parietal and pleural ganglia are paired. In many gastropods there are accessory ganglia, such as the osphradial, optic, tentacular and rhinophoral ganglia. The ganglia are linked by connectives; in terrestrial pulmonates these are short, with the result that there is a concentration of ganglia in the head region, where in some species there is, in addition, a pair of buccal ganglia supplying the buccal mass. Molluscs have a highly developed peripheral nervous system.

The **heart** is dorsally situated and comprises a ventricle and auricle. The ventricle is always unpaired, whereas the auricle may be paired or single, depending on whether there is a pair of gills or only a single one. Pulmonates usually have a single auricle. With the exception of cephalopods, the circulatory system of molluscs is an open one and comprises sinuses, spaces in the perivisceral connective tissue, which arise as a result of the widening of veins. Cephalopods have veins and arteries which are connected by capillaries. The heart is enclosed in a sack, the peri-

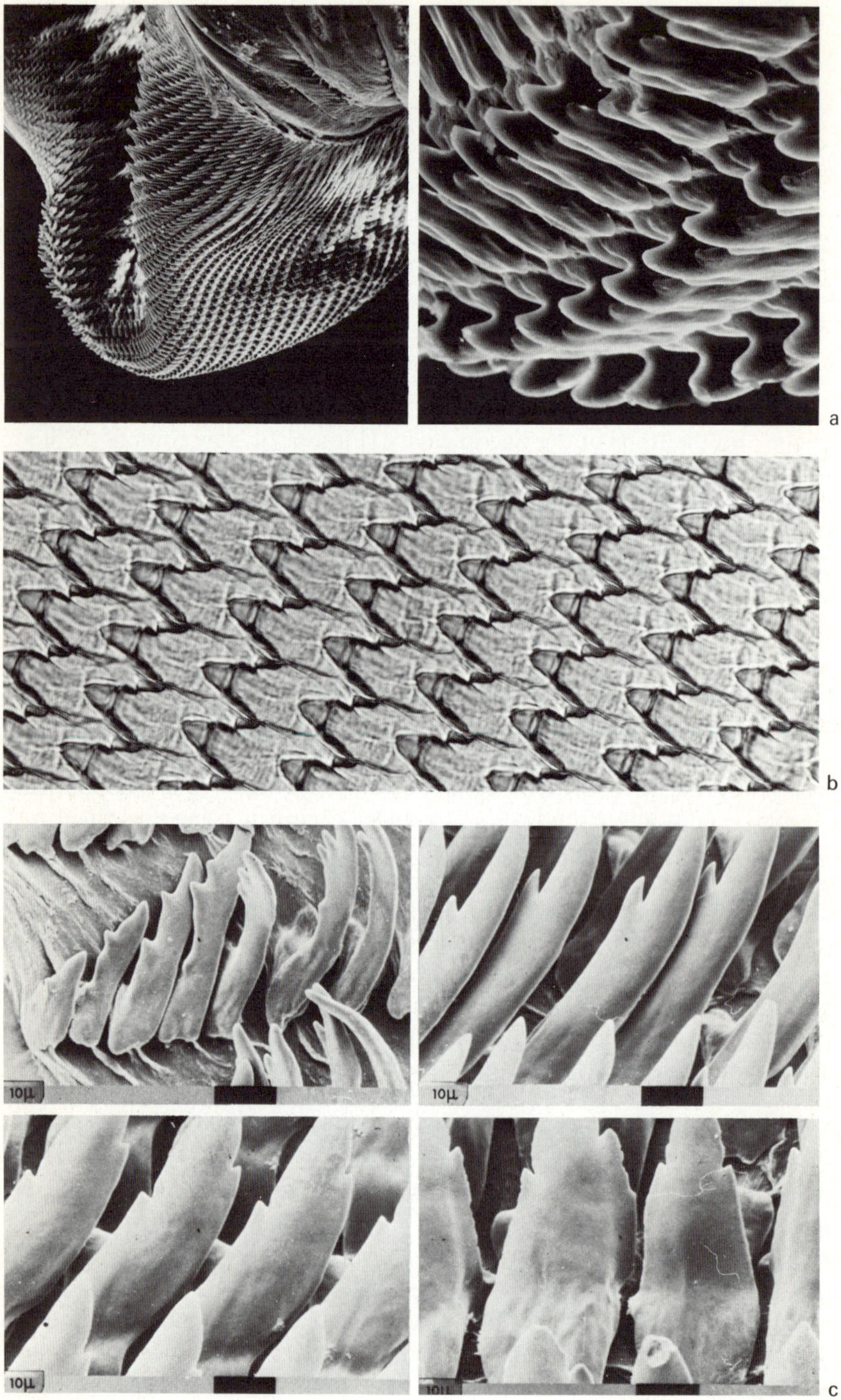

Fig. 7. Radula in feeding position (*above left*); teeth of *Deroceras reticulatum* (*above right*), *Arion rufus* (*centre*), *Boettgerilla pallens* (*below*); (Scanning electron micrographs)

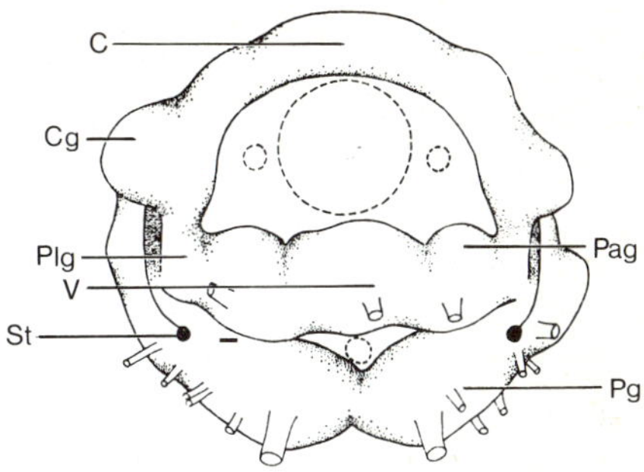

Fig. 8. *Deroceras reticulatum*, brain, posterior view. *C* cerebral connective; *Cg* cerebral ganglion; *Plg* pleural ganglion; *V* visceral ganglion; *St* statocyst; *Pag* parietal ganglion; *Pg* pedal ganglion

cardium, which is connected to the kidney by a ciliated duct, the reno-pericardial canal.

The anatomy of the reproductive organs, which undergo a transformation in one and the same animal, will be dealt with in Chapter 1.2.6, because of the special and complicated physiological processes involved.

1.2 Physiology

In recent years, interest in molluscan physiology has grown considerably. In particular, attention has been paid to the metabolism, the biochemistry of the nervous system and its endocrine function, the heart and circulatory system, and, last but not least, to reproduction and its hormonal control. The reproductive processes are of particular importance, not only because of the vast numbers of offspring which can be produced and the potential damage thus caused, but also in connection with finding new means of pest control which depend upon preventing population increase without polluting the environment (Chap. 3.6).

For this reason, the following account will deal fully with reproduction, but only briefly with other aspects of pulmonate physiology, particularly since it is the primary purpose of this monograph to present the problems of pests and their control. Despite the brevity of this account, it will be seen that snails and slugs are by no means primitive; their physiology, indeed, resembles that of vertebrates in many respects, since in both there is a complex chain of hormonal interrelations.

A comprehensive account of the physiology of molluscs is given by Wilbur and Yonge (1964/66) and more recently by Fretter and Peake (1975), while Boer and Joosse (1975) deal with their endocrinology.

1.2.1 Metabolism

The following summary is based on data from Runham and Hunter (1970).

Carbohydrate metabolism: Nutritional requirements during a period of starvation are met from reserves of carbohydrate. Glucose is a normal component of pulmonate blood and its level rises after feeding. For example, *Ariolimax columbianus*

(Gould) has 27.9–6.9 mg glucose/100 ml of blood. This represents 83% of the total carbohydrate content of the blood; the rest is made up of 3.6 mg/100 ml of glycoprotein and 1.8 mg of unidentified substances. Synthesis of polysaccharides (galactogen and mucoids) occurs via the uridine diphosphate compounds of glucose and galactose.

Protein metabolism: Molluscs appear to be capable of synthesising most, if not indeed all, the essential amino acids. In the blood of *Otala* spp., for instance, 26 different amino acids have been detected; these are incorporated into proteins, a process which probably occurs in the liver and kidneys. In the blood of pulmonates, more than 12 mg of free amino acids are found in 100 ml.

Lipid metabolism: Molluscs have unusual lipids. The fat content of the body of *Ariolimax columbianus* (Gould) is 1.12% of its wet weight. In adult *Arion ater* (L.) it is 0.8% (=8.0% of dry weight), while in young animals of the same species it is 2.3% of the wet weight the viscera alone accounting for one third of the lipid. The body fat comprises approximately 20% triglycerides (mostly in the epidermis and foot), 20% phospholipids and 40% steroids (predominantly cholesterin and its esters, found mostly in the viscera). Analysis of the triglyceride fraction revealed a high percentage of glyceryl ether derivatives; the glyceryl ether content of the phospholipids is, in fact, the highest known in any animal. It is also remarkable that only one or two glyceryl ethers occur and these in highly saturated form. Uncommon, too, is the lipid 2-aminoethylphosphonate which contains a carbon phosphate linkage, which has been found in *Lehmannia valentiana* (Fér.) and *Limax flavus* L.. It is assumed that the phospholipids of slugs are involved primarily in the structure of membranes, while the neutral lipids are involved in metabolism.

Steroids are of great interest because of their role as hormones in molluscs. Cholesterol comprises approximately 30% of the total steroids. Sterols can be synthesized from acetate in the tissues, so that when a need for them arises in the body, gastropods do not depend on plants for these compounds.

The importance of steroids in reproduction and in possible future non-chemical control measures will be discussed later.

1.2.2 The Nervous System, Neurohormones

The relative simplicity of the nervous system has led to increasing interest in molluscan neurophysiology, particularly in the neural mechanisms of memory and learning. The nerve cells are very similar in their characteristics and reactions to vertebrate neurones – at any rate, the similarity is greater than in the case of other invertebrate groups. Neurosecretion and the endocrine glands of molluscs are described by Boer (1965), Martoja (1968) and Lane (1966). The ganglia of these animals also show cholinesterase activity (Jurchenko et al. 1973; McCaman and Dewhurst 1971; Rozsa Katalin 1977). A survey of the physiology of the nervous system and neurosecretory cells can be found in Durchon (1967) and Tauc (1966). A technique for sectioning nerves has been described by Harley and Harley (1973).

The main ganglia in molluscs, the cerebral and pedal ganglia, together with the peripheral nervous system, are concerned with locomotion, while the visceral ganglion supplies the internal organs. The cerebral ganglion is primarily the co-ordinatory centre for foot, heart, and lung. In these ganglia, and also in the pedal ganglia, neurosecretory cells are found; these synthesize serotonin 5-hydroxytryptamine (= 5-HT) (YORK and TWAROG 1973; OSBORNE 1972, 1973 and others). Their rate of secretion increases during reproduction and is also higher in active summer animals than in passive snails during aestivation and in winter animals. Much work has been done recently on the exchange of Na^+, Ca^{++}, K^+, and Mg^+ ions through the neuronal membrane; work has also been done on RNA and phospholipids (CHRISTOFFERSEN 1973; KRASTS et al. 1972, OSBORNE et al. 1972). Many neurotransmitter substances are found in the brain of gastropods such as the snail *Helix pomatia* L. These include amines and free amino acids like 5-HT, tyrosine, tryptophan, tyramine, lysine, leucine, adrenalin, noradrenalin, dopamine, dopa and others (ASHER 1972; BRIEL et al. 1971; DOLEZALOVA et al. 1973; OSBORNE and NEUHOFF 1973). 5-HT and acetylcholine act as excitatory transmitters, 5-HT eliciting the prolonged and acetylcholine the brief excitatory post synaptic potentials (WALKER et al. 1972). 5-HT and dopamine stimulate the formation of adenosine monophosphate (AMP) in nerves and ganglia (KERKUT et al. 1969; CEDAR and SCHWARTZ 1972). 5-HT is found in concentrations of 2–4 μg/g in several ganglia, and also in the heart, as for example in *Aplysia* sp in concentrations of 0.6 to 1.3 μg/g (CARPENTER et al. 1971).

Water balance in pulmonates is also neurohormonally regulated (GERSCH 1964). Extirpation of the pleural ganglia in *Lymnaea stagnalis* (L.) leads to an increase in body volume as a result of water uptake, whereas destruction of the cerebral and visceral ganglia has no such effect. There are indications of a correlation between the secretory activity of various ganglia and ecological conditions such as soil humidity and osmotic concentration of soil water. This has been shown for *Helix pomatia* L. and other Helicidae, and also for the freshwater snail *Viviparus viviparus* (L.). The cerebral ganglion influences muscle tonus in the foot and can raise or lower it.

Pigments may occur in the brain; for example, two pigments were found in that of *Lymnaea stagnalis* (L.) – a red haemoprotein and a yellow carotenoprotein. These pigments occur in lipochondria within the molluscan neurones and their presence may relate to the light sensitivity of these cells (BENJAMIN and WALKER 1972).

1.2.3 Heart, Circulation

A review of the relevant literature up to 1965 is provided by GERSCH (1964) and also by HILL and WELCH (1966).

Both the volume of blood as well as the blood pressure in the open circulatory system of lamellibranchs and gastropods are high in active animals, low in periods of prolonged inactivity. The hearts of some molluscs include a pacemaker. Perfusion and stretching are necessary for the maintenance of a rhythmical beat. In most cases the heart is supplied with nerves, both excitatory and inhibitory. Storage organs for neurosecretory products (neurohaemal organs) are formed from neurites (nerve fibres) which are in close association with blood capillaries (NOLTE 1965).

The heart beat is affected by movements of the lung and foot (MENG 1960, 1962). Relaxation of the lower wall of the lung, with consequent compression of air in the respiratory cavity, and withdrawal of the foot both have an inhibitory effect on the heart beat, whereas contraction of the wall of the lung and eversion of the foot have the effect of releasing the inhibition. These inhibitory impulses reach both auricle and ventricle more or less simultaneously by way of separate nerves; this results in a certain degree of co-ordination of the two chambers of the heart.

The molluscan heart is myogenic and its beat can be reflexly modulated by tactile stimulation of different parts of the body surface, the excitation being effected by 5-HT and the inhibition by acetylcholine (MACKAY and GELPERIN 1972). Considerable concentrations of the neurohormones 5-HT and acetylcholine are found in heart muscle (IRISAWA et al. 1973; WILKINS and GREENBERG 1973). In many, though not all, molluscs the rate of the heart beat is increased by very low concentrations of acetylcholine (10^{-10} to 10^{-8} mol), whereas higher concentrations (10^{-7} to 10^{-5} mol) have an inhibitory effect on the heart rate. Amplitude and duration of the action potential are increased in concentrations over 10^{-10} mol (KISS and ROZSA 1972). Dopamine, the most common catecholamine in the nervous system of molluscs, probably also plays a role in the regulation of the heart beat. In the blood itself, however, according to OSBORNE (1972) neither 5-HT nor 5-hydroxyindol are present, while 5-hydroxytryptophan and tryptamine occur in low concentrations only.

The action potential of the gastropod heart (*Helix pomatia* L.) is so dependent on Na^+ ions that its ionic mechanism resembles more closely the excitable tissues of mammals than those of other invertebrates (KISS and ROZSA 1971, 1973).

The blood sugar level of the haemolymph is lower than the value for birds and mammals (GODDARD and MARTIN 1966). Measurements on single specimens of the snail *Biomphalaria glabrata* (Say) show that its value may vary within a period of 6 h from 1.0 to 17.2 mg/100 ml (BECKER, W. 1972).

The arterial system (for example, in *Biomphalaria glabrata*) comprises two aortic trunks, one of which supplies the intestinal tract, the digestive gland and gonads, and the other the kidneys, the anterior part of the reproductive tract, the columella muscle and the buccal region.

The **blood cells** of molluscs are white cells, amoebocytes; in rare cases erythrocytes occur (ARUN SHANKER 1973; POLEY and CHENG 1972, and others). The amoebocytes responsible for phagocytosis or incapsulation of foreign particles and for the repair of damaged tissues (SMINIA et al. 1972; SMINIA 1977, and others). They also form clots, but in contrast to the situation in vertebrates, these are reversible; the cells are simply attached to one another by their pseudopodia and can, therefore, separate again. The respiratory pigment in molluscan blood is usually **haemocyanin,** a protein containing copper which occurs in relatively high concentration. In a few species, the respiratory pigment is **haemoglobin,** an iron-containing protein. Both substances occur dissolved in the haemolymph.

Especially in the last few years, much work has been done on function, distribution and chemical properties of these two respiratory pigments (GHIRETTI 1966; READ 1966; REGER 1973; SIEZEN and DRIEL 1973 and others).

1.2.4 The Kidney, Excretion

The kidney of land pulmonates has attracted attention because of its role in nitrogen metabolism and water balance (SKELDING 1973). For a review of the literature dealing with excretory processes and the evolution of the kidney system from primitive to more highly evolved molluscs, the reader is referred to MARTIN and HARRISON (1966). More recently there have been accounts of fine structure and function by DELHAYE and BOUILLON (1972); MARTOJA (1975); NEWELL and SKELDING (1973) and others.

In gastropods the first step in the production of urine is probably ultrafiltration under pressure, not through the walls of the blood vessels, but through the kidney epithelium acting as ultrafilter (ANDREWS and LITTLE 1971, 1972; RUNHAM and HUNTER 1970). This filtrate flows into the lumen of the kidney sack, where its composition is modified by resorptive and secretory processes. The urine passes out through the ureter. Among the slugs, the Limacidae, with the exception of *Limax flavus* L., excrete large quantities of guanine (23% to 50% of total purine), while in the Arionidae thus far examined, no purine has been found in the urine (JEZEWSKA 1972).

1.2.5 Feeding, Digestion

The various methods of feeding of gastropods will be discussed in connection with the damage they cause (Chap. 2). RUNHAM and HUNTER (1970) have described the processes of feeding and digestion in slugs.

The salivary gland opens into the **buccal cavity** with its chitinous radula (PETERS 1972 and others). There is some uncertainty about the chemical nature of its secretions. KLESSEN (1973) examined the proteolytic activity of the cells; amylases were found in the saliva produced by *Arion ater* (L.) and *Deroceras reticulatum* (Müller). In these slugs there follows the oesophagus and a **crop.** The latter is a large, often yellow-, red- or brown-coloured organ, filled with digestive juice; its walls are strongly muscular and perform peristaltic movements. The crop fluid contains many enzymes such as amylase, cellulase, glucosidase, lipase, protease and others. The **stomach** is small and sack-like with two openings from the **hepatopancreas.** In this gland four types of cell are found: excretory cells, digestive cells, calcium cells and thin cells. The functional relationship of these cells to one another is not yet known. Amoebocytes are also found in the liver tissue. Many of the enzymes found in the crop also occur in the liver. In *Helix aspersa* Müller an acid, deoxyribonuclease (DNA-ase) has been found (LAVAL and PAOLETTI 1972); in *Nassarius obsoletus* (Say) glucose-6-phosphate dehydrogenase occurs in relatively high concentrations (FRIED and LEVIN 1973), while in *Arion hortensis* Fér. the lysosomes of active phagocytic digestive cells contain an acid phosphatase (BOWEN and DAVIES 1971). Furthermore, the hepatopancreas of *Helix pomatia* L., *Helix aspersa* Müller and *Cepaea nemoralis* (L.) have been found to contain large quantities of calcium and magnesium phosphate, which, remarkably, is the case also in the slug *Limax maximus* L. which has no shell (BURTON 1972).

In the liver tissue of *Helix pomatia* L. and *Helix aspersa* Müller, two isoarginases occur (BARETT et al. 1972), and in the South American terrestrial snail *Strophocheilus oblongus* (Müller), arginase has been found in association with ureolytic activity of the liver and kidney (TRAMELL and CAMPBELL 1972). VAN DER HORST (1970) has given an account of the synthesis and distribution of fatty acids in the lipids in *Cepaea nemoralis* (L.). It is also worth mentioning the occurrence of three substances with an insulin-like effect in the liver which are regarded by DE MARTINEZ et al. (1973) as insulin, proinsulin and insulin-dimer. Enzyme extracts from the stomach fluid of *Helix pomatia* hydrolyse cellulose, hemicellulose and pectin. OPEL and WOLFGANG (1963) recommend the use of the stomach fluid for maceration of plant tissues and the hydrolysis of oligosaccharides and glucosides.

The stomach is followed by the **intestine** with its blind-ending rectal caecum, and finally the anus, which lies on the right side of the mantle cavity.

According to STOREY (1970), the presence of sand in the digestive tract of gastropods, particularly in the stomach, increases the effectiveness with which food material can be crushed and ground. In experiments with *Radix peregra* (Müller), animals denied access to sand showed retarded growth. In natural conditions, gastropods take up large quantities of mineral particles. In the opinion of other authors, this serves to compensate for a deficiency of calcium in the food, which seems the more plausible explanation and will be discussed later.

The intestinal digestive juice of molluscs also contains proteolytic enzymes. Cellulases have been found in the crop and intestinal secretions of *Helix pomatia* L., *Cardium edule* (L.), *Achatina fulica* Bowdich and other molluscs (KOOPMANS 1970 and others), while in *Achatina fulica* even cellulose-digesting micro-organisms have been found (SOEDIGDO et al. 1970). Glucose, galactose, glycine and palmitic acid are absorbed by the cells of crop, hepatopancreas and intestine, while calcium is absorbed exclusively by the calcium cells (COLLATZ et al. 1975; WALKER 1972).

When there is a shortage of food, the tissue content of carbohydrates and proteins is reduced (BAYNE and THOMPSON 1970). In the mussel *Mytilus edulis* L., losses are greater from the mantle tissue than from other body tissues, while gamete formation continues. Only after prolonged, continuous starvation does production of eggs and spermatozoa stop. In experiments with the freshwater gastropod *Biomphalaria glabrata* (Say) this was found to occur after a period of starvation lasting 25 to 35 days (VIANEY-LIAUD 1973). In *M. edulis*, the ratio of oxygen uptake to nitrogen excreted falls when food is short; the rate of fall increases with increased stress either of food shortage or of unfavourable temperature conditions. In cases of prolonged starvation, the glycogen storing, vacuolated connective tissue cells, which are distributed through the whole body, play an important role in compensating for the losses (DE JONG-BRINK 1973).

1.2.6 Reproduction and Development

A survey of reproductive processes of molluscs is provided by DURCHON (1967); FRETTER and GRAHAM (1964); GRASSÉ (1968) and RUNHAM and HUNTER (1970).

1.2.6.1 The Genital Tract, Sexual Reproduction

Pulmonates are hermaphrodites. In copulation there is a reciprocal exchange of spermatozoa, and since the spermatozoa usually mature before the eggs, self-fertilization is prevented to a very large extent. Indeed, in *Crepidula*, a prosobranch, the animal is first male, but becomes female at a later stage after transformation of the genital organs and maturation of the eggs (Fig. 9). In advanced old age, the genital tract and gonads regress. Two partners are therefore necessary for copulation and each inseminates the other. Self-fertilization may, however, also ocurr (BAYNE 1973). Thus individuals of *Deroceras laeve* (Müller) isolated before the onset of sexual maturity, have nevertheless produced fertilized eggs. The basic structure of the reproductive system of pulmonates is shown in Figure 10. There is an hermaphrodite gonad or ovotestis, which leads into the hermaphrodite duct with a diverticulum (the seminal vesicle) in which the eggs are fertilized by foreign sperm. Thereafter male and female ducts separate. The male system comprises a vas deferens with associated prostate-like gland in some species, the penis and an appendage, the flagellum, which is involved in the secretion of the spermatophore. The female genitalia comprise an oviduct, an albumen gland and seminal receptacle, vagina and, in the Helicidae, a diverticulum of the vagina, the bursa copulatrix, which receives the spermatophore at copulation, and two finger-shaped glands; in the Helicidae also a blind sac of the vagina. The genital pore is situated generally on the right side of the body, just behind the head.

Recent studies have been mainly concerned with the morphology and biochemistry of the gonad and the gametes, especially in marine and freshwater gastropods, mussels and cephalopods (BOTTKE 1972; BRISSON 1968; CHOQUET 1969; LUBET and CHOQUET 1971; NORDEN 1972; QUIOT et al. 1973; STREIFF 1970; TEITELMAN 1973; VAN DEN BIGGELAAR 1971 and others).

Eggs, spermatozoa and genital tract of the following pest groups and species have been studied in recent years: the Limacidae, as exemplified by *Deroceras agreste* (L.), *Deroceras reticulatum* (Müller), *Lehmannia valentiana* (Fér.); the Milacidae, including *Milax gagates* (Drap.); and the Arionidae, such as *Arion rufus* (L.), *Arion hortensis* Fér., *Arion fasciatus* (Nilsson), *Arion circumscriptus* Johnston and the Veronicellidae, particularly *Laevicaulis alte* (Fér.) (BAILEY 1970; BAYNE 1970;

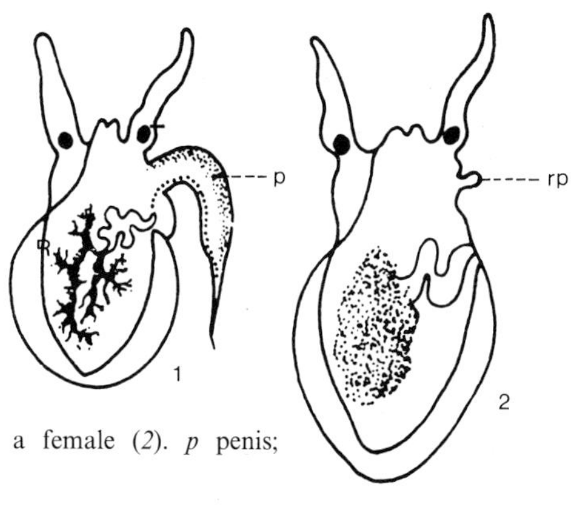

Fig. 9. *Crepidula onyx* as a male (*1*) and as a female (*2*). *p* penis; *rp* penis reduced, without function

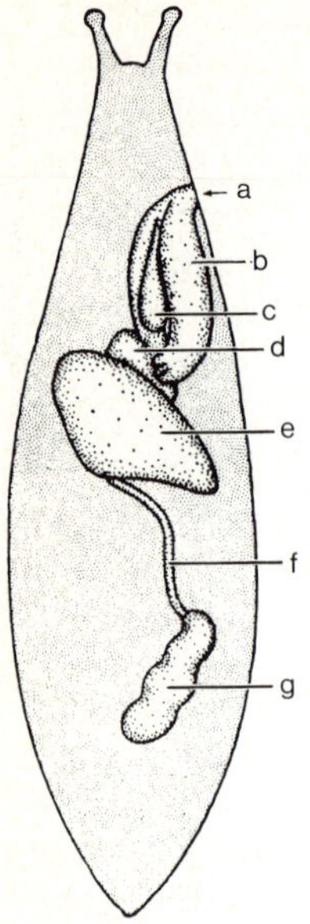

Fig. 10. *Deroceras reticulatum*, a diagram showing the position of the reproductive system within the body. *a* genital opening; *b* penial sac; *c* spermatheca; *d* common duct; *e* albumen gland; *f* hermaphrodite duct; *g* hermaphrodite gland

BOGUSCH 1972; CHEVALLIER 1969, 1971, 1974; FOCARDI and QUATTRINI 1972; KARLIN and BACON 1961; LAVIOLETTE 1950, 1951, 1954; LUCHTEL 1972; LUPU 1974; NAGABHUSHANAM and KULKARNI 1971; PACE 1971; ROUSSET-GALANGAU 1972; STERN 1975; VIANEY-LIAUD 1975). Accounts of reproduction in the snails *Achatina fulica* Bowdich, *Helix pomatia* L., *Helix aspersa* Müller, *Bradybaena similaris* (Fér.), *Oxychilus cellarius* (Müller), *Oxychilus draparnaudi* (Beck) and *Lymnaea stagnalis* (L.) are given by ANDERSON and PERSONNE (1967); BLOCK and HEW (1960); BINGHAM (1972); BREUCKER (1964); COOK (1970); EXPONDA and STOCKERT (1972); GHOSE (1965); GODDARD (1960); HERZBERG and HERZBERG (1960, 1962); KILIAS (1960); NIETZKE (1963); PLESCH et al. (1971); PINTO DE OLIVEIRA and DE OLIVEIRA (1971); RIGBY (1963); SCHMIDT (1955/56); THOMPSON (1973).

The **gonad** is embedded in the hepatopancreas; in snails it lies near the apex of the shell. In very young animals it is at first a simple sac filled with undifferentiated cells, but as it enlarges it becomes more and more lobed, while first oocytes and later spermatocytes differentiate; there are also nutritive cells. The development of the oocytes proceeds slowly, but that of the spermatocytes is rapidly completed. Only when most of the sperm has been expelled do the eggs mature (BEEMANN 1970, and others). We still know nothing about the factors which determine the differentiation of eggs, spermatozoa and nurse cells from apparently identical cells of the young gonad. Although gastropods are hermaphrodite, the male and female gametes develop in physiologically separate follicles of the gonad. STARKE (1971) has studied the fine structure of the cell types of the ovotestis of *Planorbarius corneus* (L.).

The **male copulatory organ** is a complex of muscle fibres and erectile tissue. In *Arion*, as in the Helicidae, it represents the terminal part of the vas deferens, whereas in *Limax* it is the penis sheath which can be evaginated, up to a length of 8 cm.

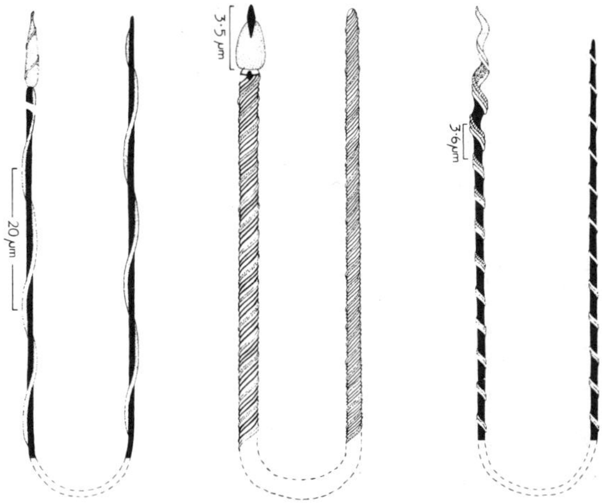

Fig. 11. Spermatozoa of *Helix pomatia* (*left*), *Lymnaea stagnalis* (*centre*), *Ariolimax columbianus* (*right*)

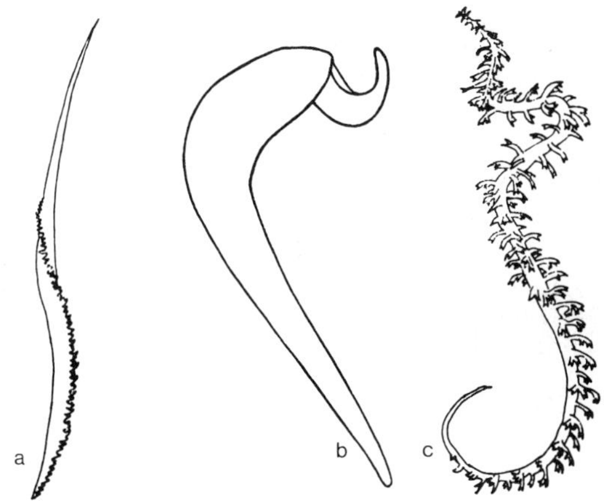

Fig. 12a–c. Spermatophores: **a** *Arion ater* rufus, **b** *Arion fasciatus*, **c** *Milax budapestensis*

In terrestrial pulmonates like the Helicinae and several other Stylommatophora, the sperm are transferred at copulation to the partner in the form of a **spermatophore,** which consists of head, sperm sack and a long tail (Figs. 11, 12).

In terrestrial gastropods the **egg** is surrounded by a proteinacious secretion of the albumen gland, which serves to nourish the developing embryo. This is in turn surrounded by a calcareous capsule secreted by glands in the wall of the uterus. This capsule may be white in appearance, as in *Helix pomatia*, or be a transparent sheat in which chalky bodies are embedded *Derocerus reticulatum* (Müller). In freshwater pulmonates, many eggs, each surrounded by a layer of albumen and a capsule, are packed together in a gelatinous mass.

Five phases can be distinguished in the reproductive cycle of terrestrial gastropods: courtship, copulation, nest-building, egg-laying and embryonic development followed by hatching of the young from the egg. Reproductive behaviour begins only when the humidity is high, as, for instance, after rain.

Courtship may be elaborate and of long duration. In its course a dart may be ejected by one or both partners; this penetrates either the sole of the foot or the

Fig. 13. Courtship in *Helix pomatia*

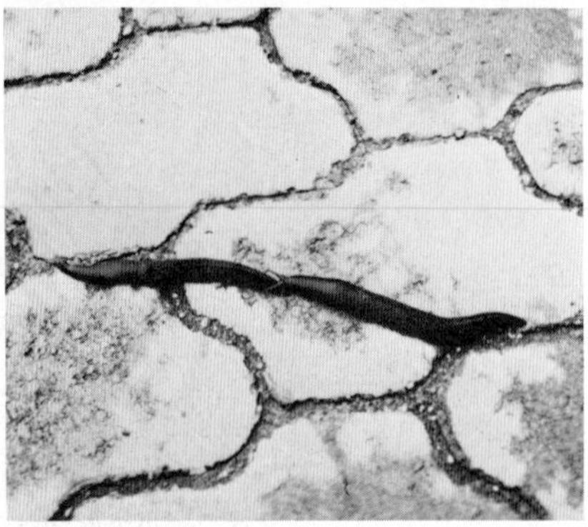

Fig. 14. Courtship in *Limax* sp.

side of the body and stimulates the partner. In *Helix pomatia*, courtship begins with the raising of the anterior part of the body and the soles of the feet of the partners are pressed together (Fig. 13). Copulation, in which reciprocal transfer of a spermatophore into the vagina of the partner occurs, lasts only 4 to 7 min. The head and sperm sack of the spermatophore, however, only reach the seminal receptacle 3 to 6 h later. In *Helix pomatia* copulation takes place several times during its life span, whereas *H. aspersa* copulates once only.

In slugs courtship begins with the animals crawling in a circle with the mouth of one animal pressed against the hind end of the other (Fig. 14). Later stages in copulation follow different patterns in the Arionidae and Limacidae respectively. In *Arion* there is a typical spermatophore; in the Limacidae, such as *Deroceras reticulatum*, on the other hand, there is no such spermatophore, the sperm being transferred in small slime balls. The dart is also lacking; instead an evertile muscular appendage of the penis sheath is used for tactile stimulation of the head and back of the partner (Fig. 15).

Fig. 15. Courtship in *Deroceras reticulatum*

Fig. 16. Stage in the courtship of *Limax maximus*, the animals dropping from a twig on a rope of mucus (length may be 45 cm)

Fig. 17. Egg laying in *Helix pomatia*

Fig. 18. Egg cluster of *Limax maximus* (*above left*), *Limax flavus* (*above right*), *Theba pisana* (*centre left*), egg cluster and newly hatched young snails of *Helix pomatia* (*centre right* and *below*)

The mating behaviour of *Limax maximus* L. is particularly interesting (KARLIN and BACON 1961; LANGLOIS 1965). The two slugs twist around one another suspended by a thick mucous thread, with the long penis-like structures everted and intertwined to form a tight spiral (Fig. 16). The sperm mass is contained in a longitudinal fold (comb) and deposited onto a glandular area on the outer surface of the partner's penis. When the penis is withdrawn, it takes with it the sperm mass which is adhering to it. From here the spermatozoa reach the seminal receptacle. Courtship may last 12 h or more in *Limax maximus* L., whereas sperm transfer itself takes only about a minute.

Oviposition in land pulmonates is a laborious and time-consuming process (Fig. 17). CADART (1955) recorded that *Helix pomatia* L. took 6 to 12 h to make a nest hole in the soil, 5 to 10 h for the ensuing rest period and 30 h for egg-laying itself. Oviposition in *Deroceras reticulatum* (Müller) takes place 8 to 10 days after copulation, in *Arion rufus* (L.) only after several weeks. The nest holes are made in loose soil and small batches of eggs are laid in them and covered with soil. The eggs are white in appearance as a result of calcium deposits in the shell. The round eggs of *Limax maximus* L. have a transparent capsule and are laid singly, while in *Limax flavus* L. they are oval and strung along a thread (Fig. 18). The terricole *Milax budapestensis* (Hazay) bores a passage 50 to 60 cm long and some 5 mm in width in a firm clod of earth, not in loose soil, and at the end of the tunnel it deposits 4–5 eggs, which are relatively large, slightly opalescent and cylindrical. The young slugs remain in the hole for 8–10 days after hatching, feeding on the egg capsules. In *Oxychilus draparnaudi* (Beck), the first young to hatch devour the other embryos in the nest (SCHMIDT 1955/56).

The *embryo* develops into a veliger larva, which is characterized by a velum. This is often divided into two parts and with its aid the larvae of marine gastropods can swim near the surface of the water before they start to crawl on the seabed and develop into the adult form. Even in land pulmonates, where the newly hatched animals already have the adult form, there is an indication of a velum during embryonic development. The embryo is rotated in the albumen within the egg by the action of cilia on the body surface. The pulmonate larva has two large protrusions – the swollen head lobe and the contractile foot lobe or podocyst; both are essential organs. The head lobe is a respiratory organ and the foot lobe a circulatory organ, making constant irregular contractions which are responsible for the circulation of body fluid (KILIAS 1960; QUATTRINI and SACCHI 1971) (Fig. 19).

Viviparity occurs in the Prosobranchia (family Viviparidae) and in the Stylommatophora (Veronicellidae, Valloniidae, Helicarionidae and Helicidae) (GRASSÉ 1968). In such cases, embryonic development starts in the uterus and the shells of the little snails can be seen through its wall. As many as seven may develop simultaneously in the uterus – a relatively large number. Such ovoviviparity may be replaced by oviparity, depending on the particular habitat of the snail or slug. Thus *Achatina panthera* (Fér.) is oviparous in the field, but ovoviviparous in captivity.

In slugs, lymnaeid snails and several others, increase in body size continues in the mature animal. Thus *Limax flavus* L. may live for 28–30 months; it is, however, sexually mature after 10–12 months and before it has reached maximal body weight. Growth continues at an almost constant rate until the end of the fourteenth to fifteenth month, by which time the slug has reached a body weight of 20 g. Sub-

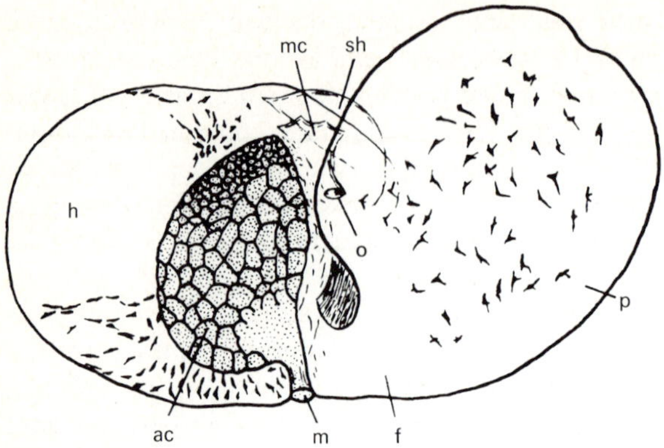

Fig. 19. Older larva (embryo) of *Helix pomatia* with head and foot lobe. *ac* albumen cells of the stomach; *f* foot; *p* foot lobe; (podocyst); *h* head lobe; *m* mouth; *mc* mantle cavity; *o* opening of mantle cavity to exterior; *sh* embryonic shell

sequently there is a loss of weight, and finally all feeding ceases, and the animal dies 8–10 days later (FRÖMMING 1950). In general, young gastropods grow at a steady and relatively fast rate after hatching; after maturation of the gonad, increase in size and weight continues more slowly. The Helicidae are an exception. Maturation of the gonad starts late and mating only occurs after completion of body and shell growth. Calcium carbonate, previously used for enlargement of the shell, continues to be deposited at the mouth of the shell, and thus gives rise to the thickened rim with the knobs and teeth characteristic of the shells of more mature members of the particular species (BOETTGER 1953).

The Limacidae are generally longer-lived than the Arionidae. Thus the life span of *Limax maximus* L. is 30 to 36 months (egg count 674–834), *Limax flavus* L. 28 to 30 months (egg count 246–342), *Lehmannia marginata* (Müller) 24 to 26 months (egg count 105–132) and *Deroceras reticulatum* (Müller) 9 to 13 months (egg count about 300). In *Arion* spp., life spans of 7 1/2 to 12 months have been recorded for *Arion hortensis* Fér. (egg count 158–205), for *A. circumscriptus* Johnson of 12 months (egg count 104–123), for *A. subfuscus* (Drap.) of 12 months (egg count 291–341) and for *A. intermedius* Normand of likewise 12 months. *Helix pomatia* L. may live for as long as 6 to 7 years (egg count 250–300). *Littorina saxatilis* (Olivi), which is male in its last sexual phase, may succumb to premature aging (INAGAKI 1972). Although this gastropod had neither reached the end of its potential life span nor achieved maximal body size, a fall in vitality, sexual potency and respiration rate was recorded.

1.2.6.2 Hormonal Control

Hormones control differentiation of the genital tract, maturation of the gametes, sex change, fertilization and oviposition.

There is much evidence that the undifferentiated germinal cells of a young gonad are potentially female; that is, in the absence of any hormonal influence, they develop into female cells only – a phenomenon referred to as autodifferentiation (GUYARD 1969; STREIFF 1966, 1967). Differentiation of germinal cells into male gametes occurs under the influence of an androgenic hormone secreted by the cerebral ganglion.

There is also evidence for the presence of gonadial hormones which stimulate the development and secretory activity of accessory glands. Thus ROHLACK (1959) found an oestrogenic hormone in the ovarian gonad of the littoral snail *Littorina littorea* (L.); this is, however, not identical with any vertebrate oestrogen. The action of such hormones may not be direct. There is, for example, a hormone which stimulates the cells of the wall of the hermaphrodite duct. According to TARDY (1973) its action is indirect, since it acts by stimulating hormone secretion from cells within the nervous system.

The interaction of individual hormones with one another is complex and the details not fully understood. Comprehensive reviews are to be found in the works by AUBRY (1959, 1962); DUNCAN (1960); GUYARD (1971) and SANCHEZ (1962).

The following brief account of endocrine control of reproduction and sex transformation may be useful, particularly in relation to the possibility of biotechnical control measures which involve self-destruction. As a result of limited space, only a part of the considerable volume of literature which has appeared in recent years can be listed here: BRISSON (1970); GOMOT (1971); GRIFFOND and GOMOT (1974); HERLIN-HOUTTEVILLE and LUBET (1975); JOOSSE (1975); LEBRETON (1971); LEGALL (1974); LEGALL and STREIFF (1975); LUBET et al. (1973); STREIFF (1970); TARDY (1967, 1969, 1970, 1971, 1973); VIANEY-LIAUD (1972, 1973). Amongst pest species, work on Arionidae, Limacidae and *Milax gagates* (Drap.) include: BAILEY (1973); GALANGAU (1964, 1969); GOTTFRIED et al. (1967); LUSIS (1961, 1966); RUNHAM and LARYEA (1968); VAN MOL (1962, 1967); WATTS (1952); WEISS (1968); YOSHIOKA et al. (1963). The Helicidae, particularly *Helix aspersa* Müller and *Helix pomatia* L. have been investigated by: BLOCK et al. (1960); FARRIS (1968); GOMOT (1973, 1975); GUYARD (1967, 1969, 1970); GUYARD and GOMOT (1969); JUNGSTAND (1962); KUHLMANN (1963); KUHLMANN and NOLTE (1967); LANE (1962, 1963); SCHLOOT (1965); VICENTE-PICCHIOCHI (1967). Finally, the amphibious gastropods (*Lymnaea* sp., *Planorbis* sp.) have been studied by: AUBRY (1954); BRAHMACHARY et al. (1972); BRINK and BOER (1967); BRISSON (1967, 1970, 1971); CUMIN (1972); JOOSSE et al. (1968); JOOSSE and REITZ (1969); LEHOUX and WILLIAMS (1971); MORRILL et al. (1973); PLESCH et al. (1971) and WENDELAAR BONGA (1972).

By means of castration and transplantation experiments, RUNHAM et al. (1973) have demonstrated conclusively the presence of two hormones in *Deroceras reticulatum* (Müller); one is responsible for the differentiation and development of the prostate gland, the other for the development of the oviducal gland (Fig. 20). The **prostate hormone** is produced during the later phase of spermatocyte differentiation and reaches a maximal level of secretion when mature spermatozoa are formed. The secretion of the hormone then diminishes, but does not cease altogether, while the **oviducal hormone** becomes dominant. It is possible that the albumen gland of the female genital tract is also influenced by this hormone. The details of the processes involved are very complex and not yet fully understood. Some factor produced by the gonad stimulates secretion of the prostate and oviduct hormones by the cerebral ganglion.

PELLUET and LANE (1961) and PELLUET (1963) found that in the Arionidae extirpation of the **optic tentacles,** or alternatively, injection of homogenates of the cerebral ganglion led to an enlargement of the oocytes in the hermaphrodite gonad. They concluded that the optic tentacles produce a hormone which inhibits oogen-

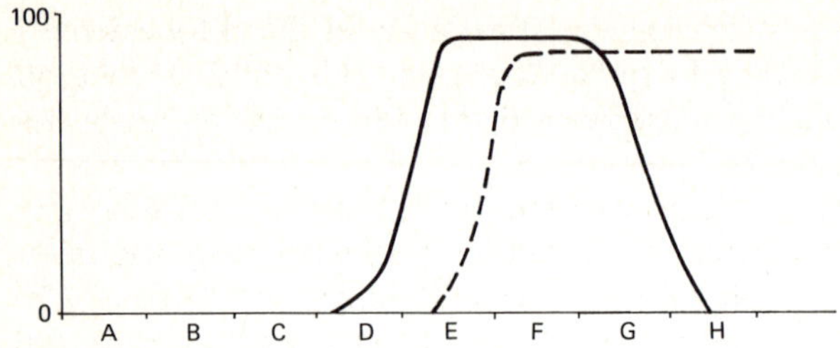

Fig. 20. *Deroceras reticulatum*. Suggested timing for the secretion of prostatic and oviducal hormones in relation to the stage of development of the hermaphrodite gland. *A* undifferentiated; *B* spermatocyte; *C* spermatid; *D* early spermatozoon; *E* late spermatozoon; *F* early oocyte; *G* late oocyte; *H* post-reproductive. —— = prostatic hormone; ---- = oviducal hormone; 0–100 = count of hormone

esis and that there is also a separate feminizing hormone. Moreover, GOTTFRIED and DORFMAN (1969) observed normal spermatogenesis in *Ariolimax californicus* (Cooper) after extirpation of the optic tentacles. The following investigations of this phenomenon have been made on economically important snails and slugs: on *Achatina fulica* Bowdich (BERRY and CHAN 1968), *Helix* sp. (LANE 1964), *Helix pomatia* L. (BIERBAUER and TÖROK 1964; BIERBAUER et al. 1965; BIERBAUER and MOLNAR 1972), *Helix aspersa* Müller (ROGERS 1971; TUZET et al. 1957), *Helicella obvia* (Hartmann) (RÖHLICH and BIERBAUER 1966), *Arion subfuscus* (Drap.) (WATTEZ 1972), *Arion rufus* (L.) and *Deroceras reticulatum* (Müller) (CHETAIL 1963), and also *Lymnaea stagnalis* (L.) (HEKSTRA and LEVER 1960; JOOSSE 1964; LEVER et al. 1961; LEVER and JOOSSE 1961).

In the cephalopods (*Octopus* sp.), a relationship between the optic glands and the testis have also been established (WELLS and WELLS 1972).

The results of experimental removal of the optic tentacles are to some extent contradictory. This is not surprising when it is borne in mind that the tentacles regenerate rapidly, that their functioning is likely to be affected by various, as yet unknown, physiological factors, as well as by interaction from other endocrine glands.

The second pair of tentacles near the mouth has the same histological appearance as the optic tentacles, but their removal in *Helix aspersa* Müller appears to have no effect on either oogenesis or spermatogenesis (SANCHEZ and SABLIER 1962).

In *Helix aspersa* Müller the effect of factors from the cerebral ganglion is both to inhibit autodifferentiation of the ovotestis to a female gonad, and to induce the differentiation of sperm cells. In the postembryonic pulmonate "brain" there must be one or more factors which inhibit oogenesis and promote spermatogenesis (GUYARD 1969, 1971). Both the relationship between the juvenile gonad and the optic tentacle and the three-way relationship of gonad, cerebral ganglion and optic tentacle have the same effect. In older animals the cerebral ganglion and optic tentacles have a masculizing effect on the hermaphrodite gonad.

In the Brown Garden Snail *H. aspersa* Müller, spermatogenesis is manifested only under the influence of an additional endocrine factor which is secreted by the cerebral ganglion and which perhaps has its effect via the optic tentacles. This androgenic hormone has a double function: inhibition of the autodifferentiation of the hermaphrodite gland to an ovary and the stimulation of the mitotic activity (cell division) in the gonad leading to spermatogenesis. This mitogenetic agent which is not sex-specific in its action, together with an androgenetic hormone, appears to be peculiar to pulmonates.

The processes involved in sex transformation in terrestrial pulmonates, have been the subject of recent experimental investigations in such pest slugs and snails as *Arion circumscriptus* Johnston, *Arion rufus* (L.) and *Deroceras reticulatum* (Müller) (BRIEN 1970; BURNETT 1967; LUCHTEL 1972 and others). Present experimental evidence suggests that the initial protandric stage continues until, under the influence of some environmental factors, the female potential of the gonad is realized. Thus isolated young *Crepidula fornicata* (L.) remained "male" and did not start to develop into females until the animals were brought into contact with other, still younger members of the species (COE 1948, 1953).

Hormonal control of the sexual cycle of pulmonates can be summarized as follows (STREIFF 1966, 1967, 1970): the optic tentacle hormone induces masculinity and inhibits femininity; the cerebral hormone initiates the onset of the female phase, and in addition there is a hormone which is responsible for the regression of the male genital organs.

According to GUYARD (1971), four hormones are involved in the regulation of **oogenesis** in molluscs (Fig. 21).

1. A mitogenic agent which stimulates cell division in the oogonia. This factor probably does not occur in the Helicidae, since in *Helix aspersa* no further cell division takes place in the oogonia after their differentiation as female cells.
2. A factor, the presence of which is suggested by the observation that a limit is set to the autodifferentiation of the oocytes. Even in prolonged in vitro culture of the gonad, full differentiation of the cells can only occur in the presence of a vitellogenic hormone.
3. A vitellogenic hormone, which is synthesized in the nervous tissue and distributed in the haemolymph. This stimulates further development but is no longer secreted during the later stages of differentiation of the female gonad. It is possible that the optic tentacles are involved as a relay point.
4. Other hormonal factors which regulate the maturation of the eggs and oviposition. Extirpation of the medio-dorsal bodies of the cerebral ganglion leads to a marked reduction in the number of eggs laid; if the medio-lateral bodies are removed as well, the frequency of oviposition is also reduced. The processes of oogenesis and spermatogenesis are not affected by these hormones.

Experimental work on the dorsal bodies and their function has been done by KUHLMANN (1966), LEVER (1958) and BOER et al. (1968).

The investigations of KUPFERMANN (1972) are particularly relevant. Oviposition in the snail *Aplysia californica* (Cooper) is regulated by neurosecretory products of the "bag cells" of the abdominal ganglion. Egg laying was found to begin 1 h after injection of an extract of the "bag cells" into the haemocoel, or following electrical stimulation of the abdominal ganglion. The protein nature of the hormone from the "bag cells" has been demonstrated by ARCH (1972).

The processes involved in **spermatogenesis** are influenced by other hormones (Fig. 22). According to GUYARD (1971), these can be summarized as follows:

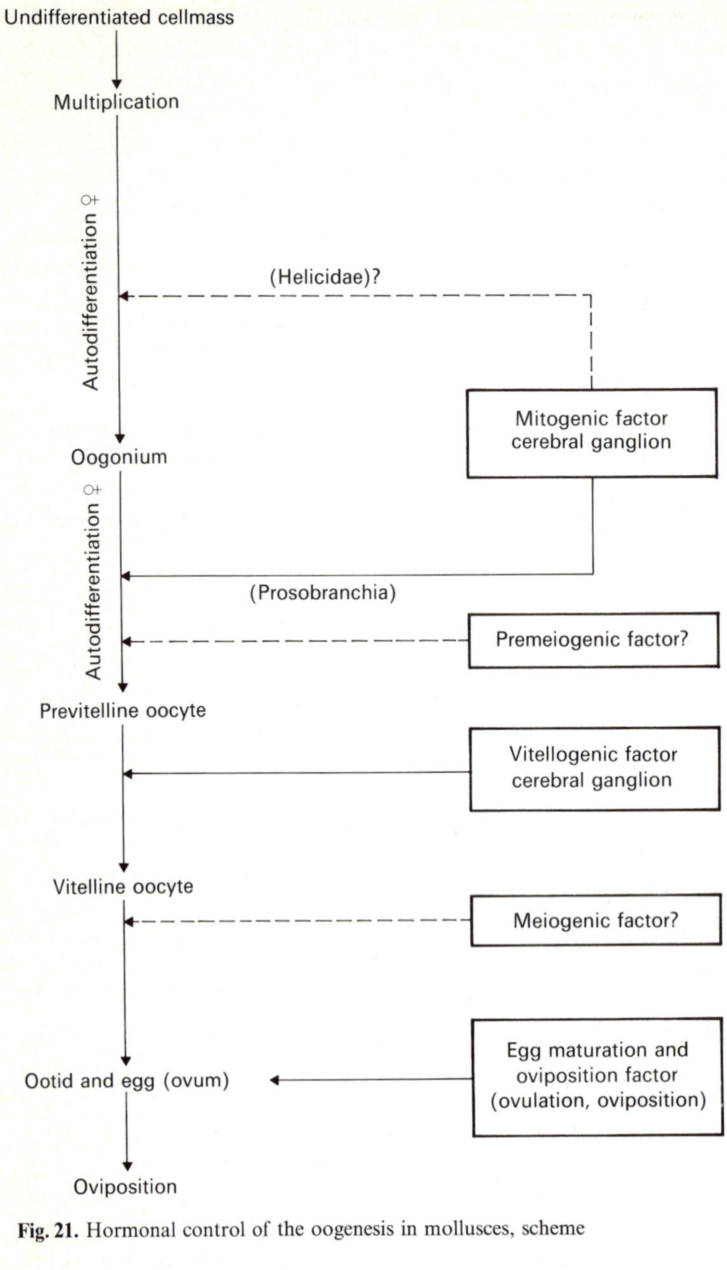

Fig. 21. Hormonal control of the oogenesis in mollusces, scheme

1. An androgenic hormone which is formed in the cerebral ganglion and stimulates both the differentiation of the ovotestis into male cells and also the mitotic activity of the spermatogonia. However, a further mitogenic factor appears to be necessary for the initiation of these two actions of the androgenic hormone.

2. A hormone, secreted by the optic tentacles, which inhibits oogenesis and the continuous effect of which is essential for the maintenance of spermatogenesis.

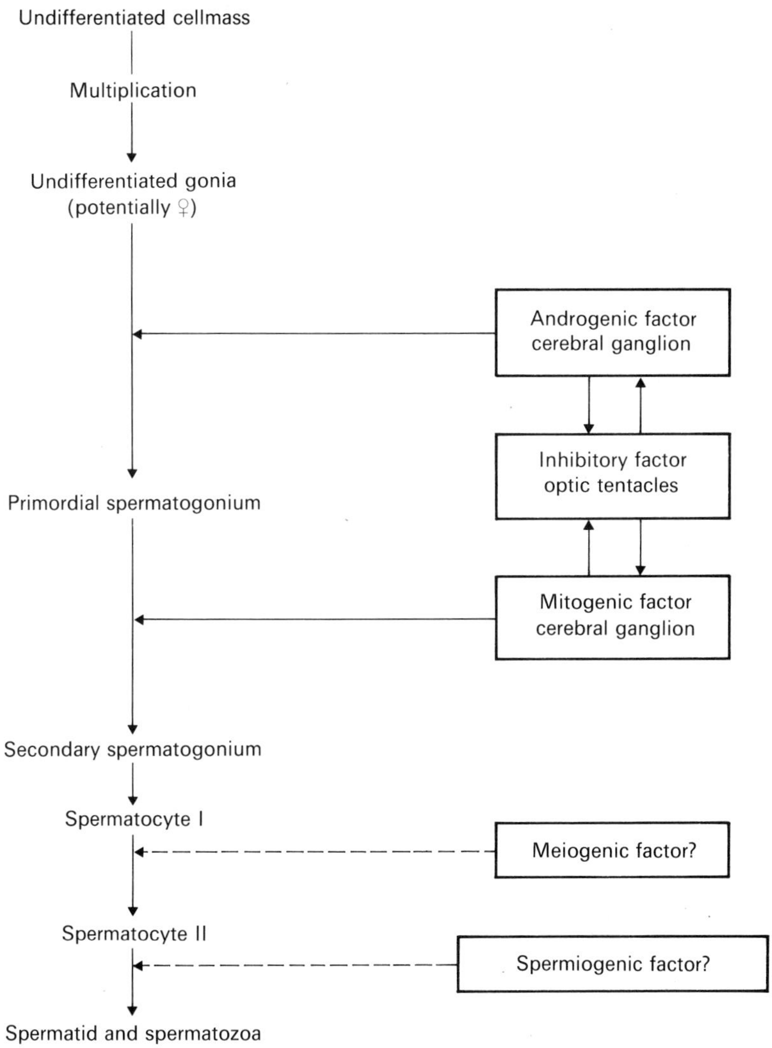

Fig. 22. Hormonal control of the spermatogenesis in molluscs, scheme

This hormone does not, however, interfere with the mitogenic function of the androgenic hormone from the cerebral ganglion. This inhibitory factor from the optic tentacles appears to be common to all pulmonates. At the distal ends of the optic tentacles, the nerve fibres terminate in a pear shaped body in which hormone-secreting, so-called "collar cells" are found (LANE 1972).

3. Other hormones may also be involved, since in *Helix aspersa* Müller, for example, the premeiotic division leading to the formation of spermatocytes and oocytes is synchronous.

The antagonistic action of the various hormones involved in differentiation of the gametes and in sex reversal are summarized below. This scheme, proposed by

GUYARD, includes both hormones whose effects have been demonstrated and others whose presence is suspected.

Only those points which apply specifically to molluscs are mentioned.

1. Feminizing autodifferentiation and the androgenic hormone

a) Androgenic cerebral neurohormone

Calyptraea sinensis L. (STREIFF 1967)

Crepidula fornicata (L.) (LUBET and STREIFF 1969)

Helix aspersa Müller (GUYARD 1969)

Viviparus viviparus (L.) (GRIFFOND 1969)

2. The androgenic hormone and inhibition of the masculinizing factor

a) The cerebral hormone and the inhibitory factor in the optic tentacles

Patella vulgata L. (CHOQUET 1969)

b) A masculinizing factor of the gonad and the inhibitory factor of the optic tentacles

Ariolimax californicus (Cooper) (GOTTFRIED and DORFMAN 1969)

3. Feminizing factor of the cerebral ganglion and the inhibitory factor of the optic tentacles.

Arion ater (L.) (PELLUET and LANE 1961)

Arion subfuscus (Drap.) (PELLUET and LANE 1961)

Arion rufus (L.) (BADINO 1967)

The chemical nature of the gonadotrophic hormones is still largely unknown. No comparable effects on the sexual process of molluscs are produced by injection of vertebrate hormones; nevertheless steroids do occur in gastropods. These are probably involved in the interaction between gonad and nervous system during the initiation of oogenesis and spermatogenesis, perhaps by modifying the release of hormone from the optic tentacles (GOTTFRIED et al. 1967; GOTTFRIED and DORFMAN 1969, 1970).

We do not yet have a detailed understanding of molluscan **steroids.** Recent research has supported the idea that in the Lamellibranchia and Pulmonata certain enzymes are required for the synthesis of androgens and oestrogens. The work of LUBET and STREIFF (1969) on *Crepidula fornicata* (L.), that of STREIFF and LE BRETON (1970) on *Littorina littorea* (L.), as well as that of GOMOT (1974) on *Helix aspersa* Müller, have shown very clearly the influence of hormones from the cerebral ganglion on the development of steroidogenesis. BARDON et al. (1971) in experiments on the mussel *Mytilus edulis* L. and on *Crepidula fornicata* (L.) using radioactively labelled steroid substances, have established the presence of an oxidoreductase which facilitates the reversible change of testosterone to androstendion. This enzyme (17α-hydroxysteroiddehydrogenase) is found in mature hermaphrodite glands of both "male" and "female" animals. After incubation of the gonad of *Crepidula*, in the absence of 17α-hydroxyprogesterone, androstendion could be isolated by chromatography. This revealed the occurrence in gastropods of a 17,20-lyase, an enzyme which has not been found in gonads of male or female mussels (also DELONGCAMP et al. 1970). The presence of several as yet unidentified metabolites has been demonstrated using radioactive markers; they are probably the reduced or hydroxylated precursors of steroids. GOTTFRIED and DORFMAN (1970) established the following sequence of steroid biosynthesis in the male phase of the ovotestis of *Ariolimax californicus* (Cooper): cholesterol → pregnenolone → 17α-hydroxypregnenolone → dehydroepirol → pregnenolone → 17α-hydroxypregnenolone → dehydroepinandrosterone → androstendione. Pregnenolone was also transformed to progesterone, but no further metabolite of progesterone could be found in the ovotestis extract. Testosterone itself was never found, since, as with BARDON et al.'s (1971) in vitro experiments, it was immediately transformed to androstendione. By incubating hepatopancreas tissue from *Buccinum undatum* (L.) with ^{14}C-labelled progesterone, TECHIMA and AKIO KANAZAWA (1971) identified the following substances: radioactive testosterone, 17α-hydroxyprogesterone, deoxycorticosterone and androstendione. The hepatopancreas probably contains the enzymes steroid 17α-hydroxylase, steroid 21-hydroxylase, C_{17}-C_{21}-lyase and steroid 17β-hydroxysteroiddehydrogenase. In the albumen gland of both *Helix aspersa* Müller and *Helix pomatia* L., the presence of an acidic β-galactosidase has been established; this can also occur in other snail tissues (BARNETT 1971).

Other investigations of molluscan steroids have been undertaken by AUBRY (1959, 1962), BAYNE (1972), COLES (1970), LEHOUX and SANDOR (1970), GOTTFRIED and LUSIS (1966), ROY and ROY JR. (1970), THIELE et al. (1964) and UHLENBRUCK et al. (1972).

The existence of hormone-like substances secreted by the **gametes** of some marine invertebrates, and facilitating the union of eggs and sperm has been known for some time. In echinoderms, such as star fish and sea urchins, at least four groups of substances occur: Gynogamone I substances, which activate the spermatozoa; this effect is antagonized by androgamone I substances; gynogamone II (G II) includes substances which agglutinate sperm, and androgamone II (A II) or "antifertilizin" substances, which inhibit the agglutinating effect and dissolve the gelatinous coat around the egg, possibly with the aid of a distinct lysin (ROTHSCHILD 1956, cited in FRETTER and GRAHAM 1964). In molluscs two types of gamone, G II and A II, probably occur; G II is found in oysters, like *Crassostrea virginica* Gmelin, for example. G II appears to be confined to the jelly in which the egg lies and to be secreted either by the follicle cells or by the egg itself while it is still in the ovary; if this jelly layer is removed, no further agglutination occurs. Both G II and the antifertilizin (A II) which antagonizes it, are proteins. The lysin which destroys the egg membrane is specific in its action and acts rapidly, dissolving the egg surface (or vitelline membrane) in 30 s if the jelly has previously been removed, but taking four to six times as long if the jelly is present.

Some molluscs which, like mussels, release their gametes into the water, have additional substances that activate spawning by other members of the same species. Thus the testicular tissue of *Mytilus californianus* Conrad has been found to excrete a substance which causes spawning in the females. The females respond only to sperm of the same species. Male oysters, however, are stimulated to release sperm by other males, by eggs or by water in which eggs have been kept, as well as by thyroxin and glutathione (GALTSHOFF 1938, 1940). This sensitivity of the males to a variety of different stimulants is of value to the species, as once initiated, the process of spawning then spreads through the whole population. The receptors are situated on the gills; their stimulation induces relaxation of the sphincter of the male genital duct and this results in spawning (NELSON and ALLISON 1940). NELSON (1936) also discovered a hormone-like protein, dianthin, in oyster sperm; this increases ventilation of the mantle cavity. Its secretion prior to spawning allows the eggs to pass easily through the gill passages into the inhalent chamber where the sperm collect.

1.2.7 The Shell

Apart from the reviews by WILBUR (1964) and TIMMERMANS (1969), there have been, especially in recent years, very many accounts of the formation, composition and regeneration of the molluscan shell and of its structure and the nature of the calcium carbonate crystals it contains, whether aragonite or calcite (COMPERSE and BATES 1973; SIEWERT 1972 and others). The periostracum, the formation of which during repair of the shell is still not understood, has also been the subject of recent investigations (KNIPRATH 1972).

The shell is secreted by the mantle epithelium and consists predominantly of calcium carbonate. There are three main layers: a thin, horny, transparent, outer layer, the periostracum, formed of an organic substance, conchin; a middle, so called prismatic layer, which consists of many lamellae of calcium carbonate embedded in an organic matrix; and an inner layer of mother of pearl, particularly characteristic of mussels.

The periostracum protects the other two underlying crystalline layers from being dissolved or destroyed by environmental factors; it is secreted by certain columnar epithelial cells of the mantle edge and undergoes an enzymatic tanning process. The protein of this outer organic layer is of a different composition from that of the proteins of the prismatic or nacreous layers. The cells of the upper surface of the mantle give rise to the calcareous middle layer by periodically secreting organic material and calcium carbonate ions into the thin layer of fluid, the extrapallial fluid, enclosed between the mantle and the inner surface of the shell. The calcium carbonate crystallizes out as aragonite or as calcite. The individual laminae are firmly bound to one another; bundles of fibrils which enclose the plates of one lamina twist together and pass into the corresponding bundles of fibrils of the next lamina. The relative calcium carbonate content of the shell of the embryo is no greater than that of the adult shell (KESSEL 1933).

Increase in size of the mollusc shell is effected by glands in the mantle collar, at the mouth of the shell; the resulting growth bands are thus parallel to its edge. The mantle collar also produces the pigments which are deposited in the shell as it is secreted. The basic colour of the shell is determined by the periostracum. In older animals this is mostly rubbed away, so that the shell has a chalky appearance. The pigments of the bands, on the other hand, are deposited within the calcareous layer.

In *Trichia hispida* (L.) the shell is covered with hairs. The distance between successive rows of hairs – particularly obvious in the middle whorls – corresponds to growth periods (SCHERF 1967) and can thus be used to determine the age of the snail. The formation of these hairs, which consist of conchin, is described by KAISER (1966).

In snails which have an operculum, it is seen as structurally analogous to the second valve of the shell of mussels (ADEGOKE 1973).

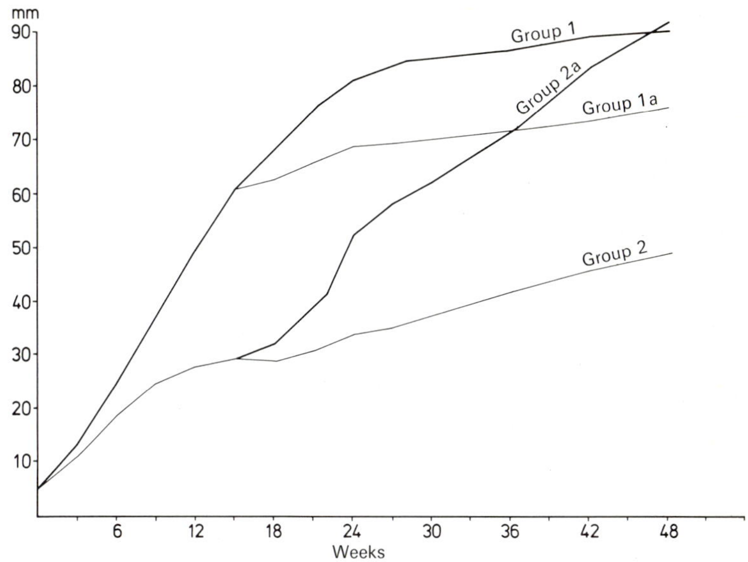

Fig. 23. Rate of increase in shell length (*Achatina fulica*) in relation to calcium supply. *Thick line* growth with calcium; *thin line* growth without calcium

Snails secrete an acidic substance from the sole of the foot; this enables them to utilize the calcium of the soil by dissolving the surface of calciferous mineral particles. Thus marble or molluscan shells show superficial etching at places where the sole of a snail has been applied. Gastropods are indeed capable of extracting Ca^{++} ions from acid or neutral soils, and they also ingest soil particles to meet their calcium requirements if the calcium content of the food is insufficient. Calcium carbonate deposited in the egg capsule is dissolved by the carbonic acid which results from respiration in the embryo, and is then used in the formation of the embryonic shell.

VOELKER (1959) studied the influence of calcium carbonate availability on *Achatina fulica* Bowdich (Fig. 23). Snails in group 1 lived on sandy soil which contained chalk particles; group 2, on the other hand, lived on chalk-free sand. At a particular point in time some animals in group 1 were deprived of chalk (group 1 a), while some of those from group 2 (group 2 a) were put on a chalky substratum. Snails of group 1, which were not deprived of calcium carbonate, laid their first eggs after 6 months; removal of chalk resulted in immediate reduction of the growth rate, delay in the onset of sexual maturity by 1 month and cessation of egg-laying for the duration of the observation period of 48 months. The relative weight increase of group 1 was 61.7%, but was reduced to 33.6% in group 1 a. The mortality of snails in group 2 was rather more than 50% and their growth practically stopped after 4 months, while after 48 weeks the shell had reached a length of only 28.5 mm, not even half the length of the shell of group 1 animals. Immediately after the addition of chalk particles to the substratum, the growth rate increased and group 2a reached the average rate of group 1. For optimal development, *Achatina* requires a calcium carbonate content of the substratum of at least

3% to 4%. Lack of calcium in the soil results in slow growth, high mortality, cannibalism and cessation of reproduction. In young animals the wall of the shell is thinner than normal because calcium carbonate is removed from the older parts of the shell so that it may be used for enlargement of the shell. Experiments on *Helix aspersa* Müller by CROWELL (1973) as well as SCHMIDT (1955) have also shown the advantages of a calcium-rich environment for health, weight, longevity and fecundity.

The significance of calcium availability or deprivation on population density and the resulting degree of damage by slugs and snails is discussed further in Chapter 2.1.2.

Regeneration of a damaged shell takes place more rapidly than does increase in size at the edge of the shell; the epithelium of the mantle plays an important role, and shows, during the process of repair, an increase in RNA and alkaline phosphatase. The increase in RNA is a clear indication that proteins are synthesized; these are used in the formation of a thin regeneration membrane. This gradually becomes calcified and the alkaline phosphatase may be associated with this process. Peroxidases in the exposed mantle epithelium and also in the newly repaired outer shell membrane are indicative of the formation of "tanned" periostracum proteins. ABOLINS-KROGIS (1968) found evidence in *Helix aspersa* Müller, 3 h after damage to the shell had occurred, of activation both of the mantle and of the hepatopancreas, amoebocytes transferring calcium and other substances from these regions to the wounded site. It is not clear whether this calcium is derived from the liver cells themselves or directly from the food. The fact that in summer months, when food is readily available, shell regeneration is much quicker than at other times suggests that the calcium is derived from the food. In some cases it is not clear how the calcium is transferred to the site of the damage; thus in *Lymnaea stagnalis* (L.), for example, TIMMERMANS (1973) was unable to detect any calcium-transporting amoebocytes.

Many pest pulmonates (Helicidae, Succineidae and the African giant snail *Achatina fulica* Bowdich) are capable of shell regeneration, while amongst the marine molluscs, 28 species of Prosobranchia, 3 species of Opistobranchia, 6 species of Lamellibranchia [including *Mytilus edulis* L. and *Cardium edule* (L.)] had still not repaired shell damage 4 weeks after it had occurred (WAGGE and MITTLER 1953).

1.3 Ecology

A characteristic feature of molluscs is the high water content of their bodies and their eggs. Thus *Deroceras reticulatum* (Müller) has a body water content of more than 80% and its eggs of over 85%; the average value for the body water content of *Arion rufus* (L.) is 86.84%, and the slug dies after loss of 20% body water. As a result there is a close relationship, especially for slugs, between population density and the moisture of the soil in a particular area. This will be described in connection with pest damage and control measures in the relevant sections of Chapters 2 and 3. As a result of evaporation of water from the skin and the secretion of mucus, body water is lost and must be replaced rapidly by water uptake from the environment; this occurs not only by drinking, but also by absorption of water through the skin itself, which is non-keratinized and consists of a single cell layer (HOWES and WELLS 1934, PUSSWALD 1948). Compared with other animals, gastropods have a relatively large volume of blood, and the blood forms the main water reservoir of their bodies; but the skin also contains much water, the relative amount varying from species to species.

Amongst the terrestrial pulmonates, slugs are found predominantly in moist, cool places, whereas there are some species of snails which can endure warmth, light and dryness over relatively long periods, as for example the mediterranean or atlanto-mediterranean *Helix aspersa* Müller, *Cernuella virgata* (Da Costa), *Theba pisana* (Müller), *Cochlicella acuta* (Müller) and members of the genera *Xerotricha* Monterosato and *Trochoidea* Brown. Such gastropods have thus the potential for being widely distributed because they can withstand long periods of dryness or fasting. Similarly the xerophilic Helicellae, like *Helicella italia* (L.), *Helicella obvia* (Hartm.) and *Helicella bolenensis* (Locard) are often transported along with dry seed from clover, lucern and the sainfoin fields in mediterranean areas.

In central, western and eastern Europe, where the climate is very moist in comparison with these warm countries, slugs are the more common molluscan pests, while in more southerly countries, snails predominate.

Among the slugs, *Deroceras caruanae* (Pollonera) prefers warm areas. This slug has a life cycle similar to that of *Deroceras reticulatum* (Müller), but development is more rapid, the optimal temperature being 20 °C. This slug is mostly found in gardens. Amongst the known pest species of Arionidae, some prefer a warm, while others prefer a cold environment. Thus according to MOENS and VAN DEN BRUEL (1960) *Arion rufus* (L.) is a "summer" slug; it increases in size in spring and summer, reproduces in August and September and passes the winter as an egg or a resting young animal. *Arion hortensis* Fér., on the other hand, is a "winter" slug, which increases in size during the warm months, reaches sexual maturity in the autumn and winter and only reproduces in the early months of the following year. *Arion circumscriptus* Johnston behaves similarly to *Arion hortensis*, reaching sexual maturity somewhat later, however. *Arion subfuscus* (Drap) is likewise adapted to a colder environment and is particularly sensitive to warmth and dryness; the egg-laying season can last from late fall right through the winter and spring. In its feeding habits this slug is adapted to wintry conditions, for besides plants, on which it feeds in the period when vegetation is available, it also devours insects and earthworms in considerable quantities.

The dependence of land pulmonates on a moist environment is, of course, greater in the case of slugs than for snails, since the latter derive a certain amount of protection from their shell. In certain circumstances they are able to withstand long periods of dryness without injurious effects. This is particularly important for those species whose general habitat is above the ground on herbaceous plants or shrubs.

In the dry summer months, many snails enter a period of suspended activity, called **aestivation;** they remain firmly attached by hardened mucus to the bark of trees, to leaves, twigs and branches, often 2 to 4 m above the ground, with the body withdrawn into the shell. In the cool of the evening, when dew falls, the snails re-

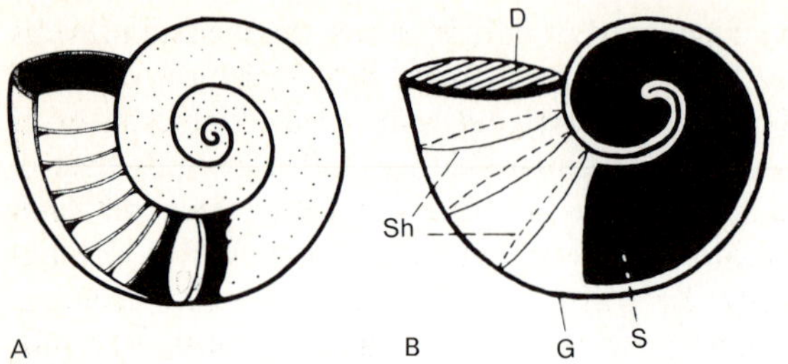

Fig. 24 A, B. Snails in aestivation (**A**) and hibernation (**B**); **A** *Helicella obvia*, showing diaphragms, often with calcareous inclusions; **B** *Helix pomatia* in hibernation; *D* epiphragm (calcareous disc); *G* shell; *S* resting animal; *Sh* partition walls of dried mucus

emerge and feed. In long periods of continuous dryness, this resting stage is uninterrupted; in this case the mouth of the shell is closed by several layers of dried mucus, each separated from the next by an air-filled space (BLOCK 1971). A snail which is particularly well-adapted to extreme dryness is the xerophilic *Helicella obvia* (Hartmann) (MAZEK-FIALLA 1933). The mouth of the shell is covered with one to nine "diaphragms", the number depending on the length of aestivation, and the membranes have inclusions of calcareous material (Fig. 24 A). Respiration and metabolism cease during this period, while the water content of the snail's body is reduced to a very low, species-specific value. When rain at last falls after several weeks of dryness, the individual mucous films dissolve, the water content of the body rises, and the normal metabolic processes are resumed.

Hygrophilic snails, whose habitat is confined to damp or wet environments, are unable to close the shell with a layer of dried mucus; of those which live in habitats in which vegetation provides permanent cover and the soil is always moist, there are some species which close off the shell and others which do not.

During aestivation several metabolic changes take place (HORNE 1973; NOPP 1972; SACCHI 1974). The osmotic pressure is also raised to three times the normal value. BRAHMANADAM and KRISHNAMOORTHY (1973) have established that in the snail *Pila globosa* (Swainson), the concentrations of calcium, magnesium and sodium in the blood rise, while those of chlorine and potassium are not affected. In *Helix pomatia* L., too, there is a rise in the calcium level during aestivation (BURTON 1971, 1972).

During **hibernation,** land gastropods creep into frost-free hides (holes in the ground, fallen leaves, dense vegetation, amongst the tree roots in woods, amongst others). The mouth of the shell of snails is then closed, not by a soluble mucous layer, but by a robust calcareous operculum, which is cast off in the spring (Fig. 24 B). Hibernation begins in autumn, even if temperature and humidity are still favourable and food plentiful. *Helix pomatia*, for example, hides itself away in the middle of October. The length of hibernation is, however, dependent on environmental conditions. Thus hibernation starts at more or less the same time in *Deroceras reticulatum* (Müller), *Deroceras laeve* (Müller) and *Arion circumscriptus* Johnston, whereas the spring re-emergence of these species takes place at different times, *D. laeve* emerging three weeks before the other species. Of course, *D. reticulatum* passes the winter mainly in the egg stage, but some adults of the species do survive and lay eggs in the spring (GETZ 1959).

It is not known what chemical substances are involved in inducing the onset of hibernation. According to MEINCKE (1972), the relative values of Na^+ to K^+ in the haemolymph is of great significance. During hibernation changes take place in the ionic composition of the blood and also in metabolism in heart tissue and nervous system (Cardot 1971; Nagabhushanam and Mantale 1971, 1972 and others).

In mid-January, at 0 °C, both osmotic pressure and ionic concentration of Na^+, K^+ and Cl^- in operculate *Helix pomatia* L. are significantly higher than in active snails in the summer. From the middle of January until the end of March at a temperature of 0 °C, both the osmotic pressure and the Na^+ and Cl^- concentrations fall, while the K^+ concentration remains unchanged. In these experiments this first rose after 24 h exposure to a temperature of 4 °C, and then after each subsequent change in illumination at a photoperiod of 8 h of light to 16 h of darkness.

During aestivation there is a rise in the amount of free Ca^{++} and during hibernation a rise in the level of protein-bound, non-excretable calcium; both these shifts in concentration are reversible. Tilgner (1958) is of the opinion that the rise of calcium ion concentration in winter is a result of the increase in protein content of the haemolymph, but in summer it is a response to an increased concentration of acidic substances which result from intermediate stages of metabolism. A similar fluctuation in phosphate concentration has not been established.

Measurement of oxygen uptake gives an indication of the rate of utilization of reserve materials which were stored before the onset of the rest period. Thus at an average temperature of 2 °C (in Prague) between the end of October and the end of March, hibernating *Helix pomatia* use 100 µl O_2 per gram weight of tissue, or 15 ml O_2 in 150 days. This is equivalent to the metabolism of 18 mg of glycogen; the observed loss of glycogen from the tissues over this period was in fact 13 mg of glycogen. Blažka (1955) therefore concludes that the reserves of polysaccharides in the snail body are sufficient to supply most of the energy needed during hibernation. In aestivation, at 25 °C, only 1.15 mg of glycogen per day are utilized. The available reserve of 15 mg/g of body tissue is sufficient for a 10–12 day period of fasting; thereafter protein is probably used as a source of energy, since considerable quantities of uric acid are excreted during prolonged starvation. Fats play no role as storage materials in molluscs.

In hibernation the rate of the **heart beat** is reduced from 10 to 13 beats/min in active *Helix pomatia* L. in summer, to 5 to 6 beats/min at the onset of hibernation on 22nd November, and even to 4 beats/min in deepest hibernation (on January 10). The reduction is, however, different from species to species; in *Helicigona arbustorum* (L.), for example, it is reduced to less than half the normal value. This possibly relates to the fact that *Helicigona* lives in loose soil and thus does not need to enter very deep hibernation; certainly aestivation is not as pronounced in these animals as in the Cepaeidae which live exclusively above the ground.

The influence of ecological factors on population dynamics (fertility, longevity, mortality rate) and on the sensitivity of slugs and snails to pesticides, will be dealt with in connection with these specific questions in Chapters 2 and 3.

1.4 Methods for Mass Rearing

Mass rearing of molluscs in mussel banks or snail farms has been practiced for a relatively long time – for food in Europe and for both food and for pearl cultivation in Japan.

1.4.1 Lamellibranchia

In bays of the shelf-sea, **oyster banks** have been established, for example, in the basin of Arcachon on the French Atlantic coast, south of the English Channel, where earlier cultivation of *Ostrea edulis* L. has been replaced today by the hardier Portuguese *Crassostrea angulata* (Lamarck). This species has a harder shell, which is elliptical in shape, in contrast to the round *Ostrea*.

Mussel farms may be set up directly on the beach or on sandbanks in shallow water. They may consist of rows of poles which are stuck into the sand, or alternatively, of frames from which lines are hung, each line carrying wooden or metal plates. These frames are sited either in non-tidal shallow water, as in the Mediterranean, or on floats in deep water, for example, off the Spanish coast. In Europe two species of edible mussel are cultivated: the thermophilic *Mytilis galloprovincialis* Lamarck, found off the Spanish coast and in the Mediterranean, and around the Atlantic sea board of northern Europe, Canada and the USA the more northerly, cold-resistant *Mytilis edulis* L. The two species are found together in the English Channel, on both the French and English coasts, and also on the coast of Ireland. Fluctuations in their geographical distribution are correlated with periods of relative warmth or cold (Lubet and Dardignac 1975).

Aquaculture, in which young animals are reared in tanks and the adults then set out on the shore area, is likely to be developed further in the future, since the demand for seafood is increasing and yields from the simpler methods used up to now are declining as a result of pollution of coastal waters; this technique will also be necessary to assist with man's growing need for sources of protein food. Aquaculture of oysters, mussels, shrimps and phytoplankton has already had success in France. On north European coastlines and in the USA, oyster culture is accompanied by cultivation of algae to use as food for the molluscs. In the Netherlands, Spain, Ireland and Yugoslavia, too, the possibilities of commercial culture of molluscs are either already being exploited or are under consideration. It is not, however, limited to oysters and mussels, but includes other species like *Pecten maximus* L., *Mercenaria mercenaria* L., *Mya arenaria* L., *Donax trunculus* L. and others. Aquaculture is being accorded ever more recognition and is growing in economic importance (Korringa 1976 and others).

1.4.2 Gastropods

Free-living *Helix pomatia* L. are collected and then kept and fed on snail farms. Shortly after the war they became an important export item from Central European countries to France. Even in the middle ages there were already snail farms in southern Germany, Baden-Württemberg and Rheinland-Pfalz. Accounts of the setting up of such "gardens" are given by Hein (1952), Kilias (1960) and Nietzke (1963). Several rules in maintenance and feeding procedures must be observed if death of the snails is to be avoided. The upper surface of the soil in the snail "garden" must not be allowed to become dry with values of less than 60% relative humidity; to avoid exposure to direct sunlight, suitable shelter, such as stones or slabs

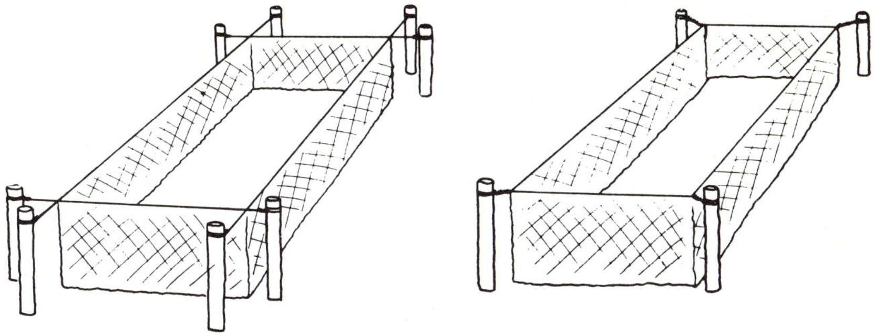

Fig. 25. Snail enclosures showing method for securing wire netting

of peat must be provided, as well as sufficient ground coverage with plants and damp moss; the soil should not be too hard and ideally should be loose, damp and loamy; both calcium deficiency of the environment, indicated by mutual rasping of shells and vitamin deficiencies, particularly of A and B resulting from feeding the snails exclusively on potatoes and carrots, must be avoided. Finally, the roof and sides of the enclosure must not be covered with airtight material such as fine gauze: most suitable is wire netting with a mesh not greater than 25 mm so that the snails cannot escape. The wire netting must be at least 50 cm high and reach 10 cm into the soil; the upper edge should be bent inwards at right angles to the sides, to prevent the animals escaping. The stakes to which the meshing is attached should, as suggested by KILIAS (1960), be on the outside of the enclosure (Fig. 25). Trees and shrubs in the enclosure should be protected by a gauze screen to a height of about 30 cm so as to prevent the animals from climbing them, thus making collection difficult. The number of snails in a "garden" should not be too large; some farmers recommend a maximum of $100/m^2$, but even this appears to be too high. Thorough and repeated watering of the enclosures is necessary, the frequency depending on weather conditions. Overhead spray irrigation should be used so that the snails are also wetted; stagnant water must, however, be avoided. The provision of a variety of food is important; this may include lettuce, cabbage, leaves, dandelions and other herbaceous plants, apples, pears, cucumber, pumpkin and now and again some baker's yeast to meet the vitamin needs. Finally, the soil should be limed frequently.

As a result of continuous, uncontrolled collection, the naturally occurring population of *Helix pomatia* L. has been reduced to such an extent that protective measures have become necessary, collection now being limited to certain times of the year only. In the Federal Republic of Germany, a Nature Conservation Regulation (Bundesartenschutz-Verordnung) of 25.8.1980 § 8 restricts collecting to certain Federal States only, and between 1 st April and 15th June each year. But only snails with a shell of more than 30 mm diameter may be collected, this being an indication that they have already laid eggs. Exporters of edible snails supply their collectors with a "measuring ring" with which to gauge shell size. Despite these protective measures, the population of *Helix pomatia* L. in the Federal Republic of Germany has continued to fall, since special exemptions are always being granted and the regulations not always observed. In each of the areas which have

Table 1. Import of fresh or frozen snails (escargots) to France. (Modified after Lecomte 1973)

Country of origin	Tonnes		
	1969	1970	1971
Turkey	1,819.4	2,150	2,052
German Federal Republic	1,804.5	1,225	864
Yugoslavia	502.7	1,028	880
Hungary	277	596	870
Greece	465.5	562	1,155

so far come under protection, the snail population has been sufficiently restored within a period of 3 years for limited collecting to be allowed once more.

In European countries, like Switzerland and the Netherlands, *Helix pomatia* L. is already included amongst the protected species, and in the Netherlands collecting, trade and import are completely forbidden. In France the situation is similar to that in the Federal Republic of Germany and a certain amount of protection is possible through action by local authorities. In two recent reports (1975 and 1976), the IUCN (International Union for the Conservation of Nature and Natural Resources) put *Helix pomatia* on the list of animals which should be protected throughout Europe.

The economic importance of molluscs as articles for **import** or **export** can be illustrated by the example of France (Lecomte 1973). In the years 1969 to 1971 60% of the imported snails came from four countries – Turkey, the Federal Republic of Germany, Yugoslavia and Greece (Table 1).

Although there has been a decrease in the number of fresh snails produced by the Federal Republic of Germany, the export of tinned snails has increased from 0.3 t in 1969 to 13 t in 1971.

The natural stock of edible snails in France is also decreasing, while demand continues not only for home consumption, but also for export to the USA and Canada. Breeding under factory-like conditions in the future will be unavoidable (Chevallier 1973, 1976). Indeed, on the heliciculture farms already mentioned, not only feeding of field-collected animals, but breeding of both *Helix pomatia* (Escargot de Bourgogne) and *Helix aspersa* (Escargot Petit-Gris) is being carried out on a large scale.

The **mass rearing** of molluscs today makes possible their use as experimental animals. Freshwater gastropods are grown so that their role as intermediate hosts of parasitic worms of man and animals can be elucidated. Terrestrial snails and slugs are reared so that their biology and physiology, and fluctuations in their population density can be studied; they are also used to test molluscicides.

There are methods for the breeding of *Biomphalaria glabrata* (Say) (Hopf and Müller 1962), *Bulinus forskali* (Ehrenberg) (Rowan 1958; Deschiens and Adetonah (1969), *Bulinus truncatus* (Audouin) (Standen 1949), *Oncomelania* spp. (van der Schalie and Davis 1968; Vogel 1948; Wagner and Chi 1959) and *Galba truncatula* (Müller) (Ricou et al. 1973). The last-named Mud Snail can be bred for relatively long periods on damp earth in containers with holes; the increase in numbers is relatively modest but, according to Kendall (1953), can be improved by feeding the snails with algae. In the natural environment, too, algae are found to be associated with the Lymnaeidae.

Recently, the mass rearing of the South American freshwater snail *Marisa cornuarietis* (L.) has been successful (RICH and ROUSE 1970); the snails are then used as biological control agents instead of herbicides to reduce the quantity of aquatic plants in canals and ditches (FERGUSON and BUTLER 1966).

Attempts to find methods for the mass rearing of pest gastropods, particularly slugs, are of particular interest for plant protection. Procedures have been described for *Limax*, *Deroceras* and *Arion* species (FRÖMMING 1954; GODAN 1958), *Milax budapestensis* (Hazay) (STEPHENSON 1966) and also for *Cepaea nemoralis* (L.) (WOLDA 1972). The methods are essentially the same as those for the maintenance of edible snails on snail farms.

For mass rearing indoors – or at favourable temperatures in the garden or open greenhouse – the slugs and snails are separated according to species and the containers isolated from one another by fine meshing. Slugs are best kept on damp, loose soil in a terrarium with low glass sides to retain moisture. For the lid of the containers, it is best to use perlon gauze of 1.0 to 1.5 mm mesh, tightly stretched across a metal frame. A tight fit between the lid and the sides of the container is very important to prevent young animals from escaping; this is best achieved by sticking thin strips of Tesamoll (adhesive-backed foam plastic) to the upper edges of the terrarium. Ideally, for ease of handling, the terrarium should have a surface area of 25 cm × 35 cm for smaller species and young animals of larger species, and 40 cm × 80 cm for adult animals (large species). Flowerpots of at least 18 cm diameter are also suitable; these can be closed at the top with perlon gauze held in place by a rubber ring; and the hole in the bottom is likewise covered with perlon gauze. A mixture of damp loam and peat (1:1) is a suitable substrate; this should be about 6 cm high, with flower pots 2/3 full. Impregnated wood should not be used to build a terrarium. Containers made of cellulose nitrate or cellulose acetate have been shown to be toxic, but not those made of polyethylene substances. About every five weeks, old, hardened, finely crushed lime or mortar is spread over the soil in the containers to provide calcium, which even slugs require. The terraria are planted with long-stemmed moss (*Sphagnum*, *Polytrichum*), green plants (*Tradescantia*) and with strongly rooted grass plants. Slugs and snails find shelter and lay their eggs in the nooks and crannies between grass roots and in moss. This applies also to slabs of peat which are kept constantly moist; they absorb and retain moisture readily, even if the container is not watered for some time. The addition of decaying but not rotten leaves, which may also have a covering of mould, is to be recommended.

Temperature is important in some cases. The freshwater snail *Biomphalaria pfeifferi* (Krauss), for example, leaves the water if the temperature is 18 °C and crawls up the sides of the container; this behaviour is not observed when the temperature is between 25 °C and 27 °C (SHIFF and GARNETT 1967). Although the mortality rate of this snail up to the onset of sexual maturity was higher at 25 °C than at 18 °C (31.7% compared with 23.3%) and fertility lower than at the lower temperature, the rate of increase of the particular population rose rapidly, since the rate of development was considerably greater at 25 °C than at 18 °C. At 27 °C the rate of increase in the population rose even further, but in this case egg production was inhibited.

A temperature of 15 °C to 17 °C appears to be optimal for the rearing of most Central European terrestrial pulmonates. For the general well-being of the animals, a relative humidity of 75% to 85% should be maintained, but pools of water must not be allowed to collect. Constant illumination, whether natural or artificial, should be avoided; light so inhibits the embryonic development of *Arion hortensis* Fér. for example, that the whole of a culture may die. Diffuse light, alternating with darkness at night, has proved optimal. Cleanliness is obviously essential; the larger slugs in particular produce large amounts of faeces, and these, as well as rotting or slimy food remains, create an unfavourable environment. Normally, weekly cleaning of the containers suffices for young animals and small species; but twice weekly cleaning is needed for adult *Limax* sp. More frequent cleaning can be disadvantageous because of the disturbance to the animals.

It is necessary to limit the number of gastropods in each terrarium. If the containers are too small, fertility and longevity decline, effects which are particularly marked in *Arion hortensis* Fér. Henderson (1968) found a sharp increase in mortality when more than 50 Field Slugs were kept in plastic boxes about 30 cm across. Values for acceptable stocking densities according to Frömming are given in the following table.

Godan (1958) kept 4 animals of the smaller but only 2 of the larger species in containers of 25 cm × 35 cm, and in terraria of 40 cm × 80 cm 6 adults of *Limax* spp. or *Arion* spp.

Container size	No. of animals	Relative size
2,000 cm^3	1	*Helix pomatia, Arion empiricorum*
1,000 cm^3	2 to 3	*Helicigona (Arianta), Cepaea, Arion subfuscus, Lehmannia*
1,000 cm^3	5 to 6	*Monacha, Helicella, Deroceras reticulatum*
500 cm^3	5 to 6	*Succinea, Arion hortensis*
500 cm^3	8 to 10	*Bradybaena, Deroceras laeve*
250 cm^3	5 to 6	*Arion intermedius*

Terrestrial gastropods make nest holes in loose soil for **oviposition.** The soil in the container must, however, be at least 6 cm deep, otherwise egg-laying may be inhibited; this was observed, for example, in *Cepaea nemoralis* (L.) when the soil was less than 3 cm deep (Wolda 1965). *Milax* spp. need firm clods of soil in which they bore passages in which the eggs are deposited. The eggs can be left in the holes to avoid damage and disturbance by their removal, but then the adults must be removed, because the container is too small for both generations and cannibalism on the part of the parents is possible; alternatively, the eggs can be removed from the nest holes and placed in another container. The advantage of the latter procedure is that the growth of the embryo can be constantly checked and if kept at a constant temperature the rate of development can be kept controlled. Two methods have been published which enable a daily check to be made of the progress of embryonic development. Schmidt (1955/56) puts the whole egg cluster into a petri dish containing sufficient damp soil; this is positioned so that the eggs can be observed through the sides of the dish and damp earth is then lightly pressed on

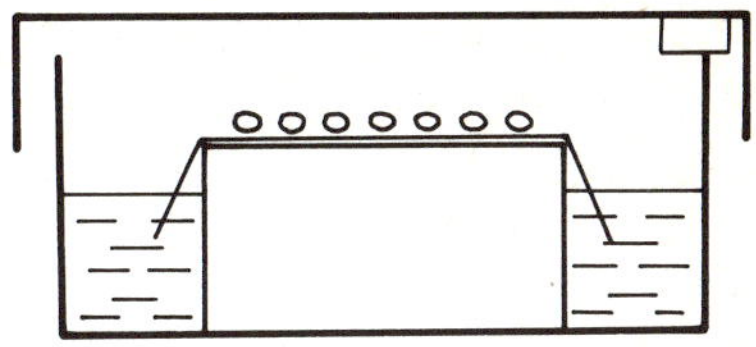

Fig. 26. Breeding dish for snail eggs

top of the little piles of eggs. A few days after hatching, the animals are removed from the petri dish.

DAXL (1970) describes a method in which eggs are separated from the egg mass on the day on which they are laid, and placed singly on wet filter paper in a culture dish (Fig. 26) and kept at a temperature of 20 °C until hatching. In both these methods the container must be kept in the dark; it may not be exposed to light, as this is unfavourable for embryonic development.

Finally, by no means the least important aspect of successful mass rearing is **nutrition.** GODAN (1958) recommends potato as the basic food, together with carrots and yeast, as sources of vitamin A and the B-complex; lettuce should be added for the phytophagous species. As a source of protein, in addition to the yeast, a casein-gelatine preparation can be used. This is prepared as follows: dissolve 1 g colourless gelatine in 15 cm^3 hot water; stir in 9 g casein powder and 100 mg Nipagin (a fungicide); knead the mass well. When it is cold, cubes or strips can be cut off. For very young animals a soft consistency is recommended; this is easily achieved by increasing the amount of water used.

The experiments of DE ANDRADE and CARVALHO (1972) on *Biomphalaria tenagophila* (Orbigny) demonstrated clearly the beneficial effects of supplementing the diet of normally phytophagous gastropods with *animal protein*. Over a period of 20 weeks they obtained the following egg counts, in each case for 10 snails with shell diameters of approx. 10 mm: for snails fed exclusively on lettuce, 219 egg batches with a total of 2,546 eggs; for snails fed exclusively on animal food (aquarium food for fishes), 648 egg batches with a total of 5,014 eggs; for snails fed on mixed food (lettuce and animal protein), 1,175 egg batches with a total of 18,024 eggs. The average per snail for each of these categories was thus 18.2, 36,0 and 128.7 eggs per day.

Winter lettuce grown in hothouses is expensive and also rots more readily than lettuce grown in the open in summer. It is therefore desirable, with mass rearing over long periods, to be independent of the supply of green leaf fodder and to use instead a substitute food preparation. In as yet unpublished experiments by GODAN, dog biscuits were used as the source of vitamins and mineral salts, since they are easily obtained and can be stored in bulk. The results thus far obtained with this method show that rearing of land gastropods on the basis of an artificial diet is possible. In these trials yeast and casein, successfully used as supplementary ingredients over a period of 20 years, were added. They could possibly be substituted by a larger quantity of dog biscuit, but this has still to be investigated. Wessonsalz, a mixture of essential mineral salts, was also added. The following recipe can be recommended: 33 g dog biscuit, 25 g casein powder, 50 g baker's yeast (dried yeast could possibly be used instead), 2 g Wessonsalz, 90 g boiled water (90 ml); this should be mixed, kneaded and formed into pellets of 3 g for larger animals, 2 g for

Table 2. Intensity of feeding on a synthetic diet (% = no. of actively feeding animals)

Species	Duration of feeding in hours		With lettuce %	Without lettuce %
Arion rufus	72		61.5	100
Adult	24	(13 animals	–	84,6
Juvenile	96	were used in	–	76.9
	72	each trial)		100
				100
	48			100
				100
				100
Deroceras reticulatum	96		–	100
		(10 animals		80
Adult	72	were used in		100
		each trial)		100
	48			90
				90
				80

smaller ones. It should always be freshly prepared and used on the same day. To prevent the pellets becoming mouldy, which was found to commence on the third day and progress rapidly thereafter, the fungicide Nipagin can be used in the proportion of 0.2 g to 200 g of the food mixture. These pellets proved, however, less acceptable to the animals than those without Nipagin, the rate of consumption of which was excellent (Table 2). They continued to be readily eaten, even when covered with a light layer of mould three days after preparation. It is thus important to replace the food pellets after 3 or 4 days.

The results of replicated experiments, set out in Table 2, give a better indication than mean values do of the attractiveness of this substitute food material, the main characteristics of which are the total absence of vegetable matter, the "tastiness" to the gastropods and the availability of sufficient vitamins and minerals.

The *increase in weight* of the individual animals was also satisfactory. After 15 days of feeding with pellets without the addition of lettuce, adult *Arion rufus* (L.) with an initial mean weight of 18.78 g, gained on average 31.4% (min. 17.7%, max. 44.9%) and for young animals of this species with a mean initial weight of 1.0 g, the average gain was 61.3% (min. 16.1%, max. 122.8%). In the same experiment, *Deroceras reticulatum* with a mean initial weight of 0.7 g, gained on average 28.2% (min. 15.4%, max. 64.2%). Mortality was normal during the 3 week observation period and did not differ from that of slugs fed with the usual plant, yeast and casein food. Egg production and fertility of the experimental animals could not, however, be assessed. This is necessary, of course, before the method can be recommended for full-scale use.

1.4.3 Parasites as Limiting Factors

Bacteria, fungi, protozoa, nematodes and mites all constitute potential threats to a mass rearing. *Fusarium* and *Arthrobotrys* were identified as fungal parasites, destroying even the eggs. Sterilization of the soil used in the terrarium helps to avoid infection. Nevertheless, contamination of the sterilized containers by the introduction of free-range snails for the purpose of complementing or improving the culture, is always possible. As a preventive measure, ARIAS and CROWELL (1963) recommend moistening the soil with water containing 200 ppm Agrimycin 100 (streptomycin 15% and terramycin 1.5%) and 200 ppm of the fungicide Captan. Dipping the eggs into this solution can protect them to some extent during embryonic development, and can also prevent the transfer via the eggs of fungi and bacteria to sterilized containers.

Adult slugs and snails may suffer infestations of protozoa, nematodes or mites. The nematodes are *Rhabditis lambdiensis* (Maupas), *Paragrolaimus* spp. and *Diplogaster* spp., while the mite is *Riccardoella limacum* (Schrank). These parasites are usually found under the mantle shield of slugs, but may occur also on the back and in the pneumostome. The mite in particular can move rapidly over the slimy surface of the body. Infection of new boxes occurs by way of the eggs, the moist outer surface of which is particularly attractive to nematodes. The eggs cannot, however, be treated with pesticides, since even low concentrations alter the protein. The elimination of nematodes is possible only after hatching, using methyl-p-hydroxybenzoate (0.025%) (ARIAS and CROWELL 1963).

Protozoa cannot be destroyed by this substance. Cultures may be infected by the highly dangerous *Tetrahymena (Paraglaucoma) rostrata* (Kahl) and also the less harmful species, *Tetrahymena limacis* (Warren), which may occur in large numbers in the Field Slug (*Deroceras* spp.) without pathogenic effects.

1.5 Molluscs as Carriers of Disease in Man, Animals and Plants

Molluscs may transmit disease, or they may serve as intermediate hosts for parasites of **man** and **animals.** Since *Deroceras* spp. and *Arion* spp. feed readily on human and animal faeces, they may ingest worm eggs and salmonellae and disperse them with their own excreta. Alternatively, these sources of infection may adhere to the slime on the body of the snail or slug, and as the latter crawl around, they may be deposited on vegetables, lettuces and strawberries. Human health is at risk when these commodities are consumed without being thoroughly washed (FRÖMMING 1952, 1953; STEINIGER 1957). *Deroceras reticulatum* and *Arion circumscriptus* Johnston ingest the proglottid stage of the tapeworms *Moniezia expansa* Rud. of sheep and *Cittotaenia pectinata* Goeze of rabbits; the eggs pass undamaged through the intestine of the molluscs. Field Slugs and the European Giant Garden Slug (*Limax maximus* L.) serve as intermediate hosts for tapeworms (Genus *Davainea* Blanchard) of fowls and that of pigeons (*Aporina delafondi* Railliet) and infect the birds when they are eaten. These parasites cause digestive disturbances and can cause substantial losses in poultry farming (WIKTOR 1973). CHEVALLIER (1973) considers that members of the families Arionidae, Limacidae, Milacidae and occasionally also Helicidae and Succineidae can all act as intermediate hosts for *Davainea proglottina* (Davaine). *Deroceras* spp. are also important as intermediate hosts for lung worms of grazing stock.

Because of its rich vitamin C content, **cress** is of great nutritional value in winter, and with the increase in its cultivation in recent years, damage caused by slugs and snails has in economic terms correspondingly increased. Control of the gastropods is necessary, not only on account of the damage they do through eating the plants, especially their roots, but also because they act as intermediate hosts for the liver fluke *Fasciola hepatica* L., the cercariae of which are deposited on the cress, with the resulting danger of infection with liver fluke when the cress is eaten (MOENS 1965; MÜLLER 1962).

In the tropics and subtropics gastropods play a major role as intermediate hosts of worm parasites of man and his domestic animals. In the wake of the expansion of agriculture and associated irrigation schemes, bilharzia, which is caused by *Schistosoma* spp. is steadily on the increase (MANDAHL-BARTH 1965; NAGUIB AYAD 1965). Infection by the parasitic larvae or cercariae, which are released into the water from the intermediate host, usually occurs in childhood. For an account of the problems and methods of combating the snails, the reader is referred to ANSARI (1973), MÜLLER (1962) and THOMAS (1973). The intermediate hosts of *Schistosoma mansoni* Sambon and *Schistosoma haematobium* (Bilharz) include aquatic snails of the genera *Biomphalaria* and *Bulinus*, while those of *Schistosoma japonicum* Katsurada include amphibious snails of the genus *Oncomelania*.

Diseases of **cultivated plants** are also carried and spread by snails and slugs. Experiments of HEINZE (1958) on *Deroceras reticulatum* (Müller), *Discus rotundatus* (Müller), *Oxychilus draparnaudi* (Beck) and *Lehmannia marginata* (Müller) have shown that Tobacco Mosaic Virus, introduced into the mouth, is still present in the digestive tract 2 days later and is thus transferable to plants by these species. Spores of *Alternaria* sp., *Fusarium* sp. and *Phytophthora* sp. have been found in the faeces and the slime on the bodies of *Arion* and *Limax* spp. and also of *Cepaea nemoralis* (L.), *Helicigona arbustorum* (L.) and *Helicella obvia* (Hartmann) (FRÖMMING 1955; HASAN and VAGO 1966). When slugs were fed on leaves infected with False Mildew (*Phytophthora phaseoli* Thaxt), and thereafter were offered healthy lima bean sprouts, the latter were found to be infected by the fungus 5 days later (WESTER et al. 1964). *Deroceras reticulatum* (Müller) spread Black Root Rot on cabbage and other crucifers in Brazil (ROBBS 1946). The snail *Achatina fulica* Bowdich distributes in its faeces spores of *Phytophthora palmivora* (Butl.), the cause of Black Pod Disease in cocoa plants in Ghana (EVANS 1973). This and other snails are responsible for the spread of Foot Rot in *Piper nigrum* (TURNER 1967). *Achatina* spreads *Phytophthora parasitica* Dast and *Phytophthora colocasiae* Rac. Spores of Rust have been found in the faeces of *Helicigona arbustorum* (L.), *Bradybaena fruticum* (Müller), *Hellicella obvia* (Hartmann), *Succinea putris* (L.) and *Arion rufus* (L.). *Zonitoides arboreus* (Say) has been identified as the disseminator of the spores of the fungus which causes Root Rot in sugar cane in Louisiana (RANDS 1924).

Freshwater molluscs can also cause damage in undertakings other than agriculture and horticulture, namely in water works. Thus MÁCHA (1971) found that the increase in numbers of *Physa acuta* Drap., *Pseudosuccinea columella* Say and *Radix peregra* (Müller) in the purification plant of sewage works could be so massive as to render the biofilters ineffective and put them out of commission. Elimination of the snails presents a problem, since the application of molluscicides may be harmful to the flora and fauna – algae, bacteria, amoebae and worms which occur in the biofilters. Recently, the freshwater mussel *Dreissena polymorpha* (Pallas) has appeared in waterworks in the Netherlands, in particular in the water of storage basins, where it clings to the walls of pipelines. Its elimination also presents problems (LEENTVAAR 1971).

1.6 Molluscs as Indicators of Environmental Pollution with Chemicals

Terrestrial, freshwater and marine molluscs are particularly sensitive to contamination with chemical substances; they store these in their bodies and are thus highly suitable as indicators of the degree of pollution of land and water.

From the point of view of plant protection, the substances which accumulate in aquatic molluscs and also in terrestrial pulmonates are of particular interest. There already exists a considerable literature on this subject. Particularly important are the chlorinated hydrocarbons. Their residues in molluscs can be remarkably high, indeed many times higher than in the affected area. A method of preparation of marine animals for determination of DDT, DDE, γ-HCH and HCB residues is described by ERNST et al. (1974, 1976), and also by PICER and AHEL (1978). A review of the literature dealing with land gastropods in regularly treated gardens and fields has been published by EDWARDS and THOMPSON (1973).

In England, residues of DDT and its metabolites up to values of 10 ppm, and in one case even 70 ppm, have been found in Field Slugs (*Deroceras* spp.) and Garden Slugs (*Arion hortensis* Fér.) from agricultural nurseries and ornamental gardens in which DDT is frequently used (DAVIS and FRENCH 1969; STRINGER and PICKARD 1964 and others). In North America, snails have been found to contain residues of chlorinated hydrocarbons from 3.5 ppm upwards; slugs were found to contain as much as 200 ppm, while earthworms from the same soils were somewhat less affected, containing only 50 and, less frequently, 100 ppm (GISH 1970). KUHR et al. (1974) report the following residues in slugs and snails collected from a regularly treated apple orchard:

	DDT	DDD	DDE	Total
	ppm/wet weight of animal			
Slugs	5.6	5.2	2.6	13.4
Snails	0.8	0.5	0.3	1.6
	ppm/dry weight of animal			
Slugs	28.1	23.9	12.1	64.1
Snails	1.8	1.5	0.7	0.4

What is significant in these results is the remarkable level of the total residue, and in particular that this is considerably higher in the slugs than in the snails. The reasons for this substantial difference are probably threefold: snails are largely protected by their shells, they go through rest periods during which they remain im-

Table 3. Residues of DDT and its metabolites at the conclusion of experiments with *Helix pomatia*. (Modified from COOKE and POLLAND 1973)

Treatment	No. of snails	Residue ppm/weight p,p'-DDT	Snail p,p'TDE	Total residue p,p'-DDT+ 1,11 p,p'-TDE)
Low dose (0.036 μg)	13	5.8 ± 1.0	1.3 ± 0.2	7.2 ± 1.1
Medium dose (14 μg)	11	84.8 ± 7.0	14.4 ± 2.3	101 ± 7
High dose (570 μg)	14	121 ± 16	35.1 ± 5.1	160 ± 20
Untreated (without $CaCO_3$)	9	0.08 ± 0.08	0.00	0.08 ± 0.08
Untreated (1.0 ml acetone)	13	0.26 ± 0.12	0.00	0.26 ± 0.12

mobile on one spot for several weeks, and they seldom make contact with the treated soil, but remain in the foliage for the greater part of their lives.

Since *Helix pomatia* L. is eaten by man, investigations of possible pesticide residues in their bodies are of special interest. DDT causes a reduction in weight of the shell and operculum of this snail. Experiments by COOKE and POLLARD (1973) have shown how high such residues can be (Table 3).

The dose was applied to lettuce leaves and these fed to the snails.

Residues of DDT do not occur in equal concentrations in all the organs of snails or slugs. The investigations of DINDAL and WURZINGER (1971) on *Cepaea hortensis* (Müller) have shown that the largest quantities were found in the gonad (29.59 ppm) and hepatopancreas (67.97 ppm) 48 h after feeding on lettuce treated with 60 μg DDT.

Earthworms and molluscs accumulate the residues of many organochlorine insecticides, such as Dieldrin, DDT, hexachlorcyclohexane, with the result that the concentration in the animals is higher than in the environment. In gastropods the residues are generally even higher than those found in earthworms from the same soil. On the other hand, the residues in both larval and adult insects are too small to be the main source of contamination of animals which prey on them. The situation in the case of gastropods is manifestly different.

Vertebrates eat molluscs and absorb the pesticide residues which they contain. Microarthropods both above and in the soil feed on the faeces of terrestrial pulmonates and thereby likewise ingest DDT-residues. Gastropods consume their own faeces and in this way they add to the already existing residues, constantly raising the level of pesticides in their bodies. This is particularly significant since the faeces of gastropods contain substantial quantities of pesticides; in the case of *Cepaea hortensis* (Müller), for example, 48 h after feeding on treated lettuce, the astonishingly high level of 648.09 ppm was found.

Given this great capacity of molluscs for accumulation of chemicals, it is not surprising that, to the extent that they feed on gastropods and mussels, vertebrates are endangered by the uptake of pesticidal residues.

Freshwater and marine molluscs also contain relatively large residues. Thus *Physa* spp. were found to contain DDT and DDE levels 115,000 times greater than the concentration in the water in which they were living (SANBORN and CHING-CHIEH YU 1973), and KHAN et al. (1972) found in *Galba palustris* (Müller) 69.15 ng of Aldrin and 13.95 ng of Dieldrin per animal. Analyses of oysters, mussels and other bivalves eaten by humans revealed residues of DDT, DDD, Dieldrin and Endrin in concentrations from 10 to 3,600 ppb (MODIN 1969), while more than 60% of the mussels contained DDT (FREDEEN and DUFFY 1970). Residues of Dieldrin have also been found in the tissues of *Rangia cuneata* (Gray) from tidal waters, which is eaten by man and represents an important economic and ecological asset. These residues, at their maximum value, were 2,000 times greater than the concentration in the surrounding water, namely 1,226 µg/kg mussel, after only 60 h sojourn in water containing 0.63 µg/l (PETROCELLI et al. 1973). This mussel, as well as the Virginian oyster *Crassostrea virginica* Gmelin, accumulates Dieldrin to a concentration of 10% to 23% of that contained in the sediment constantly washed by sea water, on which the oysters rest (PETROCELLI 1974). BUTLER (1973) cites 25 ppm DDT residue in this oyster after 96 h immersion in water containing only 1.0 ppb. It is not possible to eliminate such shellfish from the market since they appear perfectly healthy, despite the high residual contamination. Oyster tissues tolerate even 150 ppm without recognizable pathological symptoms. In this regard it should be borne in mind that in the countries in question, large quantities of oysters and other shellfish are consumed at a single meal. Oysters have, however, the capacity to eliminate pesticides and other toxic substances from their bodies if the level of pollution in the water falls (ROWE et al. 1971).

Residues of **phosphoric acid esters** can also occur in molluscs; thus in the marsh snail, *Melampus bidentatus* Say, of the salt marshes of North America, a value of 15.7 ppm/kg wet weight was measured after 10 days exposure (MARGANIAN and WALL 1972).

In the same way, **radioactive metals** can be introduced into the bodies of vertebrates by way of the food chain. For example, the snail *Viviparus ater* (Cristofori and Jan), which lives on the banks of Lake Maggiore, accumulates ^{65}Zn, which has been introduced to the lake by unnatural means. This snail forms the main foodstuff of the freshwater Sunfish, *Lepomis gibbosus* (L.), which thus ingests radioactive zinc as it feeds.

Particular attention has been paid to the pollution of the coastal waters of Japan; in 1959, after the **mercury** pollution of the mud of Minamata Bay, accumulation of the metal was found in *Katelysia japonica* (Gmelin) *(Venus japonica)*. In 1967 mercury was found in the coastal waters of Scandinavia, and in 1969 of Canada and the USA (BURTON and LEATHERLAND 1971; KOPFLER 1974). Bivalves like *Crassostrea virginica* Gmelin, *Mytilus edulis* L., *Pecten maximus* L., *Chlamys opercularis* (L.), *Mercenaria mercenaria* L. accumulate not only Hg, but also Mg, Sr, Fe, Mn, Zn, Ag, Co, Cr, Cu, Ni and Pb, and retain these metals at considerably higher concentrations than in the surrounding water (BRODTMANN 1970; BRYAN 1973; MAYER 1970; SHIMIZU MAKOTO et al. 1971; WINDOM et al. 1971). One consequence is to impede growth of the shell and the body in young animals. The freshwater snail, *Biomphalaria glabrata* (Say) is also suitable for the bioassay of heavy metals such as Cu, Cr, Cd, and Pb. Survival time of adult snails was 15 h with

1 ppm Cd, 26 h with 1 ppm Cu and 24 h with 1.4 ppm Cr; snail embryos survived for 12 h with 1 ppm Cd, 24 h with 1 ppm Cu and 4 days with 1 ppm Pb. *B. glabrata* can also be used as an indicator of radioactive ^{65}Zn (POZZI and MERLINI 1977; RAVERA 1977).

Changes in the composition of waters in the rivers and lakes in America and Europe have assumed alarming proportions. One such change results from the introduction of warm water from cooling installations, which leads to a permanent increase in the temperature and consequent reduction in oxygen content. Another source of pollution is radioactive waste. A reduction in numbers of the previously widespread cold water snail, *Lymnaea stagnalis* (L.) has occurred as a result of its high sensitivity to temperature changes as far as growth rate and fecundity are concerned. Radioactive ^{65}Zn emmitted from nuclear power stations – for example, along the banks of the Rhône – and released into the river, accumulates in the Swan Mussel *Anodonta cygnea* (L.), as does ^{90}Sr in the snail *Astraea longispina* (Lamarck). *Anodonta* can be used as an indicator of zinc contamination in water; it absorbs the metal ions very rapidly, maximum tissue content being attained within 4 days. Zinc is retained in the shell of the mussel largely as a result of physical adsorption, while the main storage occurs in the tissues of the animal, particularly the outer part of the mantle. The haemolymph plays a major role in the transport of zinc in the mussel. SHIMIZU MAKOTO et al. (1971) have drawn particular attention to *Mytilus edulis* L. as an indicator of ^{58}Co in the sea. On land, *Arion rufus* (L.) can be used for the assay of radioactive ^{54}Mn and snails for the estimation of radioactive fall-out (CAVALLERO and RAVERA 1966; SAND 1962).

As early as 1956, HARANGHY reported that carcinogens like 3,4-benzpyren cause metabolic damage in freshwater mussels.

Since molluscs store relatively large quantities of chemical residues in their bodies, many investigations of the effect of environmental pollution on their me-

Fig. 27. *Deroceras laeve*, length 20–25 mm

Fig. 28. *Helix aspersa* (shell up to 38 mm broad, 35 mm high)

Fig. 29. *Monacha cartusiana* (shell 10–12.5 mm broad, 6.5–7.5 mm high)

tabolism have been undertaken in addition to comparable investigations on water plants, crustacea, fish and insects. These include work on carbofuran (BOOTH, CHING-CHIEH YU et al. 1974), carbaryl (KANAZAWA et al. 1974), the insecticide Mirex (BORTHWICK et al. 1974; MARTIN et al. 1974) and also metoxychlor (COATS et al. 1974). The unusual capacity of gastropods to accumulate chemical residues is particularly well demonstrated in the case of metoxychlor; with a concentration of 0.0016 ppm in the water, specimens of *Physa* sp. were found to contain 15.7 ppm, while algae of *Oedogonium* sp. contained only 0.48 ppm and fish of *Gambusia* sp. 0.33 ppm. In the main, amphibious and aquatic molluscs, like *Physa* sp., are used as experimental animals; thus they were the indicators used in tests on 20 different organic substances (LU 1974), prolan (HIRWE et al. 1975), Dieldrin (SANBORN and CHING-CHIEH YU 1973), arsenic derivatives (ANSUL SCHUTH et al. 1974) and finally on the algicide Terbutryn (MACKENZIE 1974).

1.7 Important Pest Gastropod Species

1.7.1 Distribution

The molluscan fauna of continental Europe is to be found in the great **palaearctic zone,** which stretches from the Atlantic islands to the Orient and encompasses several **regions** – including the Atlantic, European, Mediterranean, Siberian, Central Asian (Iran, Irak, Northern Arabia, Turkey, Afghanistan, Tibet) and the Chinese

Fig. 30. *Cepaea vindobonensis* (shell 20–21 mm broad, 17–18 mm high)

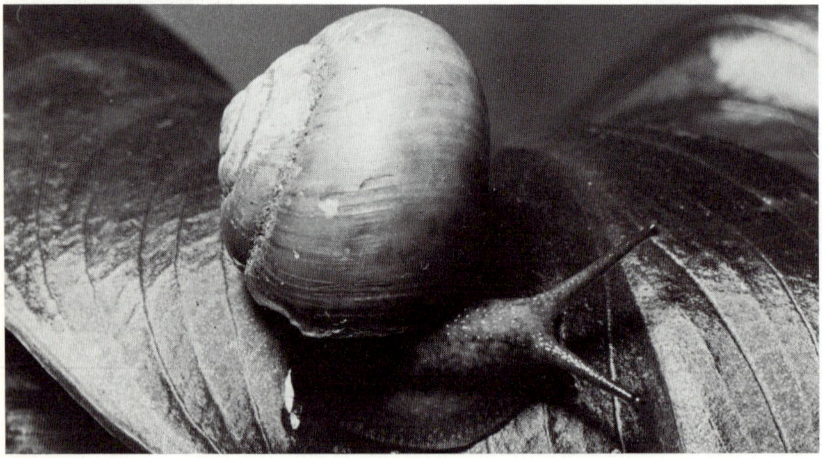

Fig. 31. *Bradybaena fruticum* (shell 18–20 mm broad, 16–17 mm high)

Fig. 32. *Oxychilus draparnaudi* (shell 12–14 mm broad, 6–7 mm high)

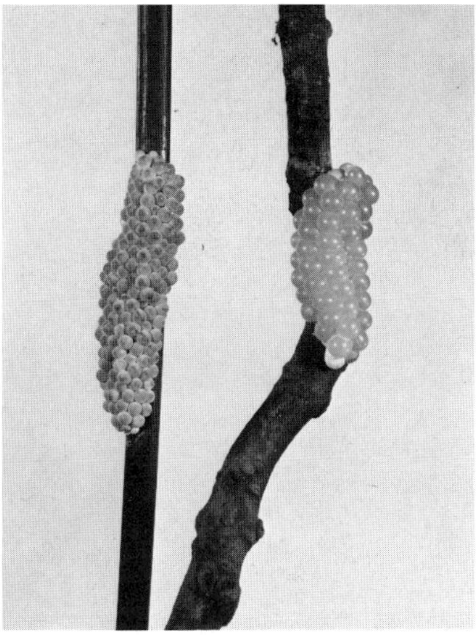

Fig. 33

Fig. 34

Fig. 33. Egg cluster of *Pomacea lineata* on a *Cyperus* stalk (*left*), egg cluster of *Pomacea glauca* on a dead branch (*right*), slightly less than natural size

Fig. 34. *Potamopyrgus jenkinsi* (shell 2.4–2.6 mm broad, 5–5.5 mm high)

and Japanese. The European pest gastropods have their origins in the Atlantic, European and Mediterranean regions. Several important terrestrial and aquatic pest snails and slugs are shown in Figures 27 to 43 and also Plates I; II; III/1–6; IV/1; V; VI; VII/1.

The pest molluscs of Europe, as well as those of other regions, have been spread in recent times by travel and trade to many countries and continents, even over oceans, so that many species are now widely distributed and no longer limited to their regions of origin. Thus the European Field Slugs *Deroceras reticulatum* (Müller) and *Deroceras agreste* (L.) as well as *Arion circumscriptus* Johnston and *Lehmannia marginata* (Müller) have spread from Europe to Australia; while the Mediterranean species *Cochlicella ventricosa* (Drap.) and *Cochlicella acuta* (Müller) have both spread northwards. The classical example is that of the African Giant Snail *Achatina fulica* Bowdich. The genus *Arion* is distributed over Europe, including the British Isles, parts of Siberia, North Africa and North America.

Representatives of the Limacidae and Milacidae are to be found in Europe, North Asia, Africa and America. *Deroceras reticulatum* (Müller) has spread to South America and *Deroceras caruanae* (Pollonera) to California and South Africa. *Milax budapestensis* (Hazay), which originated south of the Alps and Carpathians, is now found in Western and Southern Europe, USA, Canada, Hawaii and even Australia.

Of the basommatophoran pest gastropods, the following species belong to the palaearctic zone: *Galba truncatula* (Müller), *Galba glabra* (Müller), *Galba palustris* (Müller), *Lymnaea stagnalis* (L.), *Planorbarius corneus* (L.), *Gyraulus albus* (Müller); further, *Radix auricularia* (L.) is now also in Africa while *Physa fontinalis* (L.)

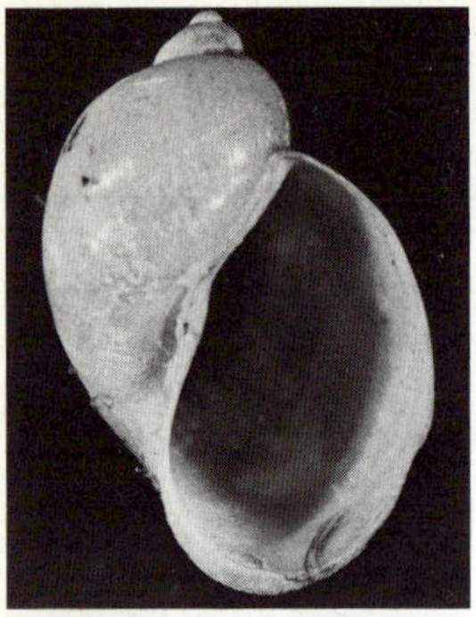

Fig. 35

Fig. 36

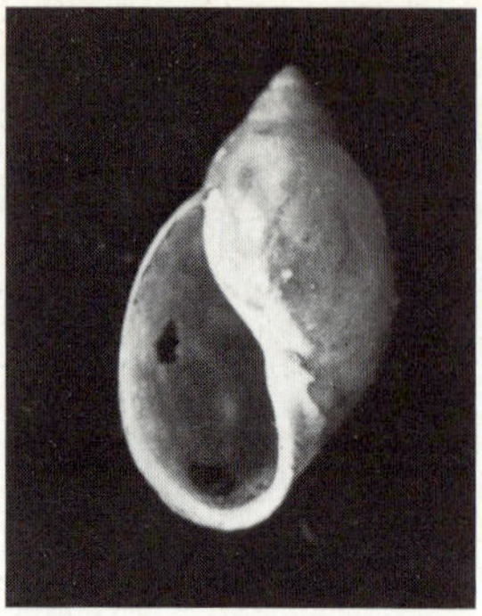

Fig. 37

Fig. 35. *Radix peregra* var. *ovata* (shell 16–17 mm broad, 24–25 mm high)

Fig. 36. *Galba truncatula* (shell 4–6.3 mm broad, 7–13 mm high)

Fig. 37. *Physa acuta* (shell 7 mm broad, 12 mm high)

Fig. 38. *Radix auricularia* (*left*), (shell 20–30 mm broad, 25–30 mm high) and *Galba palustris* (*right*) (shell 9 mm broad, 17–20 mm high)

Fig. 39. Full-grown shell of *Pomacea lineata*, 0.75 × natural size (*left*); full-grown shell of *Pomacea glauca*, slightly less than natural size (*right*)

as well as *Physa acuta* Drap. are today distributed all over the world. Amongst the stylommatophoran snails, the mediterranean *Helix aspersa* Müller has been spread world wide, and is today a pest throughout Europe, including England and Sweden, Western Asia including Israel, California and even Australia. *Theba pisana* (Müller) from the Mediterranean region, including Spain, Greece, Israel and Marocco is today a pest in Western Europe, South Africa, Saudi Arabia and America. *Helicigona arbustorum* (L.) has spread from Central Europe to Sweden

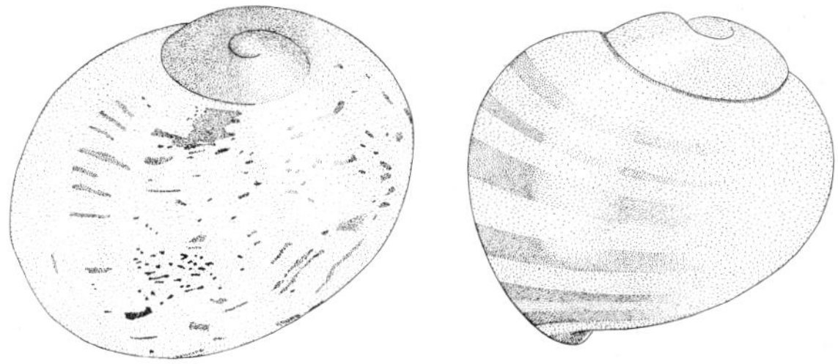

Fig. 40. Hyalinous shell of newly hatched *Pomacea lineata* (*left*) and *Pomacea glauca* (*right*)

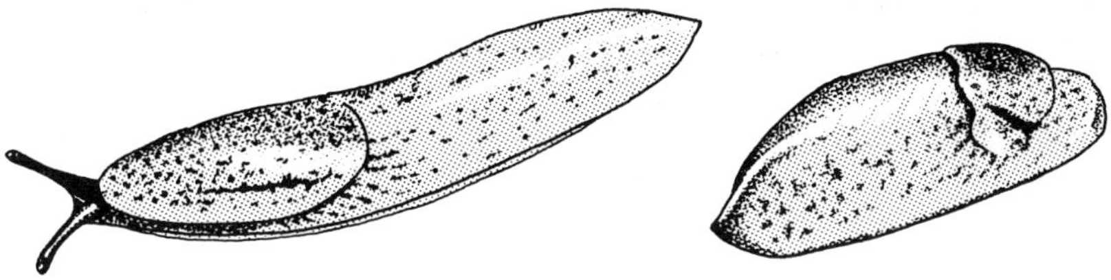

Fig. 41. *Milax rusticus*, crawling (*left*) and contracted (*right*)

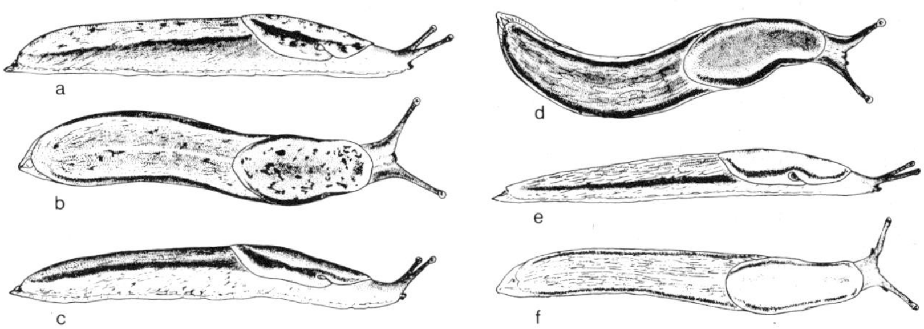

Fig. 42 a–f. Differences between *Arion circumscriptus* (**a, b**), *Arion silvaticus* (**c, d**) and *Arion fasciatus* (**e, f**)

and North America, while the mediterranean *Cernuella virgata* (Da Costa) has established itself in Australia.

Further examples, as well as the consequences of this involuntary spread, such as quarantine measures, are dealt with in Chapter 3.4.

Quite apart from human action, the spread of molluscs can result from winds (hurricanes), insects (mainly in the case of freshwater gastropods) and birds (Rees 1965). New infestation of land under cultivation has been examined by Schmidt (1955). In such areas, a species-rich gastropod fauna cannot establish itself, and the relatively small number of species can be carried to uncultivated land as a direct consequence of farming procedures. The Field Slug, in particular, has established itself on cultivated land, while *Vallonia costata* (Müller) and *Vallonia pulchella* (Müller) are frequently found in damp gardens, and on account of their small size are easily distributed unnoticed with soil removals.

Investigations of the vertical distribution of terrestrial gastropods have been undertaken on the slopes of the Jura mountains on Lake Geneva (Fabre 1927; IV

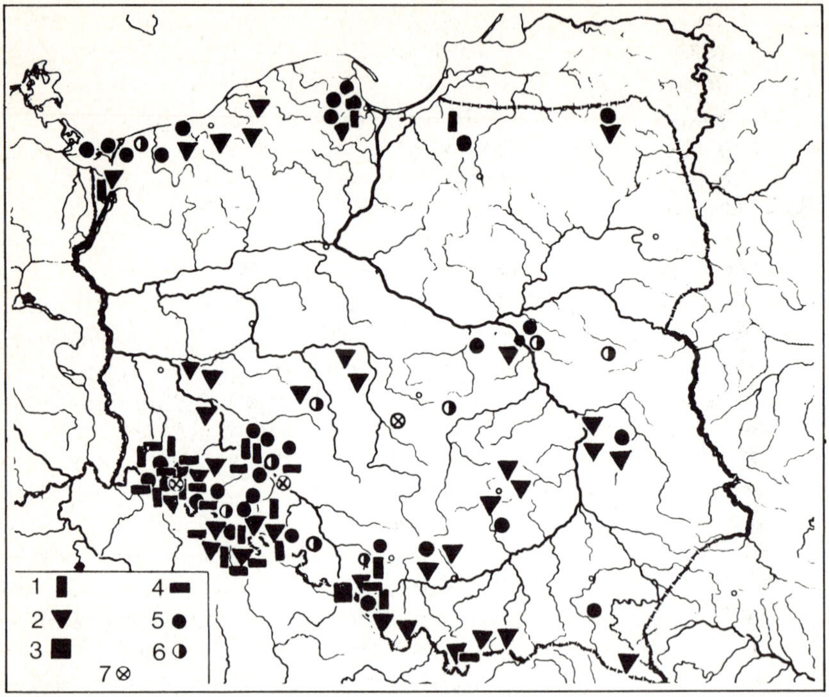

Fig. 43. Distribution of non-indigenous slugs in Poland. *1 Arion hortensis; 2 Arion fasciatus; 3 Milax budapestensis; 4 Boettgerilla pallens; 5 Limax maximus; 6 Limax flavus; 7 Lehmannia valentiana*

Cong. Malacol. Europ. 1971). The following distribution of the common species was recorded as a result: Up to a height of more than

1,700 m:	*Trichia hispida* (L.), *Fruticicola edentula* Drap., *Helicella itala* (L.), *Helicigona arbustorum* (L.), *Cepaea silvatica* (Drap.), *Cochlicopa lubrica* (Müller), *Discus rotundatus* (Müller).
1,600–1,699 m:	*Arion hortensis* Fér., *Helicodonta obvoluta* (Müller), *Cepaea hortensis* (Müller), *Oxychilus cellarius* (Müller), *Trichia villosa* (Studer).
1,500–1,599 m:	*Arion subfuscus* (Drap.), *Arion circumscriptus* Johnston, *Deroceras agreste* (L.), *Candidula unifasciata* (Poiret), *Helix pomatia* L.
1,400–1,499 m:	*Limax cinereoniger* Wolf, *Lehmannia marginata* (Müller), *Arion ater* (L.), *Galba truncatula* (Müller).
1,300–1,399 m:	*Deroceras laeve* (Müller), *Succinea pfeifferi* Rossmaeßler.
1,200–1,299 m:	*Euomphalia strigella* (Drap.), *Oxychilus helveticus* (Blum).
1,100–1,199 m:	*Cecilioides acicula* (Müller), *Succinea putris* (L.), *Succinea oblonga* Drap.
1,000–1,099 m:	*Cepaea nemoralis* (L.), *Armiger crista* (L.), *Lymnaea stagnalis* (L.), *Galba palustris* (Müller), *Radix peregra* (Müller), *Gyraulus albus* (Müller), *Bithynia tentaculata* (L.), *Zonitoides nitidus* (Müller).
900–999 m:	*Limax maximus* L.

700–899 m:	No pest gastropods recorded.
600–699 m:	*Milax rusticus* (Millet), *Bradybaena fruticum* (Müller), *Helicella obvia* (Hartm.).
500–599 m:	*Ancylus fluviatilis* Müller, *Monacha cartusiana* (Müller).
400–499 m:	*Helix aspersa Müller*, *Hygromia cinctella* (Drap.), *Physa acuta* Drap., *Physa fontinalis* (L.), *Radix auricularia* (L.).
320–399 m:	*Limax flavus* L., *Eobania vermiculata* (Müller), *Gyraulus laevis* (Alder).

The *horizontal* dispersal of pest species over large areas, particularly in **Europe**, has taken place in recent times within a relatively short period. It may be assumed that the increase in trade since the end of the 19th century, and the improvement in transport by rail and road within Europe have helped to bring this about. The case of Poland is a good illustration of this. As shown in Figure 43, all seven pest species found in this country reached it from the West (WIKTOR 1973). The species concerned are: *Arion hortensis* Fér., *Arion fasciatus* (Nilsson), *Milax budapestensis* (Hazay), *Boettgerilla pallens* Simroth, *Limax maximus* L., *Limax flavus* L. and *Lehmannia valentiana* (Fér.). The last of these species has its origin in the Iberian peninsula and has been spread over the whole of Europe. It is a pest primarily in greenhouses. In the light of current knowledge, slugs of the genus *Lehmannia* Heynemann associated with horticulture in Central Europe may all be considered as belonging to the species *Lehmannia valentiana* (Fér.) (PLATE, pers. comm. 1977; FLASAR and KROUPOVA 1976). Pest species from Central Europe have also been spread to Sweden – these include: *Limax maximus* L. since the first half of the 19th century, *Lehmannia valentiana* (Fér.), since 1921, *Deroceras caruanae* (Pollonera), since 1957, *Cecilioides acicula* (Müller), since 1822, *Oxychilus draparnaudi* (Beck), since 1871, and *Zonitoides arboreus* (Say), since 1938 (WALDÉN 1960). *Deroceras caruanae* (Pollonera) was found originally in only two places in France – in Marseilles in 1948 and in Britanny in 1958 – but by 1973 this snail had spread well beyond the coastal regions of the Atlantic and Mediterranean sea boards and was to be found in 70 different places (CHEVALLIER 1973). On account of its destructiveness *Milax budapestensis* (Hazay) was first noticed in Belgium in 1945 (MOENS 1959, 1962) and in less than 20 years spread beyond the borders of that country to infest Central Europe, England, Ireland and Switzerland as well. Pest species such as *Arion hortensis* Fér. and *Arion lusitanicus* Mabille have recently been introduced to Bulgaria (OŠANOVA 1964). *A. lusitanicus* has also recently been introduced into Germany by way of vegetable imports, and after being initially limited to the upper reaches of the Rhine and low-lying areas of Baden-Württemberg, it is now on the point of extending its territory (SCHMIDT 1971; VAN REGTEREN-ALTENA 1956, 1972), becoming in the process a very considerable pest (GÖTL, Backnang, pers. comm. 1977). Figures 44 and 45, which are based upon the studies by WIKTOR, depict the natural distribution of pest slugs in Eurasia, North America and Mediterranean countries.

Much research in recent years has been devoted to the interesting "worm slug", *Boettgerilla pallens* Simroth (≡ *Boettgerilla vermiformis* Wiktor) (ANT 1963, 1966; DAXL 1967; FLASAR 1962; GROSSU 1970; PLATE 1965; SCHMID 1962, 1963; WIKTOR 1960). This species made its appearance in 1959 in Poland, Czechoslovakia and

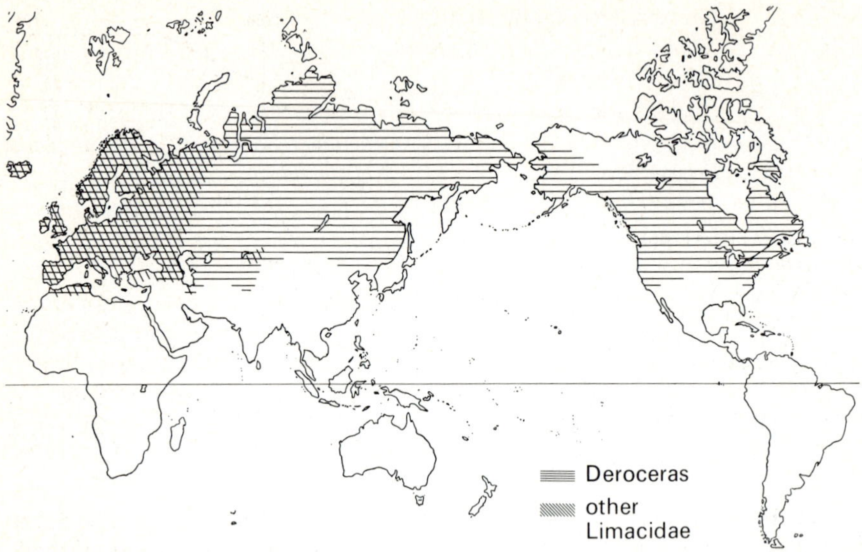

Fig. 44. Present distribution of important pest Limacidae in Eurasia, North America and the Mediterranean region

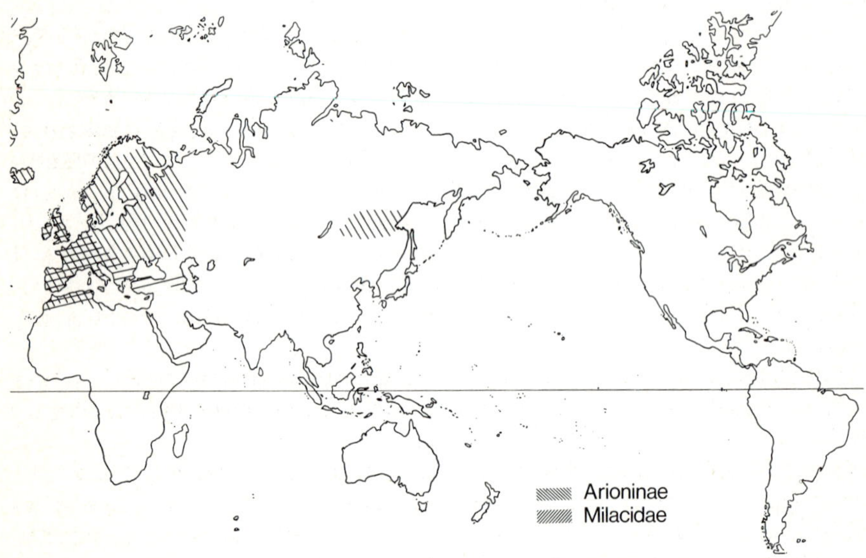

Fig. 45. Present distribution of important pest Arionidae (Arioninae) and Milacidae in Eurasia, North America and the Mediterranean region

Germany, later in Switzerland, Austria, France, Belgium, Finland and Rumania, and seems to be spreading throughout Central and Western Europe. *Boettgerilla* has a characteristic worm-like shape, unusual for terrestrial pulmonates, reaching a length of 40 to 45 mm with a breadth of only 2 to 3 mm. The light-coloured eyes, which contrast with the dark tentacles (Plate II/6) are a further good means of iden-

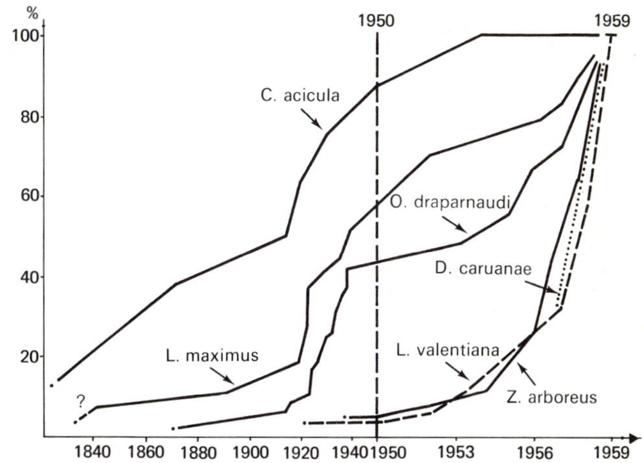

Fig. 46. Relative increase in the records of six gastropod species introduced into Sweden between 1822 and 1959. The number of record at the end of the period = 100%. Note: the time scales before and after 1950 differ

tification. The slug was probably introduced to Europe from the Caucasus in the last century. It preferentially attacks roots, buds and fruit. The holes which it makes, in contrast to those made by other slugs, are perfectly cylindrical, several millimeters deep and of constant diameter. The biology of *Boettgerilla pallens* Simroth has yet to be investigated.

At the IV Malacological Congress in Geneva in 1971, the malacologists of Europe decided to chart within the framework of the European Invertebrate Survey the distribution of terrestrial and aquatic molluscs and to prepare an **atlas** on the subject. A commission on which the various European countries were represented, was elected to co-ordinate the work (ANT 1973). Even during the preliminary stages, it became clear that a survey of this nature, on the occurrence of particular species, would have an unusually stimulating effect both on taxonomy and ecology (WALDÉN 1976). Such investigations have also enabled assessments to be made of the effect of human intervention on the gastropod fauna (ANT, Hamm pers. comm. 1977). All the molluscs of Europe are to be included in the survey, whether they are regarded as pests or not.

The first publication to appear was a distribution atlas for non-marine molluscs in Britain (KERNEY 1976). The pest gastropods listed are dealt with in Chapter 2.2.2.1.

The rate of spread of gastropod pests can be substantial. Figure 46 taken from WALDÉN (1965), shows the rapid spread of slugs, especially in recent times – in this particular case, between 1915 and 1959. The mollusc population can be expected to have subsequently declined in many centres as a result of increased industrialization and urbanization.

A further example of the speed of expansion over European countries, which applies not only to terrestrial but also to aquatic gastropods, is provided by the freshwater snail *Potamopyrgus jenkinsi* (E.A. Smith). From only a few localities in England and Ireland in 1900, this snail established itself within only 20 years throughout the whole of the British Isles, including Scotland, and by 1968 was to be found in the whole of Denmark, on the Atlantic coasts of Spain, France, Belgium and Germany and to some extent also in the interior of these countries, in the most southerly coastal regions of the Baltic, as well as in the northern parts of Sweden and Finland (REAL 1973; MOENS pers. comm. 1980).

1.7.2 Taxonomy and Nomenclature of European Pest Gastropods

1. Basommatophora: 1 pair of tentacles on the head, at the base of which the eyes are situated. The pest species live in fresh water and include the Lymnaeidae, Physidae, Planorbidae and Ancylidae. The last-mentioned family is characterized by a cupshaped shell and, like the Planorbidae, by a secondary leaf-shaped gill. The respiratory organ of the other families is a lung.

2. Stylommatophora: 2 pairs of tentacles, of which the larger pair is retractile, with eyes at the tips; the smaller pair is located near the mouth. These gastropods breathe through a lung and are either amphibious or terrestrial. In the amphibious Succineidae, the oral tentacles are either vestigial or absent altogether. There are species with, and others without a shell, while some have a small cup-shaped shell at the hind end.

The main **pest slugs** of Central Europe belong to the families Arionidae, Milacidae and Limacidae. The recent phylogenetic and morphological investigations by WIKTOR and LIKHAREV (1979) and also LIKHAREV and WIKTOR (1980) suggest that it is necessary to divide the Limacidae into five families: Milacidae, Boettgerillidae, Limacidae, Agriolimacidae and Parmacellidae.

The external characteristics which are initially important for taxonomic purposes are the position of the respiratory pore in relation to the middle of the mantle shield, the presence or absence of a keel on the dorsal surface, and the zonation of the sole of the foot (Table 4).

There are several pest species which at first sight are very similar in appearance, and can easily be mistaken for each other. Examples are: *Limax maximus* L. – *Limax cinereoniger* Wolf – *Limax flavus* L.; *Deroceras reticulatum* (Müller) – *Deroceras agreste* (L.); *Deroceras caruanae* (Pollonera) – *Deroceras laeve* (Müller) – *Deroceras sturanyi* (Simroth); *Lehmannia marginata* (Müller) – *Lehmannia valentiana* (Férussac) (= *poirieri* Mabille); finally, *Arion rufus* (L.) (= *empiricorum* Fér.) – *Arion ater* (L.) – *Arion lusitanicus* Mabille.

Among the **Limacidae,** *Limax maximus* L. is particularly noticeable on account of its size. While it is certainly not one of the major slug pests, the Field Slugs unquestionably are – at least in Central and Northern Europe. *Deroceras reticulatum* (Müller) and *Deroceras agreste* (L.) are distributed over the whole world, and both generally occupy the same habitat. Other species of the genus *Deroceras* Rafinesque found on cultivated land in Germany are: *Deroceras laeve* (Müller), *Deroceras caruanae* (Pollonera) and *Deroceras rodnae* Grossu and Lupu, as well as *Deroceras sturanyi* (Simroth) (WALDÉN, Göteborg, pers. comm. 1971; WIKTOR 1977). *D. caruanae* (Pollonera) has a similar life cycle to that of *D. reticulatum* (Müller).

The **Milacidae** mostly live in the soil and include the following species: *Milax budapestensis* (Hazay) which has brown, grey to black body colouration and darker longitudinal markings in the adult, but is a uniform white-yellow with dark

Table 4. External characteristics used to identify pest slugs

Family	Respiratory pore	Keel	Sole of foot
Limacidae	Behind midpoint of mantle shield	Present, does not reach mantle shield	Tripartite (a medial region and two lateral regions)
Milacidae	Behind midpoint of mantle shield	Present, reaches to mantle shield	Tripartite
Arionidae	In front of midpoint of mantle shield	Absent	Uniform (no longitudinal stripes)

Table 5. External characteristics for identification of small slugs

Species	Longitudinal band on either side	Creeping sole	Slime
Arion hortensis Férussac Black Field Slug	Not clear (only sharp on dorsal side); no dorsal row of tubercles	Yellow to dark yellow, orange	Body: yellow to dark yellow Sole: colourless
Arion subfuscus (Draparnaud) Dusky Slug	Clear, no sharp contrast, no dorsal line	Yellowish-white	Body: yellow to orange; Sole: colourless
Arion circumscriptus Johnston Gray Garden Slug	Clear, sharply contrasted, below a yellowish-white, dorsal row of enlarged pale tubercles	White to light grey	Body: colourless Sole: colourless

optic tentacles when newly hatched (FRÖMMING 1955); *Milax gagates* (Draparnaud) (KUHNA and SCHNELL 1963) and also *Milax rusticus* (Millet), both of which originated in the Mediterranean region and now occur also north of the Alps in Central Europe (BOETTGER 1949), and finally *Milax sowerbyi* (Férussac).

The worm-like slug, *Boettgerilla vermiformis* Wiktor, described by WIKTOR (1960) in Poland and later found in several places in Europe, proved to be identical with *Boettgerilla pallens* Simroth. *B. vermiformis* is thus a junior synonym *B. pallens*.

The family **Arionidae** comprises in the main animals which are characterized by a more compact form, a rough, wrinkled skin and an absence of the dorsal keel. In Europe, only representatives of the subfamily Arioninae are to be found. A complicating factor in their identification is the diversity of their body colouration. This is especially the case with *Arion rufus* (L.) (= *empiricorum* Fér.) which shows a colour variation from red through brown to black. The sole is light grey to yellowish and is covered with a colourless slime; the colour of the body slime, however, varies – in the case of red animals it is orange, in the case of brown or black it is colourless. *Arion lusitanicus* Mabille is something of a special case, since its classification is still uncertain. According to CHEVALLIER (1969), the external and

Fig. 47. *Testacella scutulum* (a carnivorous slug), moving (*left*) and contracted (*right*)

general features of this slug would place it with *Arion rufus* (L.) in the subgenus *Arion* (= *Lochea*), while the structure of its genitalia is such as to place it in the subgenus *Mesarion*, together with *A. (Mesarion) subfuscus* (Drap.).

Three small Arionidae are also found in central Europe, often together, on cultivated land, and are characterized by the external features listed in Table 5.

The body colour of *Arion subfuscus* (Drap.), too, can vary from ochre yellow to coffee brown.

The highly polymorphic species *Arion circumscriptus* Johnston is now separated into three distinct species, namely *A. circumscriptus* s.s., *A. silvaticus* Lohmander and *A. fasciatus* (Nilsson) (WALDÉN, Göteborg, 1971; WIKTOR 1973). On the other hand, *A. circumscriptus* has also been regarded as a synonym for *A. fasciatus* (Nilsson) (LUPU 1974).

In contrast to the Arionidae and Limacidae, the shelless **Veronicellidae** (Vaginulidae) have the anus at the posterior end of the body, and have separate male and female genital pores.

The family **Testacellidae** (Fig. 47) are carnivorous terrestrial snails; they have a strong radula which can be extruded well out of the mouth for prey-catching; in contrast to the phytophagous species, they lack an upper jaw; there is a small haliotis-like shell at the posterior end of the body. The members of this group found in south-western parts of Europe are *Testacella haliotidea* Draparnaud and *Testacella scutulum* Sowerby. Important carnivorous snails found in the Mediterranean countries, central Europe and the Caucasus are *Daudebardia rufa* (Draparnaud) of the family Oleacinidae and *Glandina algira* Bruguière of the family Zonitidae.

In the temperate climate of central Europe, **snails** are of less significance as pests than are slugs. The most important species are of the genera *Cepaea* and *Helix*. The *Cepaea* spp., in particular, are interesting objects of study on account of the extraordinary variability in the banding and basic colouration of the shell.

The Brown-lipped Snail, *Cepaea nemoralis* (L.) and the White-lipped Snail, *Cepaea hortensis* (Müller) differ not only in the colour of the shell mouth, but also the colour of the umbilical area and in the distance of the fifth band from the middle point of this area (Table 6).

Colouration and banding pattern of the shell are variable to the same degree in both species, there being yellow, reddish, brown, light and dark tones, as well as some unbanded specimens and others with one to five black or dark brown

Table 6. Distinguishing characteristics of the two *Cepaea* spp.

Snail	Edge of shell mouth	Shape of shell mouth	Umbilical area	5th band in relation to navel
Cepaea nemoralis (L.)	Brown	Height usually greater than width	Brown	Approximate
Cepaea hortensis (Müller)	White	Width greater than height	Light	Remote

spiral bands. The shell of the speckled Copse Snail, *Helicigona arbustorum* (L.) varies in colour from yellow to brownish-black with irregular yellow speckles. It occurs in the same habitat as *Cepaea* spp. but is found on or in the soil.

Of the other species which belong to the family Helicidae, particular mention must be made of *Helix pomatia* L., *Helix aspersa* Müller and *Helix aperta* Born, of which *H. aspersa* today constitutes the greatest pest, particularly in southern Europe, the Mediterranean region and the warmer parts of the USA (Chap. 2).

Species of the families **Lymnaeidae** and **Planorbidae,** which act as intermediate hosts for the bilharzia parasite (*Schistosoma* spp.) and the Liver Fluke (*Fasciola* spp.) are found virtually throughout the whole world. They live in freshwater and migrate from there to marshes and meadows, where they are often found on the grass stalks. They attach their eggs to plants and floating objects. *Galba truncatula* (Müller) is the intermediate host of the Large Liver Fluke *Fasciola hepatica* L. which infests cattle and sheep in Europe. Several species of aquatic and amphibious snails resemble *G. truncatula* in the shape of the shell and can therefore be confused with it (VAN DEN BRUEL 1968). These are: *Succinea oblonga* Drap., *Succinea arenaria* (Bouchard-Chanter.), *Galba palustris* (Müller) [syn. *Lymnaea palustris* (Müller)], *Galba glabra* (Müller) [syn. *Lymnaea glabra* (L.)], *Potamopyrgus jenkinsi* (E.A. Smith) [syn. *Hydrobia jenkinsi* (Smith)], *Radix peregra* (Müller), *Bithynia tentaculata* (L.), *Cochlicopa lubrica* (Müller) and *Aplexa hypnorum* (L.) (Fig. 48). HUBENDICK (1951), SEVO (1971) and WALTER (1968) have examined methods for distinguishing between *G. truncatula* and related species.

Many contemporary researchers are interested in the problem of **polymorphism,** but no satisfactory explanation has yet been found. The phenomenon is encountered in both snails and slugs (ALBONICA 1948; LUPU 1977; GROSSU 1977). Polymorphism in the banded snails (*Cepaea* spp.) has been described by ARNOLD (1970), SCHILDER (1953) SCHNETTER (1950) and others, and several possible causal factors have been suggested. BRUSSARD (1974) and DAY and DOWDESWELL (1968) are of the opinion that it is physiological in origin; OWEN and BENGTSON (1972) suggest that it is the result of selection by different predators in different regions; the view that it may be ecological and related to the calcium content of the soil is rejected by CAIN and CURREY (1963, 1968); finally, ARNOLD (1968, 1969) and CAIN (1968) argue that temperature is a decisive factor. WOLDA (1965) observed that in *Cepaea nemoralis* (L.) all animals with a yellow shell climbed higher up trees than those with pink shells, and this may be a reflection of the greater mobility of yellow-shelled snails. The striking red of the highly polymorphic slug, *Arion rufus* (L.) is

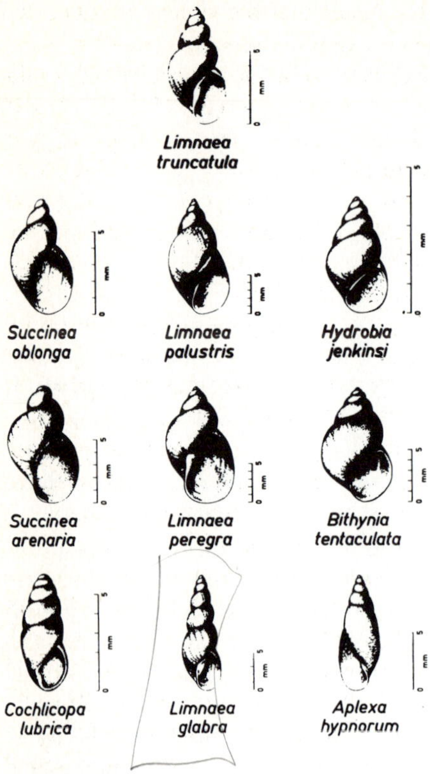

Fig. 48. Possibilities of confusing *Lymnaea (Galba) truncatula* with other species

not related to altitude, nor to the chemical composition or moisture content of the soil (FRÖMMING 1950), but according to MARENBACH (1940) is dependent on the availability of chlorophyll, carotin and the vitamins A and C in the food.

Besides various morphological features such as structure of the genitalia, the shape of the radula teeth, the protrusions and ridges found on the edge of the shell mouth in some species, or the nature of the chalk deposits in the mantle shield of slugs, **biochemical** methods have also been used in recent times for the **identification** of species and genera – for example, electrophoresis of esterases and proteins (GROSSU and TESIO 1975; GROSSU 1977; WIUM-ANDERSEN 1973). This method is quite independent of the superficial similarity that exists between animals which are now regarded as separate species, and of the polymorphism displayed by animals which are today considered to belong to one and the same species. The difference between the haemagglutinins from the albumen gland as a means of distinguishing between *Cepaea hortensis* (Müller), *Cepaea nemoralis* (L.) and *Cepaea vindobonensis* (Fér.) has also been investigated (KILIAS et al. 1972; KOTBAUER and SCHNITZLER 1972). Of the various organs tested in *Helix pomatia* L., however, the albumen gland alone was found to be a source of blood aglutinins. Finally, **cytotaxonomic** methods, involving the use of species-specific differences in chromosome number, have been used for the identification of Lymnaeidae and Rissoacea (BUTOT and KIAUTA 1966; INABA 1969; PATTERSON 1972 and others).

For taxonomic studies it is necessary to kill the snails and slugs and fix them in a relaxed and fully extended state (JAECKEL 1953). If they are killed by immersion in boiling water, the body contracts, with the result that the viscera are distorted and, particularly in the case of slugs, anatomical examination may thereafter be impossible. **Menthol** and **nembutal** have so far proved to be the only substances suitable for relaxing gastropods prior to fixation (VAN DER SCHALIE 1953; RUNHAM et al. 1965). Their action is, however, unreliable when larger species such as *Galba truncatula* (Müller) and *Physa gyrina* Say are involved. According to MCCRAW (1958), the two substances used in combination yield good results. The gastropods are first immersed in 150 ml tap water to which 1 ml of a standard veterinary solution of nembutal (60 mg/ml) has been added. The surface of the solution in which the gastropods are still immersed is then thickly sprinkled with menthol crystals. Finally the molluscs are chilled for 16 to 18 h at 4 °C. They are then ready for fixation which is usually effected with alcohol or formalin. The disadvantage of colour loss through these methods can be avoided by using the **freeze-drying** technique described by CROWELL (1973). This procedure even permits slugs to be pinned and kept in boxes like dried insects; as in the case of insects, however, they are then subject to attack by dermestid beetles, which can, in turn, be kept in check by using paradichlorbenzene. Freeze-drying is effected by maintaining the animals at −12 °C for approximately 24 h, followed by drying for 24 to 48 h according to the size of the animals; upon completion of this procedure, dried remains of mucus can be removed from the surface of the snail or slug with a paint brush. The colour of the animal is maintained by painting with a highly reflective varnish, like shellac. Extension of the molluscs is also of great importance in freeze-drying. The method of CROWELL is, however, not fully satisfactory for precise taxonomic investigation, since the viscera used for this purpose, such as the genital tract, do not always retain their original shape or position. MEIER-BROOK (1976) recommends the use of **pentobarbital** for freshwater snails.

There is not always agreement on all aspects of gastropod nomenclature. Thus in Western and Southern Europe, the Field Slugs *Deroceras reticulatum* (Müller) and *Deroceras agreste* (L.) are frequently called *Agriolimax reticulatus* Müller and *Agriolimax agrestis* L. respectively, while in USA, *Agriolimax* is the only generic name used (CROWELL, Oregon, pers. comm. 1976). In the recently published distribution atlas for the terrestrial molluscs of Great Britain compiled by KERNEY (1976), preference has been given to the name *Deroceras*, which unquestionably takes priority over *Agriolimax*.

Further examples of this are: *Limax marginatus* Müller instead of – as is customary in Central Europe – *Lehmannia marginata* (Müller); *Lymnaea peregra* Müller instead of *Radix peregra* (Müller), and *Lymnaea palustris* (Müller) instead of *Galba palustris* (Müller). In the field of plant protection it therefore seems meaningful in describing species, to give the common name in addition to the Latin name (GODAN 1974; ZATRAVKIN 1976).

1.7.3 Classification of the Most Important Pest Gastropods and some Carnivorous Snails

The following classification encompasses pest gastropods as well as a few others mentioned in connection with them. The system followed is based upon the classification of GRASSÉ (1968) and also on that of EHRMAN [1956(1933)], as augumented by ZILCH and JAEKEL (1962) and also by WALDÉN (1966).

Prosobranchia: Respiration through gills, sexes separate
Pulmonata: Respiration through lung (secondary gills in *Planorbis*), hermaphrodite.

Phylum: **Mollusca**
Class: **Gastropoda** Cuvier, 1798

Subclass: **Prosobranchia** Milne Edwards, 1848

Order: **Archaeogastropoda** Thiele, 1925

Superfamily: **Trochoidea** Rafinesque, 1815
Family: **Trochidae** Rafinesque, 1815
Subfamily: **Monodontinae** Cossmann, 1916
Genus: *Monodonta* Lamarck, 1801 (secondo Moore, 1799)
Montodonta (Osilinus) Philippi, 1847
Species: *Monodonta turbinata* (Born, 1780)

Subfamily: **Gibbulinae** Stoliczka, 1868
Genus: *Gibbula* Risso, 1826
Species: *Gibbula adriatica* (Philippi, 1844)
Gibbula albida (Gmelin, 1790)

Order: **Mesogastropoda** Thiele, 1925

Superfamily: **Viviparoidea** Gray, 1847
Family: **Viviparidae** Gray, 1847
Genus: *Viviparus* Montfort, 1810
Species: *Viviparus viviparus* (L., 1758)

Family: **Pilidae** Conolly, 1927 (= **Ampullariidae** Gray, 1824)
Genus: *Pomacea* Perry, 1810 (=*Ampullarius* Montfort, 1810)
Species: *Pomacea glauca* (L., 1758)
Pomacea lineata (Spix, 1827)
Genus: *Marisa* Gray, 1824*
Species: *Marisa cornuarietis* (L., 1758)

Superfamily: **Littorinoidea** Gray, 1840
Family: **Pomatiasidae** Gray, 1852
Subfamily: **Pomatiasinae** Gray 1852
Genus: *Pomatias* Studer, 1789
Species: *Pomatias elegans* (O.F. Müller, 1774)

Superfamily: **Rissoidea** Gray, 1847
Family: **Hydrobiidae** Troschel, 1857

* After Burch (1979)

Subfamily: **Hydrobiinae** Troschel, 1857
Genus: *Hydrobia* Hartmann, 1821
Species: *Hydrobia acuta* Draparnaud, 1805
Genus: *Bythinella* Moquin-Tandon, 1856
Species: *Bythinella viridis* (Poiret, 1801)

Subfamily: **Littoridininae** Thiele, 1929
Genus: *Potamopyrgus* Stimpson, 1865
Species: *Potamopyrgus jenkinsi* (E. A. Smith, 1889)

Family: **Truncatellidae** Gray, 1840
Genus: *Oncomelania* Gredler, 1881
Species: *Oncomelania hupensis* Gredler, 1881

Family: **Bithyniidae** Troschel, 1857 (= **Bulimidae** Guilding, 1828)
Genus: *Bithynia* Leach, 1818
Species: *Bithynia tentaculata* (Linnaeus, 1758)
Bithynia leachi (Sheppard, 1823)

Superfamily: **Atlantoidea** Orbigny, 1835
Family: **Carinariidae** Reeve, 1841
Genus: *Carinaria* Lamarck, 1801
Species: *Carinaria lamarcki* (= *mediterranea*) Péron & Lesueur, 1810

Subclass: **Pulmonata** Cuvier, 1817

Order: **Basommatophora** A. Schmidt, 1855

Superfamily: **Lymnaeoidea** Rafinesque, 1815
Family: **Lymnaeidae** Rafinesque, 1815
Genus: *Lymnaea* Lamarck, 1799
Species: *Lymnaea stagnalis* (L., 1758)
Genus: *Galba* Schrank, 1803
Galba (Galba) s. str.
Species: *Galba truncatula* (O. F. Müller, 1774)
Galba (Omphiscola) Beck, 1837
Species: *Galba glabra* (O. F. Müller, 1774)
Galba (Stagnicola) Leach, 1830
Species: *Galba palustris* (O. F. Müller, 1774)
Genus: *Pseudosuccinea* F. C. Baker, 1908
Species: *Pseudosuccinea columella* Say, 1824
Genus: *Radix* Montfort, 1810
Radix (Radix) s. str.
Species: *Radix peregra* (O. F. Müller, 1774)
Form: *Radix ovata* (Draparnaud, 1805)
Species: *Radix auricularia* (L., 1758)

Superfamily: **Physoidea** Fitzinger, 1833
Family: **Physidae** Fitzinger, 1833

Genus: *Physa* Draparnaud, 1801
Species: *Physa acuta* Draparnaud, 1805
Physa fontinalis (L., 1758)
Genus: *Aplexa* Flemming, 1820
Species: *Aplexa hypnorum* (L., 1758)

Superfamily: **Planorboidea** Rafinesque, 1815
Family: **Planorbidae** Rafinesque, 1815
Subfamily: **Planorbinae** Rafinesque, 1815
Genus: *Planorbis* O.F. Müller, 1774
Species: *Planorbis planorbis* (L., 1758)
Planorbis carinatus O.F. Müller, 1774
Genus: *Gyraulus* Charpentier, 1837
Species: *Gyraulus albus* (O.F. Müller, 1774)
Gyraulus laevis (Alder, 1838)
Genus: *Armiger* Hartmann, 1843
Species: *Armiger crista* (L., 1758)

Subfamily: **Taphiinae** Harry, 1962
Genus: *Taphius* H. & A. Adams, 1855 (=*Australorbis* Pilsbry, 1934)
Species: *Taphius glabratus* (Say, 1816) (=*Biomphalaria glabrata* (Say, 1816))
Genus: *Biomphalaria* Preston, 1910
Species: *Biomphalaria alexandrina* (Ehrenberg, 1831)
Biomphalaria pfeifferi (Krauss, 1848)
Biomphalaria smithi Preston, 1910
Biomphalaria sudanica (Martens, 1870)

Subfamily: **Helisomatinae** F.C. Baker, 1928
Genus: *Planorbarius* Froriep, 1806
Species: *Planorbarius corneus* (L., 1758)

Family: **Bulinidae** Crosse & Fischer, 1880
Subfamily: **Bulininae** Crosse & Fischer, 1880
Genus: *Bulinus* O.F. Müller, 1781 (=*Isidora* Ehrenberg, 1831)
Species: *Bulinus contortus* (Michaud, 1829)
Bulinus forskali (Ehrenberg, 1831)
Bulinus globosus (Morelet, 1866)
Bulinus senegalensis O.F. Müller, 1781
Bulinus tropicus (Krauss, 1848)
Bulinus truncatus (Audouin, 1827)*
Genus: *Indoplanorbis* Annandale & Prashad, 1920
Species: *Indoplanorbis exustus* (Deshayes, 1834)

Family: **Ancylidae** Rafinesque, 1815
Genus: *Ancylus* O.F. Müller, 1774
Species: *Ancylus fluviatilis* O.F. Müller, 1774

---

* Correct nomenclature: *B. hemprichii* (Ehrenberg, 1831) (Kilias, in litt. 1977).

Order: **Stylommatophora** A. Schmidt, 1855

Superfamily: **Veronicelloidea** Gray, 1840
Family: **Veronicellidae** Gray, 1840 (= **Vaginulidae** Gill, 1871)
Genus: *Veronicella* Blainville, 1817
Species: *Veronicella laevis* Blainville, 1817
Veronicella leydigi Simroth, 1813
Genus: *Laevicaulis* Simroth, 1913
Species: *Laevicaulis alte* (Férussac, 1821)

Superfamily: **Pupilloidea** Turton, 1831
Family: **Valloniidae** Morse, 1864
Genus: *Vallonia* Risso, 1826
Species: *Vallonia pulchella* (O.F. Müller, 1774)
Vallonia costata (O.F. Müller, 1774)

Superfamily: **Strophocheiloidea** Thiele, 1926
Family: **Strophocheilidae** Thiele, 1926
Genus: *Strophocheilus* Spix, 1827
Species: *Strophocheilus oblongus* (O.F. Müller, 1774)

Superfamily: **Succinoidea** Beck, 1837
Family: **Succineidae** Beck, 1837
Subfamily: **Succineinae** Beck, 1837
Genus: *Succinea* Draparnaud, 1801
Succinea (Succinea) s.str.
Species: *Succinea putris* (L., 1758)
Succinea (Succinella) Mabille, 1870
Species: *Succinea oblonga* Draparnaud, 1801

Superfamily: **Achatinoidea** Swainson, 1840
Family: **Achatinidae** Swainson, 1840
Subfamily: **Achatininae** Swainson, 1840
Genus: *Achatina* Lamarck, 1799
Achatina (Lissachatina) Bequaert, 1950
Species: *Achatina fulica* Bowdich, 1822

Subfamily: **Férussaciinae** Bourguignat, 1883
Genus: *Cecilioides* Férussac, 1814
Species: *Cecilioides acicula* (O.F. Müller, 1774)
Genus: *Opeas* Albers, 1850
Species: *Opeas pumilum* (L. Pfeiffer, 1840)

Subfamily: **Subulininae** Crosse & Fischer, 1877
Genus: *Subulina* Beck, 1837
Species: *Subulina octona* (Bruguière, 1789)
Genus: *Lamellaxis* Strebel, 1802
Species: *Lamellaxis gracilis* (Hutton, 1834)
Genus: *Rumina* Risso, 1826
Species: *Rumina decollata* (L., 1758)

Family:	**Spiraxidae** Baker, 1955
Subfamily:	**Euglandininae** H. B. Baker, 1941
Genus:	*Euglandina* Crosse & Fischer, 1870
Species:	*Euglandina rosea* (Férussac, 1821)
Superfamily:	**Streptaxoidea** Gray, 1860
Family:	**Streptaxidae** Gray, 1860
Genus:	*Gonaxis* Taylor, 1877
Species:	*Gonaxis kibweziensis* (Smith, 1894)
	Gonaxis quadrilateralis (Preston, 1910)
	Gonaxis vulcani (Thiele, 1911)
Genus:	*Edentulina* L. Pfeiffer, 1856
Species:	*Edentulina ovoidea* (Bruguière, 1789)
	Edentulina affinis C. Boettger, 1913
Superfamily:	**Bulimuloidea** Tyron, 1867
Family:	**Bulimulidae** Tyron, 1867
Subfamily:	**Bulimulinae** Tyron, 1867
Genus:	*Drymaeus* Albers, 1850
Species:	*Drymaeus dormani* (Binney, 1857)
Superfamily:	**Arionoidea** Gray, 1840
Family:	**Endodontidae** Pilsbry, 1894
Subfamily:	**Endodontinae** Pilsbry, 1895
Genus:	*Discus* Fitzinger, 1833
Species:	*Discus rotundatus* (O. F. Müller, 1774)
Family:	**Arionidae** Gray, 1840
Subfamily:	**Anadeninae** Pilsbry, 1948
Genus:	*Prophysaon* Bland & Binney, 1873
Species:	*Prophysaon andersoni* (Cooper, 1872)
Subfamily:	**Ariolimacinae** Pilsbry & Vanatta, 1898
Genus:	*Ariolimax* Mörch, 1860
Species:	*Ariolimax columbianus* (Gould, 1851)
Subfamily:	**Arioninae** Gray, 1940
Genus:	*Arion* Férussac, 1819
	Arion (Arion) s. str.
Species:	*Arion ater* (L., 1758)
	Arion rufus (L., 1758)
	(both = *empiricorum* Férussac, 1819)
	Arion lusitanicus Mabille, 1868
	Arion (Carinarion) P. Hesse, 1926
Species:	*Arion circumscriptus* Johnston, 1828
	Arion fasciatus (Nielsson, 1819)
	Arion silvaticus Lohmander, 1937
	Arion (Kobeltia) Simroth, 1873
Species:	*Arion hortensis* Férussac, 1819
	Arion (Mesarion) P. Hesse, 1926

Species: *Arion subfuscus* (Draparnaud, 1805)
Arion (Microarion) P. Hesse, 1926
Species: *Arion intermedius* Normand, 1952
Genus: *Geomalacus* Allman, 1842
Species: *Geomalacus maculosus* Allman, 1842

Superfamily: **Limacoidea** Rafinesque, 1815
Family: **Vitrinidae** Fitzinger, 1864
Subfamily: **Vitrininae** Fitzinger, 1864
Genus: *Vitrina* Draparnaud, 1801
Species: *Vitrina pellucida* (O.F. Müller, 1774)
Genus: *Semilimax* Agassiz, 1845
Species: *Semilimax semilimax* (Férussac, 1802)
Semilimax kochi (Andreae, 1884)
Genus: *Eucobresia* H.B. Baker, 1929
Species: *Eucobresia diaphana* (Draparnaud, 1805)

Family: **Zonitidae** Mörch, 1864
Subfamily: **Zonitinae** Mörch, 1864
Genus: *Zonites* Montfort, 1810
Species: *Zonites algirus* (L., 1758)
Genus: *Oxychilus* Fitzinger, 1833
Oxychilus (Oxychilus) s. str.
Species: *Oxychilus cellarius* (O. F. Müller, 1774)
Oxychilus draparnaudi (Beck, 1837)
Oxychilus (Ortizius) Forcart, 1957
Species: *Oxychilus alliarius* (Miller, 1822)
Oxychilus helveticus (Blum, 1881)

Subfamily: **Gastrodontinae** Tyron, 1866
Genus: *Zonitoides* Lehmann, 1862
Zonitoides (Zonitoides) s. str.
Species: *Zonitoides nitidus* (O. F. Müller, 1774)
Zonitoides (Zonitellus) H.B. Baker, 1928
Species: *Zonitoides arboreus* (Say, 1816)

Subfamily: **Daudebardiinae** Pilsbury, 1908
Genus: *Daudebardia* Hartmann, 1921
Species: *Daudebardia rufa* (Draparnaud, 1805)

Family: **Limacidae** Rafinesque, 1815
Genus: *Limax* Linnaeus, 1758
Limax (Limax) s. str.
Species: *Limax maximus* L., 1758
Limax cinereoniger Wolf, 1803
Limax (Limacus) Lehmann, 1864
Species: *Limax flavus* L., 1758
Limax (Malacolimax) Malm, 1868
Species: *Limax tenellus* O.F. Müller, 1774
Limax (Vitrinoides) Simroth, 1891

Species: *Limax monticola* Boettger, 1881
Genus: *Bielzia* Clessin, 1887
Species: *Bielzia coerulans* (M. Bielz, 1851)
Genus: *Lehmannia* Heynemann, 1862
Species: *Lehmannia marginata* (O.F. Müller, 1774)
Lehmannia valentiana (Férussac, 1823) (=*L. poirieri* Mabille, 1883)
Genus: *Deroceras* Rafinesque, 1820
Deroceras (Deroceras) s. str.
Species: *Deroceras laeve* (O.F. Müller, 1774)
Deroceras caruanae (Pollonera, 1891)
Deroceras (Agriolimax) Mörch, 1865
Species: *Deroceras reticulatum* (O. F. Müller, 1774)
Deroceras agreste (L., 1758)
*Genus: *Liolytopelte* Simroth, 1901 (=*Lytopelte* Boettger, 1886)
Species: *Deroceras* (L.) *caucasicum* (Simroth, 1901).

Family: **Milacidae** Ellis, 1926
Genus: *Milax* Gray, 1855
Milax (Milax) s. str.
Species: *Milax gagates* (Draparnaud, 1801)
Milax sowerbyi (Férussac, 1823)
Milax (Tandonia) Lessona & Pollonera, 1882
Species: *Milax budapestensis* (Hazay, 1881)
Milax rusticus (Millet, 1843)

Familiy: **Boettgerillidae Van Goethem, 1972
Genus: *Boettgerilla* Simroth, 1910
Species: *Boettgerilla pallens* Simroth, 1912 (=*B. vermiformis* Wiktor, 1959)

Superfamily: **Testacelloidea** Gray, 1840
Family: **Testacellidae** Gray, 1840
Genus: *Testacella* Cuvier, 1800
Species: *Testacella haliotidea* Draparnaud, 1801
Testacella scutulum Sowerby, 1820

Superfamily: **Oleacinoidea** Adams, 1855
Family: **Oleacinidae** Adams, 1855
Genus: *Oleacina* (Bolten) Röding, 1798
Species: *Oleacina voluta* (Gmelin, 1790)
Oleacina oleacina straminea (Deshayes, 1851)

---

* According to Wiktor (1973), *Liolytopelte* Simroth (=*Lytopelte* Boettger) should, because of the many morphological characteristics they have in common, be placed in a subgenus of *Deroceras* Rafinesque; The species *Lytopelte caucasica* Simroth would thus be synonymous with *Deroceras* (L.) *caucasicum* (Simroth).

** According to Wiktor (Wroclaw, in litt. 1977).

Genus: *Poiretia* Fischer, 1883
Species: *Poiretia (Glandina) algira* Bruguière, 1792

Superfamily: **Helicoidea** Rafinesque, 1815
Family: **Eulotidae** Möllendorff, 1898
(= **Bradybaenidae** Pilsbry, 1939)
Subfamily: **Eulotinae** Möllendorff, 1898
(= **Bradybaeninae** Pilsbry, 1939)
Genus: *Eulota* Hartmann, 1842
Species: *Eulota* (= *Bradybaena*) *fruticum* (O. F. Müller, 1774)
Genus: *Bradybaena* Beck, 1837
Species: *Bradybaena similaris* (Férussac, 1821)

Family: **Helicidae** Rafinesque, 1815
Subfamily: **Helicellinae** Wenz, 1923 (= **Leucochroinae** Westerlund, 1886)
Genus: *Candidula* Kobelt, 1871
Species: *Candidula caperata* (Montagu, 1803)
Candidula unifasciata (Poiret, 1801)
Genus: *Cernuella (Xerocincta)* Monterosato, 1892
Species: *Cernuella neglecta* (Draparnaud, 1805)
Cernuella (Cernuella) Schlüter, 1838
Species: *Cernuella virgata* (DaCosta, 1778)
Genus: *Cochlicella* Férussac, 1820
Species: *Cochlicella acuta* (O. F. Müller, 1774)
Cochlicella barbara (L., 1758)
Cochlicella ventricosa (Draparnaud, 1801)
Cochlicella conoidea (Draparnaud, 1801)
Genus: *Leucochroa* Westerlund, 1886
Species: *Leucochroa candissima* Draparnaud, 1801
Genus: *Helicella* Férussac, 1819
Species: *Helicella itala* (L., 1758)
Helicella obvia (Hartmann, 1840)
Helicella bolenensis (Locard, 1882)
Genus: *Trochoidea* Brown, 1827
Trochoidea (Trochoidea) s. str.
Species: *Trochoidea elegans* (Draparnaud, 1805)
Trochoidea (Xerotricha) Monterosato, 1892
Species: *Trochoidea geyeri* (Soós, 1926)
Genus: *Helicopsis* Fitzinger, 1833
Species: *Helicopsis hungarica* (Soós & Wagner, 1935)
Helicopsis striata (O. F. Müller, 1774)

Subfamily: **Monacheae** Fitzinger, 1833*
Genus: *Monacha* Fitzinger, 1833
Monacha (Monacha) s. str.
Species: *Monacha cantiana* (Montagu, 1803)
Monacha cartusiana (O. F. Müller, 1774)

* According to ZILCH (1962)

Subfamily: **Hygromiinae** Tyron, 1866
Genus: *Euomphalia* Westerlund, 1889
Species: *Euomphalia strigella* (Draparnaud, 1801)
Genus: *Hygromia* Risso, 1826
Species: *Hygromia cinctella* (Draparnaud, 1801)
Hygromia limbata (Draparnaud, 1805)
Genus: *Perforatella* Schlüter, 1838
Perforatella (Perforatella) s. str.
Species: *Perforatella bidentata* (Gmelin, 1788)
Perforatella (Monachoides) Gude & Westward, 1921
Species: *Perforatella rubiginosa* (A. Schmidt, 1853)
Genus: *Trichia* Hartmann, 1840
Trichia (Trichia) s. str.
Species: *Trichia hispida* (L., 1758)
Trichia striolata (C. Pfeiffer, 1828)
Trichia villosa (Studer, 1798)

Subfamily: **Helicodontinae** Hesse, 1907
Genus: *Helicodonta* Férussac, 1821
Species: *Helicodonta obvoluta* (O. F. Müller, 1774)

Subfamily: **Ariantinae** Mörch, 1864 (= **Helicigoninae** Germain, 1929)
Genus: *Helicigona* Férussac, 1821
Helicigona (Arianta) Turton, 1831
Species: *Helicigona arbustorum* (L., 1758)

Subfamily: **Helicinae** Rafinesque, 1815
Genus: *Theba* Risso, 1826 (= syn. *Euparypha* Hartmann, 1843)
Species: *Theba pisana* (O. F. Müller, 1774)
Genus: *Cepaea* Held, 1837
Species: *Cepaea hortensis* (O. F. Müller, 1774)
Cepaea nemoralis (L., 1758)
Cepaea silvatica (Draparnaud, 1801)
Cepaea vindobonensis (Férussac, 1821)
Genus: *Eobania* Hesse, 1915
Species: *Eobania vermiculata* (O. F. Müller, 1774)
Genus: *Helix* Linnaeus, 1758
Helix (Helix) s. str.
Species: *Helix pomatia* L., 1758
Helix lucorum L., 1758
Helix melanostoma Draparnaud, 1801
Helix (Cantareus) Risso, 1826
Species: *Helix aperta* Born, 1778
Helix (Cryptomphalus) Charpentier, 1837
Species: *Helix aspersa* O. F. Müller, 1774
Genus: *Iberus* Montfort, 1810
Species: *Iberus alonensis* (Férussac, 1821)
Genus: *Otala* Schumacher, 1817
Species: *Otala lactea* (O. F. Müller, 1774)

1.7.4 Key of Identification for Central European Pest Gastropods

The following key encompasses only the most important pest gastropods of Central Europe and confines itself in the main to external features, knowledge of which is necessary for quick identification of slug and snail finds. The individual species also differ from one another in their internal anatomy and the radula teeth, both characteristics which, however, can only be established upon dissection and in general remain the prerogative of specialists in taxonomy and systematics. The basis for the following key was material collated by ANT (1973), BEEDHAM (1972), EHRMANN (1956 (1933)), GODAN (1960), JAECKEL (1967), JANUS (1965), KERNEY and CAMERON (1979) and QUICK (1961).

For identification of snails, the height and breadth of the shell as well as its shape and colour are the main features (Fig. 49). The number of whorls of the shell is established by observing the shell from above and counting downwards from the

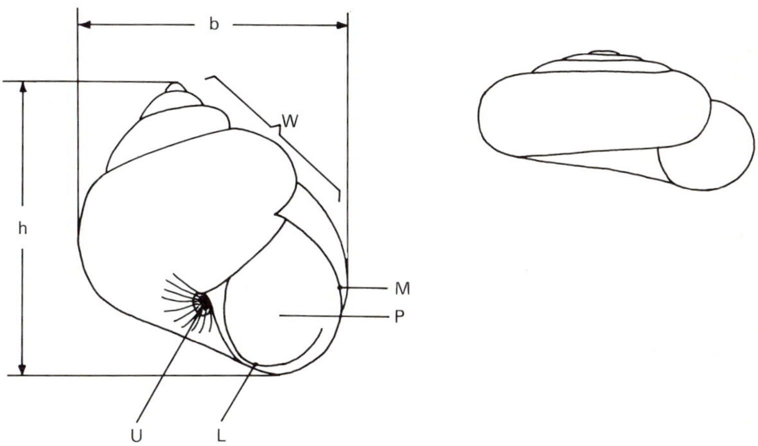

Fig. 49. *Left* shell concial to globular; *right* shell flat-conical. *L* lip; *M* mouth of shell (aperture); *P* peristome (shell mouth edge); *U* umbilicus; *W* whorls; *b* breadth; *h* height

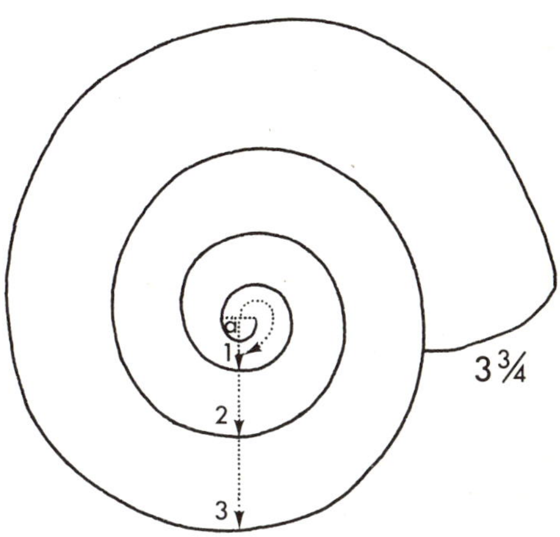

Fig. 50. Spire of the shell, seen from the top, scheme. *a* apex; *dotted line* indicates extent of first whorl

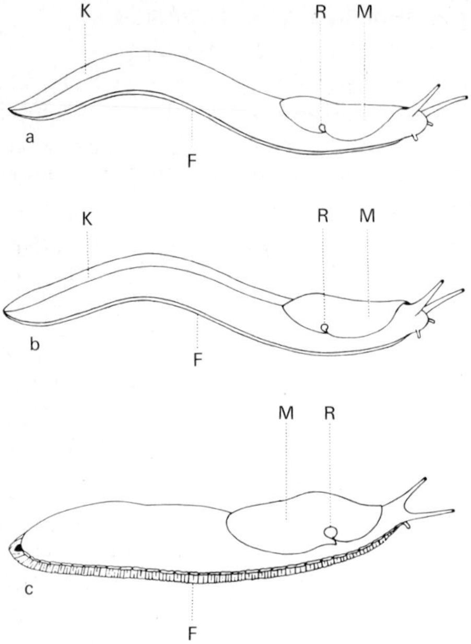

Fig. 51 a–c. External distinguishing features of slugs; *R* respiratory pore; *M* mantle (mantle shield); *F* foot; *K* keel. **a** Limacidae (Keelback Slugs); **b** Milacidae (Keeled Slugs); **c** Arionidae (Roundback Slugs)

apex, which is clearly the beginning of the spire. In Fig. 50, the dotted line indicates the extent of the first whorl. Compared with the shells of older snails, those of younger animals have a sharp edge, which is, moreover, neither thickened, folded back, nor enlarged. In those families in which these characteristics are not shown in the shells of older animals, the shell edge of younger snails is soft, flexible and without calcium deposits. The edge of the mouth (peristome) is the growing region of the shell and may be regarded as being composed of an outer lip.

The external features used for identification of the three slug families are shown in Fig. 51.

The following key of identification takes habitat as its starting point. The taxonomic characters are arranged as dichotomous types, that is, as alternatives (1–1*, 2–2*, etc.), the number on the right leading to the next pair of features.

The indications on size apply exclusively to adult animals.

Abbreviations:

M = mantle (mantle shield)
W = whorl
P = peristome (shell mouth edge)
U = umbilicus

b = breadth of shell
h = height of shell
∅ = diameter of shell or of umbilicus
$\pm$ = more or less

Class: Gastropoda

Body asymmetrical (torsion), head $\pm$ clearly defined, with eyes and tentacles, ventral creeping foot, visceral sack covered by a mantle, shell or rudimentary internal shell under mantle, respiratory pore on the right side.

I: Habitat: fields, gardens, greenhouses
II: Habitat: cellars, storerooms (also of bakeries and cheese dairies)
III: Habitat: damp meadows, ditches, river banks

To I: **Fields, Gardens, Greenhouses:** Terrestrial pulmonates, Stylommatophora (eye-spots at tips of the posterior pair of tentacles)

Ia: *Without* visible shell (animal naked), rudimentary internal shell under the mantle 1

Ib: *With* shell see. p. 79

Ia: Slugs 1

1 Body ± compact; respiratory pore *before* mid-point of M; internal shell as a mass of calcareous granules under M; back *without* keel; skin heavily wrinkled; back grooved through longitudinal rows of long coarse tubercles; sole ± uniformly coloured; animal appears hemispherical when contracted; colour of young different from adults; slow-moving animals
Family: **Arionidae,** Roundback Slugs 2

1* Body slender; respiratory pore *behind* mid-point of M; internal shell as a calcareous oval plate ("slug plate") under mantle; back *with* keel; sole tripartite (longitudinally divided into a median region with chevron-like grooves and two lateral regions); very active animals 3

3 M wrinkled; body keel only on posterior end of back (tail); with or without marking
Family: **Limacidae,** Keelback Slugs 6

3* Keel stretching from tail to mantle 4

4 M granulose; body not worm-like
Family: **Milacidae,** Keeled Slugs 5

4* Worm-like, slender form, 40–45 mm long, 2–3 mm wide, sole very narrow (1 mm wide); uniformly coloured, pale, ashen, whitish or bluish to lead grey; body side light ashen without marks; M over 1/3 body length, reaches almost to body centre; respiratory pore small, almost at mid-point of M; head dark bluish-gray, small; tentacles dark blue, eye-stalks light, transparent, knobbed at tips (eye spots); keel reaches to M, prominent, usually rather bluish at the tail; mucus colourless, clear; animals crawls actively
Family: **Boettgerillidae,** Worm-Slugs
Boettgerilla pallens Simroth
(= *B. vermiformis* Wiktor)

---

2 (1) Family **Arionidae/Arioninae**
Respiratory pore before the mid-point of M; internal shell as calcareous granules under M, M covers only the anterior part of the back; sole ± uniform 7

7 Length up to 150 mm; body sides without longitudinal bands (except young *A. lusitanicus*), very variable in colour; skin coarsely wrinkled, with small tubercles; sole grey 8

7* Length up to 70 mm; sides of body with or without dark longitudinal bands, sole uniform 9

8 Length 120–150 mm; either reddish, orange, brown (often darker on the back) or black; without marking; M coarsely granulose with elongated tubercles; foot-fringe mostly red with black alternating thick and thin transverse stripes; head dark; optic tentacles black; young animals (up tp 40 mm long) yellow-white, green-yellow or grey-green with black head and tentacles; sole light grey to yellowish; body mucus very viscous, orange (in red animals) or colourless (in brown animals); sole mucus colourless

Arion rufus (L.) (= *empiricorum* Férussac), Large Red Slug

Length 100–130 mm, usually uniform black; very similar to the above species; identification accurately only by means of anatomical examination (reproductive system)

Arion ater (L.), Large Black Slug

8* Length 70–100 mm; colour similar to the above species, variable, dark-grey, red-yellow, brown or pale greenish-grey; head and tentacles dark; sole pale or dark; young animals with dark lateral bands forming a lyre-shape on the mantle and surrounding the repiratory pore

Arion lusitanicus Mabille, Lusitanian Slug

9 Sole white or yellowish-white; dark lateral bands; M with dark lyre-like mark 10

9* Sole completely or partially strong yellow; body sides dark, but no distinct lateral band; M finely granulose, with dark rim 11

10 Length 30–40 mm (rarely up to 50 mm); light grey (especially young animals) to slate grey, lighter towards the ventral surface; prominent dark lateral band, on right side arching over respiratory pore, forming a lyre-shape on M and sometimes with a yellow stripe below; back and flanks with strong wrinkles; back with median whitish keel-line, formed of a row of light-coloured enlarged tubercles; keel-line in young animals at the tail, only faint; foot-fringe pale, posteriorly transversely lined; sole opaque, whitish; body mucus and sole mucus colourless

Arion circumscriptus Johnston*, Grey Garden Slug

10* Length 60–65 mm; ocre or red-yellow or coffee-brown; head and tentacles darker; dark lateral band on each side, faintly visible; no dorsal line; M finely granulose (small elongated tubercles); foot-fringe pale, sometimes with stripes; sole translucent, yellowish-white; body colour very variable; body mucus yellow to orange; sole mucus colourless

Arion subfuscus (Draparnaud), Dusky Slug

---

* For the accurate identification of the species belonging to the sub-genus *Carinarion* (in addition *A. fasciatus*, *A. silvaticus*) a specialist should be consulted; cf. also Fig. 42 (p. 53) and pages 16/17 of the book: WIKTOR: Die Nacktschnecken Polens, 1973, Monography Fauny Polski, Polska Akademia Nauk, Warzawa, Krakow.

11 Length 25–30 mm; grey to dark brown-grey, often with small yellow-brown dots; colour lighter towards the foot; young animals slate-grey; back darker, at the sides a dark band, that on the right enclosing the respiratory pore; sole yellow to orange; body mucus yellow to orange; sole mucus colourless

Arion hortensis Férussac, Garden Slug

11* Length about 20 mm; greenish or yellowish-grey; ±prominent dark lateral band on each side, also on M; head dark grey or black; skin covered with prominent conical tubercles which stand out with a transparent tip in the contracted animal, giving the surface a spiky appearance; sole at the posterior end of the body ± yellow; mucus golden-yellow; head and tentacles dark grey

Arion intermedius Normand, Hedgehog Slug

---

5 (4) Family: **Milacidae**

Respiratory pore *behind* the mid-point of M; keel extends length of body from tail to rear end of mantle; internal shell (calcareous "slug plate") oval and thick without skinned rim 12

12 Length 85–100 mm; sole tripartite, pale yellow at the lateral regions and with median darker or lighter region 13

12* Length less than 80 mm; sole pale; keel very prominent 14

13 Length 80–100 mm; reddish-brown (back darker) with dark spots, streaks and dots, lighter towards the foot; M with two narrow dark arch-shaped bands, internal shell (calcareous slug plate) elliptical; foot-fringe with fine rows of black dots; mucus colourless to whitish, viscous

Milax rusticus (Millet), Bulb-eating Slug

13* Length 60–70 mm; variable in colour, brown or grey to black, with a reticulum of black spots, small longitudinal rows of polygonal tubercles, with pigmented grooves between; small internal shell (calcareous plate) under M; without bands; respiratory pore grey-rimmed; back keeled; keel yellow or orange; sole grey, tripartite (middle region darker than lateral regions); mucus pale to whitish, sticky

Milax budapestensis (Hazay), Keeled Slug, Subterranean Slug

14 Length 45–55 mm (maximum up to 70 mm); yellow-brown or dark grey to black, without bands; longitudinal rows of flattened tubercles with unpigmented grooves between these; M very large; dorsal keel very prominent, often darker, abruptly truncated at tail; tentacles dark grey; behind the head on each side of the median line a yellowish-white, black-edged stripe; back with longitudinal grooves, which especially towards the tail form a network; sole tripartite, whitish; mucus white or colourless

Milax gagates (Draparnaud), Greenhouse Slug, Smooth Jet Slug

14* Length up to ±70 mm; brownish or dark brown with dark speckles; grooves between tubercles also pigmented; respiratory pore pale-rimmed;

keel prominent, lighter than the back, not truncated at the tail; sole pale; mucus yellow, sticky; resting animal laterally flattened
Milax sowerbyi (Férussac), Sowerbyi's Slug

---

6 (3) Family: **Limacidae**
Respiratory pore behind the mid-point of M; keel only at tail; calcareous internal shell ("slug plate") thin, with skin rim 15

15 Length 20–25 mm; body very soft, smooth, translucent, often covered with watery slime; light grey, brown to brown-black, either no or only slight marks; M centrally placed, extends over almost 1/2 the body length; sole pale brown or light grey; mucus thin, clear, colourless; animal crawls rapidly
Deroceras laeve (O.F. Müller), Smooth Slug, Marsh Slug

15* Length more than 30 mm; M placed behind the head, extends about 1/3 length of the body 16

16 Mucus colourless, clear, pale or yellow 17

16* Mucus milky white when animal irritated 22

17 Length ± 70 mm; whole body or at least the tail translucent; with dark longitudinal bands; prominent keel; mucus yellow 18

17* Length 80–150 mm; neither body nor tail translucent; mucus colourless, clear 20

18 Length up to 65 mm; body tender, gelatinous, waxy, completely translucent; greenish-white, pale yellow to lemon-yellow; M orange; tubercles small; head and tentacles blackish or black-brown; body sometimes with faint darker lateral bands; keel short; sole pale; body mucus yellow or orange, sole mucus colourless
Limax tenellus O.F. Müller, Slender Slug

18* Length 60–70 mm; animal soft, body with high water content; only tail translucent; skin with flat, smooth tubercles; on each side two dark longitudinal bands, the lower one towards the foot forming a network and often only faintly visible; M with 3 dark longitudinal bands, the median also faintly visible, the inner ones forming a lyre-shape on the mantle; tail with keel; sole grey, often with darker median region; mucus colourless, watery 19

19 Adult grey-reddish, with a lighter median stripe down the back, usually a large translucent water reservoir (body water) at the tail
Lehmannia marginata (O.F. Müller), Tree Slug

19* Similar to the above species; body watery and gelatinous in appearance, reddish-brown, usually with prominent dark longitudinal bands, tail translucent; mucus colourless
Lehmannia valentiana (Férussac), Canadian Slug

20 Length 100–120 mm; lighter or darker blue or blue-green, always without marks; sole blue-grey; sole mucus colourless, clear
Bielzia coerulans (M. Bielz)

20* Length 120–150 mm; whitish-grey to black, sole-mucus colourless . 21

21 Length 120–150 mm; light grey to brownish-grey; on each side two or three darker rows of spots or stripes; concentrically ridged mantle; centre-point in mid-dorsal line; M with marbled pattern; keel short, extending about 1/3 distance from tail to mantle; M and body often banded, 2 or 3 broken dark longitudinal bands; internal shell ("slug plate"); foot fringe with small black grooves; sole tripartite (in young grey, in adults pale grey-yellow); mucus colourless, sticky
Limax maximus Linnaeus, European Giant Garden Slug

21* Similar to above species; mostly ash-black, black-grey or black, except for a lighter mid-dorsal line; two faint lateral bands; sometimes with rows of flecks on the back; M dark, not speckled; keel long and pale, extending about 2/3 distance from tail to mantle; mucus colourless
Limax cinereoniger Wolf, Ash-black Slug

22 (16*) Mucus milky-white, when animal irritated 23

23 Length ± 50 mm; pale cream or yellow-white, grey to reddish brown; with a network of mottled pattern of dark brown grooves, spots and streaks, these rarely absent; M rounded at both ends with dark dots and flecks; mantle and body not banded; rim of respiratory pore slightly paler than M; centre point of the concentrically ridged mantle lies to the right of the mid-line; keel short and truncated; sole pale or light brown; mucus milky, white (lime), sticky
Deroceras reticulatum (O. F. Müller), Field Slug, Netted Slug

Very similar to above species (therefore often regarded as a synonym), in general less pigmented; sole yellowish-white; mucus less intensely milky; reliable identification only after internal anatomical examination; mucus colourless or milky-white when animal irritated
Deroceras agreste (Linnaeus), Field Slug

23* Body more transparent than previous species; light grey, grey-brown with faint small brown flecks; M lighter, a pale ring around respiratory pore; mucus colourless or milky-white; animal very aggressive
Deroceras caruanae (Pollonera)

Ib **Snails** 1

1 Shell lens-shaped, flattened. elevated, flat-conical or depressed globular, with or without bands, stripes or flecks; without or with hairs (or hair scars); P not thickened, without teeth; with or without internal rib; with U; shell up to 18 mm, small species 2

1* Shell tapered, conical or globular; with or without bands or flecks; P thick or not, internal lip (rib), without teeth; with or without U; medium to large species 13

2 Shell elevated or flat-conical, shiny and translucent (then smooth or finely striated) or not translucent (then with or without hairs); aperture elliptical or nearly circular or crescentic 6

2* Shell more lens-shaped, not translucent, limy or not; regularly ribbed or not; always without hairs; aperture round or nearly round . . . 3

3 Shell ∅ 6–8 mm, not limy, with flecks, ribbed; aperture nearly round 4

3* Shell ∅ 15–16 mm, limy, ribbed, with or without bands, stripes or flecks; aperture round or nearly round 5

4 Shell 6–7 mm b, 2.5–3 mm h; disc-shaped; strongly ribbed and finely striated (40–50 ribs on body W), thick-walled, with flecks; pale horn-coloured, with curved radial red-brown stripes at regular intervals; $5^1/_2$–$6^1/_4$ W, body W with a blunt peripheral keel; aperture nearly round; P sharp, thin with internal rib; U broad (greatest ∅ $^1/_3$ width of shell); body dark grey

Discus rotundatus (O.F. Müller), Rounded Snail

5 Shell ±16 mm b, 8 mm h; extremely depressed above, limy, opaque, white; very thick-walled; ±$5^1/_2$ W, rather flattened with shallow suture, upper W with regular ribs, lower W smooth; body W scarcely descends at mouth; 4–6 dark brown to black bands often as stripes or flecks; aperture slightly elliptical without clearly defined rib, its margins widely separated from one another; U broad, about $^1/_4$ of width of shell; growth ridges fine, irregular; body yellowish-white to brownish-yellow; eye-stalks darker

Helicella obvia (Hartmann)

5* Shell about 15 mm b, 7 mm h; very disc-shaped, less limy, slightly translucent between the bands; white to cream, fairly thick-walled; slightly glossy with fine striations; prominent pale to more usually dark brown bands, a thicker band above the mid-line and several narrower ones on the lower surface; about $5^1/_4$–$5^3/_4$ slightly convex W with shallow sutures; body W sinking below the periphery of the previous one and turns abruptly to the mouth; aperture oblique, circular-elliptical; P sharp with a very weak internal white rib; U very broad ($^1/_3$ breadth of shell); body colour similar to previous species

Helicella itala (Linnaeus), Heath Snail

6 (2) Shell elevated, up to 18 mm ∅, milky- or yellowish-white or grey-brown to red-brown; with or without hairs; U narrow or broad 7

6* Shell discoidal, always without hairs, shiny, translucent, with fine lines or finely striated; yellowish-red-brown; U broad 10

7 Shell milky- or yellowish-white; without hairs; aperture elliptical; P sharp with a thick internal rib; brownish or whitish; U narrow 8

7* Shell grey-brown to reddish; aperture oval moon-shaped, crescentic, with or without hairs; U broad 9

8 Shell 10–12.5 mm b, 6.5–7.5 mm h; depressed globular, slightly translucent, thin-walled, milky-white or creamy white, with faint transverse striations; $5^1/_2$–$6^1/_2$ convex W, not swollen, whorls rising in a flattened cone, sometimes a paler band at the periphery; aperture elliptical, rounded-crescentic; P sharp, scarcely expanded, with a strong internal rib; young animals slightly hairy; U minute, partly obscured by reflected columellar lip

Monacha cartusiana (O.F. Müller), Cartusian Snail

8* Shell larger, globular, less depressed, 16–18 mm b, 11.3–12.5 mm h; 6–$6^1/_2$ convex swollen W, in a low conical spire; yellowish-white, creamy-white; aperture nearly circular, rounded-crescentic; P sharp, scarcely expanded, internal white rib set back from the edge; body W well rounded, sutures deep; U narrow, deep, slightly obscured by reflected columellar lip

Monacha cantiana (Montagu), Kentish Snail

9 Shell about 8 mm b, 4–5 mm h; very depressed convex above with a blunt, slightly elevated spire; translucent; short curved hairs (in old specimens hair scars of fallen hairs); grey-brown to red-brown, usually with a light band on the body W; 6–7 rounded or very slightly keeled W; irregularly striated; body W (slightly blunt-keeled) sinking a little towards the mouth of the shell; less angled at the periphery than the following species; aperture oval, moon-shaped, slightly reflected lip, with prominent white internal rib; U broad, showing all whorls; wide to very wide

Trichia hispida (Linnaeus), Hairy Snail

9* Shell 11–13 mm b, 6–7 mm h; broadly depressed, convex and conical above; flattened below, with a shallow spire; hairy only in juveniles, hairless when adult; opaque, slightly translucent, grey to dark red-brown, grey to reddish-horn-coloured, creamy-yellow; 6 convex W, body W bluntly keeled, sinking a little towards the mouth of the shell; younger shells with more prominent keel; sutures moderately deep; finely ribbed; W often with pale opaque band at the periphery, rather coarse and irregular growth-lines; aperture elliptical to crescent, with a slightly reflected lip below; P sharp, with a thickened internal rib; U medium ($^1/_8$ to $^1/_5$ of width of shell)

Trichia striolata (C. Pfeiffer), Strawberry Snail, Ruddy Snail

10 (6*) Shell discoidal, ∅ greater than 6 mm, smooth with fine spiral lines; aperture elliptical 11

10* Shell discoidal, ∅ smaller than 6 mm, finely striated; aperture elliptical (as broad as tall), crescentic 12

11 Shell 6–7 mm b, 3–3.5 mm h; discoidal, convex above, flattened below, very gently convex, spire a little raised; shiny, translucent, smooth or (rarely) irregularly ribbed; very fine spiral lines (only visible with lens or microscope); pale, yellowish-brown, greenish, often whiter below; upper

side light red, lower side, around U, lighter; 4–4$^3/_4$ W, body W usually twice the diameter of previous whorl at mouth; aperture oblique, broadly elliptical, somewhat broader than high; U funnel-shaped, eccentric ($^1/_6$ of width of shell); body blue-black; smells strongly of garlic when irritated
Oxychilus alliarius (Miller), Garlic Snail

11* Shell 12–14 mm b, 6–7 mm h; discoidal with a slightly raised conical spire; thick-walled, very shiny, rather opaque, translucent, irregular striations (visible with a lens); growth-lines often rather well-defined and giving a slightly wrinkled appearance, especially at the sutures; upper side pale to dark horn-colour, lower side (around U) whitish; 5$^1/_2$–6 W, body W greatly enlarged, more than twice as broad as the previous whorl; aperture elongate-elliptical, rather higher than broad; U moderately broad, deep, ∅ about $^1/_8$ breadth of shell
Oxychilus draparnaudi (Beck), Draparnaud's Snail

12 (10*) Shell ±6 mm b, 3.5–4 mm h (seldom larger), depressed globular, convex above, flatter below, slightly elevated conical spire; glossy, rather translucent, finely striated; brown-yellow or red-brown (appearing black when animal is withdrawn); 4$^1/_2$–5 moderately convex W, body W greatly enlarged and rounded on the outside (twice breadth of previous whorl at mouth); aperture obliquely inclined, broadly elliptical crescentic; U broad, funnel-shaped, its ∅ $^1/_5$–$^1/_4$ the width of the shell; body dark grey-blue
Zonitoides nitidus (O.F. Müller), Shiny Snail

12* Shell 3.5 mm b, 2–2.5 mm h; discoidal, flatter than the previous species, spire a little raised, 4$^1/_2$ closely coiled W, suture deep; a hint of keeling at the periphery; shiny, translucent, pale yellow-brown, faint irregular growth-lines; aperture elliptical; U broad, ±$^1/_4$ the width of the shell; body grey-blue to dark-grey-blue
Zonitoides arboreus (Say), Bark Snail

13 (1*) Shell tapered, with fine striations 14

13* Shell conical or globular, medium (up to 25 mm ∅) or large (or than 28 mm ∅) 15

14 Shell 18–25 mm b, 20–21 mm h (dwarf species about 11 mm b, 10 mm h); tapered (more globular) with convex or conical spire, blunt apex; thick-walled, glossy, with irregular striations, these crossed by very fine growth-lines (visible with lens); pale to dark yellow-brown or brown-black; 5–6 convex W with rather shallow sutures; a dark band at the periphery; yellow or pale yellow flecks and streaks (sometimes missing); aperture elongate-round, a white and reflected lip; P sharp; U a tiny crescentic cleft closed by reflected columellar lip; body grey, yellow-brown to blue-black, variable
Helicigona arbustorum (Linnaeus), Copse Snail

15 Shell roundish-conical (globular, slightly depressed), U narrow (then striped with ribs) or open (then with fine spiral lines) 16

15* Shell globular or nearly globular; U reduced to a cleft or closed; completely closed by lip 17

16 Shell 10–13 mm b, 8–9 mm h; globular, slightly depressed with a convex spire; grey-white or yellowish-white, $5^1/_2$ convex W, suture deep; no bands or faint bands at and below periphery; marked regular growth-ridges; aperture somewhat elliptical, a thick white internal rib; U moderately broad, slightly obstructed by reflected columellar lip
Helicella bolenensis (Locard)

16* Shell 18–20 mm b, 16–17 mm h; globular to slightly depressed, conical or slightly convex above; thin-walled, translucent, slightly glossy, whitish, pale yellowish to reddish-brown; sometimes a faint darker band at the periphery, 5–6 convex W, slightly stepped; suture deep; body W swollen; growth-lines rather coarse and irregular, crossed by slightly spiral striations; aperture slightly elliptical with a white internal rib and a feeble lip everted basally; U medium ($^1/_5$–$^1/_6$ width of shell), open, deep, partly obscured by reflected columellar lip
Bradybaena fruticum (O.F. Müller), Bush Snail

17 Shell ∅ up to about 25 mm; aperture broad; U reduced to a cleft (only in young) or closed; umbilical region whitish or brown to dark brown 18

17* Shell ∅ 28 mm or larger; aperture large, rounded; U reduced to a cleft or closed by lip 21

18 P and umbilical region light brown, brown to black; 5th peripheral band approximate to the closed U 19

18* P and umbilical region white or whitish; 5th peripheral band at a distance from the closed U 20

19 Shell 22–23 mm b, 17–18 mm h; depressed globular with a rather conical spire; glossy, yellow, reddish or brown (variable); 0–5 brown to black bands, often faint or lacking, 5th band approximate to the closed U; $5^1/_2$ convex W; growth-ridges rather weak and irregular; aperture broad, oblique (usually higher than broad, not clearly shown in young); P brown-red to black, thickened lip dark brown; U completely closed by lip; P and umbilical region brown-red to black; shell colour and banding very variable; body grey-white, grey-yellow or greenish-yellow; wrinkles pale grey to almost black; head and eye-stalks darker, sole very light to dark grey
Cepaea nemoralis (Linnaeus), Banded Wood Snail, Brown-lipped Snail

19* Shell 20–21 mm b, 17–18 mm h; depressed globular with a conical spire; white, whitish or white-yellow; almost 5 bands, 5th band broadest, very approximate to the closed U; $5^1/_2$–6 convex W; fine, rather regular growth-ridges; P with a strong thickened lip; U completely closed by lip; P and umbilical region light brown
Cepaea vindobonensis (Férussac)

20 Shell 19–21 mm (rarely 22 mm) b, 15–16 mm h; slightly depressed globular with a low convex spire; glossy, yellow, light brown or reddish; 0–5 dark bands, often convex W; with growth-ridges, rather weak and irregular; often no bands, 5th band at a distance from the closed U; 5–5$^1/_2$ convex W; aperture elliptical, but broader than high; P white, a white thickened lip; P and umbilical region whitish to white; U completely closed by lip; body similar to previous species; very variable in shell colour and banding pattern
Cepaea hortensis (O. F. Müller), Smaller Banded Snail, White-lipped Snail

21 (17*) Shell ∅ about 28 mm; globular 22

21* Shell much larger, ∅ 35 mm or more 23

22 Shell about 28 mm b, 20 mm h; almost globular, rib striations, olive-red-brown; 3$^1/_2$–4 W, body W very large, swollen; aperture rounded; P with thin lip; body olive-brown-red; when irritiated pushes out mucus bubbles with a hissing sound and then withdraws rapidly into the shell
Helix aperta Born, Singing Snail

23 Shell 38–45 mm b and h (variable), very large, globular with more or less blunt conical spire and rounded apex, not shiny, thick-walled, deeply and coarsely ribbed and fine spiral striations; generally pale, creamy or whitish-grey to yellow-brown; 5–6 convex W, body W large, very swollen; 0–5 rather faint dark, sometimes broken, bands; aperture rounded, taller than broad, pointed above, P blunt, thickened, reflected and re-inforced by a weak white to reddish lip; U reduced to a narrow cleft by the reflected columellar lip
Helix pomatia Linnaeus, Roman Snail, Edible Snail, Vineyard Snail

23* Shell up to about 38 mm b and 35 mm h; globular, bluntly rounded spire; ± thick-walled; usually pale brown, yellow; 4$^1/_2$–5 rapidly expanding slightly convex W, body W less swollen; 0–5 dark brown, irregular, very broken bands, crossed by zig-zag streaks, often flecked with white; characteristic wrinkle-like sculpture; aperture large, roundish; P blunt, thickened, a white thickened lip; U completely closed by lip
Helix aspersa O. F. Müller, Brown Garden Snail

To II: **Cellars, Storerooms:** Terrestrial pulmonates, Stylommatophora (eye-spots at the tips of the posterior pair of tentacles)

1 *With* shell: snails 2

1* Without (visible) shell (internal shell as "slug plate"): slugs . . . 3

2 Shell 10–15 mm b, 5–6 mm h; discoidal, very flattened, spire very depressed; thin-walled, very shiny, translucent, upper side greenish-grey or yellowish-brown, underside whitish; 5$^1/_2$–6 W regularly increasing in breadth, body W almost twice as broad at mouth as previous W; feebly

and indistinctly striated (lens!); columellar lip slightly reflected; growth-lines scarcely visible, suture often presenting an illusory channelled appearance; aperture elliptical, broader than high; P sharp, symmetrical, deep, open, funnel-shaped, ∅ $^1/_6$ to $^1/_7$ of shell breadth; body dark grey-blue
Oxychilus cellarius (O. F. Müller), Cellar Snail

3 M wrinkled, respiratory pore at right side, behind mid-point of M; sole tripartitie (3 longitudinal regions)
Slugs: Family: **Limacidae**
Genus: *Limax* 4

4 Mucus colourless, body marking very dark 5

4* Mucus yellowish to dark yellow (only colourless in young), body marking faint dark 6

5 Length 120–150 mm; light grey to brownish-grey; on each side 2 or 3 darker rows of spots or stripes; concentrically ridged mantle, centre point in mid-dorsal line; M with marbled pattern; keel short, extending about $^1/_3$ distance from tail to mantle; M and body often banded; 2 or 3 broken dark longitudinal bands; internal shell (slug plate); foot fringe with small black grooves; sole tripartite (in young grey, adult pale grey-yellow); mucus colourless, sticky
Limax maximus Linnaeus, European Giant Garden Slug

6 Length 80–100 mm; greenish-yellow, yellowish-orange to reddish (young more greenish); low body-water content; on each side with ± faint dark network (marbled); M of the same colour as the marks or without pale flecks; eye-stalks dark blue to blue-green, head lighter grey-blue marbled; keel not sharp, about $^1/_2$ of the back; foot fringe dark yellow without black wrinkles; sole tripartite, median region greyish-yellow, lateral regions yellow; mucus yellow
Limax flavus Linnaeus, Cellar Slug

To III*: **Damp Meadows, Marshes, Ditches, Banks, Pools, Lakes:** Pulmonates with spirally coiled shell, without operculum on living animals

A. Stylommatophora, terrestrial pulmonates (eye-spots at the tips of the posterior pair of the tentacles) 1

B. Basommatophora, freshwater pulmonates (eyes at the base of the tentacles, reduced anterior tentacles) 4

1 (A) Shell more than 4 W or less, spire dextral, shell ovoid, very thin, horny, amber coloured, aperture wide, pointed oval; suture deep; P not thickened; no U 2

* This includes species which act as intermediate hosts for parasites and also economic pests in arable land

1* Shell less than 8 mm h, not translucent, shiny brown, more than 5 W, aperture ovoid, pointed at the top 3

2 Shell 8–11 mm b, 16–22 mm h; translucent, mostly pale yellow, grey to amber, 3–4 W, very rapidly enlarging, gently rounded; body W large, swollen; spire reduced; growth-lines not very prominent; aperture large, broad, oval, bluntly pointed above; suture deep; surface glossy, finely striated; body usually pale, yellow to grey; no U
Succinea putris (Linnaeus), Amber Snail

2* Shell 4.5 mm b, 7.5 mm h; thick, elongate pointed, opaque, pale amber to greenish-grey to horn-coloured; 3–3$^1/_2$ W; not very glossy, somewhat coarse irregular growth-lines; aperture oval, slightly pointed at the top; suture very deep; no U; body yellow
Succinea oblonga Draparnaud, Small Amber Snail

3 Shell 2.3–2.5 mm b, 5.8 mm h; small, elongate ovoid; translucent, glossy, pale brown; 5$^1/_2$–6 W; fairly convex; well-marked suture and blunt apex; aperture 2.3 mm h, oval, bluntly rounded, slightly thickened, with internal pale lip, without teeth
Cochlicopa lubrica (O. F. Müller), Agate Snail, Slippery Snail

4 (B) Shell disc-shaped, coiled in a plane spiral, tentacles long, thin
Family: **Planorbidae,** Ram's horn Snails 5

4* Shell not disc-shaped; tentacles flattened, broad, triangular or long, slender, awl-shaped 6

5 Shell large, 27–33 mm b, 11–13 mm h; dark brown, often sculptured with a network; 5 W enlarging rapidly; suture deep; spire slightly sunken above and below (above more strongly); aperture broadly kidney-shaped; P sharp; U broad, deep
Planorbarius corneus (Linnaeus), Great Ram's horn Snail

5* Shell medium-thick; 14–17 mm b, 3.5–3.6 mm h; biconcave disc-shaped; W more convex than above; body W at the periphery with a thread-like ridge (keel) towards the upper (spire) side; spire flat, slightly sunken; 5$^1/_2$–6 W, body W greatest diameter $^1/_4$ shell-breadth; aperture rhomboidal to drop-shaped; U deep
Planorbis planorbis (Linnaeus), Ram's horn Snail

6 Tentacles flattened, broad, triangular, shell dextral (coiled to the right, aperture to the right), elongate conical-ovoid; spire pointed; aperture more or less oval or ovoid
Family: **Lymnaeidae**, Pond Snails 7

6* Tentacles long, slender, awl-shaped, shell sinistral (coiled to the left, aperture to the left), elongate ovoid, smooth, shiny, living animals with lobes of the mantle turned back over the shell
Family: **Physidae**, Bladder Snails
Genus: *Physa* 16

7 Spire as long or longer than aperture, more than 5 W 9

7* Spire always shorter than aperture, 4–5 W 12

9 Shell 4–6.3 mm b, 7–13.8 mm h; small, elongate conical-ovoid; spire bluntly pointed, taller than the mouth, rather slim; usually light brown; 5–6 W, well-arched, flattened above, regularly increasing, body W large, comprising $^3/_4$ height of shell; finely striated; aperture (3.6–4 mm) bluntly pointed above; U small, open

Galba truncatula (O.F. Müller), Dwarf Pond Snail, Mud Snail

9* shell larger, up to 15 mm (then not very rounded and W not stair-like) or over 15 mm (then W stair-like) 10

10 Shell 22–30 mm b, 48–60 mm h; very large, spire elongate narrow, tapering to a point; pale to dark horn-colour; growth-striations prominent; 7–8 W, body W large, distended, ovoid, more than $^3/_4$ of height of shell; no U (closed)

Lymnaea stagnalis (Linnaeus), Great Pond Snail

10* Shell smaller, body W not distended and enlarged, spire tapered, cylindrical, no U 11

11 Shell 4–4.5 mm b, 12–15 mm h; cylindrical, spire slender, blunt; grey or brown; 7–8 W, increasing very regularly, W not stair-like, body W not distended; aperture small, pear-shaped; P with internal lip, outer and columellar lip almost parallel; U slight

Galba glabra (O.F. Müller), Mud Snail

11* Shell about 9 mm b, 17–20 mm h; elongate conical-ovoid; spire tapered, pointed and about as tall as aperture; brown, spirally and transversly striated; 6–7 W, body W not distended, about $^2/_3$ of height of shell; suture shallow, yellowish-white; aperture oval; no U

Galba palustris (O.F. Müller), Marsh Snail, Pond Snail

12 (7*) Shell 20–30 mm b, 25–30 mm h; large, broad, ear-shaped (with characteristic ear-shaped aperture); thin-walled, glossy, pale brown to whitish; spire short, sharply pointed, at most $^1/_6$ of height; 4–5 W, body W very large, greatly distended; aperture large, ear-shaped, upper edge forms a right angle with the body W; P strongly dilated; U small cleft-like or covered

Radix auricularia (Linnaeus), Ear Pond Snail

12* Shell 16–17.5 mm b, 24–25 mm h; ovoid, pointed, thin-walled; spire a short, blunt cone $^1/_3$–$^2/_5$ of height of shell; pale to dark horn-coloured; 4–5 W, body W not distended, somewhat depressed laterally; aperture medium-large, not sharp pointed above; P straight forward; U almost covered

Radix peregra (O.F. Müller)⋆, Wandering Snail

⋆ Formerly *Radix ovata* (Draparnaud) (Ant 1968)

16 (6*) Shell 6 mm b, 9.5–10.5 mm h; ovoid, sinistral (coiled to left); very thin, glossy, brittle, translucent; spire short, blunt; pale yellow-brown, horn-colour; 4 W, body W large, distended, $^3/_4$ of height of shell; aperture to the left, oval; P sharp, without lip; no U; in living animal two lobes of mantle with 6–7 finger-shaped processes turned back over the shell
Physa fontinalis (Linnaeus), Bladder Snail

16* Shell sinistral (coiled to left), 7 mm b, 12 mm h; ovoid-pointed thin-walled; spire short, conical-elongated; horn-coloured; 6 W, apex a prominent point; P with a lip; no U; in living animal only the right lobe of the mantle turned back over the shell
Physa acuta (Draparnaud), Common Bladder Snail

2 Damage Caused by Gastropods

Pulmonates can cause damage both on land and in water. Aquatic and amphibious snails are important in medical and veterinary practice because of their role as intermediate hosts for some parasitic worms of man and his domestic animals. Since such snails frequently inhabit cultivated land and there cause considerable damage to young plants and roots, they are of interest in the field of plant protection as well. This is true, for example, of *Pomacea* spp. in rice plantations of South and Central America (EFFERSON 1968; VAN DINTHER 1956, 1957, 1973), of other snails in sugar cane plantations of Tanzania (FENWICK 1970), the citrus orchards and farms of Zimbabwe (MEDENWALD and SHIFF 1965; SHIFF et al. 1973), and finally, the Lymnaeidae in cress cultivations (*Nasturtium officinale* R. Br.) of France and Belgium (RICOU and PICARD 1972; VAN DEN BRUEL and MOENS 1957, 1958). In these countries, as also in West Germany, Holland and England, many species of snail which are found in fields and pastures act as intermediate hosts for the liver fluke, *Fasciola hepatica* L., of cattle and sheep (VAN DEN BRUEL 1963, 1968, 1970).

Biomphalaria glabrata (Say), one of the vectors of *Schistosoma mansoni* Sambon, has even been found in the greenhouses of the Botanical Gardens in Berlin-Dahlem in 1950 (JAECKEL and PLATE 1967). The scale which the damage and associated control measures can assume is indicated by the fact that in some areas, for example Poland, aerial spraying with pesticides over large pastures has become necessary (TARCZYŃSKI et al. 1965), a method also recommended by STURROCK and BARNISH (1973).

Since terrestrial gastropods represent the major plant pests among the molluscs, attention will be given to these first.

The destructiveness of terrestrial pulmonates is far greater today than in former times, since limits to their spread not only from one country to another, but even from one continent to another, are practically non-existent, as a result of ever denser and faster transport and traffic. The relative degree of damage has, furthermore, also increased with the growth of intensive land utilization, while the planting of seeds at greater intervals to avoid subsequent thinning, as in the case of sugar beet, means that there are relatively fewer plants on which attack by slugs and snails is concentrated.

2.1 Causal Factors

The degree of damage inflicted upon any particular agricultural or horticultural plantation depends not only on the activity of each individual slug or snail, but also on the density of the gastropod population. Repopulation of an area either wholly or partially cleared by use of molluscicides depends on the dispersal and also the fecundity of the gastropods that may have survived. If the adjacent terrain has not been treated with molluscicides, slugs and snails will penetrate from there into the cleared area and can multiply at first unnoticed and therefore undisturbed, until the new population has reached such a density that once again considerable damage is caused to the cultivated plants.

2.1.1 Activity of Slugs and Snails

The degree of active locomotion and the intensity of feeding both have a great influence on the amount of damage caused. These two aspects of activity differ from species to species and are influenced by external factors such as temperature, humidity, light intensity, soil conditions and food supply, as well as by endogenous factors such as species, body size, age, and behaviour. The correlation between extent of damage and mobility was recognized by BARNES and WEIL (1944, 1945); since then there have been numerous investigations of this (WEBLEY 1964; ROGERS-LEWIS 1977 and others), none of which has yet produced a conclusive explanation.

As a result of **migratory behaviour,** the gastropod fauna even in adjacent gardens may differ both in the number of species present and also the number of individuals of each species. Frequently rest and food plants are situated some distance apart, and the snail *Cepaea nemoralis* (L.) has been observed to crawl back and forth between them (WOLDA et al. 1971). This "commuting" takes place along "streets" which can vary with the terrain and the season; they are not readily visible and are therefore difficult to detect (MOENS et al. 1965).

To establish the length of the trails, NEWELL (1964, 1965, 1966) marked slugs with ^{32}P. In a single night, *Deroceras reticulatum* (Müller) and *Arion hortensis* Fér. covered distances of up to 90 cm and *Milax budapestensis* (Hazay) of up to 270 cm. HUNTER and SYMONDS (1970) measured a trail of 150 cm made in one night for the Field Slug. *Helix pomatia* L. covered 69 cm in 120 min and, in one case, took only 4 min to go 12 cm (GODAN, unpublished). An example of the influence of the degree of activity on extent of damage was observed in cress cultures in Belgium (VAN DEN BRUEL and MOENS 1957). Members of a population of *Radix peregra* (Müller) are normally distributed over the whole of a ditch in which cress was being grown. In winter (February), however, the low temperatures are unfavourable for the snails which collected around a water outlet where the temperature was 10.5 °C, and damage was limited to this area.

There is also evidence that activity is dependent upon weather conditions such as temperature and humidity; this applies both to land and freshwater pulmonates (BINGHAM 1972; BURCH 1955; CAMERON 1969, 1970; HELMY 1953; LEWIS 1969; NEWELL 1966, PARTSVANIJA 1972; PINDER 1969; THEEDE 1972; WEBLEY 1964).

The migratory behaviour of slugs and snails is greatly affected by the microclimatic conditions in their habitat (CRAWFORD-SIDEBOTHAM 1972; NEWELL 1968). At the same low temperature, larger animals are much more active than smaller individuals of the same species (JUILLET 1968). Thus at 18 °C, specimens of *Arion ater* (L.), weighing 9 g and more covered up to 90 cm, while young animals with a body weight of 5 g and less moved at the most only 25 cm in an hour. At a temperature of 22 °C, however, activity was reduced and a difference between older and younger animals was no longer observed.

Field experiments were conducted by Ricou (1964) in areas where maximal temperatures lay between 15.3 °C and 25.2 °C and minimal temperatures between 8.4 °C and 15.4 °C, showed that *Deroceras agreste* (L.) and *Deroceras reticulatum* (Müller) are very active and thus cause considerable damage at **temperatures** between 17.5 °C and 20.5 °C. Low temperatures of 1 °C and 2 °C inhibited migratory activity; the animals continued to feed, but the damage they caused to clover, for example, was limited. Mellanby (1961) found that even at 0.8 °C *Deroceras reticulatum* (Müller) fed very actively on the young shoots of grain plants, so that in mild winters the damage caused to such plantations can be very considerable. This slug shows amazing resistance to low temperatures and even at 0 °C is not completely immobile, whereas *Milax budapestensis* (Hazay) is inactive and *Arion hortensis* Fér. hardly moves at 5 °C. Field Slugs have, in fact, been found to survive a temperature of −8 °C for several hours, and several days after being returned to a temperature of 18 °C, the animals laid a large number of eggs (Godan 1958).

According to Dainton (1954) the activity of slugs and snails is triggered by changes in temperature. Activity increases as a result of a fall in temperature below 21 °C and by a rise in temperature over a range of 20 °C and 30 °C; the increase in activity is, however, less marked at the higher temperatures. At a constant temperature of 18 °C and constant darkness, the Grey Field Slug, *Deroceras reticulatum* (Müller), shows a diurnal rhythm of activity, being active at night; this rhythm is lost after 4 to 5 hours.

In general, the pest gastropods of Central Europe prefer a temperature of 17 °C to 18 °C. According to Rising and Armitage (1969), the **preferred temperature** of *Limax maximus* L. is, however, not constant, but dependent on the prevailing environmental temperature to which the animal acclimatized. Oxygen consumption was used as an indicator of the degree of acclimatization. At 10 °C it was about 30% more than at 20 °C, and at the latter temperature, 25% higher than the amount used at 30 °C. The example of *Limax flavus* L. (Segal 1959) shows that gastropods can have a considerable capacity for acclimatization to the temperature conditions of their particular environment – in this particular case, over a range of 10 °C to 30 °C. There is an associated increase or decrease in oxygen uptake, which may deviate by up to 60% from the normal requirement. At a low experimental temperature of 5 °C, no increase in oxygen uptake occurs. In this case there is a temperature adaptation of the metabolism as a whole, as Blažka (1955) has shown for *Helix pomatia* L. and Roy (1963) for *Arion circumscriptus* Johnston.

Reichmuth and Frömming (1961) found, using a “temperature organ”, that slugs and snails have two preferred temperatures. The first maximum was between 19 °C and 20 °C for *Cepaea normalis* (L.), 23 °C for *Helicigona arbustorum* (L.), 19 °C for *Theba pisana* (Müller), 20 °C for *Limax flavus* L. and 21 °C to 27 °C for *Limax maximus* L. The second maxima were 32 °C, 31 °C to 32 °C, 24 °C, 25 °C and 30 °C to 34 °C. The higher preferred temperature cannot be regarded as characteristic for the particular species, since slugs, at least, cannot survive at a constantly high temperature. It is therefore to be assumed that this is a reflection of the acclimation of individual animals.

Gastropods are indeed capable of regulating their **body temperature** and can protect themselves from overheating by evaporation of water from the skin (HOGBEN and KIRK 1944; HUNTER W.R. 1964). Thus at a temperature of 33.7 °C and relative humidity of 24%, an environment which is too warm and dry for slugs, the body temperature of *Arion ater* (L.) was only 21 °C, that is 12 °C less than the environmental temperature and below the lethal temperature for slugs; such animals could survive as long as they were able to maintain this relatively low body temperature by evaporation of water. Movement of air stimulates the animals to seek out shelter so that evaporation of body water is reduced (STEPHENSON 1968). If the direction of the air flow is lateral to the slug's body, *Deroceras reticulatum* (Müller) and *Limax maximus* L. show an increase in locomotory activity, or, if the animals were at rest, movement commences. If the air stream is directed at the head, they change direction (DAINTON 1943, cited by STEPHENSON 1968). In a damp environment, slugs and snails usually crawl about singly, but as soon as the ground becomes dry they bunch together (BASINGER 1931; LUPU 1974). In this way evaporation from individual animals is reduced. Additional protection from overheating is afforded by light colours which reflect light and heat rays, and by the thickness of the shell which prevents evaporation. Thus *Cepaea* spp. with banded shells become much warmer than unbanded and yellow species (GARCIA 1973). SCHNETTER and SEDLMAIR (1953) and SEDLMAIR (1956) found that unbanded and yellow specimens were more numerous when weather conditions were dry as a result of a hot summer.

Moisture, in the form of rain or dew, also influences the activity of land gastropods. In this case the amount of precipitation is not as important as its regularity (RAINER 1962). A sudden downpour inhibits activity, while a light shower stimulates it, as RICOU (1964) and WHITE (1959) have shown experimentally (Fig. 52). Apart from any releaser effect on activity when the first raindrops fall (WELLS 1944), even a film of moisture on soil and plants stimulates both feeding and locomotion.

Activity appears also to be released by an increase in body water content after this has fallen to not less than 17% of the normal value (DAINTON 1954). However, when the body water content fell below this value, activity was reduced as a result of loss of water through the slime produced for movement (DAINTON, cited by RICOU 1964). The mucus of terrestrial pulmonates consists to 98% of water. At high temperatures of 27 °C to 35 °C, too, activity is inhibited by water loss, if the loss cannot be made good immediately by uptake of water from the environment. Gastropods are, nevertheless, capable of withstanding a loss of weight of 50% and regain full mobility when water is taken in orally, or if they are placed in an atmosphere of 100% relative humidity, since the deficit can be made good by absorption through the skin which is only one cell layer thick. Thus a specimen of *Limax maximus* L., weighing 9.1 g, was stimulated to secrete mucus heavily so that its weight dropped to 6.6 g; it had, in consequence, become immobile, but recovered within 2 h following water uptake.

Moisture attracts slugs in particular. After rain or a fall of dew, they congregate inside the heads of cabbage or lettuce, with the result that in a single cabbage as many as 20 and in a lettuce 10 slugs have been found. After prolonged overhead irrigation, even lawns which had previously been uninfested can become attractive to slugs from adjacent areas and suffer damage as a result of their feeding activity.

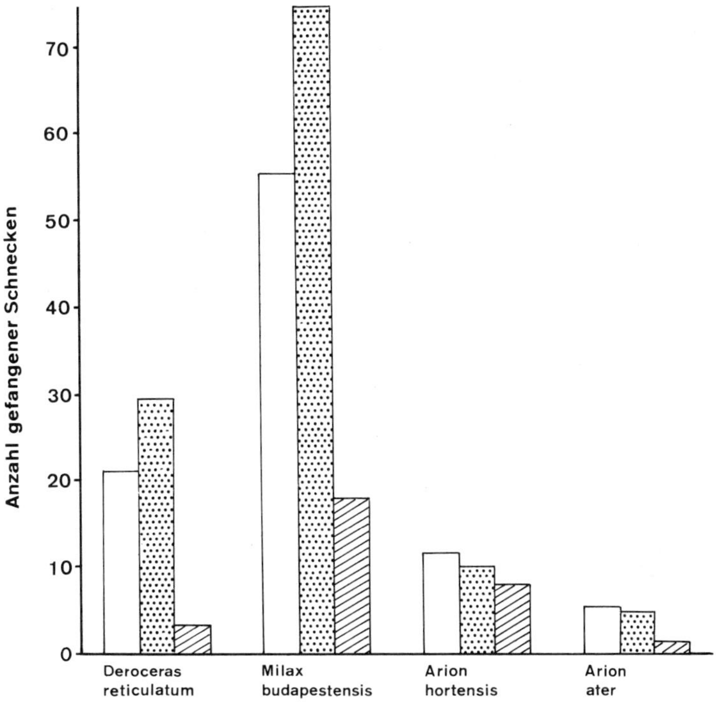

Fig. 52. Effect of falling rain on surface activity of four main species of slugs. □ = no rain; ▒ = light rain; ▨ = heavy rain

The same applies to the irrigation of areas under cultivation, particularly potato fields (STEPHENSON 1965).

Light influences the activity of terrestrial gastropods in that, except after rain, they remain in their hides during the day and only after dusk do they emerge to go in search of food (NEWELL 1966). A decrease in light intensity, together with a fall in temperature below 21 °C and a rise in humidity through dew fall at dusk, result in the animals moving onto vegetation, while in the morning, increase in light intensity stimulates the slugs and snails to retreat into the upper soil layers or between the earth clods where it is cool and shady. According to LEWIS (1969), the influence of daylight on gastropod mobility is greater than the effect of temperature changes.

The activity of slugs and snails shows a **rhythmicity** which is endogenous (LEWIS 1969); this differs from species to species and is observed only within definite temperature limits, although even outside these limits, the influence of the endogenous pattern on activity persists. The collection of slugs in the field led to the discovery by WHITE (1959) that they are most active at night, and that their activity is related to seasonal variation in the hours of darkness (Fig. 53). DAXL (1969) examined nocturnal locomotory and feeding activity of *Deroceras reticulatum* (Müller) (Fig. 54), *Deroceras laeve* (Müller), *Arion hortensis* Fér. and *Limax flavus* L. (Fig. 55). The activity rhythms of these species differ from one another. *D. reticulatum* (Müller) is periodically active, with three maxima, and intervals of greater or lesser dura-

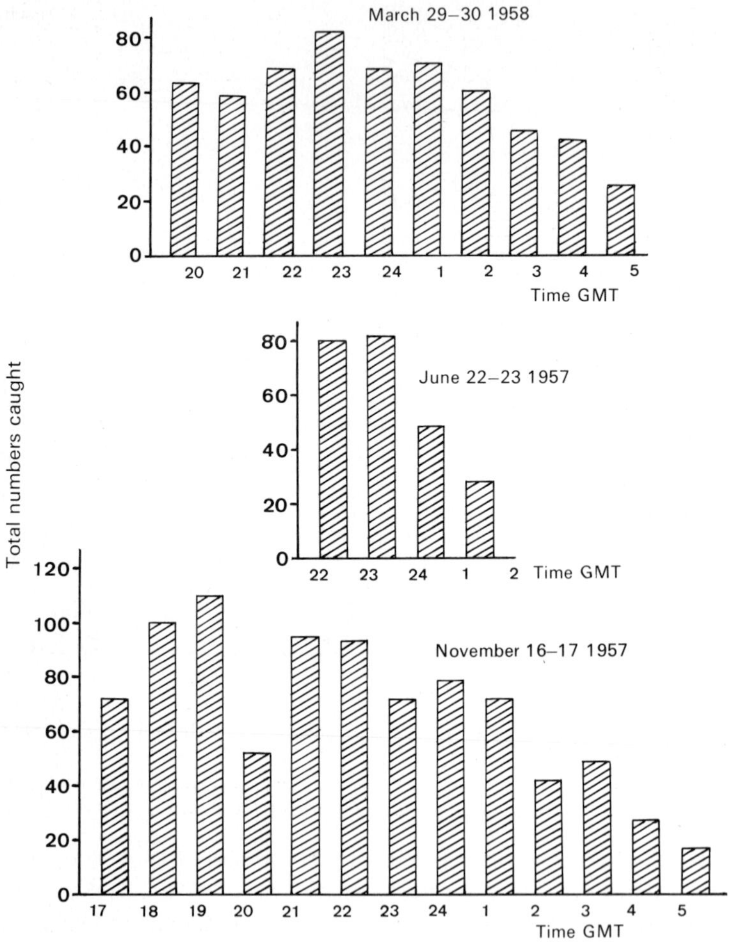

Fig. 53. Total numbers of slugs caught at hourly intervals on nights at three times of the year

tion. BARRY (1969) found that the time of greatest activity was 4 to 6 h after dark, a result which more or less corresponds with the second peak in the diagram of DAXL. The third activity peak, shown by *Deroceras reticulatum* (Müller), is especially strong and occurs in the early hours of the morning, between 03.00 and 04.00 h; by 07.00 h activity has ceased. *D. laeve* (Müller) has only one activity peak, at 06.30 h, some 2 h after sunrise; activity in this slug ceases only at about 09.00 h. The time of greatest activity for *A. hortensis* Fér. is between 05.30 h and 06.00 h. The activity of all these slugs shows a steady rise, commencing at 20.30 h in the case of *D. laeve* (Müller) and at 19.00 h in the case of *A. hortensis* Fér. The activity period of *Limax flavus* L. is on the other hand uniform and prolonged, reaching its peak soon after it begins at 22.00 h and maintaining this level for 5 h; activity ceases only after 09.00 h.

It is possible that the differences in the nocturnal activity patterns of these four species reflect adaptations to the temperature and light conditions prevailing in

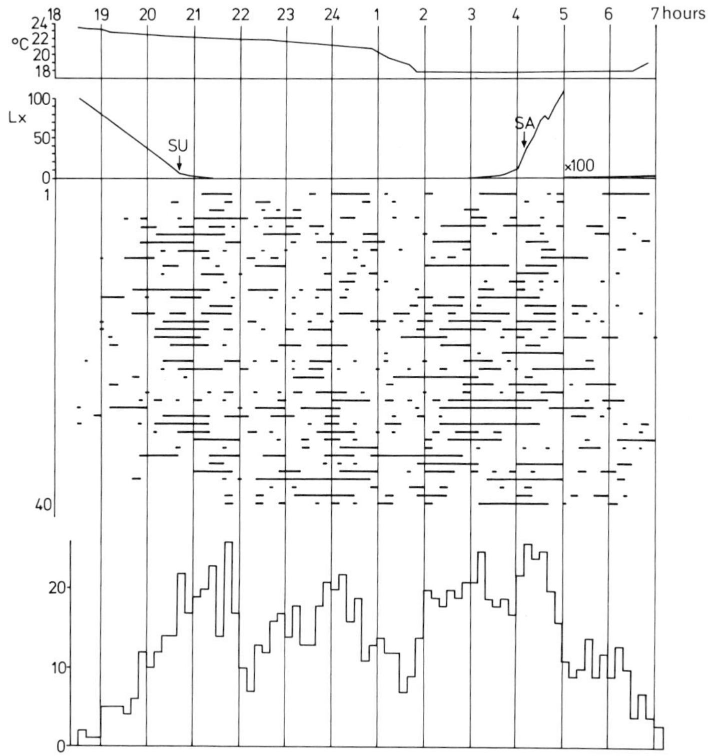

Fig. 54. Nocturnal activity of *Deroceras reticulatum*. *Horizontal lines* and *dots* represent the duration of activity of individual experimental animals; *Lx* lux; *SA*, *SU* sunrise and sunset

their different habitats. This is especially true of *Limax flavus* L. which inhabits cellars and storerooms, where daily variations in temperature, humidity and light do not occur. The fall in temperature at night between 01.00 and 04.00 h induces an increase of activity in the three outdoor species. Activity commences with a fall in light intensity, even before any appreciable drop in temperature has occurred.

The individuals of these three species differ from *Limax flavus* L. as they are not constantly active, either in locomotion or feeding, but have intermittent rest periods. The very brief periods of activity are represented as dots in the diagram of DAXL and are virtually absent in the case of *Limax flavus* L. In observing the feeding of gastropods on wheat seeds, HUNTER (1968) found that *Arion hortensis* Fér. feeds more frequently than *Deroceras reticulatum* (Müller) and the latter more frequently than *Milax budapestensis* (Hazay). These findings correspond more or less with the dots and lines in the diagrams. The weight of grain consumed by each of the three species was, however, virtually identical, since the species which fed frequently consumed less in a given time than those which fed for short periods only. Feeding activity was maximal at 20 °C and 100% rel. humidity – a further demonstration of the influence of temperature and humidity.

The activity rhythm is furthermore subject to seasonal changes such as the times of sunrise and sunset and fluctuations in temperature and humidity. The

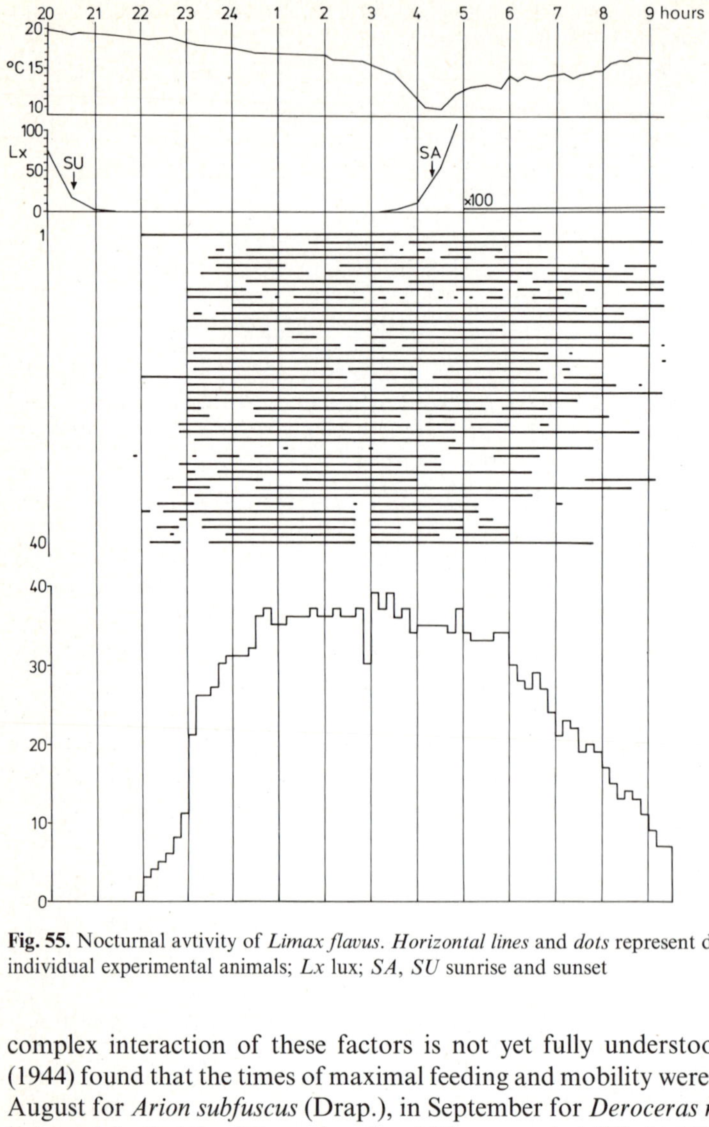

Fig. 55. Nocturnal avtivity of *Limax flavus*. *Horizontal lines* and *dots* represent duration of activity of individual experimental animals; *Lx* lux; *SA*, *SU* sunrise and sunset

complex interaction of these factors is not yet fully understood. Thus BARNES (1944) found that the times of maximal feeding and mobility were in June, July and August for *Arion subfuscus* (Drap.), in September for *Deroceras reticulatum* (Müller), and in October/November for *Milax sowerbyi* (Fér.), *Milax budapestensis* (Hazay) and *Arion hortensis* Fér.

Quite apart from climate, the nature of the **terrain** also influences the activity of terrestrial pulmonates. On heavy, wet, little-cultivated soil with large clods, snails and slugs are far more numerous than on a loose and frequently tilled soil (ANDERSSON and NILSSON 1967; DUTHOIT 1961; HUNTER 1967; WEBLEY 1962, 1963). Experiments by STEPHENSON (1975) confirm the normal preference of Field Slugs for heavy soils as a habitat, while fine, loose earth was chosen for egg-laying (Table 7).

In another set of experiments the slugs were given a choice of fine and coarse soils only: while the adult animals resided exclusively in the latter (0:15), twice as

Table 7. Distribution of young and adult *Deroceras reticulatum* (Müller) when offered a choice of four aggregate fractions. (After STEPHENSON 1975)

Slugs' age	Soil fraction			
	Fine	Medium	Medium-coarse	Coarse
3 days old	0	2	9	4
2–3 weeks old	0	14	60	44
Adult	0	2	20	28
Egg masses	8	5	3	0

many egg masses were found in the former (14:7). Coarse earth was again preferred. Experiments by DUVAL (1970) in which holes of different diameter were made in the soil, showed that *Deroceras reticulatum* (Müller) selected those which provided all-round body contact. Using adults, holes of 15 mm diameter were found to be occupied by 66% of the experimental animals, while only 6.7% of the slugs established themselves in holes of 40 mm diameter.

These slugs are thus capable of reaching mechanically planted wheat seeds by utilizing adjacent spaces in the soil. According to STEPHENSON (1975), damage caused by *Deroceras reticulatum* (Müller), *Arion hortensis* Fér. and *Milax gagates* (Drap.) to newly planted wheat seeds can be obviated by ensuring an even seed bed with fine top soil and by planting the seeds so deeply that they are thoroughly covered. Application of even a slight pressure of 390 kg/m^2 to a moderately coarse soil substantially reduced the percentage of seeds eaten. A very coarse soil, however, could not be compacted sufficiently even by nearly twice this pressure (620 kg/m^2). Spaces remained which were utilized by slugs as passages through which access could be gained to the wheat seeds.

The small *Arion* spp. as well as *Milax* spp. are ground slugs which usually avoid the surface of the soil and instead attack seeds, seedlings, bulbs, roots and tubers in the ground. Frost and dryness drive *Arion hortensis* Fér. and *Milax budapestensis* (Hazay) still deeper into the soil (RUNHAM and HUNTER 1970). Even *Deroceras reticulatum* (Müller) can penetrate several centimeters below the surface (VAN DEN BRUEL and MOENS 1958). The distribution of field slugs in the soil has recently been investigated by STEPHENSON (1975). Forty eight hours after their release on the surface, the following picture of their distribution emerged in the four types of soil mentioned above:

No. of animals in different soil types

Depth	Fine	Medium	Medium-coarse	Coarse
0– 20 mm	1	1	10	2
20– 40 mm	0	0	3	0
40– 60 mm	0	1	0	0
60– 80 mm	0	0	2	1
80–100 mm	0	0	0	1
100–120 mm	0	0	1	0
120–140 mm	0	0	1	1

This experiment confirms the observations of practical experience that coarse and very coarse soils favour movement of slugs below the soil surface and thus increase the damage they do to seeds and seedlings in the soil.

Moisture combined with a certain depth of soil are prerequisites for egg-laying by terrestrial snails. Oviposition in *Cepaea nemoralis* (L.) was inhibited when the depth of the soil was less than 3 cm (WOLDA 1965). In such cases, the snail attempted to make a nest, but regular egg-laying did not occur. Given no alternative, the snail accepted a depth of soil of only 1.5 cm, but only two batches, one of 4 and the other of 22 eggs were laid, whereas the normal egg clusters contain more than 65 eggs. This snail is stimulated to lay eggs by short dry periods with warm, damp nights; longer periods of dryness, however, cause a reduction in egg-laying.

Further effects of the nature of the soil on oviposition will be discussed in connection with population density.

2.1.2 Population Density

Apart from the migratory and feeding behaviour of an individual snail or slug, the extent of the damage they cause is also related to their fecundity. This, in turn, determines the population density, which may remain below the plague threshold, or there may be a sudden explosion in numbers. Some factors have a positive, and others an unfavourable effect on population growth, and as in the case of the two forms of activity already described, these may be endogenous or environmental in origin.

Despite being of identical age and living in comparable environmental conditions, members of the same population differ in their rate of growth, egg-laying frequency and life span (LOMNICKI 1969; WOLDA 1967, 1970). These variations are only partially genetic, however, and a final explanation of them has not yet been found.

Mild winters with rain continuing into the spring, and also reducing tilling with consequent increase of weeds and grass are, according to LEWIS and LAFOLETTE (1942), the causes of population increase of slugs and snails. RUNHAM and HUNTER (1970), on the other hand, distinguish between unfavourable external factors such as predators, frost and drought which have a direct effect on population density, and the indirect effect of adverse weather conditions on reproduction and development, and thus on the time lag between two generations (generation interval).

Furthermore, there is a close relationship between the population density of a particular area and the **water-holding capacity** of the soil (WOLDA 1965). *Deroceras agreste* (L.) prefers soil with a high water content (50% to 75%) for egg-laying (CARRICK 1942), as shown in Table 8.

It should therefore be borne in mind that measures undertaken to increase the water-holding capacity of the soil also create more favourable conditions for the reproduction of pest gastropods.

The eggs of terrestrial pulmonates are in great danger of **desiccation** (BAYNE 1968), since they contain more than 85% water. The body water content of slugs such as *Deroceras reticulatum* (Müller) is over 80%. With a fall in the water content of the soil, the eggs are no longer deposited on the surface, but are laid in the earth.

Table 8. Oviposition of *Deroceras agreste* (L.) in soils of variable water content. (After CARRICK 1942)

Day of oviposition	Saturation %				
	100	75	50	25	10
1	–	22 eggs on surface	26 eggs 1 in. deep	–	–
2	–	85 eggs on surface	59 eggs 1 1/2 in. deep	32 eggs 1 1/2 in. deep	–
3	34 eggs on surface	–	–	–	–
4	–	–	9 eggs just below surface	64 eggs 3 in. deep	–
Total no. of eggs	34	107	94	96	0

Furthermore, they are provided with a water regulating gelatinous egg membrane or with a milky-white calcareous envelope. As a rule, healthy young gastropods develop only when the water content of the soil is 60% to 85%; in dry or extremely wet earth, particularly with stagnant water, they die during embryonic development or else the young animals are too weak to hatch out.

The eggs of aquatic and amphibious gastropods are admittedly not threatened by dryness, but by low temperatures. A special mechanism enables them to survive frost: in *Physa acuta* Drap. the freezing point of the fluid in the egg capsule is raised, and depending upon the stage of development of the embryo and the salt content of the water, is about 0.1 °C above that of fresh or brackish water (STYCZYNSKA-JUREWICZ 1972).

Weather conditions such as temperature and humidity influence population size by affecting the formation of gametes, the duration of embryonic development, growth and mortality rate (Fig. 56). According to the findings of WOLDA (1972), the mortality rate of *Cepaea nemoralis* (L.) is on average 4.9% higher in winter than in summer. The extremes of weather normally encountered are admittedly not lethal in their effect, but inhibit activity and reproduction (CARRICK 1942; PAPPAS and CARMAN 1955). This has been frequently studied in terrestrial and freshwater gastropods (BERGERARD 1971; CRIDLAND 1957; DUNCAN 1959; HERZBERG and HERZBERG 1960; HUNTER, W.R. 1961; IMHOF 1973; LUSIS 1961; MCMAHON 1971; PASCUAL 1970; SEGAL 1959; VAN DER STEEN 1967; VAUGHN 1953; WILBUR and OWEN 1964).

Under unfavourable **weather conditions** the snails cease to be fertile, but longevity is increased (RITCHIE et al. 1966). Low temperatures prevent spermatogenesis; at 6 °C the maturation of eggs begins, but not that of spermatozoa. In *Cepaea nemoralis* (L.), maturation of gametes ceases between October and February (BOUILLON 1956). Between March and April oogenesis commences and somewhat later, from April to May, spermatogenesis follows. Although later in starting, the latter process is concluded before the maturation of the eggs has been completed. Higher temperatures of 23 °C promote spermatogenesis, but restrict maturation of the eggs: only a few oocytes are differentiated, but none reach maturity. Once the em-

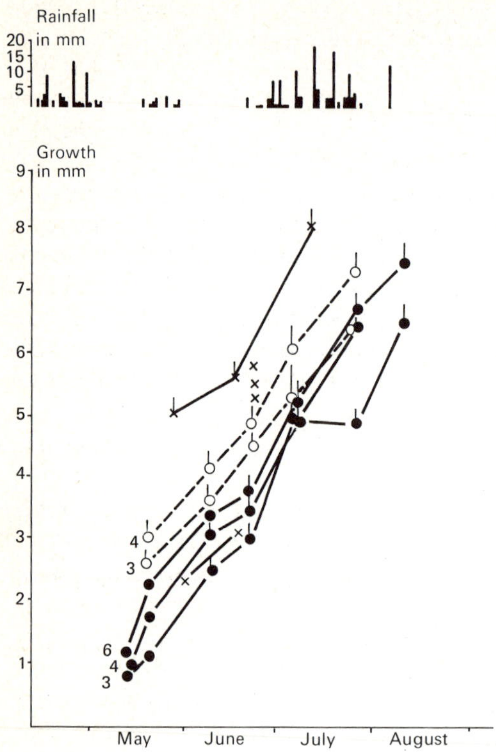

Fig. 56. Growth curves during 1970 in three experimental populations of *Cepaea nemoralis*, in two populations of *Helicigona arbustorum* living in the same area as *Cepaea* populations 3 and 4, and on some natural populations of *Cepaea*. Growth per season is different in the different populations, but during the season the growth lines are parallel. The differences are caused very early in the season. During the dry spell in June growth was retarded. ●——● = Exp. populations Cepaea; ×——× = nat. populations Cepaea; ○---○ = Exp. populations Helicigona

bryos have developed, the eggs of terrestrial pulmonates are very resistant to low temperatures. A transition from cold to warmth stimulates hatching, for which the lowest temperature limit is about 5 °C. *Deroceras reticulatum* (Müller) nevertheless manages to lay eggs at 3.3 °C.

A temperature of 17 °C to 20 °C and 100% relative humidity are optimal for terrestrial pulmonates and allow maximal growth and development (Lusis 1966).

These various experiments lead to the following conclusions: high temperatures limit the development of eggs more than that of spermatozoa; in a cool, damp environment the formation of gametes occurs normally but more slowly, and again the oocytes are more affected than the spermatocytes; lowered humidity reduces the number of spermatozoa in the gonad to half that found in "normal" animals, while the number of eggs is increased.

Warm water is not a limiting factor to population increase of **freshwater snails,** but cold water is. *Biomphalaria glabrata* (Say) tolerates a temperature range of 20 °C to 27 °C. At 25 °C the fertility is lower than at 18 °C, but the time required for development is less and the mortality rate is low (16.7%) with resulting increase in population density (Shiff and Garnett 1967). *Bulinus tropicus* (Krauss) and *Lymnaea natalensis* (Krauss) are likewise well adapted to the temperatures of the waters they inhabit, since they combine a high degree if temperature tolerance (21 °C to 27 °C) with a high rate of growth and fecundity (Prinsloo and van Eeden 1969, 1973). In *Lymnaea stagnalis* (L.), *Planorbarius corneus* (L.) and *Planorbis planorbis* (L.) the effect of a rise in temperature and lengthening of the hours of daylight is to stimulate reproductive activity. The temperature below which no oviposition occurs is species specific and ranges, for these species, between 7 °C and 12 °C. Reproductive activity and growth are linked in that arise or fall in the rate of reproduction brings about an increase or decrease in growth. A temperature threshold of approximately 10 °C is as important for *Galba truncatula* (Müller), the intermediate host for the liver fluke of cattle and sheep, as it is for the parasite itself, since its eggs can no longer develop below this temperature. Thus in Great Britain and Northern Ireland, temperatures suitable for the reproduction of the snail and at the same time of the liver fluke (10 °C and above), only occur between mid-May and mid-October (MORPHY et al. 1969). Acute outbreaks of liver fluke infestation therefore occur only in October to November, when the snail population has reached a high density, and when, as a result of favourable summer temperatures and optimal conditions for the development of the parasite, a high degree of infestation of the snail population has been established.

STURROCK (1970) has described the factors which determine survival time of the freshwater snail, *Biomphalaria glabrata* (Say) when stranded and thus suddenly exposed to the dangers of desiccation. The eggs die rapidly out of the water, but the snail itself is less sensitive. There are interactions between:

1. temperature and body size: larger snails survive better than smaller, yet this difference is less clearly marked as the temperature rises;
2. temperature and relative humidity: the chance of survival decreases with falling humidity and high temperature;
3. time and body size: larger snails survive better than smaller; the difference, however, becomes less significant the longer the time of exposure;
4. time and relative humidity: chance of survival falls with falling humidity, particularly with long exposure to air;
5. relative humidity and body size: the chance of survival decreases with falling humidity, but this applies more to large than to small animals.

A population of stranded snails is influenced by the following factors in order of decreasing importance: loss of water from the body > length of time out of the water > size and age of the animals > temperature. Maintainance of body water content is thus clearly the most important factor for survival of stranded freshwater snails.

In the case of *Galba truncatula* (Müller), the mortality of young snails in the first weeks after hatching is very high if the environment becomes dry, whereas the adult snails are far more resistant to fluctuations in moisture content of the environment (HEPPLESTON 1972).

Together with temperature and humidity, **light** also influences the maturation of the gametes and thus also the rate of oviposition (GARCIA 1975). Exposure of *Helix aspersa* Müller to 15 h illumination (60 to 80 ft. candles) alternated with 9 h of darkness, resulted in vigorous egg-laying (STEPHENS and STEPHENS 1966). High temperature in combination with prolonged hours of daylight has a markedly beneficial effect on egg-production. PELLUET and HENDERSON (1954), however, observed an inhibition of gametogenesis after continuous exposure to light of high intensity, whereas weak illumination caused more rapid maturation of spermatozoa than in snails kept under normal conditions. Complete darkness led to resorption of the gametes (SMITH 1966) in the same way as prolonged starvation (JOOSSE et al. 1968). The effect of light on the reproductive activity of gastropods has however still to be examined more fully. Day length has no influence on growth of the aquatic snails *Lymnaea stagnalis* (L.), *Radix peregra var. ovata* (Drap.), *Planorbarius corneus* (L.) and *Planorbis planorbis* (L.), but stimulates egg-laying.

Finally, there is a direct relationship between **nutrition** and population density (FISH 1972; GRIME et al. 1968, 1970; GRIME and BLYTHE 1969; MASON 1970; MOOIJ-VOGELAAR et al. 1973; MOOIJ-VOGELAAR and VAN DER STEEN 1973; VAN DER STEEN et al. 1973; WOLDA et al. 1971). Ready availability of food together with a rise in temperature can lead to a population explosion. This development is further promoted by the addition of organic material to the soil, in excess of 3% of soil content, as this provides both food and moisture for the slugs and snails. Thus young Field Slugs of 3.8 mm length, reared on soil enriched with organic material from potato fields, grew to a length of 6.5 mm, whereas others, identical in length at the outset and kept on soil lacking organic matter, reached only 4.2 mm (CARRICK 1942). Food rich in digestible protein and carbohydrate increased the number of egg masses layed by *Cepaea nemoralis* (L.) and also the number of eggs in each cluster (WOLDA and KREULEN 1973). Even snails normally regarded as phytophagous eat animal protein; thus *Cepaea nemoralis* (L.) consumes leaf lice, dead worms and dead snails (WOLDA 1972).

Just how complicated the interaction of nutrition and population density can be is shown by the fact that phytophagous gastropods need not only some animal protein but also plant protein in the form of living plant material. *Monacha cantiana* (Montagu) and *Hygromia striolata* (Pfeiffer) only grew normally when they were given living green leaves as fodder; they ceased to grow when they were offered fallen beach or oak leaves, that is, dead material, even though they consumed it until satiated (CHATFIELD 1973). Thus a population of phytophagous gastropods can increase only when offered animal protein and living vegetable matter. Living green plants are also a prerequisite for rapid population growth of fresh water snails, and are consumed in great quantities (FRÖMMING 1953; KEHDE and WILHM 1972). According to EISENBERG (1966), this nutritional requirement represents the major limiting factor to population size.

Chemical fertilization of the soil can lead indirectly to a population increase. An excess of nitrates in the soil gives rise to a tender, plasma-rich tissue in plants; this is particularly favoured by slugs and snails and causes them to flourish (NEUMANN 1954).

The **acidity** and **calcium** content of the soil, in the case of terrestrial pulmonates, or of the water in the case of freshwater gastropods, also influence population density and therefore the extent of damage caused. CARRICK (1942) reported extensive damage to potatoes in lightly alkaline and neutral soils as a result of great proliferation of slugs [*Deroceras reticulatum* (Müller), *Arion circumscriptus* Johnston, *Arion subfuscus* (Drap.) and *Arion hortensis* Fér.]. The number of individuals as well as the variety of species is greater on alkaline than on acidic soils. Increase in vitality and hence of fecundity in the snails of *Helicella* spp., *Discus rotundatus* (Müller), *Cochlicopa lubrica* (Müller), *Opeas mauritianum* (Pfeiffer) and *Oxychilus draparnaudi* (Beck), is only possible in neutral, weakly alkaline and calcium-rich soils (SCHMIDT 1955, 1959; VOELKER 1959; WAREBORN 1970). The relationship between shell growth and calcium content of the soil has already been described in Chapter 1. Lack of calcium affects the entire population as a result of the reduced vitality of the individual slugs and snails. The calcium content of the soil is of vital importance to many terrestrial snails, mainly for shell formation, and is a primary prerequisite for a large scale increase in population.

According to BURCH (1955), populations of terrestrial snails increased with a rise in the concentrations of calcium (CaO), magnesium (MgO), potassium (K_2O) and organic compounds in the soil. The greatest population densities were encountered on soil with pH values of 6.3–6.7 and P_2O_5 concentrations between 0.002% and 0.004%. Calcium, magnesium and organic constituents of the soil can be regarded as directly limiting factors, while the concentrations of hydrogen, potassium and phosphorous ions exercise an indirect effect.

LUSIS (1966) observed in *Arion rufus* (L.) a correlation between deposition of calcium in the body tissues and the onset of unfavourable conditions of temperature and humidity. The increased amount of calcium deposited in these circumstances is believed to be a consequence of impeded metabolism, which in turn results from increasing delay in development.

For **freshwater snails** the nature of the environment – here the waters in which they live – is decisive for the population density of any species (KLIMOWICZ 1962). These snails, too, are capable of absorbing calcium from their surroundings and storing it in the soft tissues of the body (HALEY and GIBSON 1971; KAPUR 1970; VAN DER BORGHT et al. 1971). SCHUTTE and FRANK (1964) have, however, found no evidence that the chemical composition of the water affects the distribution of these snails (several species of Planorbidae, Lymnaeidae and Ancylidae, as well as mussels) and suggest that a **complex of factors** is involved. Subsequently NEMENZ (1971) considered that the survival of a population of freshwater snails like *Planorbarius corneus* (L.) and *Planorbis planorbis* (L.) is primarily dependent on the salt content of the water, and that factors such as body weight, lowering of the freezing point of the blood, as well as its chloride and protein content are only of secondary importance. *Planorbis planorbis* (L.) survives better in a potassium sulphate solution and maintains a more constant weight in concentrated salt solutions (over 6% NaCl) than do the other species. The cations in the environment are less important than anions for the continued existence of this particular species. ØKLAND (1969) studied the ecology of freshwater snails in Norway. Lakes from which they are absent are poor in calcium and are often acidic, with low pH values. Similarly, the distribution of Lymnaeidae, Planorbidae and Physidae is closely related to the calcium content of the water (MACAN 1950).

The extent of damage caused as a result of large numbers is also dependent on the number of **generations** per year of the particular species, and this in turn, is determined by the length of post-embryonic development. The different phases in the life of a slug or snail are determined by the time required for development of the gonad and maturation of the gametes. CHEVALLIER (1971) has examined this in *Arion rufus* (L.), *A. circumscriptus* Johnston and *A. lusitanicus* Mabille. These species normally have an annual life cycle, in which the phases of rapid growth are dependent on climate and food availability; under unfavourable conditions growth is slow or ceases altogether. Temperatures of 17 °C to 20 °C have been shown to be favourable. There are four **phases in the life cycle:** infantile, juvenile, adult and aged (senile). In the first phase the hermaphrodite gonad is differentiated and in the juvenile phase the male genital tract develops to the point of sexual maturity. In the subsequent adult stage spermatozoa are formed and the snail is ready for copulation. Thereafter the female genital tract develops, the eggs mature and the albumen gland reaches its maximal size; this phase ends with oviposition. In the final, or senile, phase, the gonad and genital tract atrophy. Malnutrition, unfavourable weather conditions and also overcrowding are responsible for slow growth and delayed attainment of sexual maturity, of both body and gonads; it can in fact stop altogether. In such cases, mortality of infantile slugs is high and those surviving remain in the infantile phase much longer than under favourable conditions, with the result that the juvenile phase takes two years for completion.

According to HUNTER (1966), the **generation interval,** or time between successive generations, is an important factor in regulating population density in *Deroceras reticulatum* (Müller), *Arion hortensis* Fér. and *Milax budapestensis* (Hazay). As CHEVALLIER (1971) discovered in the case of the Arionidae, a complicating factor in terrestrial gastropods is that even in one and the same species the number of generations per year is not constant, but depends on environmental conditions. In *Deroceras reticulatum* (Müller), too, the number of generations varies. In Central Europe it is generally considered to be one per year. According to HUNTER (1968), however, in England this slug has two generations per year. The spring generation hatches between March and May, and reaches sexual maturity in the summer; its eggs in turn produce an autumn generation which hatches in September to October, survives the winter and lays eggs in the following spring. This pattern can be interrupted by unfavourable environmental conditions such as dryness or lack of food, and in such circumstances there are three overlapping generations in the same year (HUNTER and SYMONDS 1971). Thus one generation hatched in July 1967 produced a second in April 1968, which in turn gave rise to a third in December of the same year. It was only from the eggs of this winter generation that the new population hatched in April 1969. Thus in 1968 adult Field Slugs of three different generations were to be found instead of the usual two (Fig. 57). The number of newly hatched slugs with a body weight of less than 25 mg showed, in the years under investigation, three maxima which coincided with the appearance of a new generation (Fig. 58). There are, however, variations from year to year and from region to region. All this, of course, complicates the prediction of population growth and the related extent of damage.

There are also fluctuations in the succession of generations in the snail, *Cepaea nemoralis* (L.). As a rule, a generation requires one year to reach maturity, but sometimes, for reasons not yet known, two or three (WOLDA 1970).

Under natural conditions – as opposed to one-crop culture – some **physiological mechanism** ensures that population density remains more or less constant by inhibiting those factors which would make for overabundance. The experiments of POMEROY (1969) on *Cernuella virgata* (Da Costa) have shown that population density increases only up to a particular point, after which the number of young produced, their rate of growth and the life span of the adults all decline. This self-regulation of population density occurs even when optimal conditions pertain. According to recent investigations, discussed in Chapter 3.6.5, a so-called "density factor" is released as a secretion into the surrounding water by freshwater mol-

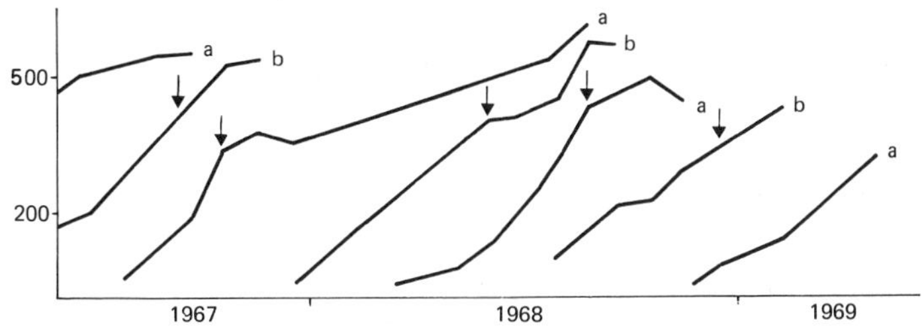

Fig. 57. Growth rates of *Deroceras reticulatum* in a natural population. *Arrow* indicates the first appearance of mature slugs in each generation. *a* a generation hatching in July 1967 giving rise to a generation in April 1968, which in turn produced a generation in December 1968. *b* a slug generation hatching in November 1967; this produced a small generation in August 1968 which in turn gave rise to a generation in April 1969

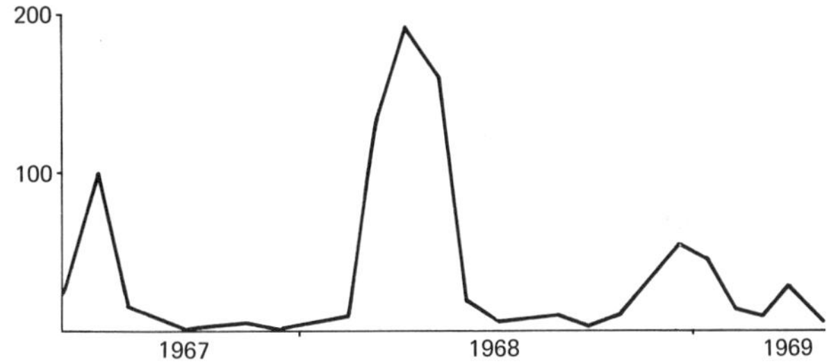

Fig. 58. Numbers of young *Deroceras reticulatum*, weighing less than 25 mg, from soil samples

luscs. This phenomenon has been very little studied in land pulmonates: it is of great importance for predictions on changes in population density when a trend towards overpopulation in an area starts to manifest itself. The enlarged population could collapse of its own accord as a result of this mechanism.

A symbiotic relationship with a **bacterium** has been shown to protect a snail population from the action of molluscicides. Bell et al. (1966) have isolated from *Biomphalaria glabrata* (Say) a bacterium which, on the basis of its morphological, biochemical, physiological and reproductive characteristics has been placed in the genus *Pseudomonas* (fam. Pseudomonadaceae). It possesses the ability to utilize the molluscide Bayluscid as its sole source of nitrogen. It is a short, thick bacillus (1.0–1.5 μm); spore and capsule production have not been ascertained. Opaque colonies appeared in a culture medium, kept at 25 °C, about 24 h after inoculation by smearing with intestinal content, faeces or slime. Maximum growth was observed between 32 °C and 37 °C, while growth stopped at 8 °C and at 42 °C (also Etges et al. 1965). The intestinal tract of the snail serves only as a passage through which the bacterium reaches the water by way of the slime and faeces. The origin of this *Pseudomonas* sp. is still unknown, but its presence in many members of a snail population treated with this molluscicide can prevent the snails from being eradicated.

The limitation of population density by disease, **parasites** and **predators** is dealt with in Chapter 3.6.

In the field of molluscan pathology, **growths** and malignant tumours have been found to occur. Snails thus afflicted have reduced vitality, so that the whole population is weakened if many of its members are affected. Surveys of tumour finds in invertebrates have been published by Scharrer and Szabó-Lochhead (1950) and LOMBARD (1972). Growths are found in annelids, in marine sipunculids and ascidians, in molluscs and arthropods, particularly insects. Abnormal cell growth occurs in these invertebrates as a result of disturbances in the endocrine and nervous systems and also through the agency of parasites. Growths were found in oysters (DINAMANI and WOLF 1973; DIX 1972, 1973; PAULEY 1972), in some slugs such as *Arion rufus* (L.), *Deroceras reticulatum* (Müller), *Limax flavus* L. (FRÖMMING et al. 1961; SZABÓ and SZABÓ 1934) and in the snail *Achatina fulica* Bowdich (MICHELSON 1973). These growths are usually neoplasms which develop in the epithelium, the mesenchyme, the pericardium or the mantle, they may also be genuine tumours which take the form of hump-like swellings on the posterior end of the body. In the lamellibranch *Mercenaria mercenaria* L., growths have been found in the gonad; histopathological examination revealed that 12 of these, taken from 316 female clams, were ovarial, and only 2 found in 233 males, were testicular (BARRY and YEVICH 1972). A granular growth from the vas deferens of *Biomphalaria* sp. was found to be the result of a rupture of the wall of the tube (KAWAZOE et al. 1971).

KRIEG (1969) induced the formation of tumours by treating the prosobranch *Ampullarius australis* (orbigny) with different concentrations (0.5% or 1.0%) and doses (0.02 ml or 0.1 ml) of 20-methylcholanthrene and carcinogenic solutions of croton oil (0.2%). After an average latency of 95 to 150 days, neoplasmic growths developed in 10% of the experimental animals in the region of the operculum. The 26 tumours were classified as 14 papillomes, 8 benign adenopapillomes, 1 malignant adenopapillome, 1 adenome and 2 adenocarcinomes. These epithelial tumours developed not only in response to injection of these substances, but also after transplantation of the epithelial growths into the dorsal side of the foot; altogether 57 epithelial tumours were induced by KRIEG (1972). These tumours correspond in type to those found in higher animals, so that the snail *Ampullarius australis* (orbigny) can be recommended for use in cancer research.

2.2 Damage to Crops and Harvests

In moist, warm summer months, in the autumn, in rainy years and of course in permanently wet regions, terrestrial gastropods cause considerable damage to agriculture and horticulture. The animals can make their appearance in any damp area, but damage can occur also during relatively dry weather, if the field is surrounded by meadows, pastures, fallow land or overgrown slopes. These areas remain moist, afford shelter for slugs and snails and provide, together with the factors already mentioned, the conditions necessary for a massive increase in numbers. From here the snails move onto any neighbouring field at dusk to feed and return at dawn. Slugs, in particular, hide under leaves, flower pots, stones, damp wooden planks, clods of earth, ground-foliage, weeds and grass, with the result that they can be detected only with difficulty. As a rule they appear by day only after a thunderstorm or prolonged rain. In wet weather and favourable temperatures a high population density is soon achieved, with total destruction of the crop. Newly sown seed can, in such circumstances, be completely devoured overnight.

Damage caused by terrestrial gastropods is, however, dependent not only on their activity and population density, but also on their feeding habits; these differ from species to species. Thus *Arion rufus* (L.) shows a preference for Solanaceae and Compositae, *Deroceras reticulatum* (Müller) for Leguminosae, Cruciferae, Compositae, carrot roots, potato leaves and tubers, as well as cereals and seedlings; *Arion hortensis* Fér. for living plants, carrot roots, tubers and bulbs. Most of the damage caused by *Arion rufus* (L.) is done to young plants in the spring, but also to bulbs and tubers. *Arion circumscriptus* Johnston is generally less in evidence (DUTHOIT 1964; MOENS and VAN DEN BRUEL 1960; VAN DEN BRUEL and MOENS 1958 and others).

FRÖMMING, who has been responsible for most of the studies on gastropod nutrition, has clearly demonstrated their significance as agricultural and horticultural pests in his book (FRÖMMING 1962). The quantity of plant material consumed by a snail or slug is very considerable. Thus an adult *Arion rufus* (L.) ate in one day 2 g lettuce, 4 g tomatoes, 4 to 5 g cucumber, or up to 6 g pumpkin, while *Deroceras reticulatum* (Müller) ate the equivalent of a quarter of its own body weight (2 g) (HERMANN 1971). One *Deroceras reticulatum* (Müller) of average size consumed in one night a leaf area of 2 cm^2 (young rape plant) which is equivalent to 60 mg (MALLET 1968). MALLET and BOUGARAN (1970) estimate that at a population density of 200/m^2, the quantity of young plants consumed per day is 60 kg. In 24 h the following quantities of lettuce were consumed by adult animals:

Deroceras reticulatum	(body weight ± 500 mg)	200 mg
Deroceras laeve	(body weight ± 390 mg)	165 mg
Arion hortensis	(body weight ± 170 mg)	95 mg
Milax sowerbyi	(body weight ± 1070 mg)	366 mg

FRÖMMING (1952) calculated that in 24 h *Arion rufus* (L.) ate more than 50 cm^2 of carline thistle leaf, or, expressed as a percentage of its own body weight:

Spinach	4.9 to 5.8%
Sugar beet	4.3 to 9%
Pear	4.3 to 9.8%
Carrot	17.2 to 29%
Tomato	11.2 to 37.2%
Lettuce	10.5 to 48.4%
Cucumber	21.5 to 70%.

Thus astonishing amounts of cucumber, lettuce, tomato and carrot roots are devoured in a 24 h period.

This applies also to *Limax maximus* L. as the following values, expressed as percentage of body weight, indicate:

Pear	6.4 to 7.5%
White radish	4.2 to 15.8%
Radish	13.9 to 17.5%
Red cabbage	3.4 to 22.2%
Tomato	34.2 to 35.6
Cucumber	41.8 to 51.7

In other experiments values for pumpkin, tomato and cucumber of as much as 55%, potatoes 20 to 30%, peach, plum and cherry of 25 to 30% were obtained.

The cellar slug, *Limax flavis* L. consumed 40% of its own weight of potato in a single night; this slug, too, prefers pumpkin and cucumber. An animal weighing 10 g ate the following amounts in a 24 h period:

Pumpkin	up to 4.6 g
Cucumber	up to 4.9 g
Potato	up to 3.4 g
Cauliflower	up to 0.2 g
White cabbage	up to 0.3 g
Apple	up to 0.5 g
Ripe pear	up to 0.6 g
Radish	up to 0.7 g
White radish	up to 0.4 g
Kale	up to 0.1 g

In one night *Helix pomatia* L. ate 50 to 70 cm^2 of *Datura stramonium* leaf and up to 5.2 g of lettuce.

Lowering of the quality and size of the harvest as well as destruction of crops in fields and gardens result not only from the feeding activities of the gastropods, but also from soiling with faeces and slime. Soiled strawberries, lettuce and cabbage, the heads of which can also harbour slugs and snails, are no longer marketable. Indeed, vegetables which have been damaged by gastropod feeding or soiled with their excreta and slime, are excluded from export in the Netherlands (DEN BOER 1969). Given the importance today of frozen vegetables to the consumer, it is essential that neither soiled nor damaged produce, nor a single solitary snail should find its way into the packages (BUNDY 1967).

Rotting agents such as bacteria and fungi usually establish themselves at places where slugs or snails have fed, so that fruit and vegetables in storage suffer further damage. Root vegetables are at risk in wet late summer and autumn weather and also in heavy soil. Fields where green manure has been ploughed into the soil are particularly endangered by fungal disease in the wake of gastropod damage; *Botrytis* in beans and cucumbers is an example of this.

Renewed sowing of the field may be necessary after the total destruction of the seed, as can happen in the autumn to germinating wheat. Rape plants grown as intercrops should be carefully observed in areas where gastropod infestation is likely to occur and, if this occurs, should immediately be ploughed into the soil. Cavities in the soil in which slugs and snails can shelter must be eliminated by intensive cultivation. Considerable losses may be incurred if, after the winter rape has been harvested, a luxuriant intermediate crop develops from fallen seed, is then ploughed in and only a coarse seed bed prepared for the winter grain crop.

Gastropod damage to winter wheat can be considerable. The optimal weather conditions for slugs and snails in Britain in 1966/67 necessitated a second drilling of 8,000 to 12,000 ha (HUNTER 1969). STRICKLAND (1965) recorded a loss of 16,000 ha of wheat. SOUTH (1973) estimated the annual loss to be 10,000 to 17,000 ha which represented 1% to 2% of all the wheat seed planted, while the loss in the case of potatoes was 30,000 to 40,000 t. In attempting to gauge the effect on

Table 9. Average values for seed consumption by four slug species. (After MALLET and BOUGARAN 1970, summarized from several tables)

Plant	Total no. of seeds offered	No. of days	*Arion hortensis*	*Deroceras reticulatum*	*Deroceras laeve*	*Milax sowerbyi*
Maize (*Zea Mays*)	10	7	2	3	2	2
Sunflower (*Helianthus annuus*)	10	7	4	7	7	9
Rape (*Brassica napus oleifera*)	20	7	3	5	12	7
Cabbage (*Brassica oleracea*)	20	7	6	7	8	14
Lucern (*Medicago sativa*)	20	2	10	20	7	20
Carrots (*Daucus carota*)	20	2	20	20	20	12

yield of thinning out of the plants by gastropods, JESSOP and READING (1969) concluded that the more marked the thinning, the greater the yield per plant, since the individual wheat plant then seeds later and more abundantly. But when 75% of the plants have been destroyed, there is no difference between the yields from thinned and non-thinned wheat fields. In this case, therefore, a marked thinning has no effect on the harvest. Indeed, EMPSON (1965) estimated that a loss of 75% of the plants as a result of gastropod damage resulted in a fall in the yield of cereal of 19%. A maize crop can also be totally destroyed by Field Slugs (MALLET 1968).

The experiments of MALLET and BOUGARAN (1970) (Table 9) clearly show the threat represented by slugs to seeds in the soil.

These results show that carrot seeds in the soil are most at risk and under favourable weather conditions 100% of the seeds sown may be eaten within two days. The same applies to lucern seed with *Deroceras reticulatum* (Müller) and *Milax sowerbyi* (Fér.). No experiments have been carried out using seed of cereals.

Even if not completely destroyed, ornamental plants attacked by slugs and snails show inhibited growth and poor flowering. The warm, damp conditions of the seedbed afford an optimal environment for slugs. Here most of the damage is caused by young animals, which may be overlooked because of their size.

In fruit farming, losses of up to 50% of the cherry harvest were recorded (NEUMANN 1954, 1955). The damaged fruit, while still on the tree, is open to attack by the fungus *Monilia* which causes rotting. Apples, pear, plums and peaches are particularly liable to disfigurement by snails. Both snails and slugs cause considerable economic losses through the destruction of windfalls, which are today not without value, being used for the manufacture of jams and fruit juices. Blackcurrents, too, are rendered unfit for marketing. Although the quantity consumed by *Monacha umbrosa* C. Pfeiffer is negligible, the fruit is rendered worthless by the subsequent contamination with fungi. The pattern of damage to the leaves resembles the holes typically left by insects, while young snails cause „Fensterfrass". Numerous species of snails are recognized pests on fruit farms, including *Cepaea nemoralis* (L.), *Cepaea hortensis* (Müller), *Candidula caperata* (Montagu), *Helix aspersa* Müller and *Euomphalia strigella* (Drap.); they represent a serious problem for fruit farmers and jam manufacturers (STRINGER 1969). STRINGER and MORGAN (1969) counted 12,000 animals per hectare. During picking they fall with the berries into the containers and later separation is difficult, particularly when the berries are kept in cold storage in a jam factory. The density of a snail population on black current bushes depends on the weediness of the ground and the cleanliness of the plantations.

2.2.1 Gastropod Damage in the Federal Republic of Germany

The main gastropod pests in the Federal Republic of Germany are undoubtedly the Field Slugs, *Deroceras reticulatum* (Müller) and *Deroceras agreste* (L.), followed, but not very closely, by members of the Limacidae. The family Arionidae is more localized. The response of these gastropods to different plants has been thoroughly investigated by FRÖMMING (1962). They eat leaves and flowers, the roots and tubers of nearly all vegetables, ornamental plants and animal fodder, aromatic and medicinal herbs, as well as fruits in field, garden and greenhouse.

2.2.1.1 Agriculture

There are two periods of maximal damage by Field Slugs. These coincide with the periods of maximal hatching of young slugs and consequent population increase and with the rainy months when the activity of the slugs is most pronounced (Fig. 59). GODAN (1973) recorded the reports of the Plant Protection Office of the German Federal Republic to the Biologische Bundesanstalt für Land und Forstwirtschaft (Federal Biological Research Centre for Agriculture and Forestry) in Berlin-Dahlem, over a ten year period (1962–1971). Most of the damage reported occurred from April to May and from September to November. Wet weather was a prerequisite for heavy damage to the seedlings of vegetables, tobacco, lettuce and strawberries by Field Slugs in the months April to August. Heavy, coarse and loamy soils enhanced gastropod infestation of winter vegetables and rape during the autumn (September to November), of winter wheat in February and of summer vegetables in May to June. Only in gardens and greenhouses, kept permanently moist through frequent watering and where optimal temperatures pertained, were population density, activity and pest damage constant over the whole cultivation period.

Figure 60 shows the relative extent to which different plants under cultivation were affected in 1,388 reports submitted. It can be seen that cabbage and winter cereals were most liable to attack, while wheat, rye and barley together constituted only 24.6% of the reports. Not only were the largest number of reports (40%) concerned with cabbage, but the economic damage was greater than for any other

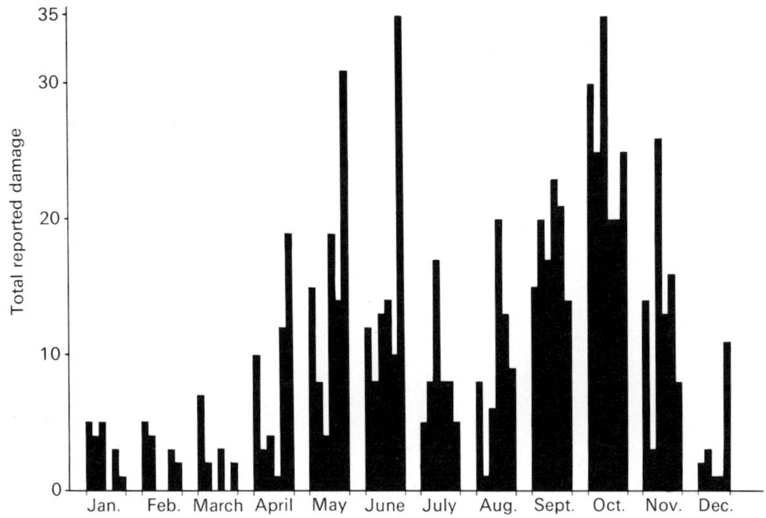

Fig. 59. Degree of damage during different months of the year according to reports from six plant protection offices

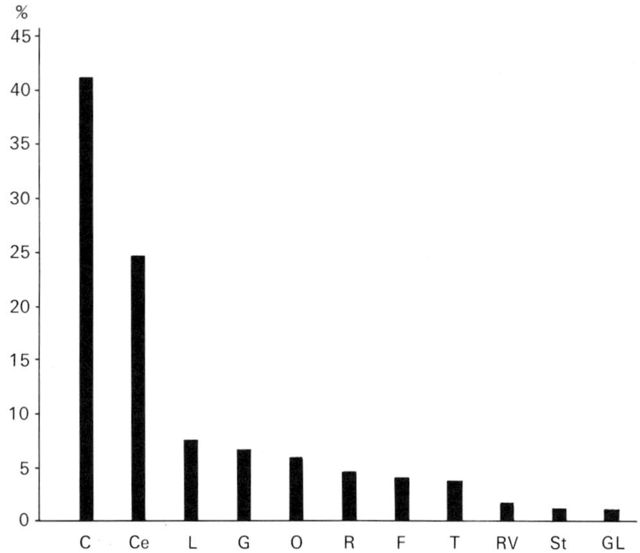

Fig. 60. Relative damage to particular crops, expressed as percentages of total reported damage, based on records over 10 years of German Plant Protection Offices. *C* cabbage; *Ce* cereals; *L* lettuce; *G* green vegetables; *O* ornamental plants and flowers; *R* rape; *F* forage plants; *T* tobacco; *RV* root vegetables; *St* strawberries; *GL* grassland

crop, since cabbage fields covered the largest area. Although reports on other agricultural plants were, at a total of 40%, fewer in number than those for cereals and cabbage, the extent of damage in each case tended to be more complete (Fig. 61). The scale of damage was often greater than in the case of cabbage or only a little less, as in the case of lettuce, green vegetables, fodder crops, ornamental

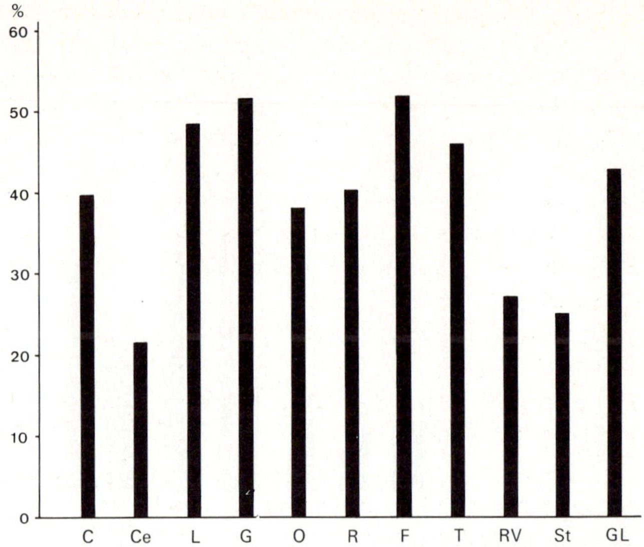

Fig. 61. Frequency of reports of heavy to very heavy damage sustained by particular crop plants expressed as a percentage of all reports of damage to each crop. *C* cabbage; *Ce* cereals; *L* lettuce; *G* green vegetables; *O* ornamental plants and flowers; *R* rape; *F* forage plants; *T* tobacco; *RV* root vegetables; *St* strawberries; *GL* grassland

plants, flowers, rape, sugar beet, tobacco and grazing land. The number of heavy to very heavy infestations expressed as a percentage of the total number of reports for each particular crop, was as follows: clover and lucern 47.8%; maize 75%; beans and peas 41.6%; carrots 66.5%; spinach 60%; cucumbers 57.2%; radish 83.3%; endives 91,7% and flowers (open field, greenhouse) 27.6%. Of the different types of brassica, brussel sprouts were the worst affected with 75%, followed by cauliflower with 50%, and cabbage (Savoy cabbage, white cabbage) with only 38.8% of the reports.

In addition pumpkin, tomatoes, spinach, cabbage seedlings, endives, late vegetables, interseed cereals as intercrops for green fodder or green manuring as well as ornamental lawns can be affected. Grasses such as *Cynosurus cristatus*, *Festuca pratensis* (except for *F. ovina capillata*) and *Phleum pratense* appear to be eaten only with reluctance by slugs and snails (Frömming and Plate 1954).

The reports of the official Plant Protection Office also provide information about factors which promote or impede the activity and population growth of terrestrial gastropods. The amount of damage they cause is enhanced by the following factors:

1. damp weather or continuous rain;
2. heavy, coarse or loamy soils;
3. proximity of the field to meadows, pastures, fallow land and slopes;
4. slow growth of the plant under cultivation, for example, beans, cucumbers, young tobacco plants, as a result of cool weather during germination and early stages;
5. late autumn sowing of seed (winter cereals in November).

Loss of young vegetable, tobacco, lettuce and strawberry plants was recorded from April to August. A second peak of damage, this time to winter cereals, occurred in September to November. With optimal soil conditions, heavy damage was inflicted on winter vegetables and winter rape from September to November, on winter wheat in February, and on summer vegetables in May to June. Fields adjacent to meadows and fallow land were seriously affected from April to June in the case of lettuce, cabbage and green vegetables as well as tobacco plants and to winter cereals in September to November.

Factors which inhibit activity and rapid population growth and thereby reduce damage are as follows:

1. dry weather, mainly in the months of June and July and again in September to October;
2. sudden falls in temperature during the summer;
3. cold snaps or light frost in early autumn, September to October.

The snails *Vertigo antivertigo* (Drap.), *Cochlicopa lubrica* (Müller), *Succinea putris* (L.) live in meadows or the banks of rivers, streams and lakes, or in the water itself. From these habitats they move onto neighbouring fields of wheat, clover and rye. The Amber Snail, *Succinea putris* (L.), causes damage to young plants of spring wheat on marshy fields adjacent to overgrown ditches, leaving these at night to invade the fields and returning at dawn.

Gastropods cause heavy damage, especially to seeds and seedlings of cereals and seed oil plants, until the first frosts occur; but even winter vegetables may be damaged so that reploughing and resowing are often necessary. Cereals such as wheat, barley and rye are particularly endangered (BECKER 1951 and others) if planted after a leguminous crop or if underseed crops are present, as these afford the snails protection and shade in dry, warm periods and provide the right nutritional conditions for a rapid growth in numbers. In general, uniform infestation with gastropods is found only at the edge of the fields, because the animals penetrate them from the unploughed perimeters. Within the field itself, however, so-called "snail plates" are found (LUTHER 1955); these are areas of particularly coarse soil where the gastropods can find shelter during the day. Tillage after drilling the seed favours rapid germination so that the damage to the seeds is less significant, as LUTHER (1955) observed in the case of winter wheat. Fields in marshland are particularly prone to attack (BREDERLOW 1975).

Field slugs consume wheat in winter even at 0.8 °C under a snow cover and destroy first the embryo and then the endosperm of the seed (DUTHOIT 1961 and others).

Experiments by DUTHOIT (1964) on the feeding preference of slugs given a choice of wheat, barley and oats showed that oats are hardly touched. *Deroceras reticulatum* (Müller) showed a preference for wheat seed feeding on wheat 85%, on barley 56% and on oats 4%; *Arion hortensis* Fér. consumed almost the same amount of wheat and barley seed (26% and 28% respectively) but again only 2% of the oats; *Milax budapestensis* (Hazay) preferred barley (barley 26%, wheat 13% and oats 2%). These experimental results are confirmed by practical experience. In 86 reports by the Plant Protection Office on slug damage to cereals over a period of five years (1948 to 1952 inclusive), oats are only mentioned on four occasions and their infestation described as "infrequent" or "average", whereas damage to wheat, barley and rye was "heavy" or "very heavy".

In experiments in which the gastropods were offered a choice of wheat seeds and wheat seedlings, the following results were obtained:

Species	Wheat seed	Wheat seedling
Deroceras reticulatum	81%	69%
Arion fasciatus	35%	28%
Arion hortensis	42%	20%
Arion ater	89%	84%
Milax budapestensis	29%	8%

The seed of cereals is preferred by all slug pests, whereas damage to seedlings is caused only by *D. reticulatum* (Müller) and *Arion ater* (L.).

In the case of **green fodder** and **silage maize,** the gastropods eat the lower leaves first and then the upper; only the light mesophyll tissue between the leaf veins is left. STAUBER (1954) observed maize fields in some of which a third of the crop and others in which a third to a half had been destroyed.

Field slugs infest **potatoes** very heavily (RAINER 1962; STEPHENSON 1965; WARLEY 1970); not only the young tubers but also the leaves and stems of the plants. Again it is the lower leaves which are affected more than the upper. Other potato pests include the *Arion* spp., *Limax maximus* L. and *Milax* spp. On neutral or lightly alkaline soils the damage they cause is particularly heavy (CARRICK 1942). Some varieties of potato may be resistant to Field Slugs. According to THOMAS (1947), the English "Arran Banner" was preferred, while "Arran Peak" was disregarded; between these two "Majestic" occupied a middle position in attractiveness. The variety "Maris Piper" was attacked more frequently than "Majestic" (HUNTER et al. 1968). According to GOULD (1965), "Ulster" was most heavily infested, while "Pentland Falcon" and "Pentland Dell" were less vulnerable than "King Edward". Few investigations have been made on German potato varieties, although RAINER (1962) mentions "Isola" as being especially prone to gastropod infestation. The pattern of damage by Field Slugs resembles that of the Colorado potato beetle, and during the rainy months of June, July and August, the foliage of potato plants can be eaten away completely. With the onset of high summer and consequent dryness, the gastropods retreat into the soil and destroy the young tubers, two or three animals sometimes boring into a single tuber. The most seriously affected potato fields are those bordering on meadows.

Young plants of **winter rape** suffer damage from mid-September. Amongst the oil plants, linseed is effectively avoided, mustard hardly eaten at all, but rape and rape seed are preferred, while poppy plants occupy an intermediate position. This was established in experiments on slugs and snails by FRÖMMING and PLATE (1955) using rape and linseed plants 10–12 cm high and poppy plants 4–10 cm high. Except in the case of winter rape and rape seed, younger plants were seldom or never affected. Rape can, however, loose its appeal for slugs and snails. Some Plant Protection Offices reported that, despite heavy gastropod infestation in October, the damage to winter rape was tolerable because of its advanced stage of development. Its large leaves form ideal ground cover from winter to spring enabling the gastropods to survive low temperatures with a consequent rise in population density. Crops following rape can thus be heavily infested; crops of winter barley and winter rye are particularly at risk. *Arion* spp. and *Milax gagates* (Drap.) are also encountered in plantations of rape and rape seed. The possibility of infestation of rape plants growing as an intercrop should be carefully watched in areas where gastropod infestation is likely to occur and in the event should immediately be ploughed into the soil. Holes in the soil in which slugs and snails can shelter must be eliminated by intensive tillage. Considerable losses may be incurred if, after the winter rape has been harvested, a luxuriant intercrop develops from fallen rape seeds, is then ploughed in and only a coarse seed bed is prepared for the winter crop.

Fodder crops (lucern, clover, lupine, sanfoin, seradella) are severly damaged by Field Slugs and, in southern Germany, also by *Helicella itala* (L.) and *Helicella obvia* (Hartmann) (Frömming 1957, 1958). In these cultivations, as also in thickly growing stubble clover and stubble seed mixture, the molluscs spend the day on the leaves nearest to the ground and on the lower part of the stalk (Anon. 1953). *Helicella itala* (L.), *Candidula unifasciata* (Poiret) and *Cochlicella acuta* (Müller) are transported with sainfoin seeds from France to Germany (Schmidt 1955). In the case of clover and turnips, seed losses of 50% often occur.

In April to May, Field Slugs can cause such damage to the newly sown shoots of **sugar beet** that a second sowing is necessary.

Field vegetables, such as the different varieties of cabbage, including Savoy cabbage, white cabbage, are attacked in wet wheather, usually in the second half of November and are poor in quality as a result of the tunnels and holes which the gastropods bore into the cabbage heads.

Young **tobacco** plants in the hotbed may be completely defoliated by Field Slugs if growth is slow and conditions very moist. *Helix pomatia* L. has also been found in tobacco plantations (Frömming 1956).

Snails seldom occur in the agricultural lands of Central Germany, because they are easily crushed during ploughing and preparation of the soil. *Helicella* spp., which can cause heavy damage to the autumn crop of maize in dry areas, have been kept in check by such soil cultivation (Empson 1965).

On the basis of written information, publications, personal observations and oral testimony, the following list of gastropod pests and the plants they attack has been prepared for quick reference (Table 10).

2.2.1.2 Horticulture

Gardens. Frequent watering in gardens and greenhouses keeps vegetables and ornamental plants constantly moist and enables slugs to multiply massively with a resultant increase in the amount of damage they cause. In the spring young animals, which are easily overlooked, do most of the damage. The warm, moist conditions in hotbeds represent an optimal environment with the result that slugs occur in considerable numbers in cultivations of ornamental plants. Heavy infestations of the snails *Helicigona arbustorum* (L.) and *Zonitoides arboreus* (Say) can cause extensive damage to young plants; *Z. arboreus* (Say) attacks the roots in particular. In horticultural undertakings the leaves and flowers of ornamental plants

Table 10. Gastropod pests in agriculture and on grassland

Feeding damage to	Slugs	Snails
Cauliflower (leaves, rosettes)		*Helix aperta* Born (south Germany) *Trichia hispida* (L.)
Beans (leaves, pods)	*Deroceras reticulatum* (Müller)	
Peas (leaves, pods)	*Deroceras reticulatum* (Müller)	
Sainfoin (seed production)		*Candidula unifasciata* (Poiret) *Cochlicella acuta* (Müller)
Fodder plants (sainfoin, lucern, clover) (leaves, young plants)	*Deroceras reticulatum* (Müller)	*Cepaea hortensis* (Müller) *Cepaea nemoralis* (L.) *Cernuella virgata* (Da Costa) *Helicella itala* (L.) *Helicella obvia* (Hartmann)
Carrots (young roots)	*Arion hortensis* Férussac *Deroceras reticulatum* (Müller) *Milax budapestensis* (Hazay) *Milax gagates* (Draparnaud)	
Potatoes (leaves, stalks, young tubers)	*Arion circumscriptus* Johnston *Arion rufus* (L.) *Arion subfuscus* (Draparnaud) *Deroceras reticulatum* (Müller) *Limax maximus* L. *Milax budapestensis* (Hazay) *Milax gagates* (Draparnaud)	
Brassica (kohlrabi, Savoy cabbage, white cabbage) (leaves, head, seedlings)	*Deroceras reticulatum* Müller)	
Lettuce (field grown) (leaves, head)	*Deroceras laeve* (Müller) *Deroceras reticulatum* (Müller) *Milax budapestensis* (Hazay)	*Trichia hispida* (L.)
Maize (leaves, young plants)	*Deroceras reticulatum* (Müller)	*Helicella* spp.
Poppy (leaves, young plants)	*Arion circumscriptus* Johnston *Arion hortensis* Férussac *Arion rufus* (L.) *Arion subfuscus* (Draparnaud) *Deroceras reticulatum* (Müller) *Milax budapestensis* (Hazay)	
Rape (winter, summer) (leaves, young plants)	*Arion circumscriptus* Johnston *Arion hortensis* Férussac *Arion rufus* (L.) *Arion subfuscus* (Draparnaud) *Deroceras reticulatum* (Müller) *Milax budapestensis* (Hazay)	
Rye (especially in limy soil)		*Cepaea hortensis* (Müller)
Brussel sprouts (leaves, sprouts)	*Deroceras reticulatum* (Müller)	
Beets (leaves, young plants, young tuber)	*Deroceras reticulatum* (Müller)	*Cepaea nemoralis* (L.)

Table 10. (continued)

Feeding damage to	Slugs	Snails
Germinating seeds and young plants		*Trichia hispida* (L.)
Spring grain (seeds, seedlings, young plants) (south Germany)	*Arion hortensis* Férussac	*Candidula caperata* (Montagu)
	Deroceras reticulatum (Müller)	*Cepaea nemoralis* (L.)
		Cochlicopa lubrica (Müller)
		Helicella itala (L.)
		Succinea putris (L.)
		Vertigo antivertigo (Draparnaud)
Asparagus (leaves)		*Cecilioides acicula*(Müller)
Spinach (young plants)	*Deroceras reticulatum* (Müller)	
Turnip (leaves, young tuber)	*Deroceras reticulatum* (Müller)	
Tobacco (young plants)	*Arion rufus* (Linnaeus)	*Helix pomatia* L.
	Deroceras reticulatum (Müller)	
Winter wheat, barley, rye (seeds, seedlings, young plants)	*Arion hortensis* Férussac	
	Arion rufus (L.)	
	Deroceras reticulatum (Müller)	
	Milax budapestensis (Hazay)	
Grassland (meadows, pastures)	*Arion fasciatus* (Nilsson)	*Helix* spp.
	Arion hortensis Férussac	*Vallonia costata* (Müller)
	Arion intermedius Normand	*Vallonia pulchella* (Müller)
	Deroceras reticulatum (Müller)	

such as orchids, summer asters, dahlias, primroses, zinnias, gladioli, chrysanthemums and others, ornamental ferns under glass, flowering bulbs and tubers are eaten by gastropods. In the case of asters, the delicate tissue of the petiole just below the flower head is rasped away, in cultivations of *Anthurium* the delicate tips of leaf and flower buds are eaten (PAPE 1964). At the end of May the lettuce in hotbeds is damaged. *Deroceras caruanae* (Pollonera), *Arion subfuscus* (Drap.) and *Arion hortensis* Fér. cause damage to greenhouse cucumbers.

MÜHLE (1941) found *Vallonia pulchella* (Müller) in addition to *Deroceras reticulatum* (Müller) in **grass seed production.** The damage generally occurs after germination, when the young grass plants are either eaten down to soil level or, if the infestation takes place at a later stage, the leaves are scraped away in places or have holes in them or the leaf tips are eaten; occasionally only the leaf veins remain.

In private **small gardens** there are considerable vegetable losses in September. Snails may damage the pods of kidney beans and runner beans or peas. Besides *Deroceras reticulatum* (Müller) and *Deroceras laeve* (Müller), *Limax maximus* L. and the smaller species, *Lehmannia marginata* (Müller) and recently also *Lehmannia valentiana* (Fér.) occur. The Great Grey Slug, *Limax maximus* L., feeds mainly on roots, tubers, carrots, potatoes, on fruit such as peaches, plums, apples and cherries (FRÖMMING 1952), but also on the green parts of the plant. The subterranian Keeled Slugs attack mainly potatoes and carrots (THOMAS 1944 and others); *Milax budapestensis* (Hazay) was first identified in the Federal Republic of Germany in 1949 (FRÖMMING 1955) and *Milax gagates* (Drap.) in 1962 (KUHNA and

Schnell 1963); both species originated in the Mediterranean region. More recently, *Milax ruticus* (Millet) has appeared. *M. budapestensis* (Hazay) causes heavy damage to lily cultivations by eating the bulbs of *Lilium davidii* and its hybrids, so that only the outer parts remain. In *Lilium regale* and its hybrids, the thick, juicy roots are often eaten through at their base or the bulb is hollowed out. On the other hand, *Lilium henryi* and *Lilium tigrinum* remained largely free of attack. Older bulbs, with dried outer leaves, are as a rule avoided. Favourite bulbs are those of *Gagea arvensis*, *G. lutea*, *G. pratensis*, *Scilla bifolia*, *S. sibirica*, *Muscari racemosum*, *Ornithogalum nutans*. The rhizome of *Anthericum liliago* and the root tubers of the orchids *Orchis latifolia*, *O. maculata*, *O. morio* and *O. palustris* are also attacked by gastropods. The green parts above the ground are not readily eaten.

The slug *Boettgerilla pallens* Simroth (= *B. vermiformis* Wiktor) first appeared in Germany in 1962 and is found in gardens and greenhouses. Together with *Deroceras reticulatum* (Müller), *Deroceras laeve* (Müller) and *Arion hortensis* Fér. this worm-slug is a pest to chrysanthemums grown in the open and of cyclamen grown in greenhouses (Plate 1965; Schmid 1962 and others). It was found in the parks of Berlin by Daxl (1967) and by 1977 had become relatively common.

Settlements bordering on forests are usually plagued by *Arion rufus* (L.), *Arion hortensis* Fér. and *Arion subfuscus* (Drap.). *Arion lusitanicus* Mabille, which resembles *A. rufus* (L.) in appearance, is found in the German Federal Republic in gardens, meadows and fields and was probably introduced through lettuce and vegetable imports (Schmid 1971). There is evidence of the occurrence of this species in the upper and lower reaches of the Rhine, in parts of Württemberg and in other areas of western Europe, including Switzerland (van Regteren-Altena 1972).

In gardens and parks, snails do considerable damage to the foliage and flowers of ornamental plants and also to fruit trees and shrubs. Thus *Helicigona arbustorum* (L.) which lives on or in the soil can be a pest of vegetables, sugar beet, tubers, fruit and ornamental plants; this snail remains active well into November, even entering greenhouses (Frömming 1958). The Banded Snails (*Cepaea* spp.) and in southern Germany the Roman Snail *Helix pomatia* L., the Brown Garden Snail *Helix aspersa* (Müller), and the Singing Snail *Helix aperta* (Born.) are pests. The latter two species are Mediterranien in origin and as a rule can only survive at warm temperatures. They frequently succeed, however, in building up large and therefore destructive, albeit localized, populations in Lorraine, the upper Rhine plateau, around Lake Constance, in the Neckar Valley, sometimes in the Lower Rhine region and, more recently, in central Germany.

Through the import of fruit, vegetables, lettuce and ornamental plants from the Mediterranean region, snails are constantly being reintroduced. *Helix aspersa* (Müller) has, however, established itself in some parts of southern Germany, even though in general the populations of this snail are destroyed in harsh winters. Furthermore, *Vallonia costata* (Müller) and *Cochlicopa lubrica* (Müller) can do damage in gardens.

Frömming (1949, 1950) listed the following **aromatic** and **medicinal herbacious plants** as food plants of *Helix pomatia* L. and *Arion rufus* (L.) which consume in 24 h over 100 cm^2 surface area of leaf of: *Allium schoenoprasum*, *Cannabis sativa*, *Crepis biennis*, *Galinsoga parviflora*, *Lactuca muralis*, *Taraxacum officinale;* from 50 cm^2 to 100 cm^2 of: *Cichorium intybus*, *Lamium maculatum*, *Levisticum offi-*

cinale, Nicotianum tabacum; less than 50 cm^2 of: *Chenopodium album, Eupatorium cannabinum, Humulus lupulus, Lamium album, Melandrium album, Petasites officinalis, Plantago major, Plantago media, Rumex acetosa, Tussilago farfara* and *Petasites* sp.

Greenhouses. Both terrestrial, freshwater and amphibious gastropods occur as pests in greenhouses. Apart from direct effect of feeding, they cause indirect damage as well, since partially eaten leaves wilt, with the result that ornamental plants are blemished and may in time die.

The following **terrestrial gastropods** are found in greenhouses: the slugs *Deroceras reticulatum* (Müller), *Deroceras laeve* (Müller), *Lehmannia marginata* (Müller), *Lehmannia valentiana* (Férussac), *Limax maximus* L., *Limax flavus* L., *Arion hortensis* Fér., *Arion rufus* (L.) and the snails *Discus rotundatus* (Müller), *Opeas pumilum* (Pfeiffer), *Oxychilius alliarus* (Miller), *Oxychilus drapanaudi* (Beck), *Zonitoides arboreus* (Say) and *Lamellaxis gracilis* (Hutton) (PLATE and FRÖMMING 1953 and others), also *Carychium minimum* (Müller), *Cochlicopa lubrica* (Müller), *Vertigo antivertigo* (Drap.), *Vallonia costata* (Müller), *Vallonia pulchella* (Müller), *Vallonia excentrica* Sterki and *Helicodiscus parallelus* (Say). *Discus rotundatus* (Müller) is said to be harmless, yet 10% of its food comprises higher plants (JAECKEL and PLATE 1967). *Oxychilus draparnaudi* (Beck) is both carnivorous and phytophagous. *Opeas mauritianum* (Pfeiffer) which usually inhabits heated greenhouses was first found in 1950 (SCHMIDT 1959). Clausiliidae and Helicidae are only occasionally carried into greenhouses, but cannot survive for long.

The following **freshwater snails** occur in greenhouses: *Physa acuta* Drap., *Physa heterostropha* (Say) and *Galba cubensis* (Pfeiffer), which lives not only in water but also on a moist substratum and in flower pots in nurseries; *Pseudosuccinea columella* Say from North America and Cuba; this is common in hothouses.

Of the **ornamental plants,** *Coleus* spp. are attacked by all the gastropod species found in greenhouses; the flowers of *Calla aethiopica* are destroyed, but the leaves are disregarded; in the case of the African violet, *Saintpaulia confusa*, the yellowed leaves are affected. Lilies, narcissi and irises in greenhouses may be damaged by *Limax maximus* L., *Limax flavus* L., *Deroceras reticulatum* (Müller), *Deroceras laeve* (Müller), *Arion hortensis* Fér., *Arion intermedius* Normand and *Discus rotundatus* (Müller) (PLATE and FRÖMMING 1951). The flowers of *Narcissus poeticus* and *Narcissus luteus* are readily eaten, but the leaves and bulbs are little affected. Experimental observations have shown that gladioli are seldom attacked. *Muscari botryoides, Scilla campanulata* and *Galanthus nivalis* are either accepted with reluctance or ignored completely.

With the exception of those plants already mentioned as well as *Tulpia, Hyazinthus, Crocus, Iris* and *Freesia*, healthy and undamaged **bulbs** are not eaten and remain, like the young plants, free from damage. Only when they have become moist as a result of the onset of germination do they become attractive to gastropods. Thus it is important to ensure that bulbs which are to be stored are healthy, dry and undamaged and are kept under dry, clean and airy conditions. Not all varieties of chrysanthemums are equally attractive to gastropods; *Deroceras reticulatum* (Müller) alone is not selective in its feeding on them (PLATE and FRÖMMING 1954). *Arion rufus* (L.) shows a liking for „Deutsche Kaiserin", "Linda Lou", "Long Island Beauty", "Mefo" and "Yellow-Isola"; it refuses or feeds less readily on "Ar-

Table 11. Gastropod pests in garden and greenhouse (vegetables, medicinal herbs)

Feeding damage to	*Garden*	
	Slugs	Snails
Beans (kidney, runner) (leaves, pods)	*Arion* spp. *Deroceras laeve* (Müller) *Deroceras reticulatum* (Müller) *Lehmannia marginata* (Müller) *Lehmannia valentiana* (Férussac)	*Bradybaena fruticum* (Müller) *Cepaea hortensis* (Müller) *Cepaea nemoralis* (L.) *Eobania vermiculata* Müller *Helix aspersa* (Müller) (south Germany) *Oxychilus cellarius* (Müller) *Oxychilus draparnaudi* (Beck)
Cauliflower (leaves, rosettes)	*Arion* spp. *Deroceras reticulatum* (Müller)	*Discus rotundatus* (Müller)
Chicory (leaves)	*Milax* spp.	*Cepaea hortensis* (Müller) *Cepaea nemoralis* (L.) *Discus rotundatus* (Müller) *Helicigona arbustorum* (L.)
Endives (leaves)	*Arion* spp. *Deroceras reticulatum* (Müller) *Milax* spp.	*Cepaea nemoralis* (L.) *Cepaea hortensis* (Müller)
Peas (leaves, pods)	*Arion* spp. *Deroceras* spp. *Lehmannia marginata* (Müller) *Lehmannia valentiana* Férussac	Same snail species as for beans
Vegetables (edge of woods, gardens)	*Arion rufus* (L.) *Arion hortensis* (Férussac) *Arion subfuscus* (Draparnaud)	
Kale (in germination)		*Helicigona arbustorum* (L.)
Cucumber, pumpkin (fruit, leaves)	*Deroceras reticulatum* (Müller) *Limax maximus* L. *Milax* spp.	*Cepaea hortensis* (Müller) *Cepaea nemoralis* (L.) *Eobania vermiculata* (Müller) *Helicigona arbustorum* (L.) *Oxychilus cellarius* (Müller) *Oxychilus draparnaudi* (Beck)
Medicinal and aromatic herbs	*Arion rufus* (Linnaeus) *Deroceras reticulatum* (Müller)	*Helix* spp. *Oxychilus alliarius* (Miller)
Potatoes (leaves, stalks, young tubers)	*Arion hortensis* Férussac *Deroceras reticulatum* (Müller) *Milax gagates* (Draparnaud)	*Helicigona arbustorum* (L.)
Brassica (cabbage, white cabbage, kohlrabi, leaves, heads)	*Arion hortensis* Férussac *Deroceras reticulatum* (Müller) *Limax maximus* L.	*Helicigona arbustorum* (L.)
Lettuce (older plants)	*Deroceras laeve* (Müller) *Deroceras reticulatum* (Müller) *Lehmannia valentiana* (Férussac)	
Carrots (leaves, young tuber)	*Arion* spp. *Deroceras reticulatum* (Müller) *Limax maximus* L. *Milax* spp.	*Helicigona arbustorum* (L.)
Celery (leaves, tuber)	*Arion* spp. *Deroceras reticulatum* (Müller) *Milax* spp.	*Bradybaena fruticum* (Müller) *Helicigona arbustorum* (L.)

Table 11. (continued)

Feeding damage to	Garden	
	Slugs	Snails
Late vegetables (in hotbed)	*Arion* spp. *Deroceras reticulatum* (Müller) *Milax* spp.	
Spinach	*Milax* spp.	*Cepaea* spp. *Helicigona arbustorum* (L.)
Tomatoes	*Limax flavus* L. (occasionally) *Deroceras reticulatum* (Müller) *Limax maximus* L.	
Feeding damage to	***Greenhouse***	
	Slugs	**Snails**
Cucumber, pumpkin	*Arion* spp. *Deroceras reticulatum* (Müller) *Limax flavus* L. *Limax maximus* L.	
Young plants in nursery beds and pots	*Deroceras reticulatum* (Müller)	*Zonitoides arboreus* (Say)
Lettuce	*Deroceras laeve* (Müller) *Deroceras reticulatum* (Müller)	*Vallonia pulchella* (Müller)
Plants (general) in greenhouses, seed growing houses and pots.	*Arion* spp. *Deroceras reticulatum* (Müller) *Deroceras sandwichiense* Eydoux & Souleyet (Botanical gardens) *Lehmannia marginata* (Müller) *Lehmannia valentiana* (Férussac) *Milax budapestensis* (Hazay)	*Galba cubensis* (Pfeiffer) *Helicodiscus parallelus* (Say) *Lymnaea columella* Say *Opeas mauritianum* (Pfeiffer) *Vallonia pulchella* (Müller) *Zonitoides arboreus* (Say) *Zonitoides nitidus* (Müller)

cadia", "Blanche Poitvine", „Irene Hofmann" and "Full White". Heavy damage has also been reported to *Salvia* seedlings, *Primula obconica*, pansies, roses and orchids (GODAN 1973). Only light infestation occurred in the case of *Gerbera*, *Amaryllis* and *Petunia*.

Datura spp. are without doubt a favourite food plant of *Deroceras reticulatum* (Müller), *Deroceras laeve* (Müller), *Lehmannia marginata* (Müller), *Limax maximus* L. and *Arion intermedius* Normand (FRÖMMING and PLATE 1951).

PAG (1957) found Field Slugs, together with *Lehmannia marginata* (Müller), *Discus rotundatus* (Müller) and *Oxychilus draparnaudi* (Beck) in **orchid** hot houses. New twigs, flowers and flower stalks are most in danger of attack. MOENS and FRASELLE (1980) have identified *Lehmannia valentiana* (Fér.) and *Zonitoides arboreus* (Say) as the major pests of cultivated *Cymbidium* and *Cattleya*, while *Deroceras laeve* (Müller), *Arion hortensis* Fér., *Discus rotundatus* (Müller) and *Oxychilus draparnaudi* (Beck) are more localized in their effect, feeding on the parts of the orchid which are in the soil. *O. draparnaudi* (Beck) is to some extent carnivorous, feeding on the other gastropods. *Opeas pumilum* (Pfeiffer) which has been transported

Table 12. Gastropod pests to ornamental plants and lawns

Feeding damage to	Slugs	Snails
Aconitum	*Deroceras reticulatum* (Müller)	*Cepaea hortensis* (Müller) *Cepaea nemoralis* (L.) *Helicigona arbustorum* (L.)
Adiantum	*Arion hortensis* Férussac *Milax budapestensis* (Hazay) *Milax gagates* (Draparnaud)	*Oxychilus cellarius* (Müller)
Aesculus	*Deroceras reticulatum* (Müller)	*Helicigona arbustorum* (L.)
Amaryllidaceae (leaves, bulbs)	*Arion hortensis* Férussac *Arion intermedius* Normand *Deroceras laeve* (Müller) *Deroceras reticulatum* (Müller) *Lehmannia marginata* (Müller) *Limax maximus* L.	*Helicina subfusca* Menke
Anthurium	*Arion* spp. *Deroceras* spp. *Limax maximus* L.	
Aster (summer and winter plants) (seedlings and young plants)	*Arion hortensis* Férussac *Deroceras reticulatum* (Müller) *Lehmannia marginata* (Müller)	
Calla aethiopica (flowers)		*Vallonia pulchella* (Müller)
Chrysanthemum (field grown)	*Boettgerilla pallens* Simroth *Deroceras reticulatum* (Müller)	
Chrysanthemum indicum (leaves) (in greenhouse)	*Arion hortensis* Férussac *Deroceras laeve* (Müller) *Deroceras reticulatum* (Müller) *Lehmannia marginata* (Müller) *Lehmannia valentiana* (Férussac)	
Coleus (flowers, leaves, germinating bulbs)	*Arion hortensis* Férussac *Arion intermedius* Normand *Deroceras laeve* (Müller) *Deroceras reticulatum* (Müller)	*Opeas mauritianum* (Pfeiffer) *Vallonia pulchella* (Müller)
Cornus	*Arion ater* (L.)	
Cyclamen (young plants) (in greenhouse)	*Boetgerilla pallens* Simroth *Limax maximus* (L.)	
Datura stramonium (leaves, flowers) (in fields)	*Arion intermedius* Normand *Deroceras laeve* (Müller) *Deroceras reticulatum* (Müller) *Lehmannia marginata* (Müller) *Limax maximus* L. *Limax tenellus* L.	*Cepaea nemoralis* (L.) *Helix pomatia* L.
Fittonia verschaffeltii	*Deroceras laeve* (Müller) *Deroceras reticulatum* (Müller)	
Gerbera (leaves, flowers)		*Oxychilus draparnaudi* (Beck)
Gladiolus (flowers)	*Arion* spp. *Deroceras* spp.	
Hedera		*Helix aspersa* Müller
Hyazinthus (bulbs)	*Limax flavus* L. *Limax maximus* L. *Milax* spp.	
Iris, lilies (flowers, leaves, brood bulbs	*Arion hortensis* Férussac *Arion intermedius* Normand *Deroceras reticulatum* (Müller)	*Cepaea hortensis* (Müller) *Helicigona arbustorum* (L.) *Helicina subfusca* Menke

Table 12. (continued)

Feeding damage to	Slugs	Snails
Narcissus	*Lehmannia valentiana* (Férussac) *Limax maximus* L. *Milax budapestensis* (Hazay)	
Nicotiana tabacum (leaves)	*Lehmannia valentiana* (Férussac)	
Orchids (roots, young shoots, leaf stalks, flowers)	*Arion hortensis* Férussac *Deroceras laeve* (Müller) *Deroceras reticulatum* (Müller) *Lehmannia valentiana* (Férussac)	*Carychium minimum* (Müller) *Discus rotundatus* (Müller) *Opeas pumilum* (Pfeiffer) *Oxychilus draparnaudi* (Beck) *Zonitoides arboreus* (Say)
Papaver	*Deroceras reticulatum* (Müller)	
Plants in general in a *Victoria regia* house		*Lymnaea columella* Say *Physa acuta* Draparnaud *Physa heterostropha* (Say)
Plants in parks and gardens	*Arion* spp. *Boettgerilla pallens* Simroth *Deroceras reticulatum* (Müller) *Limax maximus* L.	*Cepaea* spp. *Cochlicopa lubrica* (Müller) *Helix* spp. *Vallonia costata* (Müller)
Rhododendron	*Arion fasciatus* (Nilsson)	*Helix aspersa* Müller
Rosa (young plants)	*Deroceras reticulatum* (Müller)	
Salvia splendens		*Discus rotundatus* (Müller) *Oxychilus draparnaudi* (Beck)
Solanum tuberosum (leaves, flowers) *S. dulcamara* *S. lycopersicum* *S. nigrum*	*Arion intermedius* Normand	
Spathiphyllum harveyanum		*Vallonia pulchella* (Müller)
Tradescantia albiflora	*Deroceras reticulatum* (Müller) *Lehmannia marginata* (Müller) *Lehmannia valentiana* (Férussac)	
Tulipa (germinating bulbs, flowers, leaves)	*Arion hortensis* Férussac *Arion intermedius* Normand *Deroceras reticulatum* (Müller) *Lehmannia valentiana* (Férussac) *Limax maximus* L.	*Cepaea hortensis* (Müller) *Helicigona arbustorum* (L.) *Helicina subfusca* Menke
Viburnum		*Helix aspersa* Müller
Water plants (in general)		*Armiger crista* (L.) *Australorbis camerunensis* C. R. Boettger *Discus rotundatus* (Müller) *Galba cubensis* (Pfeiffer) *Helisoma tumida* Pfeiffer *Isidora proteus* Sowerbyi *Physa acuta* Draparnaud *Physa heterostropha* (Say) *Lymnaea columella* Say *Radix auricularia* (L.) *Vallonia costata* (Müller) *Vallonia pulchella* (Müller)
Ornamental ferns under glass	*Arion hortensis* Férussac *Arion subfuscus* (Draparnaud) *Deroceras laeve* (Müller) *Milax gagates* (Draparnaud)	*Helicodiscus parallelus* (Say) *Oxychilus cellarius* (Müller)

Table 12. (continued)

Feeding damage to	Slugs	Snails
Ornamental plants (in general) in greenhouses (leaves, flowers)	*Deroceras laeve* (Müller) *Deroceras reticulatum* (Müller) *Lehmannia marginata* (Müller) *Lehmannia valentiana* (Férussac) *Limax maximus* L. *Milax* spp.	*Discus rotundatus* (Müller) *Oxychilus alliarius* (Miller) *Oxychilus cellarius* (Müller) *Oxychilus draparnaudi* (Beck) *Retinella pura* Alder *Vertigo pygmaea* (Draparnaud) *Vitrea cristallina* (Müller) *Zonitoides arboreus* (Say) *Zonitoides nitidus* (Müller)
Grass plants in grass seed production	*Deroceras agreste* (L.) *Deroceras reticulatum* (Müller)	*Vallonia pulchella* (Müller)
Lawns	*Deroceras reticulatum* (Müller) *Lehmannia valentiana* (Férussac)	

from as far afield as South America, eats orchid bulbs. In Hawai, *Deroceras reticulatum* (Müller) is also found on orchids (ALICATA 1950).

Field Slugs also blemish **Opuntii** and **Cacti** (SAINT PIE and BEAUCHARD 1963). The molluscan pests of vegetables, medicinal herbs and ornamental plants in gardens and greenhouses are listed in Tables 11 and 12.

2.2.1.3 Fruitgrowing

In the **fruit-growing** the most important gastropod pests are snails, in particular – and mainly in southern Germany – *Cepaea nemoralis* (L.), *Cepaea hortensis* (Müller), *Euomphalia strigella* (Drap.), *Monacha umbrosa* C. Pfeiffer, *Helix pomatia* L., *Bradybaena fruticum* (Müller), *Cochlodina laminata* (Montagu), *Lacinaria biplicata* (Montagu), *Trichia hispida* (L.) and *Trichia striolata* (C. Pfeiffer). Sweet and sour cherries, apples, pears, plums and peaches are all subject to snail damage. The pattern of damage caused by these snails, which includes complete defoliation, resembles that resulting from heavy caterpillar infestation (NEUMANN 1954, 1955 and others). Sour cherries are particularly subject to attack and often all that remains is the stone attached to the fruit stalk. During periods of aestivation, when the Banded Snails remain firmly attached to the trunk and branches by day and feed in the damp early morning hours, they may eat through the bark, even reaching the fibre of the trunk; this occurs particularly in those parts of the tree which were damaged by frost during the previous winter. PLATE (1964) observed that Banded Snails caused feeding damage to both fruit and bark of apple trees of the variety "Ontario". Even young fruit of walnut size, was attacked, and in consequence fowling by the fungus *Monilia* followed. In southern parts of the Black Forest *Cepaea hortensis* (Müller) is widespread in gardens, lawns and fruit orchards, while *Cepaea nemoralis* (L.) has only a very localized and sporadic distribution, but is more common in North Germany. Plums and peaches are eaten also by the slug *Limax maximus* L. (FRÖMMING 1952). The substantial damage to windfalls is of economic importance where the fruit is used in the manufacture of jams and fruit juices.

Table 13. Gastropod pests in fruit growing and viticulture

Feeding damage to	Slugs	Snails
Berries		
Strawberries (fruits)	*Deroceras laeve* (Müller) *Deroceras reticulatum* (Müller)	*Trichia striolata* (C. Pfeiffer)
Redcurrants (leaves) Blackcurrants (leaves, fruit on bush) Gooseberries (leaves)		*Bradybaena fruticum* (Müller) *Cepaea hortensis* (Müller) *Cepaea nemoralis* (L.) *Cochlodina laminata* (Montagu) *Euomphalia strigella* (Draparnaud) *Helix pomatia* L. *Lacinaria biplicata* (Montagu) *Monacha umbrosa* C. Pfeiffer
Pomaceous fruits		
Apples (leaves, fruit, bark) Pears (leaves, fruit)	*Limax maximus* L.	*Bradybaena fructicum* (Müller) *Cepaea hortensis* (Müller) *Cepaea nemoralis* (L.) *Euomphalia strigella* (Draparnaud) *Helix pomaria* (L.) *Trichia hispida* (L.)
Stone fruits		
Sweet and sour cherries (especially south Germany) (leaves, fruit on tree mainly by snails) Peach (leaves, fruit) Plums (leaves, fruit)	*Deroceras reticulatum* (Müller) *Limax maximus* L.	*Bradybaena fruticum* (Müller) *Cepaea hortensis* (Müller) *Cochlodina laminata* (Montagu) *Helicigona arbustorum* (L.) *Helix pomatia* L. *Lacinaria biplicata* (Montagu) *Monacha umbrosa* C. Pfeiffer
Grapevines (sprouts, green shoots and leaves)	*Deroceras reticulatum* (Müller) *Deroceras agreste* (L.)	*Cepaea hortensis* (Müller) *Cepaea nemoralis* (L.) *Helix aspersa* Müller *Helix pomatia* L.
Fallen fruits	*Arion rufus* (L.) *Deroceras reticulatum* (Müller) *Limax maximus* L.	*Helicigona arbustorum* (L.)

Strawberries are rendered unfit for sale and rot rapidly as a result of damage by *Deroceras* spp. and also by the snail *Trichia striolata* (C. Pfeiffer). Even the foliage of gooseberries and red currants does not escape damage and black currents are frequently hollowed out from the side.

In **viticulture** *Deroceras reticulatum* (Müller) and *Deroceras agreste* (L.) can become a plague because they destroy the leaves and green shoots. *Helix* spp., *Bradybaena fruticum* (Müller) and *Cepaea hortensis* (Müller) eat the bark of the new green shoots and developing buds in spring, as has been observed for example in the case of Müller-Thurgau vines (HERING 1966). As a result of intensive tillage of the soil and chemical weed control, these gastropods no longer play the dominant

Table 14. Gastropod pests in forestry

Feeding damage to	Slugs	Snails
Beech (soft leaves of seedlings, young plants)	*Lehmannia marginata* (Müller) *Limax maximus* L.	*Cepaea hortensis* (Müller) *Cepaea nemoralis* (L.) *Helicigona arbustorum* (L.)
Crataegus-standard (leaves)		*Cepaea hortensis* (Müller)
Alder (bark)		*Cepaea hortensis* (Müller) *Cepaea nemoralis* (L.)
Larch		*Vallonia costata* (Müller)
Conifers (especially seedlings)	*Arion circumscriptus* Johnston *Arion rufus* (L.) *Arion subfuscus* (Draparnaud) *Limax cinereoniger* Wolf *Limax tenellus* Müller	*Cepaea nemoralis* (L.) *Cochlodina laminata* (Montagu) *Helicella itala* (L.) *Helicella obvia* (Hartmann) *Helicigona arbustorum* (L.)
Pines (bark)		*Oxychilus alliarus* (Miller)
Poplars (young, bark)		*Cepaea nemoralis* (L.)
Juniper berries		*Helicella obvia* (Hartmann)
Woodland berries	*Arion rufus* (L.) *Arion subfuscus* (Draparnaud) *Deroceras reticulatum* (Müller)	*Bradybaena fruticum* (Müller) *Cepaea nemoralis* (L.) *Helicigona arbustorum* (L.)

role they once did. Plagues are, nevertheless, still possible; for example, exceptional damage by *Deroceras reticulatum* (Müller) in vineyards at Bernkastel-Kues on the Moselle River was reported by GÄRTEL in 1977.

The snails and slugs affecting fruit growing and viticulture are listed in Table 13.

2.2.1.4 Forestry

While slugs and snails certainly contribute to the formation of humus on the forest floor (FRÖMMING 1956, 1958), they can also be important pests in forestry (Table 14). In beech woods, where the undergrowth is dense, snails of the genera *Cepaea* and *Helicigona* occur in large numbers, mostly amongst nettles. *Cepaea* spp. cause damage to the bark of young poplars and alders, in the same way as they do to fruit trees. As the trunk increases in diameter, cracks appear in which fungi can become established, with the result that the tree ultimately dies. KOTTHOFF (1962) found *Cepaea nemoralis* (L.) on stems of 4- to 6-year-old poplars; these were only 10 cm in diameter and although the bark was still smooth, many showed scrape marks of 1 cm length. At the beginning of August, the snails were found along the whole length of the trunk, right up to the crown of the tree. GODAN reported damage to young alders at the end of July. During aestivation, however, most of the Banded Snails were found attached to the tree trunks up to a height of 3 m. Slug damage to beech seedlings has been observed. REINMUTH (1938) reported complete defoliation of the upper branches of *Crataegus* by *Cepaea hortensis* (Müller).

Slugs which are known pests to agriculture and horticulture also occur in coniferous forests. Thus *Arion subfuscus* (Drap.) is found mainly in coniferous forests where it can inflict heavy damage to seedlings (BOBACK 1952). This applies also to

Oxychilus alliarius (Miller) which, as has been reported by RAEHLE (1970), lives under the bark of pine trees, although it is found in the open in western and northern Europe and in the alps.

Herbaceous plants and forest undergrowth may be attacked by *Helix aspersa* Müller. Even woodland fruits, such as juniper berries, do not escape damage (FRÖMMING 1952).

2.2.1.5 Agricultural Warehouses, Storerooms and Cellars

The Cellar Slug, *Limax flavus* L. is of particular importance as a pest to agricultural products in storage (FRÖMMING 1951). It is found in cellars and storerooms occasionally in greenhouses and, south of the alps, in fountains. This slug is encountered in caves in Hungary being found in the open only in thinly populated areas of Eastern Europe. Its food consists of stored fruit such as apples and pears, as well as potatoes, carrots, sugar beet, pumpkin, cucumber and tomatoes. In the case of green leaf vegetables, it attacks the middle vein of the leaves which is poor in chlorophyll but contains a high percentage of water, in preference to the green parts of the leaves. Bulbs which are infected with fungi or bacteria are readily eaten. FRÖMMING regards as significant the spread of fungi and bacteria by slugs moving about at night; a single slug can cover 20 to 40 m in one night, feeding on several different potatoes. This may result in the destruction of 1 $^1/_2$ kg of potatoes within 1 month, some of the damage arising from the partially eaten tubers becoming rotten.

The European Giant Garden Slug, *Limax maximus* L., makes its way into cellars and storerooms from gardens and farms. It originates from Southern and Central Europe, yet is most common in Western Europe; it is found in almost all parts of Germany, most frequently near human habitation. *L. maximus* L. feeds on green plants but prefers roots, tubers and rhizomes; in the case of leaf vegetables it eats the juicy midrib.

The Cellar Snail, *Oxychilus cellarius* (Müller) is also found in cellars and storerooms. Further examples are given in Chapter 3.3.

2.2.2 Gastropod Damage in Other Countries

In the preceding section damage caused by gastropod pests in the German Federal Republic was considered. Accounts of the gastropod fauna and in a few cases, their effect as pests, have been given for Belgium by ADAM (1960), Bulgaria by HÚDEC and VASÁTKO (1971), PINTER (1969) as well as PINTER and PINTER (1970), Finland by KOLL (1961) and VALOVIRTA (1967), France by CHEVALLIER (1973, 1974), GERMAIN (1930/31), MALLET (1973) and also RITTER (1955), Greece by JAECKEL and PLATE (1961) and by KLEMM (1962), Great Britain and Ireland by CHATFIELD (1972), ELLIS (1926, 1951, 1969), EVANS (1968) and QUICK (1949, 1961), Holland by BENTHEM-JUTTING (1933), BENTHEM-JUTTING and VAN REGTEREN-ALTENA (1962), VAN BRUGGEN (1957), BRULINS et al. (1959) and GITTENBERGER et al. (1970), Italy by ALZONA (1971) and SACCHI (1957, 1971, 1974), Yugoslavia by JAECKEL and MEISE (1956), Canada by LAROCQUE (1953), North America by PILSBRY (1939/40,

1946/48), Norway by ØKLAND (1925, 1930, 1969), Austria by KLEMM (1960, 1974), East Africa by VERDCOURT (1972), Poland by URBAŃSKI (1957) and WIKTOR (1973), Portugal by NOBRE (1930), Rumania by GROSSU (1955), Switzerland by MERMOD (1930), Spain by PARDO (1943), SACCHI (1965) and SACCHI and NOS (1958), Hungary by PINTER (1970, 1974), the USSR by LIKHAREV and RAMML'-MEIER (1952), VASIL'EV (1973) and ZHADIN (1952), the northern part of the Balkan peninsular by S.H. JAECKEL (1954), exotic molluscs in the USA by BRANSON and BATCH (1969) and in an account of the prosobranchs by GRAHAM (1971).

2.2.2.1 Western Europe

In the countries bordering on the Federal Republic of Germany the same pest slugs and snails occur, although their relative economic importance of any particular species differs from country to country. The Field Slugs are, however, the most important gastropod pests in all these countries.

In **France,** Brittany and Normandy have a rich pulmonate fauna (CHEVALLIER 1970; LUCAS 1969; RICOU 1961). In contrast to Central Europe, where the Limacidae (*Deroceras* spp.) cause most damage, the Arionidae are here the most important gastropod pests. Those species which are regarded as most dangerous are *Deroceras agreste* (L.), *D. reticulatum* (Müller), *Deroceras laeve* (Müller), *Arion hortensis* Fér. and, with a localized distribution, *Milax budapestensis* (Hazay) (CHEVALLIER 1973; MALLET 1968, 1973). *Arion rufus* (L.) occurs in the north and west, while *Arion subfuscus* (Drap.) which is less significant as a pest, is distributed over the whole country, with the exception of high mountain areas in the Pyrenees. *Geomalacus maculosus* Allman (Family Arionidae) is found not only in France, but also in Ireland, Spain and Portugal. *Deroceras caruanae* (Pollonera) was first found in France twenty years ago (CHEVALLIER 1973; HAMEURY 1958). *Limax maximus* L. has occurred in France since about 1925 (LEMEE 1928).

Although the occurrence of *Milax budapestensis* (Hazay) is still sporadic, its further spread into France from Belgium, Germany, England and Switzerland is only a question of time. It has already attracted attention as a pest in sunflower fields, where up to 50% of the cultivated area may be destroyed (AGUILAR and PACQUÉTIAN 1963). In May, sunflower seeds planted in the soil are eaten, later the cotyledons of the seedlings, before they break through the upper surface of the soil, as well as the hypcotyl and the young roots. As a preventive measure, it is recommended that the sunflower seeds be planted deeper than usual (5 to 7 cm) so that access to them is difficult for the slugs.

Other recognized pests are: *Helicigona arbustorum* (L.), *Helix aspersa* Müller, *Helix aperta* Born, *Helix pomatia* L., *Hygromia limbata* (Drap.), *Cepaea nemoralis* (L.), *Cepaea hortensis* (Müller), *Helicella itala* (L.), *Hygromia cincttella* (Drap.), *Eobania vermiculata* (Müller), *Theba pisana* (Müller), *Zonites algirus* (L.).

Otala punctata (Müller) is found in southern France. CHEVALLIER (1975) found *Helix lucorum* L. in the area around Lyon. The amphibious *Succinea putris* (L.) is found in meadows and pastures, but feeds at night in adjacent fields of clover, vegetables and cereals. The freshwater snails *Radix ovata* (Drap.) (=*Limnaea limosa* L., *Radix peregra* (Müller) (=*Limnaea peregra* Müller, *Lymnaea stagnalis* (L.) and *Galba palustris* (Müller) destroy beds of cress (LEIMBACHER 1973;

Ricou 1961, 1966; Ricou and Marre 1960; Ricou and Picard 1972; Ricou et al. 1973).

The following agricultural crops are the most subject to gastropod attack: cereals (mostly winter varieties, following rape), oil plants, maize, lettuce, peas, beans, fodder crops (lucern, clover), cauliflower, cabbage, potatoes, tobacco, strawberries, medicinal herbs, sunflower seedlings and ornamental plants. *Deroceras reticulatum* (Müller) and *Arion hortensis* Fér. are the most common gastropod pests of cereal crops (Ricou and Mallet 1977). In the case of maize, total destruction of young plants in south eastern regions of France is common. Rape is planted at the time of greatest activity of these slugs, with the result that often all the seed is eaten in September to October. In winter and early spring the ground cover of rape leaves favours an increase in numbers of the gastropods, with the result that the succeeding crop suffers great damage (Mallet and Bougaran 1970).

Theba pisana (Müller) causes considerable losses to the lucern harvest, particularly in south-east France. This species is one of the most mobile snails, surpassing the other members of the Helicidae in this respect, and can also increase in numbers very rapidly. It leaves the ditches and hedges where it spends the winter and appears in the fields in mid-April. The considerable damage which it causes results not so much from its feeding activity, as from the slime with which the lucern flowers become soiled; this causes them to stick together, so that pollination by Hymenoptera becomes impossible (Cairaschi and Lecomte 1973). Damage to the harvest is also significant, since the sieves of the harvesting machines are rendered unusable by squashed snails; furthermore, lucern seeds soiled in this way are less marketable. Cairaschi (1970) mentions also *Hygromia limbata* (Drap.). Clover and lucern grown in autumn on soil free of weeds are particularly endangered, since the gastropods are then entirely dependant on these crops as a source of food.

Amongst ornamental plants, flowers such as asters, dahlias, gladioli and carnations are mainly eaten; so, too, are bulbs and cacti. In this connection, *Deroceras reticulatum* (Müller), *Deroceras agreste* (L.), *Arion hortensis* Fér., *Deroceras laeve* (Müller), *Milax sowerbyi* (Fér.) and, though less frequently, *Milax budapestensis* (Hazay) are regarded as "enemies" (Mallet 1968, 1973 and others). *Cecilioides acicula* (Müller) has become a pest to *Asparagus plumosus* (Pussard 1951).

In **Belgium,** too, the Field Slugs are the most important gastropod pests and attack the same plants as in Germany and France. The pests of grain, including rye, are *Deroceras reticulatum* (Müller) and *Deroceras agreste* (L.), to potatoes and carrots *Arion hortensis* Fér., the latter is also a pest in parks and gardens, together with *Arion circumscriptus* Johnston, *Arion subfuscus* (Drap.), *Deroceras laeve* (Müller), *Milax budapestensis* (Hazay), *Milax rusticus* (Millet) and *Milax sowerbyi* (Fér.) (Moens 1959, 1960, 1962; Moens and van den Bruel 1958, 1960). The three *Milax* species, as soil dwellers, eat carrots, tubers and bulbs. *Limax maximus* L. and *Limax flavus* L. cause great damage in mushroom beds. Amongst the snails the most noteable are *Cepaea hortensis* (Müller), *Cepaea nemoralis* (L.), *Bradybaena fruticum* (Müller), *Euomphalia strigella* (Drap.), *Helicigona arbustorum* (L.), *Helix pomatia* L. (van den Bruel and Moens 1957) and since 1972, *Cernuella neglecta* (Drap.) (Sitva 1973). In Belgium cultivation of cress is of economic importance and is mainly plagued by *Galba truncatula* (Müller), *Galba palustris* (Müller), *Radix peregra* (Müller) and *Potamopyrgus jenkinsi* (E.A. Smith) (Moens 1965,

1967, 1973; SEVO 1971 and others). *Galba truncatula* is also of importance as the intermediate host of the Liver Fluke of cattle and sheep.

The pest gastropods of Holland, which offers a favourable environment for terrestrial pulmonates (MÖRZER BRUYNES et al. 1959; GITTENBERGER et al. 1970), include *Deroceras reticulatum* (Müller), *Arion ater* (L.), *Arion hortensis* Fér., *Arion circumscriptus* Johnston, *Arion subfuscus* (Drap.), *Milax gagates* (Drap.), *Milax budapestensis* (Hazay) and *Milax sowerbyi* (Fér.) (BACKHUYIS and PIPKEN 1966; VAN DINTHER 1973; SCHOUTEN 1949; VAN REGTEREN-ALTENA 1970; VISSER 1966). *Milax* spp. attack the seedlings of cereals, rape and rape seed, especially in autumn and mild winters; they further destroy the young plants of cabbage, lettuce, peas, beans, clover, carrot roots, sugar beet, strawberries as well as ornamental plants, flowers and bulbs, which are of economic importance particularly in Holland. *Geomalacus maculosus* Allman has been found in Holland since 1913.

The following snails occur in Holland: *Helix aspersa* Müller, *Helix pomatia* L., *Cernuella virgata* (Da Costa) which reached Holland from southern Europe in 1874 (MIENIS 1969), *Cochlicopa lubrica* (Müller), *Discus rotundatus* (Müller), *Oxychilus draparnaudi* (Beck), *Candidula intersecta* (Poiret), *Helicigona arbustorum* (L.), *Cepaea nemoralis* (L.), *Vallonia pulchella* (Müller), *Vallonia excentrica* Sterki and, particularly important, *Trichia hispida* (L.). *Helix pomatia* L. attacks hedges which serve as windbreaks. Although this snail has been found in Holland since 1874, it has made its presence felt to a far greater extent since 1970 (BUTOT 1970, 1972). The mediteranean *Cochlicella ventricosa* (Drap.) had spread as far as the North Sea coast only 20 years ago (BOULANGE et al. 1958). Areas which were previously dry but have since 1955 been under irrigation have, with the exception of the newly identified *Gyraulus laevis* (Adler), the same gastropod fauna of Lymnaeidae and Planorbidae as the rest of Holland (KUIJPER 1973). In this country, too, intermediate hosts of the Liver Fluke, like *Galba truncatula* (Müller), play a role (OVER 1967). *Radix ovata* (Drap.) attacks cress cultivations. New species like *Opeas mauritanum* (Pfeiffer) may be added to MEUSE and HUBERT'S (1949) list of gastropod fauna of greenhouses.

In **England** gastropod pests are a serious problem, since the climate, with its high relative humidity, moderate temperatures and mild winters, affords ideal environmental conditions for terrestrial pulmonates. Every year, damage involving considerable financial losses is inflicted on cereals, potatoes, vegetables, celery, lettuce, maize, carrots, beetroot, peas, clover, cabbage, turnips as well as other agricultural and horticultural crops. Ornamental plants, such as chrysanthemums and tuberosa, schrubs and flowers also suffer. Highly valuable hothouse plants are particularly endangered, since in the wake of damage by slugs and snails fungal plant diseases can establish themselves (BARNES and WEIL 1944, 1945; BARRY 1969; EYSTER 1951; HARTOG and SANDEE 1962; MARTIN and FORREST 1969; PINDER 1969; QUICK 1947; RUNHAM and HUNTER 1970; SOUTH 1965; WARLEY 1970; WINFIELD et al. 1967 and others). Winter cereal crops are particularly endangered if they are planted after a crop which affords particularly favourable environmental conditions for gastropods. GOULD (1961) estimated the destruction of wheat seed sown after a harvest of peas or cabbage to be 90%, of clover and cereals 70%, of beans up to 35%, but of potatoes only up to 5%; by contrast, there was no loss of seed sown after a harvest of sugar beet, nor of that sown on fallow land.

The most important pests of potatoes, celery and root vegetables such as carrots, include *Milax budapestensis* (Hazay), *Milax gagates* (Drap.), *Milax sowerbyi* (Fér.), *Deroceras reticulatum* (Müller), *Deroceras agreste* (L.), *Arion hortensis* Fér., *Arion circumscriptus* Johnston, *Arion subfuscus* (Drap.), *Arion intermedius* Normand (CARRICK 1942; STEPHENSON 1967). Plants in both field and garden are also attacked by *Arion ater* (L.), *Arion fasciatus* (Nilsson), *Arion lusitanicus* Mabille, *Lehmannia marginata* (Müller), *Lehmannia valentiana* (Fér.), *Limax cinereoniger* Wolf, *Limax maximus* L., *Limax tenellus* Müller and *Prophysaon andersoni* (Cooper), while *Limax flavus* L. attacks food in storage (CHATFIELD 1972, 1973; ELLIS 1964; HAYWARD 1954). *Geomalacus maculosus* Allman is found in south west Ireland and eats mainly seedlings of sugar beet. *Deroceras caruanae* (Pollonera) reached Ireland shortly before 1959 (MAKINGS 1959).

The following snails should be mentioned: *Monacha cantiana* (Montagu), *Helicella* spp., *Cochlicopa acuta* (Müller), *Helicigona arbustorum* (L.), *Cepaea hortensis* (Müller), *Cepaea nemoralis* (L.) – the latter being found not only in gardens but also in clover, lucern and pastures – and since 1970 *Helicodonta obvoluta* (Müller) (CAMERON 1972; CAMERON and PALLES-CLARK 1971 and others).

BEEDHAM (1972) and WALDÉN (1976) have published lists of British terrestrial pulmonates, including those which are pests.

According to the distribution atlas of KERNEY (1976), the following pest gastropods have a singularly wide occurrence in the British Isles: *Limax maximus* L., *Lehmannia marginata* (Müller), *Deroceras laeve* (Müller), *Deroceras reticulatum* (Müller), *Arion ater* (L.), *Arion subfuscus* (Drap.), *Arion circumscriptus* Johnston, *Arion hortensis* Fér., *Arion intermedius* Normand, *Oxychilus cellarius* (Müller), *Oxychilus alliarus* (Miller), *Oxychilus helveticus* (Blum) (only in England), *Monacha cantiana* (Montagu) (occurs only in east England, but in very large numbers), *Trichia striolata* (Pfeiffer), *Trichia hispida* (L.), *Cepaea nemoralis* (L.), *Cepaea hortensis* (Müller) (less numerous in Ireland), *Helix aspersa* Müller (only in Scotland less numerous), *Potamopyrgus jenkinsi* (Smith), *Physa fontinalis* (L.), *Galba truncatula* (Müller), *Galba palustris* (Müller), *Lymnaea stagnalis* (L.) (not in Scotland), *Radix peregra* (Müller), *Gyraulus albus* (Müller), *Ancylus fluviatilis* Müller, *Cochlicopa lubrica* (Müller), *Discus rotundatus* (Müller).

Freshwater snails also serve in England as intermediate hosts for the Liver Fluke of cattle and sheep. The most important are *Galba truncatula* (Müller), *Radix peregra* (Müller), *Planorbis* spp., *Physa fontinalis* (L.), *Galba palustris* (Müller) and *Lymnaea stagnalis* (L.) (HEPPLESTON 1972 and others).

2.2.2.2 Southern Europe

In **Spain** and **Portugal** attention was formerly directed mostly at the intermediate hosts of the trematode parasites of man and domestic animals (SANCHEZ 1965). More recently the focus of interest has changed. In the northern and north eastern parts of Spain, central European species which also occur in France are found; these include *Lymnaea stagnalis* (L.) (since 1966), *Deroceras laeve* (Müller), *Arion rufus* (L.), *Arion hortensis* Fér., *Arion intermedius* Normand, *Arion lusitanicus* Mabille, *Geomalacus maculosus* Allman, *Deroceras reticulatum* (Müller), *Lehmannia valentiana* (Fér.), *Limax maximus* L., *Limax flavus* L., *Helicigona arbustorum* (L.) with one sub-species, *Trichia hispida* (L.), *Cepaea nemoralis* (L.) and *Oxychilus draparnaudi* (Beck) (ALVAREZ 1972, pers. comm. 1968; ALONSO and IBAÑEZ 1975; HARVEY 1971; RIEDEL 1972). In most cases these are mediterranean species which prefer a warm environment; their distribution extends into Italy, which is also true of *Helicella* and *Helix (Trochoidea)* species, *Helicella itala* (L.), *Helicella neglecta* (Drap.), *Cernuella virgata* (Da Costa) and others which up to now have not been regarded as pests (ALONSO 1975).

Helix pomatia L. is mostly found in vineyards. Other gastropods from this region are *Eobania vermiculata* (Müller), *Iberus alonensis* (Fér.), *Iberellus* spp., *Cecilloides acicula* (Müller) and *Carinaria lamarcki* Peron & Lesueur. DEL RIVERO and ROCA (1969) describe *Helix aspersa* Müller and *Theba pisana* (Müller) and FIGUEIREDO and LEPAGE (1953) describe snails as particularly important pests on orchids. Other possible pests are *Leucochroa candissima* Drap., *Ancylus hispanicus* Guénée (found also in Western Europe and North Africa), *Rumina decollata* (L.) and *Otala* species.

In **Italy** the situation is similar to that in Spain and it is only recently that attention has been directed to terrestrial gastropods on cultivated land, major emphasis being placed on the snail vectors of worms parasitising man and animals. Indeed, snails hitherto only regarded as pests in agriculture and horticulture have been found to act as intermediate hosts for trematodes. These include *Candidula unifasciata* (Poiret), *Helicella itala* (L.), *Cochlicella acuta* (Müller), *Cochlicopa lubrica* (Müller), *Euomphalia strigella* (Drap.), *Abida frumentum* (Drap.), *Zebrina detrita* (Müller). In addition there are the usual intermediate hosts like *Galba truncatula* (Müller), *Galba palustris* (Müller), *Lymnaea stagnalis* (L.), *Radix peregra* (Müller), *Planorbarius corneus* (L.), *Bulinus truncatus* (Audouin) and *Bithynia leachi* (Sheppard) (PAMPIGLIONE and TOFFOLETTO 1971).

In damp places the following species occur: *Deroceras* spp., *Arion rufus* (L.), *Geomalacus maculosus* Allman, *Limax maximus* L., *Limax cinereoniger* Wolf, *Limax flavus* L. and the well-known *Milax* species. In the vegetable-growing area of Sardinia, *Milax gagates* (Drap.) causes considerable damage to artichokes under cultivation *(Cynara cardunculus* subsp. *scolymus)*, feeding on both the leaves and heads (PROTA 1960).

The most important pest snails are *Helix aspersa* Müller, *Helix aperta* Born, *Theba pisana* (Müller), *Helix lucorum* L., *Helicella itala* (L.) and *Helicella obvia* (Hartmann) which is transported to other countries with fruit and vegetable exports. Other species which are found in Italy are *Helix pomatia* L., *Helicodonta obvoluta* (Müller), *Helicigona arbustorum* (L.), *Abida frumentum* (Drap.), *Trichia striolata* (C. Pfeiffer), *Carinaria mediterranea* Fér., *Cepaea hortensis* (Müller), *Oxychilus alliarius* (Miller), *Oxychilus cellarius* (Müller) and *Helisoma anceps* (Menke) (since 1967) (ALZONA 1971; GIROD 1973; GUISTI 1968; HENRARD 1968; SACCHI 1974). *Bradybaena similaris* (Fér.) and *Subulina octona* (Bruguière) are pests on *Hibiscus* spp. in gardens.

In **Greece** and in **Turkey** *Theba pisana* (Müller) occurs, which, together with *Helicella itala* (L.), *Helix aspersa* Müller and *Helix aperta* Born, is also important as a quarantine pest in the USA.

CURTIN (1962) also mentions, amongst others, *Helicopsis striata* (Müller) and *Oxychilus cellarius* (Müller).

2.2.2.3 Austria and Eastern Europe

In **Austria**, as in West Germany, France and England, gastropods are regarded as very important plant pests, and in general the same species are involved in all these countries. Again *Deroceras reticulatum* (Müller) is the most important pest in field, garden and greenhouse (Böhm 1957).

Oxychilus alliarius (Miller), which is found in Switzerland and is often confused with *Oxychilus helveticus* (Blum), also occurs here (Wüthrich 1963).

In **Czechoslovakia,** as in central Germany, species of the families Arionidae und Limacidae are found, and the same plants are attacked.

Again the Field Slug, *Deroceras reticulatum* (Müller) is important. In addition to the gastropods commonly found in nurseries and greenhouses, the following species occur: *Boettgerilla pallens* Simroth, *Lehmannia valentiana* (Fér.), *Opeas pumilum* (Pfeiffer), *Lacinaria biplicata* (Montagu), *Lacinaria plicata* (Drap.), *Carychium tridentatum* (Risso), *Carychium minimum* (Müller), *Gyraulus albus* (Müller), *Trichia hispida* (L.), *Cochlicopa lubrica* (Müller), *Physa heterostropha* (Say), *Cochlodina laminata* (Montagu), *Perforatella incarnata* (Müller). *Melanoides tuberculata* (Müller) has been introduced from Indonesia and India into greenhouses in which *Victoria regia* is being cultivated; here *Helisoma trivolis* (Say) is also found (Flasarova and Flasar 1962, 1965; Flasar and Kroupova 1976; Mácha 1971). Dirlbeck (1969) stated that in hotbeds seedlings of *Brassica pekinensis* (Lour.) Rupr. were destroyed by *Arion rufus* (L.) and *Deroceras laeve* (Müller). In greenhouses *Discus rotundatus* (Drap.), *Aegopinella nitidula* (Drap.), *Arion circumscriptus* Johnston and *Deroceras laeve* (Müller) destroyed, amongst other things, leaves of *Gerbera jamesonii* (Dirlbeck 1970).

The gastropod fauna found on cultivated land in **Poland** comprises, in the western parts of the country, species which have been introduced from elsewhere (Wiktor 1967, 1973). These include the garden pests *Arion hortensis* Fér., *Arion fasciatus* (Nilsson), *Boettgerilla pallens* Simroth, *Limax maximus* L., *Limax flavus* L., *Lehmannia valentiana* (Fér.), *Milax budapestensis* (Hazay). The main pests are of course *Deroceras reticulatum* (Müller) and possibly also *Deroceras agreste* (L.). In gardens *Bielzia coerulans* (M. Bielz), *Deroceras sturanyi* (Simroth), *Cepaea vindobonsensis* (Fér.) are found, while *Lehmannia valentiana* (Fér.), *Limax flavus* L., *Deroceras laeve* (Müller), *Limax maximus* L. and also *Arion hortensis* Fér. occur in hothouses. *Limax maximus* L. and *Limax flavus* L. can cause complete destruction of mushrooms under cultivation.

Galba truncatula (Müller), *Planorbis planorbis* (L.), *Radix peregra* var. *ovata* (Drap.) and *Bithynia leachi* (Sheppard) live in ponds which are subject to periodic drying up (Klimowicz 1962).

In the case of **Rumania,** material is a available on the family Limacidae, in particular the genus *Deroceras* Rafinesque and also the genera *Milax* Gray and *Helicigona* Férussac as well as the family Zonitidae (Grossu 1955, 1970; Grossu and Lupu 1959, 1960, 1962, 1963; Grossu and Tessio 1975; Lupu 1971, 1972, 1974). The following representatives of the Limacidae occur: *Deroceras reticulatum* (Müller), *Deroceras agreste* (L.), *Deroceras laeve* (Müller), *Limax cinereoniger* Wolf, *Limax maximus* L., *Limax flavus* L., *Bielzia coerulans* (M. Bielz), *Lehmannia marginata* (Müller), *Lehmannia valentiana* (Fér.), while of the Arionidae: *Arion* *ater* (L.), *Arion subfuscus* (Drap.), *Arion hortensis* Fér., *Arion circumscriptus* Johnston, *Arion fasciatus* (Nilsson) (LUPU 1970, 1974).

In **Bulgaria** there are, according to WIKTOR (1977), 34 slug species, 13 of which are widely distributed throughout central and western Europe, 7 species over the Balkan area, and partly also in Rumania, while 2 species are found in the Mediterranean area. On cultivated land, Limacidae and *Milax* species, *Deroceras transcaucasicum* Simroth, *Arion rufus* (L.), *Arion hortensis* Fér. occur, and also snails of the family Helicidae, like *Cepaea vindobonensis* (Fér.) and *Euomphalia strigella* (Drap.) (HÚDEC and VASÁTKO 1971, OŠANOVA 1964). *Cecilioides petitiana* Benoit is a pest to lettuce grown in greenhouses (PALFI 1973). Other gastropod pests in this region are *Monacha cartusiana* (Müller), *Monacha cascaloides* (Bourgs), *Ena obscura* (Müller), *Helix pomacella* Mous, *Helix lucorum* L., *Helicella obvia* (Hartmann), *Helix lutescens* (Rossm.), *Helicella krynickii* (Andrz.) and the freshwater snail *Radix peregra* var. *ovata* (Drap.).

2.2.2.4 Northern Europe

In **Denmark** many of the gastropod pests common in Germany are also found. The most important are *Deroceras reticulatum* (Müller) and *Limax maximus* L. *Helicigona arbustorum* (L.) is distributed over the whole of Denmark, while *Bradybaena fruticum* (Müller) is found only in the north and west (PETERSEN 1964). Several have spread as far as southern Sweden, including, since 1965, *Limax flavus* L. (NILSSON 1962).

The gastropod fauna of **Sweden** also includes the Field Slugs, particularly in plantations of rape and rape seed. The same species are found here as in central Europe: the Arionidae (*Arion* spp.), Limacidae (*Deroceras* spp., *Lehmannia* spp., *Limax* spp.) and Helicidae [*Helix* spp., *Cepaea nemoralis* (L.), *Helicigona arbustorum* (L.) and others]; also *Lehmannia valentiana* (Fér.), *Deroceras caruanae* (Pollonera) (since 1857), *Arion silvaticus* Lohmander, *Cecilioides acicula* (Müller), *Oxychilus* spp., *Zonitoides arboreus* (Say) (since 1950) (ANDERSSEN and NILSSON 1967; BORG 1962; WALDÉN 1954, 1960, 1961, 1965, 1969; WAREBORN 1970). Although *Limax maximus* L. originated in southern and central Europe, this snail was introduced to northern Sweden with fruit imported from California.

Helix pomatia L. can even survive the winter in **Norway** (VIK 1968). ØKLAND (1969) studied the ecology and distribution of freshwater snails in Norway, where they play a role as the intermediate host of liver flukes. The snails are *Galba truncatula* (Müller), *Galba glabra* (Müller), *Lymnaea stagnalis* (L.), *Radix peregra* (Müller), *Galba palustris* (Müller), *Radix auricularia* (L.) and *Radix glutinosa* (Müller).

In the extreme north of Europe, in **Finland**, *Deroceras reticulatum* (Müller) is once more the main pest in fields, gardens and greenhouses, where it causes damage especially to cucumbers, tomatoes and lettuce, as well as chrysanthemums, carnations, roses and tulips (MARTII MARKKULA 1969). The same types also occur of the Arionidae, Limacidae, Helicidae as well as *Oxychilus* spp., *Succinea* spp., *Zonitoides* spp., *Bithynia tentaculata* (L.), *Physa fontinalis* (L.) and others (LÄHDESMÄKI 1970; VALOVIRTA 1967). Finland represents the most northerly limit of *Limax* *cinereoniger* Wolf (RAJALA 1957). *Milax gagates* (Drap.) was probably spread to Finland with roses from Denmark and with begonias and azalias from Belgium (VALOVIRTA 1969).

2.2.2.5 America

The gastropod fauna of north America consists of many species which had their origin in Europe or Asia. Thus many common names begin with the word "European" or end with "of Europe". Some examples of snails and slugs involutarily introduced are: *Arion ater* (L.) (European Black Slug), in 1963, *Arion circumscriptus* Johnston (European Slug), *Limax maximus* L. (Great Slug of Europe), *Lehmannia valentiana* (Fér.) (Iberian Slug), *Prophysaon andersoni* (Cooper) (British Garden Slug), *Milax* spp., and further *Cepaea nemoralis* (L.) (European Snail) shortly before 1966, and also Radix auricularia (L.) (DEXTER 1966; VAN DINTHER 1973; JULLIET 1968; O'NEIL 1964 and others).

In **Canada,** *Deroceras reticulatum* (Müller) and *Deroceras agreste* (L.) are important pests in field, garden and greenhouse. After these two, the most important are *Arion circumscriptus* Johnston, *Arion fasciatus* (Nilsson), *Arion silvaticus* Lohmander, *Limax maximus* L., *Lehmannia valentiana* (Fér.) and *Milax gagates* (Drap.) which spread from Europe to Australia, Canada, Hawaii, the Canary Islands and Tristan da Cunha (FOX 1962); finally *Prophysaon andersoni* (Cooper) is a special pest to gladioli and clover (GLENDENNING 1952; LANGLOIS 1965; WILKINSON 1964).

In the **USA** the Field Slugs are once again the most important gastropod pests. They cause considerable damage to maize, especially to young plants at the stage at which adventitious roots are developing (BARRY 1969). Damage is done at a later stage, too, through tunneling into the stake root, if the field is unploughed and there is a heavy coverage of mulch (MUSICK 1972). Brussel sprouts are destroyed by *Deroceras reticulatum* (Müller) and also *Milax gagates* (Drap.) with an infestation of 20,000 slugs per $^1/_2$ hectar. *Arion ater* (L.) caused great damage to narcissus cultivations by rasping the bulbs (DOUCETTE 1954). In the damp coastal regions of California considerable economic losses are incurred as a result of gastropod damage to artichoke heads (LANGE and MACLEOD 1941). Mention must also be made of *Deroceras laeve* (Müller), *Arion hortensis* Fér., *Arion circumscriptus* Johnston, *Prophysaon andersoni* (Cooper), *Limax maximus* L., *Milax* spp., *Veronicella* spp. and species of the Philomycidae (BRANSON 1962; BRANSON and BATCH 1969; VAN BRUGGEN 1955; GETZIN and COLE 1964; RISING and ARMITAGE 1969), as well as the indigenous slug, *Ariolimax columbianus* (Gould) found in gardens (SMITH, San Francisco, pers. comm. 1968), *Parmarion* sp., *Vaginulus* sp. and *Mariaella dursumierei* Grey which is a pest on leguminous plants, and finally the snails *Bradybaena similaris* (Fér.), *Subulina octona* (Bruguière) and *Xestina striata* Gray (Anon. 1967). Cabbage, beans, peas and tomatoes are attacked by the snails *Helix aspersa* Müller and *Theba pisana* (Müller) which have been introduced from the mediterranean region, and on dry sloping ground by *Helicella obvia* (Hartmann) (BRANSON 1962; VAN DINTHER 1973). The two former species constitute a threat to orange groves in California. According to FISHER et al. (1976), owners of gardens paid as much in combating *Helix aspersa* Müller as they did for all other pests together. MEAD

(1961) found 3,000 specimens of *Theba pisana* (Müller) on a single citrus tree. *Drymaeus dormani* (Binney) is, in fact, called the "citrus tree snail" (KNUTSON, Washington, pers. comm. 1968). *Cepaea hortensis* (Müller) and *Cepaea nemoralis* (L.) cause damage to vegetables, lucern and pastures. *Opeas mauritianum* (Pfeiffer) lives in hot houses.

Lamellaxis gracilis (Hutton) is a pest in greenhouses of **Central America.**

The paddy fields of **South America** harbour *Pomacea (Ampullarius) glauca* (L.), *Pomacea lineata* (Spix) and *Pomacea dolioides* Reeve; these are spread over Surinam, Brazil, Columbia, Bolivia, Venezuela, Trinidad, Guadeloupe and Barbados (VAN DINTHER 1956, 1957, 1973). Their egg batches enclose the thin stalk of the rice plant and contain 200 to 300 eggs (Fig. 33). The young snails can crawl around relatively quickly in the water shortly after hatching and can spread at the rate of 2 to 5 cm/min. *Strophocheilus globosus* (Müller) is found in the north-west of Uruguay. KLAPPENBACH and OLAZARRI (1965) have described a new species, *Strophocheilus parodizi* n. sp. from the Argentine.

2.2.2.6 Africa, Asia, Pacific, Australia

The genus *Otala* Schumacher is indigenous to the **western Mediterranean** region (Marocco, western Algeria, the Iberian peninsula). *Limicolaria kambeul* Bruguière is found in the peanut and maize plantations of the southern Sudan.

The African Giant Snail, *Achatina fulica* Bowdich is amongst the most dreaded terrestrial gastropods in the tropics and sub-tropics, mainly because of its voracity and also the speed with which it spreads (MEAD 1961, 1979). This snail kills young banana plants, eats the leaves and fruits of plants under cultivation in fields and gardens, including cereals, destroys the seedlings of *Phaseolus* sp., *Asplenium nidus*, the fruits of the bread fruit tree (*Artocarpus* sp.), seedlings, flowers, seeds and young pods of cocoa *(Theobroma cacao)*, Papaya, Citrus, young *Hevea* plants, cacti *(Cactus, Cereus, Opuntia)*, the leaves of *Amaranthus* sp., *Bryophyllum* sp., *Canna* sp. amongst others. Remarkably, sugar cane, Yam-beans *(Pachyrhizus tuberosus)* and Betel *(Piper betle)* are not attacked. Of vegetables, Cruciferae, Leguminosae and Cucurbitaceae are preferred. Amongst the other *Achatina* species *A. zanzibarica* Bourg. causes damage to sisal and cotton plants, while *A. craveni* Smith attacks sesame seed and coffee in Tanzania.

Freshwater snails (*Bulinus*, *Biomphalaria* and *Oncomelania* spp.), play an important role as vectors of parasitic worms (*Schistosoma* sp., *Fasciola* sp. and others); more is said about this in Chapter 3.6.

In **South Africa** representatives of the Arionidae, the Limacidae and the Milacidae are found (VAN REGTEREN-ALTENA 1966). Here as in other sub-tropical countries (Oases in Saudi Arabia, Israel), the snail *Theba pisana* (Müller) is a dangerous pest in vegetable gardens and also on young leaves and fruit of citrus plants (ANON. 1969; VAN DINTHER 1973; HARPAZ and OSERI 1961; MIENIS 1970). While *Helix aspersa* Müller does occur in Israel, it is not found in such large numbers as *Theba pisana* (Müller).

In the **USSR** three species of the Mediterranean genus *Cernuella* Schlüter are found – in the Crimea, the Caucasus and Trans-Caucasus (SHILEIKO 1967). SHAPIRO (1975, 1976) investigated gastropod infestation in fruit growing, in marked gardens and among cultivated strawberries and gooseberries. He lists the following pests: *Arion circumscriptus* Johnston, *Arion hortensis* Fér., *Arion subfuscus* (Drap.), *Deroceras reticulatum* (Müller), *Deroceras laeve* (Müller), *Deroceras agreste* (L.), *Limax* spp., *Milax gagates* (Drap.) as well as *Cepaea* spp. and *Eulota* spp. *Limax (Vitrinoides) monticola* Boettger attacks tobacco seedlings in hotbeds (AKRAMOVSKII and BABAYAN 1977). In the European part of the USSR, *Deroceras caucasicum* (Simroth) is found on winter wheat and tomatoes (IZZATULLAEV 1977). MEKHRALIEV (1978) discovered that *Acroloxus lacustris* (L.) (Ancylidae) acts as an intermediate host for trematodes. Mention must also be made of the books of LIKHAREV and RAMML'MEIER (1965) and by LIKHAREV and WIKTOR (1980).

QUICK (1953) has given an account of Helicidae which have been introduced into **Australia.** *Helix aspersa* Müller is a pest on wheat fields; the snails attach themselves to the ears of wheat during aestivation and clog up the machines during harvesting. As many as 400 snails have been found per square meter. The same high population density may be encountered on pastures (ANON. 1969); this snail is also a threat to green vegetables. *Helix aspersa* Müller and *Cochlicella acuta* (Müller) have become pests in the citrus plantations along riverbanks of southern Australia (FURNESS 1977).

In the **tropics,** *Helicarion salius* Bens. damages the seedlings of tea plants of Ceylon; *Vaginulus hedleyi* Simroth and *Veronicella leydigi* Simroth are pests on coffee, tobacco, sweet potatoes and vegetables in the West Indies and India; *Vaginulus bleekeri* Keferst is found on young tobacco plants, *Mimosa* and *Pueraria* in Indonesia; *Parmarion pupillaris* (Humbert) (a slug which is 5 cm long and has a small dorsal shell) and another *Parmarion* sp. attack young tobacco, maize, cabbage and *Crotalaria* plants; *Lamellaxis gracilis* (Hutton) and *Subulina octona* (Bruguière) are pests on seedbeds of tobacco in Indonesia and Sumatra; and last but by no means least, *Achatina fulica* Bowdich is a pest on vegetables (cabbage, beans, peas, tobacco and young plants of *Ipomoea* and *Hevea)* (VAN DINTHER 1973).

In **Malaysia,** *Mariaella dursumierei* Grey, *Parmarion martensi* Simroth and *Vaginulus* sp. are considered to be particularly harmful, mostly to Leguminosae, *Pomacea dolioides* Reeve and *Pomacea glauca* (L.) to young germinating rice plants, and finally *Bradybaena similaris* (Fér.), *Subulina octona* (Bruguière) and *Xestina striata* Gray are also found in this region (ANON. 1967).

In Table 15 the pest gastropods of selected crops of countries other than Germany are listed.

2.2.3 Pattern of Damage and Pest Identification

There is no *typical* pattern of damage caused by gastropods. Small leaves are completely consumed, while larger ones may have holes in them or be eaten around the edge. Thicker leaves are mostly rasped on the lower surface, while the upper surface remains undamaged („Fensterfraß"). Rasping damage to thick, fleshy leaves and stalks is possible. Mine-like identations, holes and tunnels are bored into tubers, roots, bulbs and cacti *(Cactus, Opuntia, Cereus)*. Snails damage the bark of young apple trees, poplars and alders, eating down to the fibres of the trunk. The swollen

Table 15. Gastropod pests of particular crops in countries other than Germany

Crops attacked or pattern of damage	Country	Slugs	Snails	Author
Apple tree *Malus*/Rosaceae, (wild and cultivated plants)	India (northwest)		*Pila globosa* (Swainson)	REDDY and KARUMURI (1968) pers. comm. from Indian experts
Artichokes *Cynara scolymus*/ Compositae (rasping on leaf buds)	Italy Sardinia California (coastal region)	*Milax gagates* (Draparnaud) *Deroceras agreste* (L.)		PROTA (1960) LANGE and MACLEOD (1941)
Bananas *Musa*/Musaceae (leaves, stem, fruit, young plants)	USA	*Ariolimax columbianus* (Gould)	*Achatina fulica* Bowdich	MEAD (1961) SMITH (San Francisco, pers. comm.)
Cotton *Gossypium*/ Malvaceae (seedlings)	Tanzania Mauritius		*Achatina zanzibarica* Bourguignat *Achatina fulica* Bowdich	VAN DINTHER (1973) MEAD (1961)
Arrowroot *Canna edulis*/ Cannaceae	Philippines		*Achatina fulica* Bowdich	MEAD (1961)
Bread fruit tree *Artocarpus*/ Moraceae (fruits)	Saipan Pacific islands		*Achatina fulica* Bowdich	MEAD (1961)
Yew tree *Hibiscus*/ Malvaceae	Italy USA (Florida)		*Bradybaena similaris* (Férussac) *Subulina octona* (Bruguière) *Achatina fulica* Bowdich	VAN DINTHER (1973) MEAD (1961)
Strawberries *Fragaria*/ Rosaceae (fruits)	USA	*Deroceras* sp.	*Helix aspersa* Müller *Helix rufescens* Pennant	HOWITT (1960) HOWITT and COLE (1962)
Peanuts *Arachis hypogaea*/ Leguminosae	Sudan (southerly)		*Limicolaria kambeul* Bruguière	VAN DINTHER (1973)
Hemp, East Indian *Crotalaria*/ Leguminosae	Indonesia	*Parmarion* sp. *Parmarion pupillaris* (Humb.)	*Achatina fulica* Bowdich	VAN DINTHER (1973) MEAD (1961)

Table 15 (continued)

Crops attacked or pattern of damage	Country	Slugs	Snails	Author
Medicinal herbs (in general)	France	*Deroceras reticulatum* (Müller) *Arion* spp.		MALLET (1973)
Raspberry bushes *Rubus*/Rosaceae	England		*Cepaea nemoralis* (L.)	
Coffee tree *Coffea*/Rubiaceae	South America West India India	*Veronicella leydigi* Simroth *Vaginulus hedleyi* Simroth	*Strophocheilus oblongus* (Müller)	VAN DINTHER (1973)
	Tanzania		*Achatina fulica* Bowdich	MEAD (1961)
			Achatina craveni Smith	VAN DINTHER (1973)
Cocoa tree *Theobroma cacao*/ Sterculiaceae (seedlings, flowers, young pods, seeds)	Ceylon		*Achatina fulica* Bowdich	MEAD (1961)
Cacti *Cactus*, *Opuntia*, *Cereus*/ Cactaceae (feeding holes)	Bohemia France	*Deroceras reticulatum* (Müller)		FLASAROVÁ and FLASAR (1965) SAINT PIE and BEAUCHARD (1963)
	USA		*Achatina fulica* Bowdich	MEAD (1961)
Rubber tree *Hevea*/ Euphorbiaceae (in nurseries)	Indonesia	*Parmarion pupillaris* (Humbert)		VAN DINTHER (1973)
(young plants, flowers, pods)	Ceylon Malaysia Ceylon	*Mariaella dursumierei* Grey	*Achatina fulica* Bowdich	REHM and ESPIG (1976)
Rattlebox *Crotalaria*/ Leguminosae	Indonesia	*Parmarion* sp.	*Achatina fulica* Bowdich	VAN DINTHER (1973)
		Parmarion pupillaris (Humbert)		MEAD (1961)
Clover *Trifolium*/ Leguminosae	Pacific islands USA	*Deroceras agreste* (L.) *Arion ater* (L.)		HOWITT (1961)
Kopou beans *Pueraria*/ Leguminosae	Indonesia	*Vaginulus bleekeri* Keferst		VAN DINTHER (1973)

Table 15 (continued)

Crops attacked or pattern of damage	Country	Slugs	Snails	Author
Cress *Nasturtium officinale*/ Cruciferae (leaves, sucking roots)	Belgium Holland France		*Lymnaea stagnalis* (L.)	MOENS (1960)
			Galba palustris (Müller)	MOENS (1965)
			Galba truncatula (Müller)	VAN DEN BRUEL and MOENS (1957)
			Radix ovata (Draparnaud)	RICOU and MARREE (1960)
			Radix peregra (Müller)	
			Candidula caperata (Montagu)	
			Potamopyrgus jenkinsi (Smith)	MOENS (1967)
Leguminosae (generally)	Malaysia	*Mariaella dursumierei* Grey *Parmarion* sp. *Vaginulus* sp.	*Achatina fulica* Bowdich *Bradybaena similaris* (Férussac) *Subulina octona* (Bruguière) *Xestina striata* Gray	Anon. (1967)
Lucern *Medicago sativa*/ Leguminosae	France		*Theba pisana* (Müller)	CAIRASCHI and LECOMTE (1973)
Maize *Zea mays*/ Gramineae (young plants, main root)	Sudan (southerly)		*Limicolaria kambeul* Bruguière	VAN DINTHER (1973)
	Indonesia	*Paramarion* sp.		VAN DINTHER (1973)
	USA	*Deroceras reticulatum* (Müller)		BARRY (1969)
(seed in soil)	France (southerly, westerly)	*Arion circumscriptus* Johnston *Arion hortensis* Férussac		MALLET (1973)
Mimosa *Mimosa*/ Leguminosae	Indonesia	*Vaginulus bleekeri* Keferst		VAN DINTHER (1973)
Orchids (in greenhouse, young shoots, flowers, flower stalks, roots) (especially slugs)	England Europe	*Lehmannia marginata* (Müller)	*Discus rotundatus* (Müller) *Oxychilus draparnaudi* (Beck) *Zonitoides arboreus* (Say)	PAG (1957) BLOWERS (1963)

Table 15 (continued)

Crops attacked or pattern of damage	Country	Slugs	Snails	Author
Cymbidium/ Orchidaceae	Belgium	*Arion hortensis* Férussac	*Discus rotundatus* (Müller)	MOENS and FRASELLE (1980)
Cattleya/ Orchidaceae		*Deroceras laeve* (Müller)	*Oxychilus draparnaudi* (Beck)	
		Lehmannia valentiana (Férussac)	*Zonitoides arboreus* (Say)	
	Portugal		Snails	FIGUEIREDO and LEPAGE (1953)
Orchids	Israel North America	*Lehmannia valentiana* (Férussac)	*Zonitoides arboreus* (Say)	MIENIS (1980)
	Hawaii	*Deroceras reticulatum* (Müller)		ALICATA (1950)
Phalaenopsis/ Orchidaceae	Philippines Java		*Achatina fulica* Bowdich	MEAD (1961)
Rape *Brassica napus*/ Cruciferae (young plants)	France Sweden	*Deroceras reticulatum* (Müller)		MALLET and BOURGARAN (1970) BORG (1962)
Rice *Oryza sativa*/ Gramineae (seedlings)	India Burma		*Pila globosa* (Swainson) *Pila polita* Deshayes	VAN DINTHER (1973)
	Africa Sierra Leone Swaziland Nile Valley		*Lanistes ovum* Trosch.	
	Surinam		*Pomacea lineata* (Spix) *Pomacea glauca* (L.) *Pomacea dolioides* Reeve	VAN DINTHER (1973)
	South America Puerto Rico		*Marisa cornuarietis* (L.)	VAN DINTHER (1973)
Lettuce *Lactuca sativa*/ Compositae	Hungary	*Milax budapestensis* (Hazay)	*Ceciliodes petitiana* Benoit	PALFI (1973)
Blackcurrant bush *Ribes*/ Saxifragaceae (leaves, fruit)	England		*Trichia striolata* (C. Pfeiffer)	STRINGER (1969)
			Helix aspersa Müller *Candidula caperata* (Montagu)	STRINGER and MORGAN (1969)

Table 15 (continued)

Crops attacked or pattern of damage	Country	Slugs	Snails	Author
			Cepaea hortensis (Müller) *Cepaea nemoralis* (L.)	
Sesame *Sesamum indicum*/ Pedaliaceae	Tansania		*Achatina craveni* Smith	VAN DINTHER (1973)
Sisal *Agave sisalana*/ Agavaceae	Tansania		*Achatina zansibarica* Bourguignat	VAN DINTHER (1973)
Sunflowers *Helianthus annuus*/ Compositae (seedlings)	France	*Milax budapestensis* (Hazay)		AGUILAR and PACQUÉTIAN (1963)
Asparagus *Asparagus officinalis*/ Liliacea	France		*Cecilioides acicula* (Müller)	PUSSARD (1951)
Bird's nest fern *Asplenium nidus*/ Polypodiaceae	Saipan		*Achatina fulica* Bowdich	MEAD (1961)
Sweet potatoes *Ipomoea batatas*/ Convolvulaceae	West Indies India	*Vaginulus hedleyi* Simroth *Veronicella leydigi* Simroth		VAN DINTHER (1973)
Tobacco *Nicotiana tabacum*/ Solanaceae (leaves, young plants)	Sumatra		*Opeas gracilis* Hutton *Subulina octona* (Bruguière)	VAN DINTHER (1973)
	West Indies India	*Vaginulus hedleyi* Simroth *Veronicella leydigi* Simroth		VAN DINTHER (1973)
	Indonesia	*Vaginulus bleekeri* Keferst *Parmarion pupillaris* (Humbert)		VAN DINTHER (1973)
	France	*Deroceras reticulatum* (Müller) *Arion hortensis* Férussac		MALLET (1973)

Table 15 (continued)

Crops attacked or pattern of damage	Country	Slugs	Snails	Author
		Arion circumscriptus Johnston		
	Tropics and sub-tropics of America	*Veronicella laevis* Blainville		
	Oriental Africa Pacific islands Indonesia	*Laevicaulis alte* (Férussac)		
	Pacific islands	*Semperula idae* Semper		
Tea plants *Thea sinensis*/ Theaceae (plants in seedbed)	India	*Helicarion salius* Benson		VAN DINTHER (1973)
Victoria regia and plants in Victoria-regia houses	Czechoslovakia		*Melanoides tuberculata* (Müller) *Helisoma trivolvis* (Say)	FLASAROVÁ and FLASAR (1962, 1965)
Grape vines *Vitis vinifera*/ Vitaceae (sprouts, young plants)	Mediterranean countries South Europe Sub-tropics		*Bradybaena fruticum* (Müller) *Helix aspersa* Müller *Helix pomatia* L.	VAN DINTHER (1973)
Citrus trees *Citrus*/Rutaceae (leaves, fruits)	USA (California)		*Helix aspersa* Müller *Drymaeus* dormani (Binney)	MUMA (1955) VAN DINTHER (1973) ARNOLD (1962) NORRIS (1952)
	Israel Libya Mediterranean countries South Africa Oases in Saudiarabia		*Theba pisana* (Müller)	MIENIS (1970) SHARAF and GHALEB (1978) VAN DINTHER (1973)
Sugar cane *Saccharum officinarum*/ Gramineae	Tropics	*Veronicella leydigi* Simroth		VAN DINTHER (1973)
Sugar beet *Beta vulgaris*/ Chenopodiaceae	Ireland	*Geomalacus maculasus* Allman		

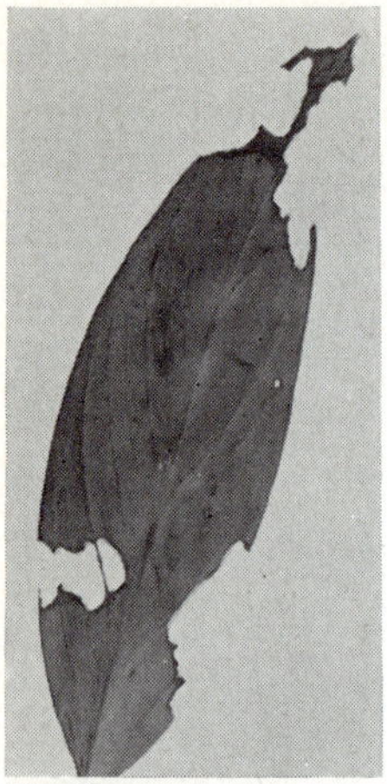

Fig. 62

Fig. 63

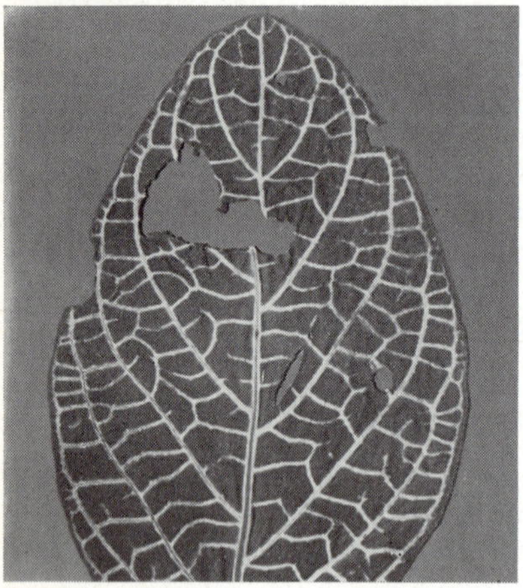

Fig. 64

Fig. 65

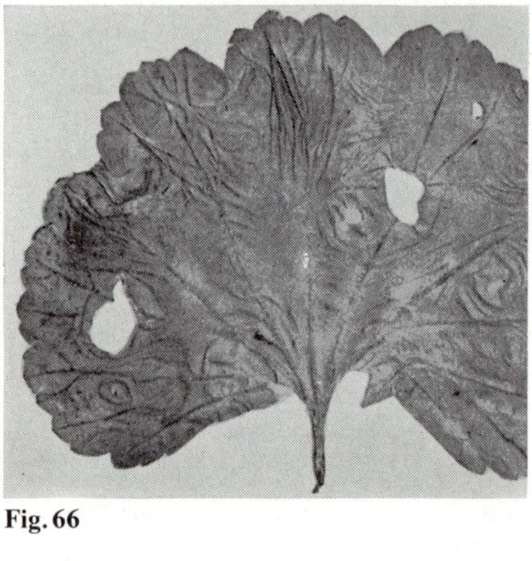

Fig. 66

Fig. 62. Feeding damage by *Deroceras reticulatum* to *Spathiphyllum harveyanum*

Fig. 63. Feeding damage by *Cepaea nemoralis* to *Coleus verschaffeltii*

Fig. 64. Feeding damage by *Deroceras reticulatum* to *Fittonia verschaffeltii*

Fig. 65. Feeding damage by *Lehmannia marginata* to *Tradescantia viridis*

Fig. 66. Feeding damage by *Deroceras reticulatum* to Zonal Pelargonium

Fig. 67. Slug damage to leaves and flowers of *Gerbera jamesonii*

Fig. 68. Lemons from the grove of Nipomo Farming Co., California, showing destruction by *Helix aspersa*

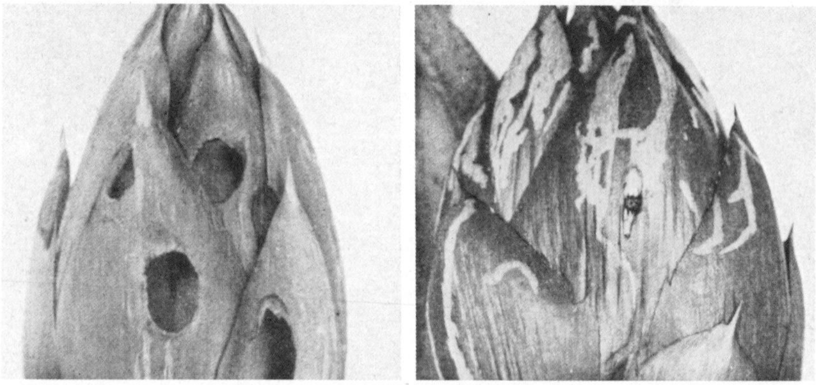

Fig. 69. Artichokes showing feeding damage by *Milax gagates*

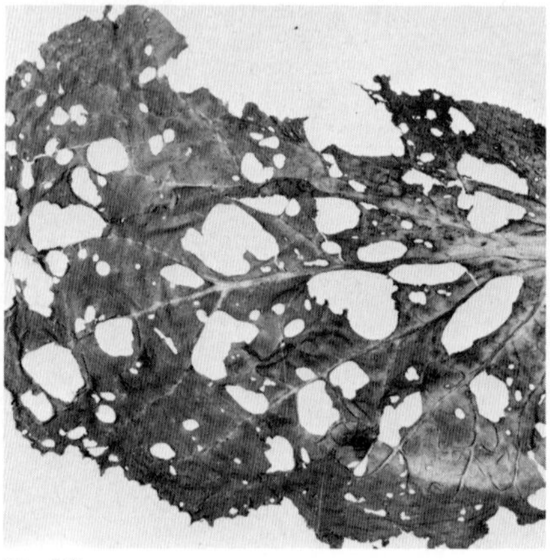

Fig. 70

Fig. 71

Fig. 70. Feeding damage by slugs to chinese cabbage (*Brassica pekinensis*)

Fig. 71. Feeding damage by Lymnaea to the root of cress

Fig. 72. Feeding damage by slugs to young maize plants

Fig. 73

Fig. 74

Fig. 73. Scarred areas of damage by *Deroceras reticulatum* to ripening wood of grapevine

Fig. 74. Feeding damage by *Deroceras reticulatum* to the bark of green vine shoots

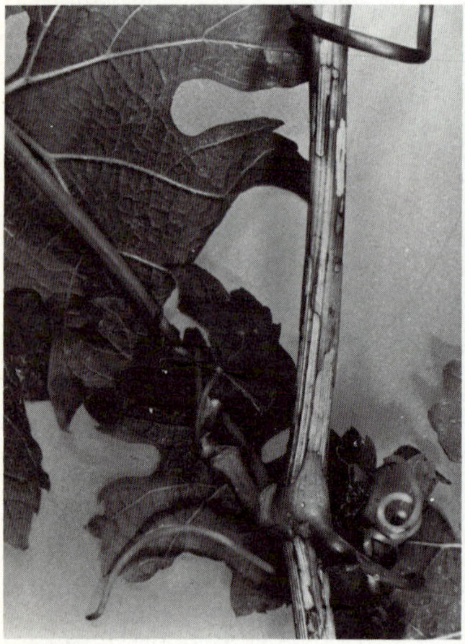

Fig. 75

Fig. 76

Fig. 75. Feeding damage by *Deroceras agreste* to 1-year-old Müller-Thurgau shoots

Fig. 76. Rasping damage by *Deroceras agreste* to Müller-Thurgau leaves

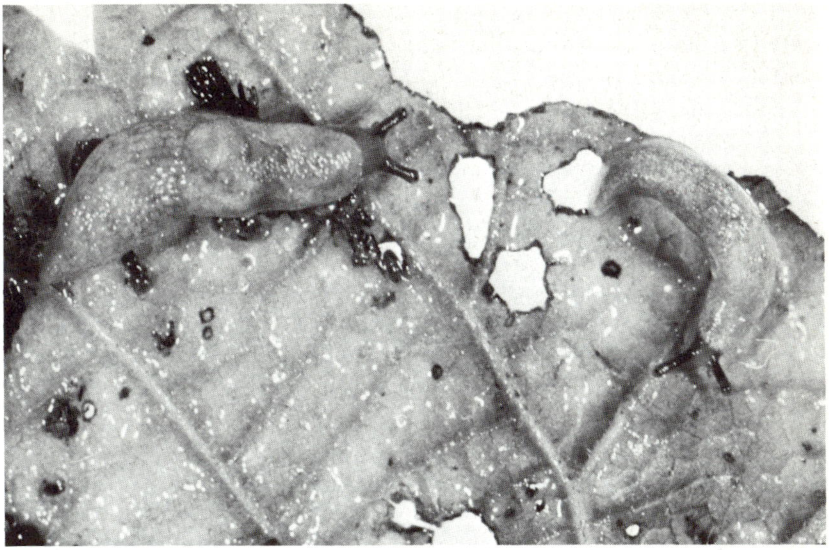

Fig. 77. Grey Field Slug on vine leaf

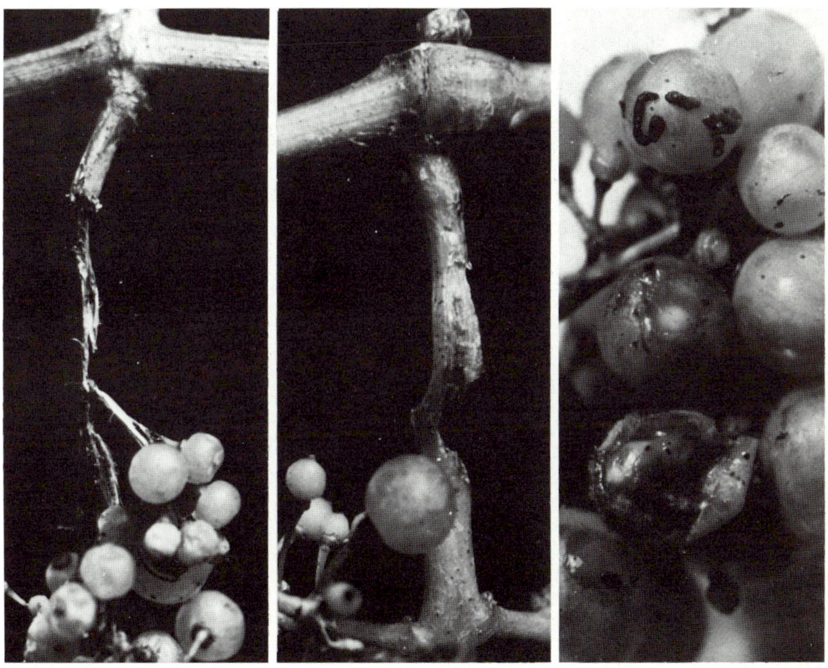

Fig. 78. Feeding damage by slugs to vine stalks and grapes

seeds of winter wheat lying in the earth are hollowed out, starting from the embryo, and either the embryo, or at a later stage of germination, the tip of the shoot is destroyed. In many cases, such as rice, the young plants are eaten, in others, like cress, the roots are destroyed.

The **patterns of damage** are thus diverse. Some impression of this is provided by the photographs of agricultural and horticultural plants shown in Figs. 62–78 and Plate VII.

An **indication** of **infestation** by gastropods is provided by the shiny silver slime trails left behind by slugs and snails as they crawl about. These trails may, however, disappear after a short time if the plants and soil are moist. Finally, there remain the small, irregular, sticky faeces at the sites of damage. Identification of the pest involved is made more difficult by the fact that the snails and slugs are usually not to be found at the site of damage when this is inspected, because they hide away during the daytime.

Since slugs and snails have no characteristic pattern of damage, that caused by other pests may mistakenly be attributed to them. The pattern of damage to winter wheat often resembles winter or seedling damage when the little leaves are eaten down to the soil surface (LUTHER 1955). The following data are based on field observations of the German Plant Protection Office over a period of 10 years (1962–1971). In spring, the pattern of gastropod damage resembles winter damage to winter grain and seedling damage in the case of vegetables; in April to May it resembles fungal damage in beans and young tobacco plants in the field as well as other crops in hotbeds. Pest identification in May is made difficult by the feedling of wireworms on young tobacco and sugar beet plants, by Pygmy Mangold Beetles, cockchafer grubs, isopods, millipedes and spring tails on vegetables. Cutworms, wireworms and *Deroceras reticulatum* (Müller) may all feed at the same time on mustard plants. Similar damage is caused to flowers by isopods and millipedes. Sugar beet and turnips are attacked by larvae of cockchafers, by millipedes, isopods and spring tails. In August lettuces may be attacked simultaneously by gastropods and caterpillars. Vegetables are attacked in June and July by cutworms, wireworms and larvae of cockchafer grubs. After heavy rain, tomatoes which have been damaged by slugs and snails become infected with Brown Rot. Growth in young plants can be disturbed both by aphid infestation and by gastropods. In November confusion may arise between the damage caused to the sprouts of winter rape by gastropods with that caused by Field Mice. In this month, too, damage to ornamental plants by both gastropods and isopods is to be feared. Damage following late sowing of winter cereals, turnips and lettuce may be due to Field Slugs, cutworms, wireworms, millipedes and isopods.

2.3 Prognosis, Critical Number

Prognoses demand knowledge about the prerequisites for large scale population increase of a pest. Methods of prognosis and knowledge of the critical number are of particular importance in the case of Field Slugs, the major pests of newly sown winter cereals, of oil plants and young vegetable plants. BUHL and SCHÜTTE (1971)

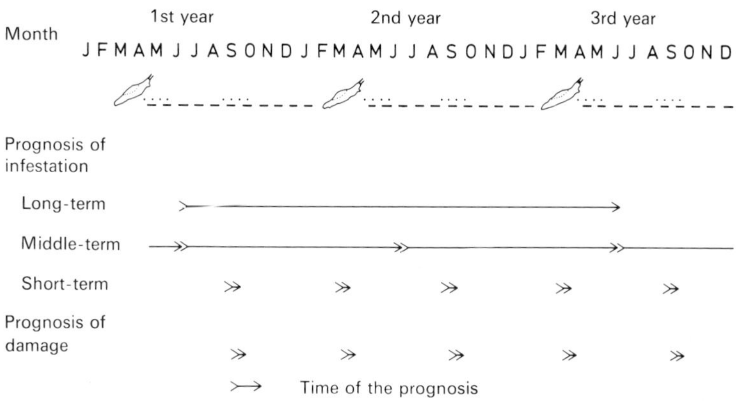

Fig. 79. Scheme for the prognosis of damage by Field Slugs

give data for *Deroceras reticulatum* (Müller), *Deroceras agreste* (L.) and *Arion hortensis* Fér. (Fig. 79). For long- or middle-term prognosis, in which at least one generation is involved, the gastropod population infesting the preceeding crop (rape, clover, lucern, *Dactylis* and others) is estimated if these crops have previously been grown for at least two successive years on the same field. In the case of short term prognosis, the use of traps (damp cloths, pieces of sacking, planks, tiles for example) is recommended. Even collection of a single specimen per trap over 1 to 5 nights is indicative of a need for constant vigilance.

The "danger level" for any gastropod species is taken as the number of healthy specimens capable of effecting a 10% destruction of seed or seedlings.

In the field of medicine, prognoses of the expected incidence of bilharzia are based on determination of the population density of the intermediate hosts of the parasite (*Schistosoma* spp.). The rate of infection of man and animals with *Schistosoma* larvae is directly proportional to the number of freshwater snails in the area concerned. Further, the aquatic plant, *Salvinia auriculata*, creates optimal environmental conditions for *Bulinus africanus* (Krauss) and *Biomphalaria pfeifferi* (Krauss) and thus promotes a high population density. *Galba truncatula* (Müller) is the most important intermediate host of Liver Flukes and its population fluctuations have been studied by HEPPLESTON (1972). In making a prognosis about mass increases in its population, it may be of importance that at temperatures below 10 °C, egg-laying ceases and there is no increase in numbers.

2.3.1 Problems Encountered in Prognosis

To make a prognosis, it is essential to have *precise* knowledge, not only of population density, but also of causal factors affecting it. The latter are difficult to determine because they are different for different gastropod species. Besides the mortality rate, the development periods of eggs and juveniles, and the longevity of adult and senile slugs and snails must also be taken into account when establishing for any particular species the life span and number of generations per year, which can moreover vary for one and the same species. Thus the Field Slug, *Deroceras reticulatum* (Müller), may have either one or two generations per year. When there are two, the scheme for prognosis shown in Figure 79 has to be adjusted with respect to "1., 2., 3. year". In other cases, for example *Cepaea nemoralis* (L.), development of a generation can be delayed by one or two years.

It is relatively easy to make prognoses about population explosions in slugs. Moderate summer temperatures with a constantly high humidity, mild winters, dense covering of weeds and grass and coarse soil all provide optimal environmental conditions, the effects of which are augmented by the increasing tendency to scanty tilling (Chap. 2.1). After dry spring and summer months, no population explosion is to be anticipated. The number of juveniles which hatch in late summer is significant for later damage. For example, they eat the potato tubers in the ground in September and October. If the summer has been dry and this dryness continues throughout August, the reproductive activity of gastropods is reduced to a minimum, so that no damage takes place in September. If, however, there is heavy rain in August, a large number of young gastropods develop, and cause considerable damage to the tubers in September.

Snails can, however, survive a period of dryness by going into aestivation and are even able to endure several months without food (Chap. 1.3).

The prospects for survival of eggs, young and older animals are all critical for a **population explosion.** CAIN and CURRY (1968) estimate the average survival rate of animals of one generation of *Cepaea nemoralis* (L.) to be 55%; later they put it at 70% under normal weather conditions; however, WOLDA'S (1972) estimate was 50%. This illustrates the difficulty of obtaining data on which a prognosis can be based on only one particular aspect of even a single species.

In general, in the case of Field Slugs, two periods of maximal damage can be anticipated. These correspond with rain fall in particular months of the year and usually occur in the period April to June and again September to October (GODAN 1973). These periods of maximal damage are preceded in each case by a peak period of infestation, the cause of which is attainment of adult size by young slugs, which thus swell the numbers of the existing adult population. Subterranean slugs like *Arion hortensis* Fér. and *Milax budapestensis* (Hazay) show only one peak in population density each year, in July/August/September in the case of *Arion hortensis* Fér. and in June/July for *Milax budapestensis* (Hazay) (RUNHAM and HUNTER 1970). There are two annual peak periods of population density for *Cepaea nemoralis* (L.), from May to the beginning of June and again September to the beginning of October, while the peak periods for *Theba pisana* (Müller) are in July and November (CARAISCHI and LECOMTE 1973; WOLDA and KREULEN 1973).

There is indeed a relationship – at least in the case of *Cepaea nemoralis* (L.) – between the number of eggs in a cluster and the mortality rate of the eggs. The fewer the eggs, the higher the mortality rate and, in the case of very small egg clusters with less than 13 eggs, no young snails hatch. If the number of eggs in the batch is over 30, there is no longer any relationship between mortality rate and egg number. Experimental observations revealed that on average there were 103 juveniles to each adult. There are, however, differences in this respect from one population to another and from one year to another; thus for eight different populations studied over two years, the number of juveniles produced varied from 25 to 150 per adult snail. The relative number of juveniles to adults decreases with in-

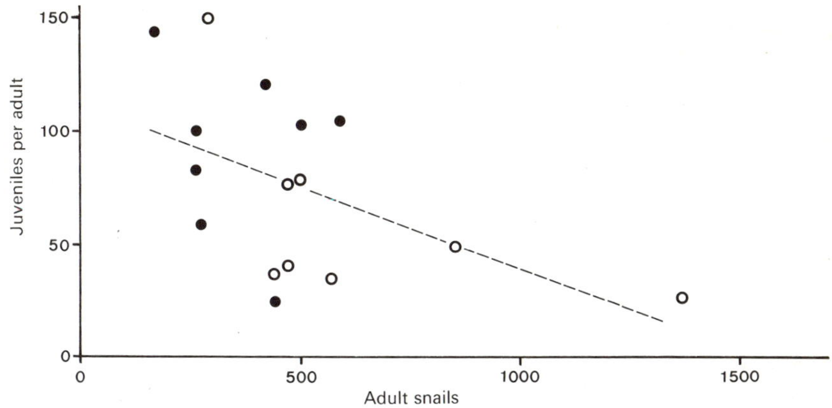

Fig. 80. The number of juveniles produced per adult as a function of adult density, *Cepaea nemoralis*. ● = 1967; ○ = 1968

crease in population density (Wolda and Kreulen 1973) (Fig. 80). Herein lies another problem for prognosis, for it can happen that a snail plague is predicted, in consequence of which control measures are instigated and costs incurred, whereas the population would later have declined of its own accord (Chap. 2.1.2).

Wolda and Kreulen (1973) compared the process of expansion of a healthy population of *Cepaea nemoralis* (L.) with one with a tendency to decline. Egg production in the increasing population was twice that of the declining population and the survival rate of the eggs 25% greater, with the result that the juveniles represented 46% of the expanding population; that is 20 times their relative number (2.3%) in the declining population. The key factor in population dynamics of *Cepaea nemoralis* (L.) is thus not the mortality rate of the adults but the survival rate of the young snails in the first year after hatching (Wolda and Kreulen 1973). Bayne (1968) regarded the survival rate of the eggs as more important. Naturally dryness, predators and parasites are important factors limiting population size (Chap. 3.6).

To judge the likelihood of a future population explosion of a gastropod pest, a knowledge of the **number** of **eggs** laid is desirable. Frömming (1954) regards the following as critical numbers for slugs: *Limax maximus* L. – 676 to 834; *Lehmannia marginata* (Müller) – 105 to 132; *Deroceras reticulatum* (Müller) – 300 to 342; *Arion hortensis* Fér. – 150 to 203. It must, however, be borne in mind that by no means all the eggs hatch successfully as the example of *Cepaea nemoralis* (L.) has shown. Dryness or stagnant water may kill them before they hatch. Field Slugs lay the greatest number of eggs at temperatures between 10 °C and 20 °C when the mortality of adult slugs is 0%, while their mortality rate increases to 100% after 1 h at 35 °C and 80% after 4 h at 30 °C. The slugs survive at 25 °C, but lay no eggs at this temperature (Carrick 1942). A brief period of cold – for example, 5.8 °C during the day and 2.0 °C at night – in no way inhibits egg-laying. In winter, at 5 °C, embryonic development was observed to take 105 days, whereas at 20 °C it took only 18 days. During development, the embryo becomes increasingly cold-resistant, so that shortly before hatching, it is able to survive a temperature of 0 °C for several days or even −2 °C for 2 h. Thus in frosty conditions in winter, many eggs are found with fully developed embryos; if these are brought under cover at 18 °C to 20 °C, they hatch within several hours or days. Temperatures around 0 °C inhibit hatching, but it is triggered by a change from cold to warm. Large quantities of organic substances and dung in the soil are favourable for the survival of eggs and young animals. Eggs may develop in such soil, even when the ground is frozen.

Unlike Field Slugs, *Cepaea nemoralis* (L.) is very sensitive to low temperatures (WOLDA and KREULEN 1973). An entire experimental sample of 594 eggs laid in October and maintained at 10 °C died; at 13 °C 61% survived and hatched into healthy young snails; at 16 °C this rose to 69%. Some eggs do, admittedly, survive the winter, but the number of juveniles which hatch in the following spring is relatively small, and in two experiments only 12.6% and 13.6% of young snails reached adulthood.

To make a prognosis it is necessary also to establish the number of juveniles (and thus indirectly the mortality rate of the eggs), the growth rate and mortality rate of the adults and the ratio of juveniles to adults. According to SCHNETTER (1950), only about half the adult population of *Cepaea nemoralis* (L.) is replaced by juveniles each year, while SPIEKERMANN (1976) found that the percentage of the total population which represented young snails was 40.6% for *Helix pomatia* L., 51.5% for *Oxychilus cellarius* (Müller), 55.5% for *Perforatella incarnata* (Müller) and 68.9% for *Lacinaria biplicata* (Montagu).

A distinction can be made between juvenile and adult slugs according to their **phase** in the **life cycle** (BETT 1960). Thus the largest number of juvenile *Arion hortensis* Fér. are found from May to August, mature, unmated slugs in September, and impregnated slugs with eggs from February to April the following year. The eggs usually hatch in January and February; growth occurs in spring and summer and sexual maturity is attained in September; the life span is only one year. Young *Arion subfuscus* (Drap.) emerge in October, grow slowly during the winter, rapidly in April; adult slugs are found only during the summer, egg-laying occurs in June and July, and the adults die at the end of the summer when the next generation hatches; the life span is one year. Juveniles are found in late summer, winter, spring and early summer. The number of young of *Deroceras reticulatum* (Müller) rises considerably in May and June, but they are to be found throughout the year; they reach sexual maturity in June, July or at the latest in August. The colder the autumn, the longer the period of embryonic development and the higher the mortality rate of the eggs. It has now been established that there is a spring and an autumn generation of Field Slugs (Chap. 2.1.2). The life cycle lasts about one year. Young *Milax budapestensis* (Hazay) are few in number from January to June, but numerous from August to November; they become adult and reach sexual maturity between September and December, but do not yet mate. Only in the following year, from January to early May, does the number of mated snails reach a peak; the adults die after oviposition during July. *Milax sowerbyi* (Fér.) is very active in August and September, copulates in September. The young hatch in early spring, grow to adult size in the summer, become sexually mature and ready to mate in September to November the following year; they lay their eggs in late autumn and die soon afterwards. Information is also available for other Arionidae: *Arion intermedius* Normand – egg-laying in July, August; hatching in September, October; sexual maturity reached in spring the following year; life span about 12 months;

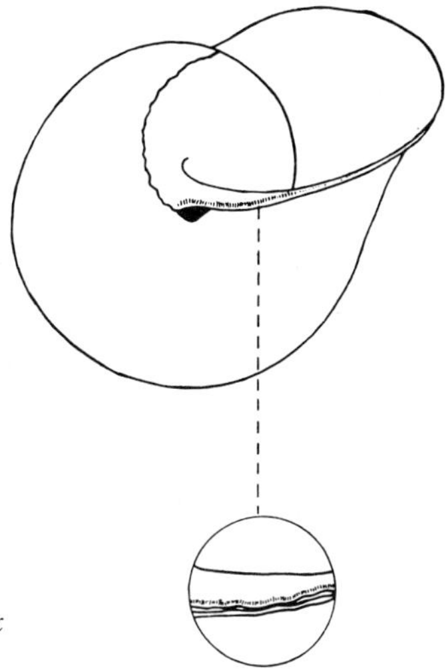

Fig. 81. Growth lines on the edge of the shell mouth of *Helix pomatia*

Arion rufus (L.) – egg-laying in August, September; juvenile in October, November; sexually mature in the summer of the following year; life span 12 to 14 months; *Arion circumscriptus* Johnston – egg-laying in May, June, July; juvenile in July, August, September; adult in the spring of the following year; life span about 12 months.

It is relatively easy to **determine** the **age** of pest snails from external features. There is usually a sharp edge to the mouth of the shell in young snails. After cessation of growth, which continues in some species after the attainment of sexual maturity, a thickened lip is formed; this indicates that the snail is adult. POLLARD (1973) found that in *Helix pomatia* L. there are **growth lines** in this thickening (Fig. 81); from these the age of the snail can be judged, so that a small population can be subdivided into age groups. Snails with many growth lines, that is older snails, are more mobile than are the younger adults which have few lines. Senile animals which have reached the end of their reproductive phase are sluggish; in their case the shell shows no increase in the number of growth lines, but distinct signs of wear and tear. It is desirable that other species which have a thickened shell edge should be examined for the presence of such growth lines.

Differences in the normal life span between different species also makes general prognosis difficult, since damage is often caused by the action of several species at once.

The **"gradation type"** of a gastropod pest is also important for prognosis. This is difficult to determine in the case of slugs, because in an area of damage there are usually several species, which have different life cycles, habits and feeding patterns; even the places in which they hide are different. Thus *Arion circumscriptus* Johnston prefers potato fields where the soil surface is not covered with dense vegetation, and lives on the young tubers and roots of the potato plants in the open field. The Field Slug lives on the ground or in crevices and under earth clods and prefers areas where there is good ground cover. Thus in any one area, the size of the population, particularly of the smaller species, is usually underestimated because of their varied and secluded ways of life.

The following points are important and must be taken into account when making **prognoses,** particularly where **Field Slugs** are concerned:

1. There are, as a rule, two peaks of population density each year, possibly corresponding to two generations, and coinciding with periods of maximal damage. They occur in April to May and in September to October.
2. Factors which favour a population explosion are mild winters, evenly distributed rainfall in the spring, summer and autumn and an absence of extremes of temperature; optimally, this should lie between 15 °C and 21 °C.
3. In damp summer months, heavy infestation with slugs can be expected from the middle to the end of August.
4. A reduction in population density of slugs follows dry conditions in June and July or September and October, or sudden cold periods in summer, or light frost in autumn (September, October).
5. Damage is greater if crop growth is slow as a result of cooler weather in spring.
6. Damage is increased by frequent irrigation of the fields.
7. Damage is also greater if tilling is scanty and also if seeds are spaced out to avoid later thinning, so that damage is concentrated on a smaller number of plants.
8. Dense weed cover furthers the development of slugs. Nevertheless, even complete removal of weeds by no means completely rids an area slugs.
9. Heavy, coarse soil increases the likelihood of damage.
10. The location of the field is important, since adjacent meadows, pastures, weed-covered slopes or overgrown, fallow land considerably promote increases in population.
11. Limy soil increases the fecundity of both slugs and snails; population density is greatest with a soil pH of 6.3 to 6.7.
12. Large quantities of organic material, such as vegetation which is ploughed in, remains of straw or cattle dung, favour survival and reproduction of slugs in particular.
13. Over-fertilization with nitrogen increases fecundity of gastropods and thus the damage they cause.
14. Important for egg-laying is the water-holding capacity of the soil; the largest egg masses are found in soils with 75% saturation.
15. The preceding crop is of importance: damage to winter wheat is usually heavier after peas, beans, clover, vetch and rape.
16. In the case of snails, a large number of empty shells on the ground is an indication that a population explosion has already occurred and has come to an end; it is unimportant whether predation or some control mechanism in the population itself has brought this about.

2.3.2 Critical Number

The **critical number** is the number of slugs or snails which a plant can carry, or the maximal intensity of infestation at which no economic losses are incurred. Such numbers, on which prognoses could be based, have not yet been published for gastropods. Values of the critical number which might apply to healthy crops and pests (FRANZ 1970), vary according to the "degree" of healthiness, the capacity for resistance and for regeneration of the plants as well as the mobility and fecundity of the gastropods.

In the literature few results are to be found which correlate the intensity of infestation of an area and a high level of damage. CARRICK (1942) reported significant damage to potato fields with an infestation of *Deroceras reticulatum* (Müller), *Arion circumscriptus* Johnston, *Arion hortensis* Fér. and *Arion subfuscus* (Drap.) of 100,000/ha. According to MOKRZECKI (1931), total destruction of a 2,400 ha wheat field occurred with a density of 7 to 10 Field Slugs per m^2; the young shoots were destroyed and bare areas of up to hectare resulted. In other cases, counts of 30,000, 500,000 or 600,000 Field Slugs per hectare have been reorded. Heavy damage to winter rape was incurred with an infestation of 200/m^2 (MALLET and BOUGARAN 1970). STEPHENSON (1968) recorded 15,000 slugs per hectare in the centre, but 700,000 on the periphery of cultivated fields, 26,000 to 275,000 per hectare on strawberry fields in England and 13,370/ha in the case of sweet potatoes in Trinidad. STRINGER (1967) recorded heavy damage to black currents with infestations of 2,800/ha of *Helix aspersa* Müller, 200/ha of *Cepaea nemoralis* (L.), 900/ha of *Cepaea hortensis* (Müller), 5,000/ha of *Helicella caperata* (Montagu) and 800/ha of *Hygromia strigella* (Pfeiffer).

Preliminary investigations have been made by GODAN (unpublished) of the critical number of wheat and rye seedlings. Seeds were sown by conventional methods in damp soil (85% R.H.) and conditions were optimal for both seeds and slugs (18 °C to 20 °C). An observation period of 48 h following slug infestation showed the following: *total* destruction of the embryo in 55% of the experiments with *Deroceras reticulatum* (Müller), where the embryo was still in the soft, swollen seed or had just broken through the seed coat; but no destruction of shoots which were more than 2 to 3 mm long. The root was destroyed in 23% of the seeds and root hairs in 17.3%. In older plants the embryo itself remained free of attack, but the soft contents of the seed were eaten (12.4% of the experiments); the point of penetration of the slug was usually near the base of the intact wheat germ. In many cases the shoot nevertheless died as a result of lack of nutrition, since the nutritive reserves of the seed were lacking and the root not yet sufficiently developed. Values for *Lehmannia marginata* (Müller) were 50% destruction of the embryo up to a length of 0.5 mm, 32% destruction of the radicle. 50% destruction of root hairs and 0.7% destruction of the seed contents when the shoot was 1 to 5 mm long. *Limax maximus* L. and *Limax flavus* L. destroyed the embryo at the stage of breaking of the seed coat in 30% of the seeds respectively, while if the wheat germs were 4 to 5 mm long, only the radicle was eaten.

In field experiments using caged snails and the wheat variety "Carstens" (2 animals per seed, maximum temperature 14 °C, minimum 7 °C, 72% to 97% R.H.), the following incidence of attack was obtained: on wheat germ less than 0.5 mm with *Deroceras reticulatum* (Müller) (100 animals) 23.5%, *Lehmannia marginata* (Müller) (100 animals) 12.8% and *Limax maximus* L. (50 animals, and on account of their size, only one slug per seed) 0.5%. With 475 older seedlings (40 to 140 mm long), the relative number attacked by *Deroceras reticulatum* (Müller) was 26%, *Lehmannia marginata* (Müller) 21.5%, *Limax maximus* L. 0% and *Arion rufus* (L.) 20%.

Deroceras reticulatum (Müller) represents the greatest threat to the embryo of wheat and rye seeds when the seed lies swollen in the earth; seeds which are still hard are avoided. The seed embryo is mostly eaten after the stage of breaking through the seed coat until it reaches a length of about 0.5 mm.

The newly emerged **radicle** is always endangered. It was usually totally destroyed so that a quarter of the embryos died. This radicle has, however, very great regenerative capacity, as is shown by the fact that destruction by slugs can result in the production of many new side rootlets. Even when there were 5 or 6 *Lehmannia* per seed, the embryo rootlet regenerated, and such seedlings had, four weeks later, a green leaf 4.5 to 6 cm long. Admittedly these young plants grew more slowly than those which had not been attacked in this way, but only in a few cases were the plants deformed so that the leaf tips remained in the leaf sheath. Feeding at the leaf base usually took the form of rasping off of the epidermis, and this had no effect on the plant.

Loss of a quarter or even less wheat or rye seed through the action of slugs, mainly *Deroceras reticulatum* (Müller) represents an unacceptable economic loss, even when it can be expected that the plants will regenerate and the gastropod pests reduced in number as a result of dry weather and other unfavourable environmental conditions. Dry weather is, however, not good for the crop and reduces both growth and hardiness of the plants. The experimental results presented here must, however, be checked by thorough investigations and observations, to obtain reliable data on the critical number, which can then be used in making prognoses. It is, however, clear that the embryo of a plant is in a particularly critical stage just at the time at which it breaks through the seed coat.

There is, as yet, no method for establishing the **threshold limit** of damage caused by gastropods. Random samples of soil should be examined, as is done for insect pests in forestry (Franz 1970), to establish the density of egg batches, juveniles, adults (particularly of subterranean slugs), the number of small species and of *Helicigona arbustorum* (L.). The upper surface of the soil should also be examined for larger species of slugs and for snails, while low bushes, trees and shrubs should also be searched for other snail species, since in dry summers, in June, July and August, many snails [*Cepaea* spp. and *Theba pisana* (Müller)] are attached during aestivation to trunks, branches and leaves three or four meters or more above the ground. As soon as the weather becomes damp, these snails become active and return to the ground. The number of empty shells found on the ground may be an indication that a population increase has come to a halt; the reason for this, whether unfavourable circumstances or predators, is of no importance; in this case no warning is necessary. Since the nests of land pulmonates are usually in hollows in the upper 5 cm of soil, this should be examined to a depth of 10 cm and over an area of at least one square meter. The areas selected for investigation should be covered with weeds and grass, the soil should be loose and rich in humus and contain 50% to 90% water relative humidity. The species and age of both young and old animals should be listed separately. If no gastropods are to be found, from observation of both the slime trails they leave behind as they crawl around, and of visible damage to vegetation some information about the gastropod population can be obtained. This method is, however, not reliable since the trails remain visible only for a significant length of time on relatively dry ground and on leaves.

2.3.3 Methods

2.3.3.1 Estimating Population Density

An important prerequisite for making forecasts about possible future population explosions to obtain an estimate of the number of pest gastropods present in an area. Determination of the population density both before and after control measures have been taken is also important in order to check the effectiveness of such measures. As has already been stated, there are many biotic and abiotic factors which influence population density. It is difficult to estimate the exact size of a gastropod population, since slugs in particular hide away during the day. Subterranean slugs like *Milax budapestensis* (Hazay) live mostly in the earth, and even *Deroceras reticulatum* (Müller) may remain largely in the spaces in the soil and under or between clods of earth. Terrestrial gastropods are usually active after dusk and at night, but even then not all individuals of a population are active (Chap. 2.1.1). Collecting must therefore be done at night, or at least in the evening and at dawn, and preferably in moist weather conditions. Collection is made difficult by the fact that not all slugs crawl about at the same time or to the same extent, since there are species specific differences in times of maximal activity; these have already been discussed.

Furthermore, the distribution even of the members of one species is not uniform. The periphery of fields tends to be more heavily infested than the middle, but here earth clods can be of advantage for the gastropods. Even on a uniformly moist lawn, slugs and snails are not evenly distributed. According to SOUTH (1965), the Field Slug is commonly found in the vicinity of orchard grass, *Dactylis glomerata*. This grows in early spring and affords protection for any eggs laid at its base should the weather in the following months be dry. At these places there is also a larger number of slugs than elsewhere on a lawn.

It should also be borne in mind that a field which is regarded as infested does not necessarily harbour gastropods by day or even by night, since these may live in adjacent grassland, overgrown slopes, ditches and weedy fallow land and move onto the neighbouring field only at night. Fields with very leafy crops can also serve as a reservoir of gastropod pests which hide there and from where they can move onto other fields. The survey should therefore include neighbouring fields, irrespective of whether they are under cultivation or not.

Many problems arise, and these can only be solved when an estimate of the true density of a population has been made. Examples from the literature make clear just how difficult this is. Estimates of the population density of slugs around clumps of *Dactylis* range from 67.8/m^2 to 115.2/m^2 for *Deroceras reticulatum* (Müller), from 11.8/m^2 to 102.3/m^2 for *Arion hortensis* Fér and from 67.8/m^2 to 59.2/m^2 for *Milax budapestensis* (Hazay) (SOUTH 1965). CRAWFORD-SIDEBOTHAM (1970) regard 80 to 161 slugs per square meter in a natural environment as normal. The main difficulty in making these estimates lies without any doubt in determining the activity patterns which differ in the different gastropod species and depend on many factors including climate, species and age as has been shown in the preceeding chapters.

The methods used at first for **capturing** slugs and snails were traps or bait (DUTHOIT 1961; HUNTER, W.R. 1961). These are usually effective only in damp weather, when the gastropods are sufficiently mobile. The use of traps allows an estimate to be made only of those gastropods which happen to be active at the time (GOULD 1962). This does not give an indication of the true population density, since many gastropods, particularly slugs, live under the ground or may be resting on it. VAN DEN BRUEL and MOENS (1958) have shown that traps are totally unsuitable for measuring the population density of subterranean slugs. Furthermore, both these methods are very time-consuming when large areas are involved. It is really only in gardens that a gastropod population can be estimated by such a method as putting out damp cloths or wooden platters with fresh potato or carrot slices as bait; this must, however, be done in the evening or late afternoon. Collection of animals is made the following morning.

THOMAS (1947) reports on a **sampling method** for determining the gastropod infestation of potato fields, and this could still be used today. In this procedure, 10 plants in a field – or 20 if the field is very large – are dug up and examined for slug damage and soil is also sieved and examined for slugs. This is done first when the young tubers have just formed – in this particular case on 7th September – and sampling is then repeated at fortnightly intervals thereafter. The proportion of potato tubers which have been affected by slugs give an indication of the degree of infestation, which often increases as the autumn progresses.

Many different methods for estimating gastropod populations have been published, indicating the importance of finding a solution to this problem.

Good results are obtained with the **flotation method** if it is done with thoroughness, yielding catches of 94% in the case of *Arion fasciatus* (Nilsson) and 81% with *Arion intermedius* Normand. HUNTER (1968) considers washing through of the soil to be suitable for determining the actual population density related to species and age of *Deroceras reticulatum* (Müller), *Arion hortensis* Fér. and *Milax budapestensis* (Hazay). This method is 100% sure, except for young, newly hatched slugs, when the catch is only 63% to 86%; in the case of eggs, values of 91% to 100% are obtained. The soil samples should be collected to a depth of 10 cm in meadows or 20 to 25 cm in the case of agricultural land or gardens, while the surface area encompassed by each sample should be 0.1 m^2 or 0.05 m^2; 12 to 18 such samples should be taken to obtain statistically valid results (SOUTH 1973). Almost the full compliment of small slugs such as *Deroceras reticulatum* (Müller), *Arion fasciatus* (Nilsson) and *Arion intermedius* Normand and their eggs are removed from soil samples taken from fields and pastures and then washed under a fine jet of water over sieves (SOUTH 1964). Slugs and eggs pass through the top sieve which has a gauge of about three meshes every 2.5 cm, and collect in the lower sieve, with 10 to 20 meshes to 2.5 cm. Once the residues of organic matter have taken up water and float on the water surface, slugs and eggs become visible. This method is said to be 99% effective for *Deroceras reticulatum* (Müller), 94% for *Arion fasciatus* (Nilsson), when the soil samples are washed very carefully several times and all the slugs and eggs removed. The simple method of **collecting by hand** always results in an underestimate of population density. Over-estimation of the relative number of one species in relation to others may however arise when collecting is done at night, as the lighter coloured species are more noticeable than the darker and are thus collected in greater numbers, together with other light-coloured objects such as bits of wood, sticks and pebbles. It is therefore advisable, before starting collecting at night, to spend 5 min getting accustomed to the distinction between light and dark species; very thorough searching is also essential (CRAWFORD-SIDEBOTHAM 1972).

SPIEKERMANN (1976; also SCHORER 1973; cited by SPIEKERMANN) describes a method for both qualitative and quantitative estimation of terrestrial molluscs this involves two complimentary methods of operation: searching large areas (about 6 m^2) marked out at regular intervals and also the sieving of detritus and soil from samples of uniform size (each 50 by 50 cm). The residues on the sieves are dried out to enable the shells of very small species to be picked out. To distinguish empty shells from those containing living snails, they should all then be wetted thoroughly, as this causes any live snails to become active.

It is desirable to mark out any particular area (at least 1 m^2) in which the gastropod population is to be sampled for species composition and numbers. This can be done with mechanical wooden frames or other types of **barrier.** Ammonium sulphate can be used as a chemical barrier (THOMAS 1944). For molluscs which are sensitive to electric shocks, an electric barrier can be used – for example, for *Helix aspersa* Müller on citrus trees (CARMAN 1965; WOLDA 1970). MOENS et al. (1967) invented a metal barrier; this was actually intended to protect valuable garden plants (Chap. 3), but can also be used to study the population density within a large area.

There are also methods for the estimation of population density of fresh water gastropods like *Galba truncatula* (Müller) (HEPPLESTON 1972; MOENS 1973; ROSS and O'HAGAN 1968).

2.3.3.2 Marking

Marking individual gastropods may be necessary to obtain data on the degree of infestation either for making a prognosis, for studying the fluctuations in populations in their natural habitat (MCCLELLAND 1965; LOMNICKI 1964; MOENS et al. 1966) or for establishing the population density *before* and *after* control measures have been undertaken, so as to estimate the success or otherwise of the methods used. There are marking procedures for both terrestrial and freshwater gastropods.

To mark **snails** is relatively easy because the shell has simply to be painted, as was done, for example, with *Theba pisana* (Müller) (CAIRASCHI and LECOMTE 1973). Alternatively, plastic strips of different colours may be used. Such methods are, however, time consuming and the marking not always permanent. HEPPLESTON (1972), for example, painted *Galba truncatula* (Müller) with nail polish, but 10% of the marks came off after 2 weeks and 15% to 20% after 4 weeks.

Marking of **slugs** is difficult, since no marker will hold on the slimy body. Attempts have been made at radioactive tagging by way of the food, as was later tried with other small soil dwellers, like collembolans (MÜLLER and BIERINGER 1971). Experiments using radioactive substances like ^{32}P and ^{131}I were carried out early on *Deroceras*, *Arion* and *Helix*. According to FRETTER (1952), radioactive phosphorus taken in with the food passes by way of the digestive glands to the haemocoel, and from there to the other tissues; it accumulates in the nervous system, mucous and salivary glands, odontophore and some of the cells of the mantle epithelium. ^{32}P is also stored in the calcium cells of the digestive gland. In the nervous system, radiophosphorus is found mainly around the nerve fibres, to a lesser extent in the cytoplasm. In mucous cells, which are especially abundant in the foot gland

Table 16. Counting rates of ^{32}P during the two first stages. (After FRANÇOIS et al. 1965)

μCi/g	Distance in cm	Days after treatment									
		0	1	2	3	4	5	6	7	8	9
		Counting rates (in counts per second)									
600	15	–	185	313	310	293	283	267	253	275	222
450	15	64	210	248	273	245	253	258	231	243	220
300	15	100	270	300	338	335	345	325	348	320	295
150	15	25	45	78	105	107	118	113	97	104	93
	10	65	130	160	230	225	255	220	215	225	208
30	5	–	180	200	255	228	225	225	215	235	198

and salivary glands, the phosphorus may be bound to calcium. The isotopes are also incorporated into the structure of the developing radula. Radio-iodine, on the other hand, is readily absorbed from the gut lumen into the digestive gland, but is stored only in the kidney; it is eliminated with the faeces. MOENS et al. (1965) and FRANÇOIS et al. (1965) mixed radioactive solutions to rice flour in the proportion 2:5 and fed this to isolated slugs. The resultant radioactivity could be detected with a geigercounter at distances of 5, 10, 15, and 30 cm. Field experiments were conducted using ^{32}P, ^{198}Au, ^{86}Rb, ^{131}I and ^{58}Co, of which ^{32}P and ^{58}Co gave the most satisfactory results. ^{32}P is not suitable for studying the movements of individual slugs, but owing to its relatively long half-life (14.3 days), it is very useful for determining population density and movement of the slug population as a whole. The slugs were still found to be radioactive 45 days later. Labelling with ^{58}Co is readily detectable and long lasting (half-life 71 days), so that even on uneven ground, individual slugs could be easily found every day over a whole month. Mortality rate is normal compared with non-tagged controls, and feeding is not affected. The results of some such experiments are listed in Table 16.

The method is highly sensitive since a readily detectable count is obtained even at distances of 10 and 15 cm and with a dose of 150 μCi/g.

Using this radio-isotope method in heavily infested clover fields, between 23% and 61% of a known population of Field Slugs was recovered (FRANÇOIS et al. 1968).

Another method of marking involves insertion of a radioactive metal rod (0.008 mm diameter, 1.5 mm length) with an activity of 25 to 30 μCi ^{182}tantalum under the tegument of the chalk platelet in the mantle (FRANÇOIS et al. 1968).

Radioactive tagging of those freshwater snails which are intermediate hosts of worm parasites of man and animals is easier than in the case of terrestrial gastropods, since this does not have to be done by way of the food. Simply adding the radioactive isotope to the water suffices. It is even possible to tag the parasites while they are in the vector, but for this to occur it is necessary that the snail be already radioactive before the sporocytes and cercariae of the trematodes develop (JEWSBURY and HOMEWOOD 1971). In addition to ^{32}P, ^{131}J, ^{64}Cu, ^{85}Sr, ^{59}Fe, ^{86}Rb, ^{22}Na and ^{14}C, ^{75}Se has been found to be suitable. Because of its markedly long half-life of 127 days, ^{75}Se can be used in field investigations. Specimens of *Biom-*

phalaria glabrata (Say) labelled by application of $H_2\,{}^{75}SeO_3$ or Se-methionine gave a count of more than 100,000 cpm. The radioactive selenium is incorporated mainly into the soft body tissue, less into the shell. Cercariae developing in the snails were radioactive from the third day, reaching a maximum after three weeks. The rate of infection of mice by radioactive cercariae was not significantly less than by non-radioactive larvae, and they developed in the mice to normal worms. Labelling can be effected in two ways: either the snails may be kept in water to which has been added $H_2\,{}^{75}SeO_3$ in 1 N HCl, with an activity of up to 5 μCi per snail per litre of water; or the snails may be fed on a substitute diet containing 10 μCi of ^{75}Semethionine/250 mg food, representing a total innoculation of 70 to 100 μCi.

As this chapter has shown, the importance of gastropods as pests in agriculture and horticulture should not be underestimated, and their control must constantly be borne in mind.

3 Control of Pest Gastropods

Molluscan pests can be destroyed in several different ways: by chemical means (molluscicides), through the use of biotechnical measures, by mechanical methods and by interference with the environment. These different methods can, and sometimes have to be used simultaneously. VAN DINTHER (1973) would add to this list legislative measures, which are concerned with quarantine regulations, as is the case in the United States and Canada.

The control of pulmonates by whatever method is not simple, and each of the above-mentioned methods has its associated problems. A prerequisite for success is knowledge of habitat and behaviour, but this differs from species to species even in the same area. It is, furthermore, essential to take into account the physiological effects of the method used; these depend not only on the chemical substances administered, but also on the particular species involved and on the physiological state of the population. It is also important that the campaign of control should be conducted over a wide terrain and not be limited to a restricted and relatively small area; for it must always be remembered that gastropods from adjacent land may move onto a treated and thus pest-free field, and there give rise to a new pest population.

In general, different molluscicides must be used for control of freshwater snails. Ecological factors are of importance in determining the success or otherwise of all the various means of control.

3.1 Problems Encountered in the Control of Pest Gastropods

It has gradually become clear through practical experience that the effectiveness of molluscicides formulated as spray, dusting powder or bait are greatly influenced by weather and soil conditions and also by the behaviour of the slugs and snails. Furthermore it varies from species to species, is affected by the physiological condition of the animals, and is made difficult by the variety of places in which gastropods can hide.

Let us first consider **terrestrial gastropods.** Slugs are, as a rule, active only during or after light rain, in warm moist weather and in the damp evening and morning hours; it is thus only under these conditions that slugs can come into contact with molluscicides or be attracted to them. It is difficult to effect successful control measures in greenhouses, because daily watering gives rise to a constant high humidity. Metaldehyde and other substances which depend for their effect on desiccating slugs and snails are rendered less effective because the animals can quickly recover in the damp and optimal environment. If molluscicide becomes absorbed by the sheath of mucus which covers the body, it can (like a coat) be cast off within a short time. Slugs can, in this way, rid themselves not only of molluscicides which have come in contact with the skin, but even if they have been ingested (DAXL 1968, 1970). Control measures against snails are fundamentally more difficult since they can withdraw into their shells, thus avoiding contact with molluscicides.

It is almost impossible to destroy any significant number of terrestrial gastropods by use of bait; probably no more than 10% of a population can be destroyed in this way (FRÖMMING and PLATE 1952). For species that live in the soil (*Milax* spp.) baiting results in catches which are in no way commensurate with the actual population density (VAN DEN BRUEL and MOENS 1958; GOULD 1961). The coarseness of the soil and the opportunities it thereby offers for slugs to remain hidden plays a role here.

Furthermore, the toxicity of any molluscicide depends on the manner in which it is taken up by the gastropods. Thus the effects of metaldehyde, the molluscicidal carbamates and herbicides differ according to whether they have come into contact with the skin, or are eaten and absorbed through the intestine. Mortality rate is always greatest for *Deroceras reticulatum* (Müller) after body contact with metaldehyde, and not following oral uptake, whereas precisely the opposite applies to *Helix aspersa* Müller (CRAGG and VINCENT 1952; THOMAS 1948).

As these examples show, failure of control measures may not be due to ineffectiveness of molluscicides, but a consequence of ignoring the biological and physiological requirements of the gastropods and also the influence of the environment.

3.1.1 Influence of Biological and Ecological Factors

Control measures must be adjusted to suit both the autecological and synecological factors and their interrelationships (GODAN 1972). The following compilation shows what these factors are and the way in which they indirectly influence the effect of a molluscicide (Table 17).

The toxicity and attractiveness of a molluscicide prepared in the form of **bait,** today the usual method of application in dealing with **land gastropods,** depend on the species and its sensitivity, on its behaviour, memory and activity rhythm and habituation to the substance. The method of feeding, the nature and quantity of available food and also the plant cover which affords shelter all have an important bearing on the activity and behaviour of slugs and snails and thus indirectly on the effectiveness of any bait. The structure and texture of the soil are also important, since clods of earth, stones, weeds and leaves close to the soil afford ideal shelter and thus make contact with the molluscicide more difficult. Demecological factors, such as age, composition and variation in size within a pest population influence the success of control measures, because the sensitivity of a snail or slug changes during its life span.

Abiotic factors such as temperature, humidity, moisture content of the soil, illumination (sun, shade) promote or impair activity, sensitivity and recovery of gastropods; since all these factors influence water-regulation, one of the most important problems in the control of terrestrial gastropods. Thus high humidity favours recovery after treatment with metaldehyde preparations, while in dry conditions

Table 17. Ecological factors and effectiveness of a molluscicide

		Gastropod	Molluscicide
1. Autecological factors			
Abiotic factors			
Temperature, Humidity, Light →	Water regulation	Sensitivity recovery (metaldehyde) increased sensitivity (carbamates) →	Toxicity
Soil (earth clods as shelter) →	Activity →	Behaviour →	Attractiveness
Biotic factors			
Nutrition, Plants as shelter →	Activity →	Behaviour →	Attractiveness
2. Demecological factors			
Age groups, Individual size →		Sensitivity →	Toxicity
3. Species →		Sensitivity —	Toxicity

the animals die. The converse is true of treatment with carbamate. Furthermore, the molluscicide itself is affected by the climatic conditions mentioned, decomposing rapidly or more slowly, with consequent diminution of effectiveness.

The mode of life is an important aspect, since species which live on, in or above the soil may all be pests in the same area. Thus *Milax* spp., *Arion hortensis* Fér., *Deroceras* spp., the terricolous *Helicigona arbustorum* (L.), and *Cepaea* spp. which are mainly found on leaves, twigs and branches, can all occur together. When spraying or dusting is employed, the molluscicide must reach the soil surface. If the leaves and upper parts of the plant only are reached, there is no chance that slugs which live on or in the soil will come in contact with the molluscicide (Howitt and Cole 1962), and the substance appears to have no effect. On the other hand, control of snails like *Helix aspersa* Müller and *Theba pisana* (Müller) is inadequate using one of the normal baits (Cairaschi and Lecomte 1973).

Further investigation of many aspects of the interaction between the above mentioned factors is necessary. The aim in gastropod control should be to increase its reliability and effectiveness. There are various ways in which this can be achieved as part of an integrated plant protection scheme. A gastropod population may be so weakened by sub-lethal environmental factors, not only weather conditions, that smaller quantities of molluscicide than usual are sufficient to keep the population density below the threshold limit. It is also possible that a single application of herbicide to destroy weeds in a monoculture will also eliminate the molluscan pests.

As has been shown in Chapter 2, **freshwater** and **amphibious snails** must also be controlled. Besides biological and ecological factors, the effect of water on the molluscicide, as well as the composition of the water in which the snails are living, are of great importance. In determining the suitability of the molluscicide, consideration must be given to its speed of sedimentation, its rate of degradation, its dispersion, solubility and also its toxicity for aquatic plants and animals, which may constitute a danger for inundated crops. As far as the habitat is concerned, the following are all important: the size and depth of the water body, the nature of the bed (sediments, mud, bog-like base), turbulence, flow and clarity of the water, its pH value and salt content, as well as the depth of penetration of UV light in continuous sunshine. The poison must be kept above the lethal dose for snails for a long time. For successful control either the molluscicide must remain in the water for sufficiently long to ensure that young snails which hatch after application will also be killed, or, if the molluscicide decomposes rapidly or is washed away, two applications must be made within a few days of one another; the second should be made directly after hatching of the eggs laid by animals which survived the first treatment by finding sufficient protection in the muddy bed of a lake, pond or river. Young snails may not yet be capable of reproduction by the time the second application is made. Since the onset of sexual maturity is partially determined by temperature, which cannot be predicted with any accuracy, it is difficult to judge precisely the best time for the second application. Using the first of these methods, the control of *Biomphalaria* spp. is successful if the molluscicide Niclosamide is maintained at a concentration of 0.25 ppm in water for 30 days.

The interaction of the factors which relate to the pest itself are best known in the case of **land gastropods,** and are dealt with in the following sections.

3.1.1.1 Species-Specific Factors

In general, snails are more sensitive than slugs. Thus the molluscicide Carbaryl in a 10% mixture with bran resulted in 100% mortality of *Cepaea nemoralis* (L.), but failed to affect slugs, even at twice this concentration. According to THOMAS (1948), *Helix aspersa* Müller is very sensitive, while the black *Milax gagates* (Drap.) is extremely resistant.

Since the sole of the foot of land gastropods comes into contact with the treated soil, the sole size should be of considerable importance for the uptake of molluscicides. This is true not only when spraying and dusting are employed, but also when the molluscicide is in granulated form, since the snails and slugs may crawl over these and poison themselves without having eaten them. In this regard it might be assumed that larger animals are more sensitive to treatment than smaller, since the large surface area of the sole should enable more molluscicide to be taken up. As the following account shows, however, this is true only when different species are being considered, but not for animals of different size of the same species.

GODAN (1966) made comparisons between slugs of the same genus (experiment series A), between slugs of different genera (series B) and between different families (series C). The food comprised potatoes, carrots, casein and yeast; in the case of *Arion* and the smaller species lettuce was added.

Experimental series A:	Equal in age Limacidae:	*Limax maximus – Limas flavus*
Experimental series B:	Equal in	1. body size but not age 2. body size and age
	Limacidae:	1. *Limax maximus – Lehmannia marginata* *Limax flavus – Lehmannia marginata* *Limax flavus – Deroceras reticulatum* 2. *Lehmannia marginata – Deroceras reticulatum*

Experimental series C: Equal in body size and age
Limacidae and Arionidae
Limax maximus – Arion rufus
Limax flavus – Arion rufus

In experimental series A we are dealing with the two largest species of Limacidae, *L. maximus* L. being somewhat larger than *L. flavus* L. In experimental series B 1, young animals of a large species *(L. maximus* or *L. flavus)* are compared with adults of smaller species *(Lehmannia* or *Deroceras)* so that the sizes of body and creeping sole correspond. Only the small species in series B 2 were of corresponding age as well as body and sole size and food. In experimental series C, comparisons were made with the larger species of the families Limacidae and Arionidae. Table 18 shows the results of such comparisons, using a 2% Isolan preparation.

Table 18. Species-specific sensitivity of slugs. (After GODAN 1966)

Experimental series	Species	Average age (months)	Average weight (g)	Average sole (mm^2)	Mortality in %	P	Duration of test (h)
A	*Limax maximus*	9	7,1	626.7	35.5	<0.01 sign.	6
	L. flavus	9	3,6	464.3	85.0		
	Limax maximus	3	2,1	299.6	10.0	<0.01 sign.	5
	L. flavus	3	1,7	288.4	75.0		
B 1	*Limax maximus*	2–3	1,2	177.1	14.0	<0.01 sign.	6
	Lehmannia marginata	6–7	0,9	133.0	78.0		
	Limax flavus	3–4	1,1	217.0	46.7	=0.05 sign.	6
	Lehmannia marginata	6–7	0,9	133.0	78.0		
	Limax flavus	3	0,5	108.4	39.0	<0.01 sign.	6
	Lehmannia marginata	5	0,5	118.0	61.0		
	Limax flavus	3–3 1/2	1,0	146.5	44.0	<0.02 sign.	1 3/4
	Deroceras reticulatum	6	0,6	92.0	77.5		
B 2	*Lehmannia marginata*	6	0,8	125.3	40.0	<0.02 sign.	1 3/4
	Deroceras reticulatum	6	0,6	92.0	77.5		
	Lehmannia marginata	1	0,2	68.2	98.0	>0.05 sign.	1 3/4
	Deroceras reticulatum	1 1/2	0,2	54.2	88.0		
C	*Limax maximus*	10	13,4	1,070.6	44.0	<0.01 sign.	6
	Arion rufus	10	15,3	1,177.4	0.0		
	Limax flavus	6–7	4,1	516.0	78.0	<0.01 sign.	6
	Arion rufus	6–7	3,6	436.1	0.0		

P=probability of error (MUDRA 1958)

Large species are more resistant than smaller. Furthermore, the effect of the molluscicide differed from species to species, even when body size and age were the same. Thus *Arion rufus* (L.) (Arionidae) is more resistant than *Limax maximus* L. (Limacidae).

Further data were obtained from experiments using molluscicide bran mixtures and 24 h exposure: the mortality of the small Brown Slug was 40% after application of 5.0% promecarb or metaldehyde (commercial bait pellets) and 70% after treatment with 2.5% mexacarbate or 0.6% isolan: the corresponding mortalities of the large Cellar Slug, on the other hand, were 4%, 23%, 20% and 17% respectively. Spraying with 0.5% solution of mexacarbate which then came into contact with the sole, resulted in death of only 50% of experimental adult *Limax maximus* L., but mortality was 100% in the case of *Limax flavus* L. CRAWFORD-SIDEBOTHAM (1970) also established a species dependant sensitivity of slugs to metaldehyde and methiocarb. The investigations of DAXL (1971) with metaldehyde, isolan and the herbicide ioxynil showed *Arion hortensis* Fér. to be more resistant to all three preparations than *Deroceras reticulatum* (Müller). The Brown Slug *D. laeve* (Müller) which has a high body water content was most susceptible to the carbamate isolan.

The following order of decreasing resistance to molluscicides has been compiled with the aid of information from the literature:

Milax budapestensis → *Milax gagates* → *Arion rufus* → *Limax maximus* → *Limax flavus* → *Arion hortensis* → *Lehmannia marginata* → *Deroceras laeve* (except in the case of carbamates) → *Deroceras reticulatum.*

This series relates to adult gastropods and applies also to the young of the larger species. In the small species, differences between individuals of the same body and sole size and the same age were found amongst adults, but not young animals: thus adult *Lehmannia marginata* (Müller) were more resistant than adult *Deroceras reticulatum* (Müller), whereas young *L. marginata* did not differ in their resistance from young *D. reticulatum.*

Since the resistance of young snails in greater than that of older, reproductively active snails – a matter which is discussed further in Chapter 3.1.1.2 – young *Limax flavus* L. and *Limax maximus* L. were as resistant as adults of the species which rank one step above them in the rating of resistance. Young specimens of the larger species *Limax maximum* L. and *Limax flavus* L. have shown themselves more resistant than equally sized adults of the smaller species, *Lehmannia marginata* (Müller) and *Deroceras reticulatum* (Müller).

These investigations have important implications in **practice.** They demonstrate the need for careful survey in an infected area, the more so since mixed gastropod populations are the rule. They also show that in any control campaign the sensitivity of the individual species must be taken into account. Amongst the larger species of slugs, *Arion rufus* (L.) is more resistant than *Limax maximus* L. The latter, in turn, is the only one of the larger species of Limacidae which is more resistant than the smaller *Limax flavus* L. although both species feed on roots and tubers. Amongst the smaller Limacidae, *Lehmannia* spp. and *Deroceras* spp., *Deroceras reticulatum* (Müller) is the most sensitive.

Since the young of *Limax* spp. are more resistant than older individuals of the same species, their sensitivity compared with that of adults higher up the sensitivity

list has been investigated. This showed that young *L. maximus* were much more resistant than adult *L. flavus* twice their age and size:
L. maximus: 3–4 months old– 1.3 g weight = 14.0% mortality.
L. flavus: 8 months old – 4.9 g weight = 78.0% mortality.
The following results were obtained from three series of experiments (5 to 6 h contact with isolan), in which young slugs were compared with the adults of a more resistant species:

1. *L. flavus*
   (3–4 months old, weight 1.7 g) = 37.5% mortality.
   L. maximus
   (7–8 months old, weight 7.1 g) = 35.0% mortality.
2. *L. flavus*
   (3–4 months old, weight 1.1 g) = 46.7% mortality.
   L. maximus
   (6–7 months old, weight 13.4 g) = 44.0% mortality.
3. *L. maximus*
   (3–4 months old, weight 1.2 g) = 14.0% mortality.
   A. rufus
   (6– 7 months old, weight 6.3 g) = 0% mortality.

The mortality of young animals of a sensitive species is about the same as that of older individuals of a more resistant species. *Arion rufus* (L.) alone is an exception, demonstrating in this experiment, too, extraordinary resistance. The results were confirmed in practice.

Species-specific differences in sensitivity have also been demonstrated for snails subjected to control by spraying or using pellet bait (GODAN 1967) (Table 19).

Helix pomatia L. was found to be the most resistant to both these methods of application of molluscicides. This is remarkable since the sole of adult *H. pomatia*, with a surface area of 1,516 mm^2, has a far greater area for contact with molluscicides than have the smaller, roughly equally sized, *Cepaea* and *Helicigona*, with sole sizes of 108 mm^2 and 151 mm^2 respectively. The body weight of *Helix pomatia*, too, with an average of 25.6 g, is 10 times that of the two smaller species which are of about the same size and have an average weight of 2.9 g (50 animals). The two species of Banded Snails are of roughly equal size, but the darker species, *Helicigona arbustorum* (L.) was more resistant than the lighter *Cepaea nemoralis* (L.). This observation is interesting in view of the fact that REICHMUTH and FRÖMMING (1960, 1961) found that heavily **pigmented** slugs are more resistant than lighter species. SEDLMAIR (1956) observed that longevity is greater in darker than lighter specimens of the two species *Cepaea nemoralis* (L.) and *C. hortensis* (Müller). GODAN found that, following contact of the sole with 0.2% metaldehyde, the first deaths amongst light-coloured *Cepaea nemoralis* (L.) occurred within 24 h, whereas all the darker individuals of this species survived for at least 3 days. The final result of an experiment with 20 light and 20 dark specimens of approximately the same average body weight (3.42 g and 3.56 g) showed mortality rates of 20% and 10% respectively. In the field, the more resistant members of a *Cepaea* population are the more abundant, which is disadvantageous for control measures. In a wooded parkland near Berlin, the numbers of light, dark and grey *Cepaea* col-

Table 19. Species-specific effect of molluscicides on snails. (After GODAN 1967)

Active ingredient concentration	Species	Number	Mortality %
As spray			
Metaldehyde	*C. nemoralis*	100	90
0.04%	*H. arbustorum*	100	100
	H. pomatia	50	55
Carbaryl	*C. nemoralis*	90	95
1.0%	*H. arbustorum*	45	60
	H. pomatia	73	75
Isolan	*C. nemoralis*	170	98
0.12%	*H. arbustorum*	74	60
	H. pomatia	40	10
Mexacarbate	*C. nemoralis*	100	100
0.05%	*H. arbustorum*	80	40
As bait pellets (active ingredient-bran mixture)			
Metaldehyde	*C. nemoralis*	40	65
(bait pellets)	*H. arbustorum*	40	80
	H. pomatia	40	25
Carbaryl	*C. nemoralis*	40	60
5.0%	*H. arbustorum*	40	50
	H. pomatia	40	10
Isolan	*C. nemoralis*	40	95
0.6%	*H. arbustorum*	40	40
	H. pomatia	40	0

lected were 14, 59 and 47. Interesting in this connection is the case of the carnivorous Butter Clam, *Saxidomus giganteus* (Deshayes) which captures its prey by producing a paralysing poison (PRICE and LEE 1971) and the distribution of melanin in the clam shows that it plays an important role both in retaining the poison and preventing the clam's own body tissues from being affected.

Further examples of species-specific differences in the effect of molluscicides on snails are provided by the experiments of BÖHM (1961) with Slugit (metaldehyde spray, dose 30 cm^3/5 l water /10 mm^2 sole contact in which the two smaller species, *Helicigona arbustorum* (L.) and *Bradybaena fruticum* (Müller) were found to be more susceptible than the larger *Helix pomatia* L. Furthermore, RICHARDSON and ROTH (1963) found that *Cochlicella barbara* (L.) is more resistant to methyl bromide than *Theba pisana* (Müller). According to MAYER (1957) treatment with metaldehyde spray is adequate in controlling *Oxychilus draparnaudi* (Beck), but not *Discus rotundatus* (Müller).

In conclusion it can be said that the species-specific differences in sensitivity to molluscicides are quite independent of the size of the sole. Species-specific differences in the effect of molluscicidal poisons occur not only in land gastropods, but also in freshwater and amphibious pulmonates. *Bulinus (Physopsis) globosus* (Morelet) is very resistant, *Biomphalaria pfeifferi* (Krauss) less so, while *Lymnaea natalensis* (Krauss) is extremely sensitive (SHIFF and WARD 1966).

3.1.1.2 The Effect of Age

Complications exist, however, even when only a single species is involved. In this case the age and associated size of both body and sole which are dependent on age can be important. These two factors are, however, only related to one another in young, growing specimens, while in the adults, the reproductive phase and later on aging affect sensitivity to molluscicides.

With the exception of *Deroceras*, young animals only a few months old have been found to be more resistant than sexually mature adults of the same species reared on the same diet. This was shown in trials involving sole contact for several hours to isolan (2% in acetone and 7.1 μg/cm^2 of glass surface) and in which the slugs were fed with potatoes, carrots, casein and yeast (GODAN 1966) (Table 20).

Experiments in which the effect of metaldehyde on Cellar Slugs was tested point to the same conclusion (GODAN 1961).

The result indicate that **within** a particular *species* mortality is very much higher in slugs with a relatively larger sole, in other words, the adults, than in younger specimens. This may depend on the amount of molluscicide taken up, or, on the physiological condition of the gastropods, which can change in the course of time.

The studies of GODAN (1966) (Fig. 82) have shown that there are different sensitivity levels at different times in the life span of a gastropod and have determined the length of each of these phases. The columns in the graph indicate age groups of a total of 300 Cellar Slugs (*Limax flavus* L.) tested. The age of group I was 2 to 3 weeks, group II 2 to 3 months, group III 4 to 5 months, group IV 8 to 9 months, group V 12 months and group VI 24 months. Tests were made using a 2% solution of isolan in acetone, and 7.1 μg/cm^2 glass surface for 5 h.

As a result of the low mortality of the 4- to 5-months age group, the number of unharmed, active juveniles was high.

The **physiological state** of the slugs changes in the course of their life span. Very young, newly hatched specimens are extremely susceptible; from the second month, sensitivity to molluscicides diminishes, so that three month old juveniles

Table 20. Influence of body size (weight, sole size) and age on mortality in slugs of the same species. (After GODAN 1966)

Species	Average age (months)	Average Weight g	Average sole size (mm^2)	Mortality %	P	Difference (time for effect) (h)
Limax maximus	9	12.5	918.9	46.7	<0.02	sign.
	4	3.0	305.9	6.0		(6)
Limax flavus	8	4.9	553.6	78.0	<0.05	sign.
	3–4	1.1	217.0	46.7		(6)
Limax flavus	9	3.6	464.3	85.0	<0.01	sign.
	3–4	1.7	232.2	28.0		(5)
Lehmannia marginata	6–7	1.4	205.1	98.0	<0.01	sign.
	2	0.2	56.9	64.0		(24)
Deroceras reticulatum	6	0.6	92.0	77.5	>0.05	no sign.
	1$^1/_2$	0.2	54.2	88.9		(13)

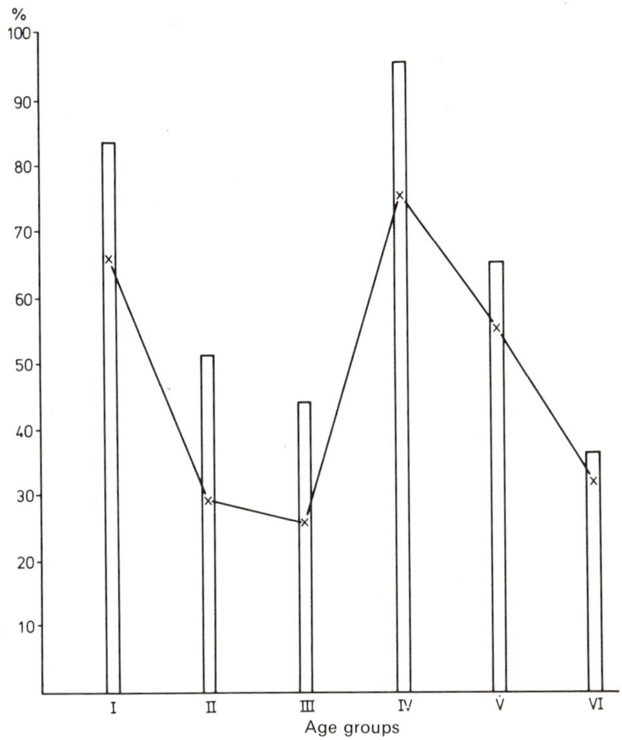

Fig. 82. Sensitivity and age, 5-h contact with a 2% isolan preparation; *Limax flavus* (300 test specimens); *columns* percentage strongly affected slugs ("side position"), ——— = mortality (%)

have a relatively high degree of resistance. They become once more very sensitive as adults during the reproductive phase – in the case of Cellar Slugs, from the 8th to the 12th month. At a later stage, when oviposition has ceased, susceptibility decreases until, at the onset of senility, gastropods are as resistant as juveniles. The age periods given above apply to *Limax flavus* L. The phases of greater or lesser sensitivity are different for other species with shorter or longer life spans, but the general pattern of changing sensitivity with age and physiological condition remains.

The periods of greatest sensitivity, which are important for control measures, thus occur when the gastropods are very young and again when they are adults, at the peak of their reproductive activity. There is a progressive decline in sensitivity with increasing age. Resistance is greatest amongst juveniles before the onset of sexual maturity and old gastropods which have ceased or almost ceased oviposition. As a result, young, as yet immature, gastropods survive the control measures in sufficient numbers to ensure that a new generation develops after control has been completed, giving rise to a new infestation. This is a problem which has not yet been overcome; it applies equally to terrestrial and freshwater gastropod pests.

The mean body weight and sole size respectively for each of the groups of *Limax flavus* L. were as follows: Group I – 0.15 g and 75.2 mm^2; II – 0.85 g and 158 mm^2; III – 2.16 g and 272.6 mm^2; IV – 5.25 g and 620.2 mm^2; V – 6.39 g and 608.0 mm^2; VI – 6.5 g and 617.9 mm^2. The following values have been established for body weight and sole size for various age groups of other species:

Arion rufus	3 months old (∅ weight 6.7 g)	– 645.6 mm^2
	10 months old (∅ weight 16.6 g)	– 1205.4 mm^2
Limax maximus	7 days old	– 26.4 mm^2
	2 months old (∅ weight 1.1 g)	– 177.1 mm^2
	4–5 months (∅ weight 3.1 g)	– 305.8 mm^2
	11 months (∅ weight 13.5 g)	– 918.9 mm^2
Limax flavus	7 days old	– 30.1 mm^2
	12 months (∅ weight 7.12 g)	– 608.0 mm^2
Lehmannia marginata	3 months old	– 68.2 mm^2
	6 months	– 125.3 mm^2
	14 months (∅ weight 0.8 g)	– 148.2 mm^2
Deroceras reticulatum	3 months old	– 54.2 mm^2
	5 months	– 91.9 mm^2

3.1.1.3 Water Regulation, Activity

As already indicated, the various species of land gastropods do not all have the same body water content, and this may be the basis of species-specific differences in reactions to molluscicides. Slugs have an equally high water content, and even brief exposure to intense sunlight or warm sunny weather combined with high temperatures, indeed dryness in general, are all fatal to them. The **humidity** of **air** and **soil** represent limiting factors for slugs and the success of chemical control measures is to a large extent dependant on these factors. The water slugs lose through secretion of mucus or through evaporation from the skin must continually be replaced by drinking or by uptake through the skin, so that a correct water balance is maintained. If, as a result of dryness of the soil or low relative humidity of the air, this does not take place sufficiently rapidly, slugs die even before the molluscicide has had time to take effect. If, on the other hand, rain or heavy dew fall follows immediately on control measures, the water loss caused by the molluscicide (metaldehyde, but not carbamate) can be made good and the animals recover. Thus application of dehydrating compounds, such as metaldehyde, in damp environments, which is precisely where control measures are necessary, has to be repeated two of three times for the method to be successful.

Temperature, in combination with humidity, also affects water balance. Slugs which had fed on metaldehyde bait pellets died after only 1 or 2 h exposure to a temperature of 30 °C and direct sunlight (Rammos 1962). Although this represents a very welcome result from the point of view of control, it can lead to false conclusions when testing molluscicides – a subject which has still to be discussed – since such high temperatures are unlikely to occur for any length of time when control measures are being put into practice. In other experiments conducted in damp and mild weather conditions, which are more likely to be encountered in practice, the gastropods only died after 24 h. Experiments should therefore be made – and this applies also to tests on molluscicides – during damp, mild weather. The most suitable temperatures for this purpose lie between 15 °C and 20 °C.

In natural conditions, reduced fitness as a result of lack of moisture can be excluded as a factor affecting control, since a population explosion and associated large scale damage would not occur if environmental conditions were not optimal. Sprinkler systems should not be used once control measures have been initiated. HOWITT (1966) reported that after irrigation of a clover field, the population density of slugs increased within 18 h to about 600% of that originally recorded. This was due to the migration of a large number of animals from neighbouring areas, attracted by the excessive moisture.

The **activity** of gastropods is influenced by both temperature and humidity (VAN DEN BRUEL and MOENS 1958; DAINTON 1964; GODAN 1961; HOWITT 1960, 1961; MELLANBY 1961 and others) (Chap. 2.1). Success of control depends, in turn, on their activity. In light rain, the number of slugs caught by using bait pellets is greater (BARNES and WEIL 1942, 1944, 1945) than in heavy rain when the catch is smaller because the activity of slugs is inhibited (THOMAS 1948). Temperatures between 12 °C and 20 °C and a total rainfall not greater than 15 mm elicit particularly marked activity (RICOU 1964).

Of all the forms of gastropod activity elicited by mild, wet weather dispersal and feeding are of especial importance, because they co-determine the outcome of control measures; since the most usual method is to use bait pellets, feeding activity is particularly important. It must also be borne in mind that gastropods will usually remain by day in their damp hideaways, emerging to feed only at dusk or at night, and thus only encountering the bait at these times. The activity of gastropods increases when it is dark. The bait should thus be put out in the evening, and not in the morning, so that during the first night it is still fresh and attractive. After application to the upper surface of the soil, a molluscicide only affects those gastropods which are on and not those which are in the soil. It is, however, under beans, cereals and the remains of oil plants on heavy soils, as well as in the soil itself, that large concentrations of gastropods occur.

The shape and size of the plants may furthermore influence the effectiveness of the molluscicide: for example, the large plants of Brussel sprouts afford shelter for slugs above the soil on which a molluscicide is strewn. Even during the day, the slugs remain in the moist centre of the plant and do not descend on the soil. LANGE and SCIARONI (1952) regard dusting as more effective than the use of bait pellets; the powder should, however, reach the soil under the leaves.

There are, furthermore, species-specific differences in the **speed** with which gastropods-**creep** over the ground, and on this, too, the effectiveness of the bait depends. Thus *Theba pisana* (Müller) and *Deroceras laeve* (Müller) are characterized by particularly great mobility, while *Arion hortensis* Fér. is very sluggish and creeps very slowly. *Arion rufus* (L.) is also very mobile. In this case it suffices to spread protective barriers across the area, between hideaways and the plants which are to be protected. By contrast, in the case of the relatively sedentary and inert Field Slug, *Deroceras reticulatum* (Müller), the whole area must be treated, because this slug covers significant distances in search of a new source of food only when the old supply has been exhausted. Before control measures are undertaken, it is necessary to establish which species are present in an area of damage. If the dominant species is *Milax budapestensis* (Hazay), control measures should only be carried out on warm, damp nights because this species lives in the soil and rarely emerges at the surface where the bait pellet clump lies. This rule does not have to be applied so strictly in the case of the Field Slug, which lives above the soil.

3.1.1.4 Season (Hibernation, Aestivation)

In the damp months of late summer and autumn, increase in numbers and thus also of damage reach a peak, so that detecting the animals in the field presents no difficulties. Control measures and also tests on molluscicides should be carried out during this time.

There are, however, other factors which are dependent on the humidity of the environment and which affect the sensitivity of gastropods to molluscicides. These include hibernation and aestivation.

Let us first consider hibernation. In the spring, before egg-laying has begun, snails are more resistant than in autumn, despite the fact that the latter season is preceded by several months of active feeding (GODAN 1967). The mortality of *Helix pomatia* L. is considerably greater in autumn than in spring when hibernation has just ceased and the epiphragm cast off (75% compared with 6.7% when exposed over 24 h to 2% carbaryl). In these experiments, mortality at 75% was just as high from May onwards as in autumn. In late summer the body water content of these snails reaches a maximum. This is reduced when the snail goes into hibernation, closing the shell mouth with an epiphragm and the lower value is retained until the end of the cold period (DEXHEIMER 1951). During hibernation and also aestivation the body water content in the case of the now passive *Helix pomatia* L. is considerably reduced. The blood is thus greatly concentrated and its protein content almost doubled (WILBUR and YONGE 1964/66); its freezing point is thus lower than in an active snail.

Many species of snail spend the dry months of the year in a state of aestivation. If this is of long duration, the body water content falls to a lower value, and this is characteristic for each species (GEBHARDT-DUNKEL 1953). The mouth of the shell

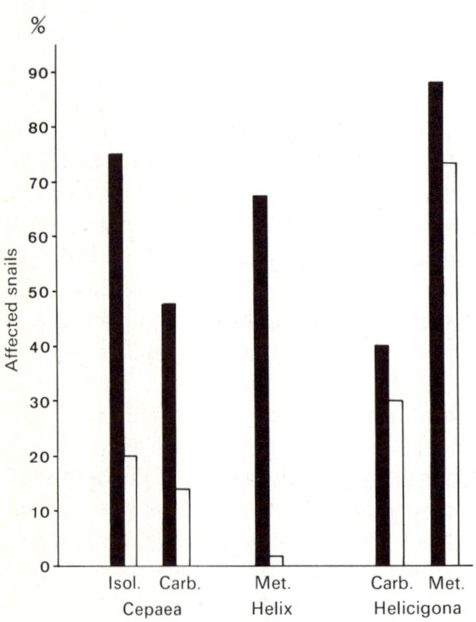

Fig. 83. Affected snails of the active group (*black columns*) and passive group (*white columns*); 24 h after application (μ/cm²) of isolan 10.7, carbaryl 177.9 and metaldehyde 7.0 (100 snails per trial)

is then closed by several layers of dried mucus, separated from one another by insulating layers of air (Chap. 1.3).

Probably as a result of the lower body water content during aestivation, sensitivity to carbamate is reduced in comparison with that of active snails, as it is also during hibernation.

Figure 83 compares the sensitivity of active and passive experimental animals of the species *Cepaea nemoralis* (L.), *Helicigona arbustorum* (L.) and *Helix pomatia* L. (100 animals in each group); the amount of active ingredient applied (μg/cm^2) for isolan, carbaryl and metaldehyde was 7.1, 59.3 and 23.7 respectively; the period of treatment was 24 h. The average body weight of adult *Helix pomatia* L. during the passive phase in aestivation was 23.9 g (max. 27.7 g; min 19.2 g) while that of active animals was 31.3 g (max. 33.9 g; min 30.0 g).

Active snails came immediately into contact with the molluscicide; several animals quickly withdrew into their shells for protection. The passive animals came into contact with the molluscicide later, when in response to being wetted by the fluid, they extended the foot from the shell; this occurred about 25 min after spraying.

Cepaea nemoralis (L.) and *Helix pomatia* L. which are able to endure long periods of dryness, were less sensitive during the passive state than during constant activity. Interestingly, this difference is not found in the terricolous *Helicigona arbustorum* (L.). This species occupies a quite different environment which is damp and shady as a result of dense vegetation, and where it lives in the upper layers of the boggy soil. Periods of aestivation in order to survive dry weather conditions are thus unnecessary. The other two species, by contrast, particularly *Cepaea*, inhabit relatively dry areas, often above the soil up to several meters high in shrubs and trees, where these snails spend the period of aestivation firmly attached to bark or leaves. As a result of its permanently damp surroundings, *Helicigona* is not adapted to long periods of dryness and its capacity for physiological regulation of water content consequently not very well developed. Clear differences between active and passive animals are not found in this species (Godan 1972). Other examples of the greater difficulty of control of aestivating snails are provided by the experiments of Lewis and LaFollette (1942) on *Helix aspersa* Müller and by Richardson and Roth (1963) on *Cochlicella barbara* (L.) and *Theba pisana* (Müller). The latter case involved fumigation of a shipment with a mixture of ethylene oxide (10%) and carbon dioxide (90%). A higher dose and longer period of treatment than usual were necessary for aestivating snails with a thick-walled, tightly closed shell.

Thus aestivating snails frequently survive control measures and give rise to a new pest population.

3.1.1.5 Nutrition

Godan (1961, 1966) tested the effect of nutrition on the resistance of gastropod pests to metaldehyde and carbamate. In each case specimens of *Limax maximus* L. and *Limax flavus* L. were separated into three groups immediately after hatching from eggs of the same cluster, so that the starting material was the same. The three groups were fed on different diets over a period of nine months: group I on carbohydrate-rich food (potatoes); II on protein (daily fresh-scraped meat or wet casein-paste) and III on a mixed diet (potatoes, carrots, meat, casein, yeast). The criteria used to estimate the degree of poisoning were production of mucus and resumption of feeding as an indication of recovery.

Surprisingly, in tests with metaldehyde at concentrations of 0.1%, 0.01%, and 0.02% and 10 or 30 min of treatment respectively, gastropods reared on a mixed diet proved to be far more sensitive to the molluscicide than those on a pure potato or protein diet. There was initially no detectable difference between the other two groups, but over a long period, a diet of potato alone proved to be unsuitable, since the animals weakened and finally died. This can, however, be remedied by the addition of yeast as a source of vitamins (Chap. 1.4.2).

There were clear differences in the outward appearance of slugs of groups I and II on the one hand, and those of group III which had been fed on a mixed diet, on the other. Thus at the end of the nine month period, the average body weight of *Limax maximus* L. from group I was 15.38 g and group II 4.97 g, while for group III it was 9.28 g. The average weights of *Limax flavus* L. from group I were 1.80 g, group II 1.46 g and group III 5.41 g.

In the test with metaldehyde, secretion of mucus remained normal in groups I and II, but in the case of group III was strong to very strong. The corresponding loss of weight resulting from increased production of mucus and consequent loss of water was greater than for groups I and II. After only 10 min treatment with metaldehyde, a weight loss of up to 46% was recorded for *Limax maximus* L. from group III, whereas for groups I and II this was 27% and 23% respectively (environmental humidity 85% to 95% RH). After 30 min treatment, the value for the "carbohydrate group" approached that of the "mixed diet group" while the "protein group" was still unaffected, or at least showed no visible symptoms. At this point the loss of weight was 53% for group III, 43% for group I and 23% for group II.

A similar result was obtained for the Field Slug, *Deroceras reticulatum* (Müller). The three diet groups into which the animals were divided immediately after hatching were fed as follows: group I on lettuce; group II on casein and yeast; group III on lettuce, potatoes, casein and yeast. The body length of the animals fed on a mixed diet (plant and protein food) was at 2.5 cm measured in an extended, creeping position, greater than in the case of the other two groups.

The influence of diet on the effectiveness of carbamate is shown in Table 21. Here too, *Limax maximus* L. reared on an exclusively plant diet (group I) were

Table 21. Influence of nutrition on sensitivity to isolan. (After GODAN 1966)

Species	Average age (months)	Average weight (g)	Average sole size (mm²)	Mortality %	P	Difference (treatment in h)
Limax maximus	9	I–2.6 III–4.2	209.5 504.2	13.3 46.7	<0.01	sign. (18)
Limax flavus	6–7	II–1.4 III–3.1	168.0 322.8	53.3 93.3	<0.01	sign. (5)

more resistant than animals of the same age which had been fed for nine months on a pure protein diet (group II). Furthermore, when exposed to a preparation of 2% isolan in acetone (7.1 μg/cm^2 of glass surface) over a 5-h period of treatment, *Limax flavus* L. of diet group II were more resistant than slugs of the same age fed on a mixed diet of potatoes, casein and yeast (group III).

The slugs of the "protein group" were, in consequence of the unbalanced diet, only half as large as those of group III, yet they were the most resistant to the effects of molluscicides.

That a protein diet does indeed decrease sensitivity of gastropods to molluscicides is shown by further experiments with 7-month-old Cellar Slugs, in which mortalities after 5 h exposure to isolan (2%) were as follows: for animals fed on potatoes it was 84.4% (weight 2.1 g; sole 23.2 mm^2), for animals fed on casein and yeast 53.3% (weight 1.4 g; sole 168.0 mm^2; $P < 0.01$).

It can be reasonably concluded that a diet consisting exclusively of animal protein increases the resistance of gastropods to molluscicides. Such a diet, provided from immediately after hatching, contributes to the health of the animals, as has been demonstrated also for Mole Crickets (GODAN 1964). Slugs reared on a mixed diet were admittedly larger and therefore appeared better fed, but they were nevertheless more susceptible to the action of molluscicides than the smaller group II. As a result of over-feeding, the chances of survival are smaller than those of animals on a protein-rich diet.

The greater sensitivity of the slugs which were larger as a result of being raised on a mixed diet may be due not only to the type of nutrition, but be related also to body size and the correspondingly larger surface area of the sole. Larger animals with a greater area of the sole must take up more molluscicide than smaller specimens and this could account for the greater sensitivity of these slugs compared with the other nutritional groups. Indeed, the sole area of group III with an average value of 700 mm^2 was more than double that of animals of groups I and II with an average area of 280 mm^2.

If this assumption is correct, young animals of group III would be expected to have the same degree of sensitivity as animals of the same size from groups I and II, since, despite different diets, the sole sizes are then the same and amounts of molluscicide absorbed will be the same in both cases. This was, in fact the case. *Limax*

Table 22. Comparison of the sensitivity of young *Limax flavus* slugs of group III with that of equal-sized older slugs of groups I and II. (After GODAN 1966)

Metaldehyde concentration %	Exposure time (min)	χ Young	χ Old	Difference between young and old	P
0.1	10	3.6	4.4	None	<0.05
0.02	10	4.1	2.4	None	<0.02[a]
0.01	10	3.0	3.0	None	<0.05
0.01	30	5.8	4.6	None	<0.05

[a] = estimated according to WAERDEN and NIEVERGELT (1956).
χ = arithmetic mean of the number of days for recovery.
P = probability of error (MUDRA 1958).

flavus L. reared on a mixed diet had attained, after three months, the same body weight (average for 50 animals was 1.5 g) and therefore the same sole size as adults of groups I and II. These young slugs were found to be just as resistant to metaldehyde as the equal-sized adults of groups I and II. Mucus production, too, was not greatly increased in the sublethally affected experiment animals (Table 22).

While there is a relationship between species and body or sole size, nutrition and age also determine the surface area of the sole and as a result the effectiveness of molluscicides. However, this relation only applies to members of the same species, not to members of a pest population which comprises several different species.

The investigations of GODAN (1961) provide further insight into the interaction of body size and weight, nutrition, age and metaldehyde sensitivity. As the above example of *Limax flavus* L. shows, the following three patterns between these variables can be established, but they apply only to members of a single species:

A

Body size	Different	(Weight about 1.5 g and over 5.0 g)
Nutrition	Different	(Groups I, II, III)
Age	Identical	(9 months)

In experimental animals of the same age, sensitivity is affected by nutrition which in turn affects body and sole size. Slugs fed on a carbohydrate and protein rich diet (group III), which have a higher growth rate and reach a larger size, require a longer recovery period and are therefore more sensitive than the animals reared on an unbalanced diet, which remain smaller in size (groups I and II). The surface area of the sole is here the determining factor.

B

Body size	Different	(Weight about 1.5 g)
Nutrition	Identical	(Group III)
Age	Different	(3 and 9 months)

With identical nutrition, the younger and therefore smaller slugs are less sensitive than the older and thus larger, sexually mature animals. Here age plays a role.

C

Body size	Identical	(Weight about 1.5 g)
Nutrition	Different	(Groups I, II, III)
Age	Different	(3 and 9 months)

With identical body size, the younger, better fed gastropods of group III are just as sensitive as the older specimens of groups I and II. Nutrition here is important.

These interrelationships indicate anew the difficulties with which control of pest gastropods is confronted. They also explain the apparently contradictory practical observations. BÖHM (1957), for example, speaks of the failure of metaldehyde in treating small, young slugs, while FRÖMMING (1955, 1957) regards juveniles

Table 23. The effect of nutrition before exposure on the attractiveness of bait pellets. (After GODAN 1966)

Molluscicide	Nutritional group	% Slugs on bait	P	Attractiveness of bait
Isolan 0.9 mg	I	75.7	<0.01	Attractive
	II	63.0	>0.05	Indifferent
Mexacarbate	I	69.2	$=0.01$	Attractive
3.75 mg	II	40.0	>0.05	Indifferent
Metaldehyde	I	64.7	$=0.01$	Attractive
(Pellets)	II	62.5	>0.05	Indifferent

P = probability of error

as more sensitive than adults. The composition of a population of pest gastropods in an infested area also leads to differences in the effectiveness of molluscicides.

There are further problems connected with nutrition. The food preference of gastropods in a particular area can change, so that the reputed attraction of some commercial molluscicide preparations may be questioned. Change in behaviour of gastropods with regard to food can occur before and during control measures (Table 23). Older *Limax flavus* L. were fed for five weeks with potatoes (group I) or with ripe tomatoes (group II) (GODAN 1966). In "single animal choice" tests, the slugs were allowed to choose between this food and a molluscicide (a carbamate-bran mixture with 0.12% isolan or 0.5% mexacarbate or a trade preparation of metaldehyde). The concentration of the active ingredient was 0.75 g per pellet.

In the case of the experimental animals fed exclusively on ripe tomatoes, the molluscicide-bran mixture was no more attractive than the fruit. Only those slugs which had been fed for weeks exclusively on potatoes preferred the bait, irrespective of the poison used. This result confirmed the observations of THOMAS (1948) that metaldehyde bait was without effect in fields of tomatoes. Succulents and strawberries as well as lettuce are usually preferred by phytophagous gastropods to most baits containing starch as an attractant. Molluscicide pellets applied in clumps, as was the earlier practice, were only effective when eaten by the gastropod pests; in this case the type of crop may have an adverse effect on the outcome of control measures.

Finally, it must be emphasized that feeding mainly with animal protein improves the condition of the gastropods and increases their capacity to resist the action of molluscicides. Thus adult gastropods reared for months on casein and yeast were more resistant than animals of the same age and roughly the same size which had been reared on a purely vegetable diet.

Rapid growth on a normal diet (plant and protein) leads to an increase in surface area of the sole and thus to greater sensitivity to molluscicides. This applies, however, only to members of the same species.

The type and amount of preferred food available in an area influences the attractiveness of the bait (in tests using bran); when the available food is juicy (fruits, tomatoes and others), the attractiveness of the bait is lost; when the food is dry (potatoes, amongst others) the bait remains attractive.

3.1.1.6 Behaviour, Learning Capacity, Habituation

Formerly, bait pellets were usually applied in clumps, a practice which has not, however, stood the test of time. For behaviour, learning capacity and habituation of gastropods can all affect the success of chemical control (GODAN 1961).

With regard to **behaviour**, unprotected, exposed areas are avoided in the search for food. The pellet bait should therefore not be laid out in an open area, but placed close to a wall, the borders of beds, near fences, or the edge of ditches, under an overhanging structure or leaves, stones and similar objects which serve as hide-aways. Specimens of *Limax maximus* L. were placed singly under a shelter, in dishes with 0.25 m^2 surface area, containing damp garden soil and 24 h later the position of slime trails of the slugs which had been active during the night were recorded. If the animal was in a strange environment for the first time, the mucus trails ran mostly around the edge of the experimental dish and the bran clumps and metaldehyde bait in the middle were not noticed (Fig. 84). If the bait was situated at the edge of the dish, it was inevitably found and eaten by the slug as it crept along the wall of the dish. Only after repeated trials with the same animal in the same dish were the trails found to run crisscross over the soil. Better results were obtained when the bait was placed on the edge of the dish, irrespective of whether it consisted of clumps of pellets (each 1.5 g), of potato slices, pure bran or a bran-metaldehyde mixture. The same behaviour was observed in *Deroceras reticulatum* (Müller), *Deroceras laeve* (Müller), *Lehmannia marginata* (Müller) and *Helix pomatia* L.

The following series of experiments on **learning** behaviour also demonstrates that the success of control measures depends to a large extent on the location of the bait clumps. To test learning capacity, each slug was offered an unpoisoned bran bait, placed on 3 successive days at the edge of the dish; on the 4th day the position of the bait was altered, being placed in the middle of the dish. The maze of slime trails found next morning at the edge of the dish show that the slug had searched for the bait at the old site. After unsuccessful searching the slug returned to its hide, and the bait in the middle of the dish remained undiscovered (Fig. 85). This result, based on tests with over a hundred experimental animals, shows that slugs are able to learn (also ALLOWAY et al. 1973).

The bait must therefore be laid out at the same place if the application is to be repeated. If this is not done, the success of the control measures is jeopardized for the simple reason that the behaviour of the slug or snail is such that the bait cannot be found.

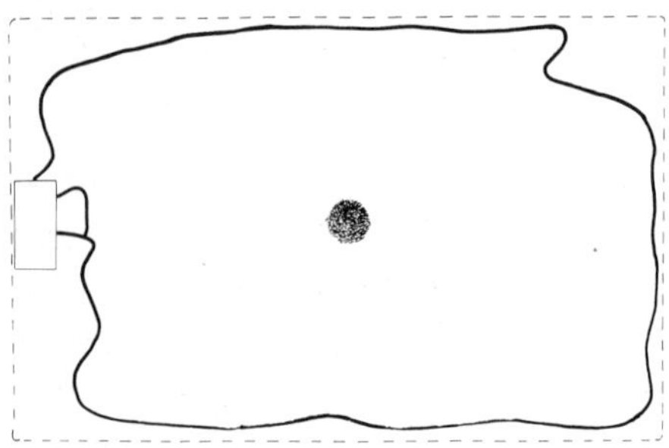

Fig. 84. Trails made by crawling *Limax maximus* in strange environment; *rectangle* shelter; *mass of dots* bran bait

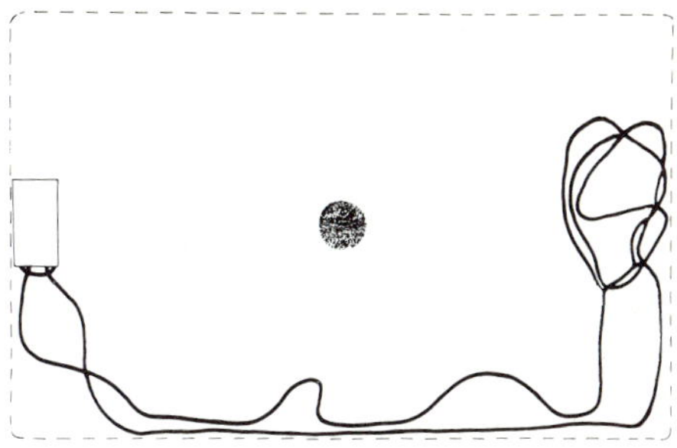

Fig. 85. Trails made by crawling *Limax maximus* searching for bait at the old position

Now to the question of **habituation.** In the field, slugs and snails can recover from sublethal doses of molluscicides, either as a result of high humidity during the application of metaldehyde, or of a relatively dry environment when carbamate has been used, so that it is necessary to make repeated applications, particularly of metaldehyde, to the infested area. The gastropods therefore come into contact with the poison several times and take up small quantities of it. The possibility of habituation to the molluscicide must therefore be borne in mind. One and a half year-old *Limax maximus* L. were fed with metaldehyde-bran paste (20 g wheat bran moistened with 200 ml of a 0.0001% metaldehyde solution in distilled water). This extremely low molluscicide concentration had been shown in preliminary tests to be completely harmless; there were no signs of poisoning. Each slug was presented with a clump of 0.5 g of the mixture; this was checked and renewed every day in order to confirm that it had been ingested (group I). Untreated animals were given daily the same weight of fresh bran moistened only with water. Mortality of the two groups was normal and equally low during the five weeks of the experiment. The test to establish the occurrence of habituation was conducted in the form of a choice experiment. Each of the slugs was given a choice between moistened bran and the same amount of metaldehyde-bran mixture (in proportion of 2:1). After 30 min almost all (90.3%) of the untreated animals (group II) were found on the metaldehyde bait, while of the slugs which had been exposed to metaldehyde over several weeks, only 63.3% chose the bait. They had become accustomed to the poison and did not react to it as anything particular.

Practical observation shows that molluscicide pellets achieve a satisfactorily high mortality rate only during the first three days of application as a result of great attractiveness; thereafter their effectiveness dwindles rapidly, as Barnes and Weil (1942, amongst others) established in the case of a metaldehyde-bran bait.

The problems thus far discussed relate to the gastropods themselves. There are, however, problems associated with the molluscicides.

3.1.2 Repellent Effect of Molluscicides

Many chemical molluscicides are repellent at high concentrations but are attractive at very low concentrations; between these two extremes the response to them is indifferent (Godan 1958, 1959). In choice experiments (at 20 °C to 22 °C), 1-year-old *Limax flavus* L. displayed the following behaviours (Godan 1961): the slug

Table 24. Repellent effect of molluscicides on slugs: choice experiments. (After GODAN 1965)

Molluscicide	Species	Number	% on "treated" places	P	Evaluation
Isolan	*L. flavus*	36	26.4	<0.01	Repellent
Mexacarbate	*L. flavus*	36	0.0	<0.001	Repellent
Promecarb	*L. flavus*	12	0.0	<0.01	Repellent
Promecarb	*Lehmannia marginata*	12	20.8	<0.05	Tendency repellent
Promecarb	*Cepaea nemoralis*	10	10.0	=0.01	Repellent

avoided potato slices which had been treated not only with the heavy dose (4% to 6% poison) already present in the preparation, but had also been moistened with concentrations of 1% to 0.1%. By contrast this slug showed a preference for slices treated with a 0.0001% metaldehyde solution. This exceptionally low concentration was so attractive to the slugs that they used the treated slices as resting places and for oviposition. At concentrations between 0.01% and 0.0001% metaldehyde, the reaction of the slugs was indifferent. Pure metaldehyde is repellent (SPEYER 1954). This repellent effect could be used in practice to inhibit feeding. According to MORETON (1953), application of 1 kg/ha protected the plants from being eaten by gastropods for up to 10 days, although this concentration was not lethal; an amount of 5 kg/ha was even more strongly repellent and afforded protection for up to 14 days from the surviving gastropods (total mortality 90%).

A repellent effect of high concentrations and attractiveness at low concentrations has also been established for carbamate. Table 24 shows the result of choice tests with treated and untreated halves of the experimental dish (GODAN 1965). It was not possible to use a stronger dose of the wettable powder than 2% because the gastropods no longer reacted normally as a result of the onset of poisoning. Since each half dish was wetted with 2.5 ml of a 2% concentration, the following actual amounts of molluscicides were applied to each half dish:

Isolan – preparation with 6% active ingredient = 3.0 mg
Mexacarbate – preparation with 25% active ingredient = 12.5 mg
Promecarb – preparation with 50% active ingredient = 25.0 mg.

Amongst the carbamates tested, mexacarbate had the strongest repellent effect. *Lehmannia* spp. totally rejected lettuce treated with a 2% concentration of carbaryl and zectran and to a large extent also that treated with isolan.

Confirmation of the repellent effect of carbamate has been provided by PAPPAS and CARMAN (1955) (8% isolan-bran bait) with *Helix aspersa* Müller, by GETZIN and COLE (1964), HARPAZ and OSERI (1961), PROTA (1960) and WILBUR and YONGE (1964/1966). DAXL (1968), on the other hand, describes the attractiveness of low doses. In this case, again using choice tests, bran bait with either 0.6% or 0.3% isolan attracted 81.8% of Field Slugs, a significantly larger proportion than was found on the poison-free bait clumps (18.2%).

The repulsion of gastropods by a relatively high concentration of active ingredient in the molluscicide preparation is so strong that the attractiveness even of the

Fig. 86. *Lehmannia marginata:* repellent effect after application of isolan: the slugs crawl onto one another

otherwise preferred protein substances (Chap. 3.1.3) is lost. Thus carcass meal bait lost its attractiveness when 5% or even 2% metaldehyde of 1% carbamate (isolan) was added. According to DAXL, the percentage of gastropods attracted to the 2% metaldehyde-mixture was 47.1%, to the pure carcass meal bait it was 52.9%, which represented an indifferent reaction, whereas only 33.3% were attracted to bait containing 1% isolan as compared with 66.6% attracted to the pure carcass meal bait; the poisoned bait thus had a repellent effect.

GODAN (1961) observed a shock reaction of gastropods on contact with metaldehyde. The animals at first crawled around energetically, to escape from the treated area; in the experimental dishes they collected along the molluscicide-free edge. They subsequently tried to avoid contact with the metaldehyde by using their own mucus trail or that made by another member of the same species. Lifting of the anterior part of the sole and withdrawal of the head were observed in *Limax* and *Helix*. Gastropods thus try to avoid areas which have been treated with high concentrations of metaldehyde. Similar behaviour was observed in response to carbamates (GODAN 1965). About 10 to 15 min after the onset of the effect, flight reactions became evident. If escape is impossible, slugs will creep over one another and form clumps, with the animal lowest in the clump trying to get on top of the others (Fig. 86). Snails climb on to each others' shells or direct the shell opening upwards to avoid contact with the poison.

The mortality rate provides indirect evidence for the altered behaviour of gastropods in response to high concentrations of molluscicides: mortality is less at high concentrations of the molluscicide than at lower. The repellent effect is responsible. This was clearly shown by feeding experiments with mexacarbate-bran mixture which contained either 2.5% or 0.5% molluscicide and were used in the form of pellets. At the higher dose, only 10% of Cellar Slugs were found on the bait and mortality was, at 25%, correspondingly low. By contrast, the mixture with the lower dose was preferred and attracted 95% of the slugs, and resulted in a high mortality (100%). Similarly, with *Arion ater* (L.) a 4% carbamate preparation caused a lower mortality than one containing 2% (CROWELL 1967).

This apparent discrepancy can be explained as follows: the repellent effect of the strong molluscicide-bran mixture means that only a few gastropods approach the bait and feed from it, and then consume only a small, non-lethal amount. Bait with a lower concentration of molluscicide causes a higher mortality because it does not repell the animals; on the contrary, it attracts them, so that large quantities are consumed and a lethal dose reached.

Instead of the repellent effect which acts when the gastropods are at some distance from the molluscicide a **deterrent** effect may come into play, since some animals were observed to taste the bait. In this case the animals are not repelled from a distance, but only after making contact with the treated surface or tasting the poisoned bait. This deterrent effect can, however, result in inhibition of feeding; the gastropods are driven away and seek their food elsewhere.

3.1.3 Bait and the Attractiveness of Bait Pellets

Since only high concentrations of poison offer any guarantee of a lethal effect on gastropods, an attractant must be added to molluscicide pellets. Metaldehyde bait contains 4% to 6% of the poison; such high doses do not attract gastropods, but rather repell them. Since the different species display preferences for many different types of food, so various different attractants must be added to the preparation. It is, however, uneconomic for the manufacturers of these products to alter the attractant from case to case. For the moment this problem remains unresolved and is a further complicating factor in gastropod control.

At first carbohydrates such as wheat, bran, fruit pulp, boiled potatoes, and mollasses were considered, and their attractiveness in mixtures with molluscicides was tested (BARNES and WEIL 1942). *Helix pomatia* L., for instance, was offered a choice, at a distance of 15 cm, between wheat and rye flour, wheat bran and crushed grain, sucrose and dextrose. The snails crawled in various directions, but never directly to any of the substances. It was only by chance that they encountered one of the baits and ate from it. Sucrose and dextrose were always avoided, even when strewn immediately in front of the animals. The addition of flour to dextrose did not attract the animals for long. This is the more surprising since earlier sugar had often been added to bait to make it more attractive. Rye and wheat flour, wheat bran and crushed grain were accepted by *Helix pomatia* L., yet a typical attraction response was not observed. Nevertheless these substances, too, are included in molluscicide pellets available today. Furthermore, dry milk powder was ignored, but moist casein-paste was eagerly eaten (GODAN 1958). In addition to the very large amount of carbohydrate [according to RICOU and FERRET (1973) 85%] in the form of wheat or rice bran which is added to act as attractant and as carrier base, a fungicide is also necessary to prevent pellets becoming mouldy. This too can be repellent in high concentrations (WEBLY 1965) and can reduce the effectiveness of the bait pellets.

It now appears that so-called attractants used in baits do not appeal to all gastropod species, but only to those which feed on a starch-rich diet (potatoes, bulbs, tubers, et cetera), like *Limax maximus* L. and *Limax flavus* L. or to omnivorous species such as *Deroceras reticulatum* (Müller). The attractiveness of bran has been questioned even in the case of Cellar Slugs and Field Slugs (DAXL 1968; FRÖMMING 1957).

Wheat bran is at best only slightly attractive to purely phytophagous species like *Helix aspersa* Müller (CROWELL 1967). For this snail, which is a serious pest to citrus plantations in California, freshly prepared orange pulp (a by-product of citrus farming) has proved to be an effective bait (LANGE and MACLEOD 1941; LEWIS and LAFOLLETTE 1942 and others). The use of carbohydrate baits is also uncertain of success in the case of the terricolous species like *Milax gagates* (Drap.) and *Milax budapestensis* (Hazay) (VAN DEN BRUEL and MOENS 1958), as well as the phytophagous Arionidae. According to THOMAS (1948), good catches and therefore high mortality [*Deroceras reticulatum* (Müller)] is achieved using metaldehyde-bran bait; this is not the case with *Arion hortensis* Fér., though better results are obtained when casein is added. Some gastropods also eat animal matter (dead members of the same species, earthworms, carrion and faeces) and these are attracted by bait which contains dried blood meal or has a meat odour.

Thus when carbohydrates, formerly regarded as *attractive*, are used, mostly in the form of bran as a carrier base, it is mostly a matter of chance whether the gastropods find the bait or not. Most molluscicides are used today in the form of bait.

The attractiveness of the bait is increased by the addition of protein (carcass meal) as a carrier base. DAXL (1968) tested the effect of dried blood, fish meal, casein, bran and vegetable matter on the following species, and arrived at the following order of preference:

Deroceras reticulatum:
1. vegetable diet
2. carcass meal
3. casein
4. bran
5. fish meal
6. blood meal

Lehmannia marginata:
1. vegetable diet
2. casein
3. carcass meal
4. bran
5. blood meal and fish meal

Arion hortensis:
1. vegetable diet
2. casein and fish meal
3. bran
4. carcass meal
5. blood meal

Limax flavus:
1. carcass meal
2. fish meal
3. casein
4. blood meal
5. bran
6. vegetable diet

Apart from vegetable matter, protein concentrates must be recognized as being more attractive than bran. Casein is also attractive to gastropods, although its effect is limited to certain species; according to THOMAS (1948), casein increased the catch of *Milax budapestensis* (Hazay) and *Deroceras reticulatum* (Müller), but not of *Arion hortensis* Fér.

These investigations show clearly the difficulty of formulating a bait which competes effectively with the much preferred plant food, which is normally abundantly available. The Cellar Slug, *Limax flavus* L., is an exception, feeding readily on casein and flour in store-rooms. Surprisingly, bran results in a better catch of *Deroceras laeve* (Müller) – which was not examined by DAXL – than does carcass meal. The explanation of this lies in the feeding habits of this slug, which attacks with preference sugar-rich fruits, like strawberries.

It seemed relevant to compare the attractiveness of carcass meal with that of wheat bran in a mixture with a molluscicide. In tests with isolan (0.3% or 0.6% active ingredient), 36 Cellar Slugs chose the carcass meal bait, but only 4 the brain bait; with metaldehyde, 90% of the animals were attracted to the carcass meal bait and only 10% to the bran. For both metaldehyde formulations, bait containing carcass meal attracted a greater number of slugs and resulted in higher mortality than did the bran bait.

The problem seemed at last to have been solved, as has already been mentioned, by the use of carcass meal. Unfortunately it has become clear first, that the required high concentration of molluscicide either reduces or completely destroys the attractiveness of the protein additives, and, secondly, that after being fed for several days on carcass meal, the slugs become habituated to this, lose interest and increasingly return to their preferred food, the cultivated plants. Evidence that gastropods become habituated to the bait is provided by choice experiments of DAXL (1968) with *Limax flavus* L., which had previously been fed for about four weeks on carcass meal. After the 1 st day, of this diet, 90.6% of the animals were attracted to the carcass meal, 9.4% to the plants. After the 3rd day, these values were 37.9% and 62.1% respectively. After the 14th day, of this uninterrupted diet of protein, the result obtained after the 1 st day was reversed, with values of 18.5% and 81.5% respectively. This shows that animal protein is only initially more attractive than plant food. Furthermore, the replacement of carcass meal with another protein substance, which is known to be an attractant, does not lead to recovery of the lost effectiveness of the carcass meal bait.

According to RAMMOS (1962) oils from the fir tree *(Abies alba Mill.)* and its cones are attractive. Oils are, however not suitable for use in bait formulations unless perhaps in the form of pastes, but these would give rise to problems in distribution and in durability. Slugs avoided mustard oil, camomile oil and castor oil; they accepted, but in no way showed a preference for Palmin, olive oil, margarine and butter.

It may be mentioned that beer is attractive to *Deroceras reticulatum* (Müller) and *Vaginulus plebeius* Fisher, and moistening metaldehyde-bran bait with bear is recommended to increase its attractiveness (SELIM 1974; SMITH and BOSWELL 1970). According to MOENS and FRASELLE (1980), fermenting barley malt is very attractive to slugs.

The use of furfurol (α-furfurylaldehyde) has been considered as a means of making bait attractive. This substance, however, when mixed in different concentrations with sea sand, to which gastropods have a neutral reaction, released various responses. The 0.1% concentration was repellent to *Deroceras reticulatum*

(Müller), an effect which wore off in time due to the slug becoming habituated to the substance, while the 0.05% concentration had no effect on *Deroceras* or on *Helix pomatia* L. At 0.01% it was attractive to *H. pomatia*, but still did not attract *Deroceras*. The extremely low concentration of 0.005% had no effect on the behaviour of the experimental animals, although furfurol pellets were only 5 to 10 cm apart.

It is thus clear that the problem of finding a substance as carrier base of the molluscicide and which is really attractive to most, even if not all the gastropods in an infested area has not been solved.

3.1.4 Reasons for Failure of Chemical Control

Failure in the application of "slug pellets", and also false assessments in their official testing, leading to an over- or under-estimate if the efficacy of the commercial product, can arise in a number of ways.

Success of control measures resulting in possible over-estimation of a molluscicide in official tests may be due to:

1. High temperatures combined with dry weather *before* control is started. This leads to a disturbance in the water balance of gastropods which means that the animals are in a less than healthy condition. This is because in warm conditions with low humidity of soil and air, not only is water continuously lost through mucus production and respiration, but it cannot be replaced. They may thus die without even making contact with the molluscicide.
2. Direct sunlight which weakens and damages gastropods, even without any action of the molluscicide, in the same way as the climatic factors mentioned in point (1).
3. Disregard of variation in sensitivity of the species comprising the gastropod fauna in an infested area; this may include several different species, each covering a considerable age span: in general, it can be said that smaller species are more sensitive than larger ones. Tests on molluscicides should therefore not be confined to smaller species or to those with a high water content, as these are particularly sensitive.
4. Disregard of age: before the onset of sexual maturity gastropods are particularly resistant, so that they survive the control measures, and later, when they reach reproductive age, give rise to a new pest population. Tests should therefore not be carried out on gastropods in a particularly sensitive phase, such as the height of the reproductive phase or immediately after hatching from the egg.

Failure of control measures resulting in possible under-estimation of a molluscicide preparation may be due to:

1. Low temperatures: these inhibit gastropod activity, the animals remain in their retreats, do not creep about and so do not make contact with the molluscicide.
2. Heavy rain or irrigation during or shortly after completion of control measures, but before assessment of the result: gastropods move from neighbouring, non-irrigated areas into the optimally damp area, with the result that in the testing area,

more living gastropods are found at the end of the experiment than were present at the outset.

3. Disregard of the activity rhythm: gastropods usually creep about only during the twilight hours of late evening and early morning, when dew has fallen; continuous rain, however, inhibits activity. It is advisable, therefore, to lay the bait only in the late afternoon and to check the results next morning.

4. Results may be affected by the habitat of the particular species being controlled: species which live *in* the soil are more difficult to control than those which live *on* the soil; those species which live *above* the soil on leaves, shrubs and trees are equally difficult to reach.

5. Snails seem to be more resistant than slugs since they find a certain amount of shelter in their shells.

6. Application in early spring after hibernation and, in the case of snails, during aestivation, both periods when resistance is particularly high.

7. Disregard of the behaviour, learning ability and habituation of gastropods when using pellets: the distribution of bait over open, unprotected areas which are generally avoided by gastropods, changing the place at which the bait is placed when it is renewed, and allowing too great a distance between the pellets so that they are not within the olfactory range of the gastropods, a distance which anyway varies from species to species.

8. Disregard of the food preference *before* and *during* control measures: in the case of juicy fruits, like tomatoes, the enticing effect of attractants (mostly bran) is reduced.

9. Influence of type and quantity of available food: plant food is mostly preferred to bait; bran is not attractive to all species.

10. Disregard of feeding habits of experimental animals reared in the laboratory: gastropods fed exclusively on animal protein (caracass meal), albeit smaller, are nevertheless more resistant than those on a mixed diet (plant food and protein), although the latter grow faster and are therefore larger. In tests only gastropods reared on a mixed diet, which is normal for gastropods in the field, should be used.

11. Disregard of the species-specific effect of molluscicides so that tests are made only on larger specimens of gastropods: in comparison with smaller species, these have considerable powers of resistance. Tests should be made on large and small species.

12. Disregard of age: juveniles before sexual maturity and senile slugs past the reproductive phase are extremely resistant. Thus to arrive at a fair assessment of a formulation, animals of different ages should be used, including some which are at a very susceptible stage, during oviposition, for example, or shortly after hatching.

3.2 Chemical Control Methods

Control with chemical substances is today still regarded as the most useful method, particularly over large areas. Since terrestrial gastropods are more important in plant protection than freshwater snails, substances used against them will be considered first. Before the toxicity of metaldehyde had been recognized, bait containing compounds of arsenic, such as calcium arsenate, was used; this was still recommended in 1953 in California for the control of *Theba pisana* (Müller) and *Helix aspersa* Müller (BASINGER 1953). Other baits used were Paris Green, calcium cyanamide, copper sulphate, and later, sodium pentachlorophenolate (VAN DEN BRUEL and MOENS 1960; FRÖMMING and PLATE 1952; GETZIN and COLE 1964; MEAD 1961). The last three compounds mentioned can also be used against freshwater snails.

In the Federal Republic of Germany molluscicidal products can only be placed on the market with the approval of the Federal Biological Research Centre for Agriculture and Forestry after registration.

Of the many substances for control of land gastropods which have been officially tested, those with the most useful practical application are metaldehyde and mercaptodimethur (methiocarb), and they are the active ingredients in commercial formulations. Since the problems associated with the use of metaldehyde have not yet been solved, further chemical compounds are being tested for their possible molluscicidal effects, even substances with other uses, such as fertilizers, fungicides, insecticides and herbicides.

JUDGE (1969), for example, tested the effect of 74 substances on Field Slugs, including niclosamide, chlorinated hydrocarbons (DDT, dieldrin, chlordan, endosulfan); carbamates, such as aryl-N-methylcarbamate (carbaryl, mesurol, amongst others), oxim-carbamate (aldicarb, amongst others), N-acetyl-methylcarbamate (Hercules and others) and N,N-dimethylcarbamate (isolan, pyramat); phosphoric acid esters such as phosphate, phosphorthioate, phosphordithioate (phorat, amongst others), pyrophosphate, phosphonate, phosphonothioate, phosphonodithioate, phosphoramidate and phosphoramidothiate. Of these, 23 have proved to be really useful molluscicides when tested under comparable conditions (spraying young pea plants); in particular, from the series of insecticidal phosphoric acid esters, the carbamate mercaptodimethur (mesurol), temik (as pellets with 10% aldicarb) likewise a carbamate, as well as phorat (10% in pellets). These substances will be mentioned again in discussion of the particular groups of molluscicides.

3.2.1 Molluscicides Used Against Terrestrial Gastropods

In the Federal Republic of Germany, commercially available molluscicides, together with the application rate in each case, are listed in the Catalogue of Approved Plant Protection Substances published by the Federal Biological Research Centre for Agricultural and Forestry in Brunswick; this Index appears regularly with supplements. Official testing of molluscicides before they are put on the market is a legal obligation.

The aim is to achieve a "uniform, environmentally oriented system of evaluation and registration of plant protection substances in Europe" (HERFS 1974). Methods of testing as a basis for registration of molluscicides have been discussed in France and Belgium (MALLET 1974; MOENS 1971, 1972; RICOU andFERRET 1973). The authorities responsible for the control of gastropod pests in the different countries of Europe, and in North America (USA and Canada) should improve their cooperation so that control of gastropod pests and the development of molluscicides can be correlated (WEISS 1973).

In the Federal Republic of Germany, the molluscicidal properties of commercial products are tested in both feeding and contact experiments. Among the methods published by the Federal Biological Research Centre for Agriculture and Forestry (Guide Lines for Official Testing of Plant Protection Products) there are specific "Guide Lines for Testing the Efficacy of Molluscicides on Slugs"; these are conducted both in the field and in the laboratory (GODAN, MOSEBACH and PLATE 1965). The experimental methods described in the literature (HENDERSON 1968, 1969; JUDGE 1969; SHARAF and GHALEB 1978 and others) are not suitable for tests on molluscicides.

Furthermore, methods are available for testing molluscicides on freshwater snails, which involves a molluscicidal formulation which floats on the water surface (STRUFE et al. 1965 and others).

The **evaluation** of a molluscicide is based on tests of its toxicity, persistence of effectiveness, attractiveness to gastropods (in the case of bait) and also the chances of recovery of affected slugs or snails. With preparations in the formulation of pellets, swelling capacity and capacity to withstand mould and decay are also considered. Laboratory tests are usually made on slugs reared in the laboratory (Chap. 1.4).

A molluscicide formulation must not only be safe for warm-blooded animals (mammals including man, birds) but it must also be effective against several different gastropod species with different habits, and also remain active with high soil humidity and in cool weather conditions. Bait should remain attractive and toxic for at least one week, and should swell up readily, because gastropods feed only on damp, swollen pellets, not on dry ones.

A series of tests is made, using for each molluscicide three small plots (test product, alternative formulation of the same ingredient applied in the same way, untreated area) which must be largely uniform with respect to the density and type of plants, nature and humidity of the soil, and light conditions (shade). The experimental areas should ideally be enclosed by hedges, slopes, pastures and lawns or ditches, because these serve as refuge areas for slugs and thus ensure a continuous supply of experimental animals to the areas under observation.

In evaluating the results of tests, only gastropods of one species and of approximately the same age should be considered together; this presents no difficulties when laboratory reared animals are used.

A method which differs from the normal laboratory procedures is described by MOENS (1972). This is based on assessing not the mortality of gastropods, but their feeding intensity some days after application of molluscicide pellets (Fig. 87). Five rows, each with 10 wheat seeds (altogether 50) were laid out in a terrarium containing clods of soil. The damp clods and spaces between them closely resembled the natural habitat of slugs in the field. The required humidity (100% RH) was ensured by capillary action from below. The temperature was allowed to fluctuate between 10 °C and 25 °C, because the activity rhythm of slugs is partly determined by rise and fall in temperatures. The molluscs were kept for 72 h in containers which were arranged in six sets each of three terraria. Of these three terraria, the first contained

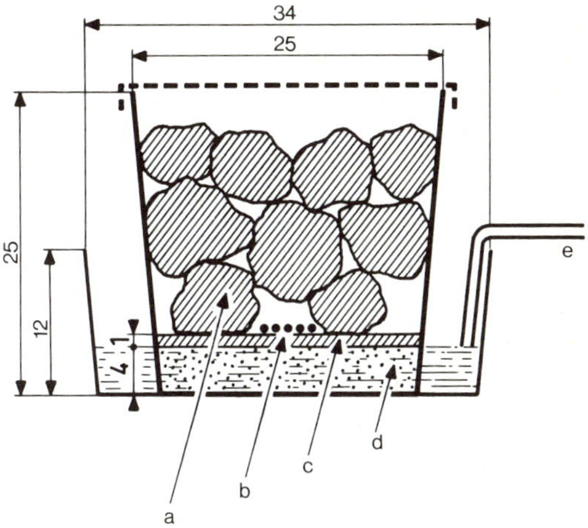

Fig. 87. Test to determine inhibition of feeding activity by molluscicides. *a* sterile earth clods; *b* 5 rows of 10 wheat seeds each; *c* sterile soil; *d* fine sand; *e* water reservoir with pipe

100 mg of the unknown test product, the second 100 mg of a comparative product of unknown effect, while the third was untreated. The six sets were further handled differently: in three, the molluscicide granules were applied mixed with the grains, and in the other three at a distance of 3 cm away from the grains.

In both the application of molluscicidal formulations in practice and their assessment in official tests, the aut- and demecological factors mentioned earlier must be borne in mind. If these are overlooked, the preparation could fail in practical application, despite a positive evaluation in the test; it might thus be regarded as inadequate, even though the active ingredient is highly toxic to gastropods. This unavoidably incorrect conclusion leads to failure in practice, and this in turn leads to the unfortunate condemnation of a molluscicide which is in fact effective. Under-estimation of the efficacy of a molluscicide can also occur, if it rains during, but not before the start of the test period. In this case it is possible that in the initial estimation of gastropod infestation, taking into account both species and total number, a relatively low population density in the experimental area is arrived at, but that this increases, despite the positive effect of the molluscicide. An initial population estimate should therefore be made only after thorough and prolonged irrigation of the area; stagnant water, which is avoided by slugs and snails, should, however, not be allowed to collect.

Metaldehyde will be considered first, since it is used as the active ingredient in most commercial molluscicide formulations.

3.2.1.1 Metaldehyde

Metaldehyde $[(CH_3-CHO)_4]$ is synthesized by polymerization of acetaldehyde in suspension in ethanol in the presence of acidic catalysts. Metaldehyde crystallizes in needle form and also as tetragonal prisms; in its pure state it is tasteless and odourless. It is almost completely insoluble in water, a suspension containing only very small amounts of metaldehyde in solution: the amount will vary with temperature: 0.018% at 0.5 °C; 0.020% at 17 °C and 0.026% at 30 °C (Cragg and Vincent 1952).

Metaldehyde is effective either through ingestion or as a result of dermal contact. It was used as early as 1934 in South Africa, while in Europe it was known only as a solid fuel in the form of meta-tablets. Its molluscicidal properties were discovered in Europe purely accidentally, in 1936, when farmers in southern France found dead and dying gastropods on meta-tablets left behind by picnickers. As a result, tests were made in France, England and later in other countries in Europe, in the USA and in Australia. For reviews of the main literature and critical discussions of the pros and contras of the use of metaldehyde, the reader is referred to MAYER (1957) and MEAD (1961).

The molluscicidal effect of metaldehyde is based on two properties: first, its **irritant** effect, which causes gastropods to secrete large amounts of mucus, with resulting desiccation, and secondly on its **toxicity** (nerve poison) at high concentrations (STRINGER 1946).

Metaldehyde initially ellicits excitement, then an over-production of mucus and later **loss of activity,** muscular contraction and finally paralysis and death. Following the heavy secretion of mucus, the movements of gastropods poisoned by metaldehyde are inhibited, inhibition of mobility and mucus production being related (DAINTON 1954). Thus *Limax maximus* L. becomes motionless, but is not yet dead, after a 50% weight loss as a result of mucus production. CRAGG and VINCENT (1952) investigated the loss of body weight in Field Slugs after 3-h contact with a 0.02% metaldehyde solution (dose 6.3 $\mu g/cm^2$); this was compared with animals treated only with distilled water (Table 25).

Compared with the "distilled water slugs", the loss of weight caused by metaldehyde through water loss is considerable. Other experimental results obtained by DAXL (1970) on the influence of the molluscicide on water balance of gastropods will be discussed in another connection (Chap. 3.2.1.3).

Desiccation leads to death of Field Slugs within 7 to 17 h (70% RH, 15 °C). When, however, the body water lost as a result of excessive mucus production can be replaced within a short time as a result of rain, dew or already existing high humidity, slugs recover; they also recover when washed off with water, while other animals, just as strongly poisoned, but not brought into contact with water, die. Control measures are thus more likely to succeed in dry, warm weather, because the water deficit cannot be made good rapidly enough when the air is dry; rainy, cool weather, on the other hand, can lead to failure of control measures. In these

Table 25. Comparison of percentage loss of body weight after contact with metaldehyde and with distilled water. (Modified from CRAGG and VINCENT 1952)

Treatment (h)	Metaldehyde treatment		Water treatment	
	Immediately after contact	24 h later	Immediately after contact	24 h later
1	−21.99±3.52	−24.29±6.77	−1.98±4.24	−4.66±3.77
3	−26.93±5.24	−32.60±6.68	+8.30±6.60	+6.75±4.41

− = loss of weight; + = increase in weight

conditions mortality is low; it was only 30% and 35% respectively in the case of *Limax flavus* L. and *Lehmannia marginata* (Müller); 15% and 25% respectively for *Cepaea nemoralis* (L.) and *Helicigona arbustorum* (L.).

Slugs died within an hour of eating metaldehyde-bran bait in warm weather (30 °C and direct sunlight), but after only 24 h in damp, mild weather (RAMMOS 1962). The molluscicidal effect of metaldehyde is thus dependant to a very large extent on temperature and humidity. MOENS (1970) established the following relationship between toxicity and temperature (day and night rhythm) for slugs: *high* toxicity when the maximum temperature lay between 20 °C and 25 °C, and the minimum between 10 °C and 15 °C; *moderate* toxicity when the maximum temperature lay between 15 °C and 20 °C, and the minimum between 5 °C and 10 °C, and *no* toxicity when the temperature was constantly low (maximum 4 °C to 7 °C, minimum 0 °C and 5 °C).

Efficacy of metaldehyde bait, which is dependant on the animals being very active, is optimal when rainfall during the control period does not exceed 15 mm and the prevailing temperature lies within the following limits: the minima between 9 °C and 16 °C and the maxima between 15 °C and 20 °C (RICOU 1964). MALLET (1968) achieved the best results after application of metaldehyde dust at maxima of 18 °C and 20 °C and minima between 10 °C and 12 °C. The following result shows that mortality depends not only on the dose of the molluscicide, but also on the temperature: the mortality of slugs injected with a suspension of metaldehyde (20 µg/slug) was 28% at 20 °C at a dose of 20 µg per slug (CRAGG and VINCENT 1952), while three times this dose (66 µg per slug) was needed to achieve this rate of mortality at 10 °C (HENDERSON 1969). The influence of humidity is demonstrated by the fact that mortality of Cellar Slugs was 85.7% higher at 44% RH than at 75% RH (DAXL 1970).

Lethal dose: According to MALLET (1968), for Field Slugs this is 0.01 mg metaldehyde applied to 1 cm^2 surface area of sole and over a 1-h period; according to MÜNCHBERG (1950) this is 2.60 to 3.33 mg/g body weight for *Limax maximus* L. (17 °C and 95% to 100% RH); CRAGG and VINCENT (1952) put it at 0.06 mg pure metaldehyde per 400 to 800 mg slug body weight after 1 h contact, corresponding to 6.3 $\mu g/cm^2$; according to MALLET (1968), the lethal dose for an average sized Field Slug is 0.05 mg. Since one slug usually consumes 2 cm^2 of plant tissue ($\approx$ 60 mg), for example of young rape leaves, during a night, it will take up 0.02 mg metaldehyde applied as a powder, a dose which, although dangerous to the slug, is not lethal. MALLET, therefore, considers the poison to be more effective following dermal contact than after ingestion.

Metaldehyde was at first administered as **toxic bait** mixed with wheat bran, rice flour or maize meal as attractant. In 1950 it was acclaimed as an infallible control measure against terrestrial pulmonates. Later, however, disadvantages became apparent (PLATE and FRÖMMING 1952 and others) such as its great weather dependance and the resulting rapid recovery of slugs which at first appeared lethally poisoned. Metaldehyde was also considered, at the outset, to be attractive; later experiments, showed, however, that this is not the case. The pure substance is, in fact, repelling; only when very diluted is metaldehyde attractive (GODAN 1961), but this is at such low concentrations as to be totally unsuitable for gastropod control, since they are not lethal.

Gastropods are not usually killed immediately by metaldehyde bait, but remain alive for several days, secreting mucus heavily and in a state of paralysis. Thus members of the larger slug species which have a low body water content, often move a certain distance from the bait after they have been poisoned and before they become immobile as a result of excessive production of mucus with resulting loss of water. Furthermore, when too little poison has been taken up, they may recover after rain or irrigation or when air humidity is high. It is therefore advisable on the day after application of the molluscicide to search the terrain for immobile gastropods and kill them, a procedure which is practically feasible over small areas such as gardens and greenhouses.

The degree of **inhibition of activity** as a result of excessive production of mucus differs for different individuals and different species. Thus *Limax maximus* L. have been found motionless close to metaldehyde bait and also at distances of up to 85 cm from it. Even if recovery is possible, the plants under cultivation are spared, at least for a limited time, damage from the feeding activities of the inactivated gastropods.

For applications in the field, the **total quantity** of pure **metaldehyde** per hectare seems to be more important than the concentration in the bait: 5 kg/ha is regarded as necessary by ARNOUX and RITTER (1954) and others. However, the recommended quantities vary considerably, evidence of the uncertainties coupled with the use of this very weather-dependant poison. Thus both low and very high doses are recommended: for control of Field Slugs, MEAD (1961) recommends 10 kg/ha metaldehyde spray (50% metaldehyde powder in suspension in 500 to 1,000 litres of water/ha); and 50 kg/ha dusting preparation (15% metaldehyde powder); LANGE and SCIARONI (1952) recommend 40 to 45 kg/ha (5% metaldehyde powder); according to MAYER (1957), 50 to 100 kg/ha of bait pellets should be used; VAN DEN BRUEL and MOENS (1958) recommend 100 kg/ha giving 100 pellets/m^2, or 140 kg/ha; according to HUNTER and SYMONDS (1970) the required amount is only 10 kg/ha; the Federal Centre for Plant Protection, Vienna (1970, 1971), recommends 300 g in 50 l water/100 m^2 for spray application; 3 to 5 kg/ha for bait in the form of egg-sized clumps at distances of 25 cm, laid out at night, or 6 to 7 kg/ha in fields and gardens or 5 kg/ha on arable land; 50 g/ha in gardens and 10 pellets/m^2 in hot beds and greenhouses; finally also 10 to 20 kg/ha in fields under cultivation and 10 to 20 pellets/m^2 in gardens or 7 kg/ha; according to the Federal Biological Centre for Agriculture and Forestry in Brunswick, 10 to 15 kg/ha of the bait officially registered for cereals, 20 to 30 kg/ha for vegetables. Very much larger amounts are needed in the case of gastropod infestation of lettuce and half the dose

Table 26. Mortality at different temperatures. (Modified from MALLET 1968)

Species	6 °C %	8 °C %	10 °C %	16 °C %	20 °C %	22 °C %	25 °C %
Deroceras reticulatum	0	15	15	40	80	100	100
Arion hortensis	0	10	20	50	50	–	–
Milax sowerbyi	0	10	20	20	20	–	–

should be applied at the time of germination and the other half when the seedlings are thinned out. Further treatment would normally be carried out after formation of the head and shortly before the harvest, since this is the time at which infestation with slugs is first noticed. This second application should therefore not be too generous. Since 1970 slug pellets have been available which have a very low application rate of 3 kg/ha and a pellet size of only 2 mm.

MALLET (1968) established that the most favourable temperature for a high mortality rate (80% to 100%) of Field Slugs using 65 metaldehyde (5%) pellets/m^2 is 20 °C or a little higher, while this temperature and density of pellets caused mortality of only 50% in *Arion hortensis* Fér. and 20% in *Milax sowerbyi* (Fér.) (Table 26).

Metaldehyde is ineffective in cool weather (MOENS and VAN DEN BRUEL 1956). Species-specific sensitivity has also been demonstrated; *Deroceras reticulatum* (Müller) is the most susceptible and *Milax* spp. the most resistant of the three species considered, which included *Limax*.

Cleared **barriers,** 3 to 5 m in width, around fields have proved to be very effective in keeping slugs in check, particularly where a cereal cultivation borders on a field of newly ploughed-in clover, or pastures, meadows, hedgerows and ditches.

Snail infestation of fields and gardens from neighbouring ditches and marshes can be kept in check by treating the areas surrounding fields. Since about 1980 "bait ribbon" formulations (3.5 cm broad, 4% metaldehyde) have been available.

In general, molluscicidal pellets are applied by **broadcasting.** In horticultural establishments they are spread between the plants or pots (in cold frames and greenhouses). Here treatment is carried out throughout the year, the bait constantly being renewed, and with such success that the gastropod population in fact diminishes in the course of time. In this way tulips, chrysanthemums, freezias, calla lilly and ferns can be satisfactorily protected from gastropod damage. In greenhouses, where the humidity is constant (over 70% RH) as well as the temperature (16 °C to 22 °C), metaldehyde bait pellets (2 to 3 g/cm^2) is sufficient for the control of slugs (KOSOGLAZOW 1974).

Few data are available on the effect of metaldehyde on snails. The problems involved in their control, which are due to variations in sensitivity associated with aestivation, species-specific factors and the habit of living on leaves and branches some distance above the ground, have already been mentioned. In addition, technical factors play a role (CAIRASCHI and LECOMTE 1973).

As has already been said, damp warm nights, which increase the activity of gastropods, are optimal for the effectiveness of metaldehyde if followed by a warm dry day, during which slugs are unable to correct their water content. Sunlight rapidly depolymerizes metaldehyde, and thus inactivates it. Whether applied as a spray, dusting powder or bait pellets, the molluscicide should therefore as far as possible be spread *under* the leaves. Bait, in particular, should not be exposed to direct sunlight, first to protect the metaldehyde from the effects of sunlight and secondly, because gastropods are mostly located under leaves and in the soil, if they do not happen to be feeding on leaves above the soil.

Metaldehyde has also been tested in water on freshwater snails (Lymnaeidae, Planorbidae); its only effect was to stimulate heavy secretion of mucus (RAFFY 1939).

Table 27. Rate of heartbeat of 15 day embryos on the second day after immersion for 20 min in a 0.05% molluscicidal suspension. (Modified from DAXL 1968)

Eggs	Heartbeats/min	
	Maximal number	Minimal number
Untreated	77	76
Metaldehyde	68	36
Isolan	24	30

The **ovicidal effect** of metaldehyde was for a long time a matter of dispute. At the beginning of the metaldehyde era, an ovicidal effect was attributed to it, but this resulted from dusting with metaldehyde mixed with talcum powder or powdered basic slag and the observed effect was due to desiccation of the eggs by the latter two substances, not by the metaldehyde (LANGE and MCLEOD 1941; LANGE and SCIARONI 1952; THOMAS 1948).

DAXL (1970) tested the toxicity of metaldehyde and isolan on the eggs of the Cellar Slug, *Limax flavus* L. at 20 °C and 100% RH. For periods of 8, 15 or 25 days, commencing immediately after oviposition, the eggs were dipped for 10 min at a time in a solution or suspension of a molluscicide. Metaldehyde as a spray preparation (Slugit*) was used, and isolan at a concentration of 0.025% or 0.05%. The oxygen uptake of the eggs (measured with a Warburg apparatus) was used as a measure of the ovicidal effect. The experiments demonstrated the ovicidal effect of both metaldehyde and the carbamate, isolan. With metaldehyde the effect is very small, but nevertheless present, while isolan has a highly toxic effect from the 8 th day of embryonic development.

The various stages of development of the embryo differ in sensitivity which increases as development proceeds. While the oxygen consumption of 15- and 25-day-old eggs started to decline immediately after treatment, embryos poisoned at an earlier stage (on the 8 th day) continued to develop for some time, oxygen consumption increased for 5 days after treatment and only then started to decline. Eggs treated on the 25 th day after oviposition, when embryonic development was already completed, were particularly sensitive and died about 3 days later. For example, at the start of the treatment, the following values for oxygen consumption of 15-day-old eggs were recorded: on day 1 after treatment with metaldehyde (0.025%) it fell from 10.5 µl 0_2/h to 4 µl 0_2/h; following treatment with isolan (0.025%) it fell from 9.5 µl 0_2/h to 3 µl 0_2/h.

The two molluscicides induced characteristic symptoms and morphological changes in the embryos: in the case of metaldehyde, embryos, like both young and old slugs, made spasmic contractions; with isolan, the embryo swelled up until it almost completely filled the space enclosed by the egg capsule.

The influence of the molluscicides is also reflected in the frequency of the **heart beat** of the embryos (Table 27).

* This product is no longer available in the Federal Republic of Germany

Both molluscicides slowed down the heart beat of treated compared with untreated embryos; metaldehyde furthermore, induced arythmicity.

While in earlier investigations metaldehyde was said not to be volatile, HENDERSON (1970) has found that a **gaseous phase** does occur, as a result of decomposition into acetaldehyde. Experiments by DAXL on the toxicity of the gaseous phase of both molluscicides to gastropod eggs show that metaldehyde vapour inhibits hatching so that the embryo dies in the egg (100% mortality of 100 treated eggs). The mortality of eggs in the gaseous phase of isolan was 72%, that of untreated control eggs 0%. Interesting in this connection are the results of experiments by GODAN (1965) on the effect of a likely gaseous property of carbamates and of metaldehyde on adult Cellar Slugs, not their eggs. The mature slugs are not as easily killed by gaseous molluscicides as are the embryos. Even when the slugs were kept for as long as 6 days at a height of 1 cm above the molluscicide, Isolan (2%) had no effect, while with metaldehyde, mortality was only 10%.

3.2.1.2 Carbamate

Carbamate compounds have a wide range of effects. They are used as insecticides, herbicides and fungicides (KENAGA et al. 1962; METCALF 1961; RICH and HORSFALL 1961; SHOREY et al. 1962; UNTERSTENHÖFER 1962 and others). Now they have also won a place as molluscicides.

Investigations on the effect of carbamates on gastropods were first made in the USA, then in Austria and Germany. These have shown that these compounds are just as toxic to molluscs as metaldehyde. Their advantage lies in the fact that, quite unlike metaldehyde, they are little affected by environmental conditions, and their toxicity increases in humid surroundings, which are optimal for gastropods.

Molluscicidal properties are shown particularly by the methylcarbamates (CROWELL 1967, 1977). These are esters or salts of carbamic acids and toxic mainly to terrestrial pulmonates. They cause metabolic disturbances in insects and warm blooded animals, including man, and inhibit acetylcholine esterase in the nervous system (CASIDA 1963 and others). It is to be assumed that a similar effect occurs in molluscs.

Valuable molluscicidal properties were established first for carbaryl, isolan and mexacarbate and later mercaptodimethur (methiocarb). The chemical substances (GETZIN and COLE 1964; KLIMMER 1971; MARTIN and FORREST 1969) in each case are:

Carbaryl:	1-Naphthyl-N-methylcarbamate
Isolan:	1-Isopropyl-3-methyl-5-pyrazolyl-N,N-dimethylcarbamate
Mexacarbate:	4-Dimethylamine-3.5-xylyl-methylcarbamate
Mercaptodimethur:	3.5-Dimethyl-4-methylthiophenyl-N-methylcarbamate

The superiority of these carbamates over other molluscicides was first established for **carbaryl,** which was found to be toxic to freshwater snails at a concentration of 20 ppm, causing death within 24 h (ASPÖCK and AN DER LAAN 1963). At 10 ppm, however, *Lymnaea* sp. and *Ancylus fluviatilis* Müller were little affected and even oviposited. This substance was not very effective against large terrestrial species like *Cepaea* sp., *Helicigona* sp. and *Helix pomatia* L., but it was very effec-

Table 28. Percentage mortality of *Limax flavus* L. after spraying with carbamates. (Modified from GODAN 1965)

Active ingredient	Series I Concentration				Series II Concentration		P=0.01	
	0.1%	0.2%	1.0%	2.0%	0.2%	2.0%	l_1 %	l_2 %
Carbaryl	0	0	0	0	0	52.5	32	73
Isolan	0	0	5	10	0	92.5	75	99
Mexacarbate	–	0	25	30	10	100	88	100
Promecarb	0	0	0	0	0	100	90	100
Metaldehyde	–	85	100	–	70	100	88	100

P = probability of error
l_1 and l_2 = upper and lower values at a 1% fiducial limit for the final column of figures

tive against the small species, *Discus rotundatus* (Müller), even at concentrations of 0.1% and 0.2% (DIRLBECK 1969). RUPPEL (1959) and WEBLEY (1965) claimed that this molluscicide was of value against *Deroceras reticulatum* (Müller), but this favourable effect has not been confirmed by other authors (CROWELL 1967; GETZIN and COLE 1964; GODAN 1965, 1966). In mixtures with metaldehyde, however, the success rate of carbaryl in molluscicide-bran bait was improved by 15% when used against the slugs *Deroceras reticulatum* (Müller), *Arion hortensis* Fér. and *Milax budapestensis* (Hazay) (ASPÖCK and AN DER LAAN 1963).

Isolan and **mexacarbate** were found to have significant molluscicidal properties when formulated as sprays or as bait pellets (GODAN 1965, 1966) (Table 28). In experimental series I, the molluscicides were dissolved or suspended in water and sprayed on to damp soil on which the gastropods were kept for 6 days; in series II they were dissolved in acetone in glass dishes, in which the gastropods were kept for 48 h.

On 1 cm^2 of treated glass surface the following amounts were deposited:

Carbaryl, Promecarb each	59.3 µg
Isolan	7.1 µg
Mexacarbate	29.7 µg
Metaldehyde	23.7 µg.

The experimental conditions used in series I approximated field conditions. The gastropods could get away, just as they can in the field, and often came into contact with the carbamate film for short times only while feeding, with the result that contact with it was interrupted. In series II escape was not possible: the slugs were in constant contact with the molluscicide. In this case a 2% concentration of isolan, mexacarbate or promecarb caused high mortality of Cellar Slugs.

When applied as bait pellets laid out in clumps, mexacarbate caused 100% mortality even at a low dose and was thus considerably better than the normally available metaldehyde bait. Isolan was, however, effective only at a high concentration.

These experiments provided further evidence for the dependance of mortality rate on the attractiveness of the bait (Table 29). Both the 2% and 10% carbamate formulations proved attractive. This is a consequence of the presence of bran,

Table 29. Attractiveness and toxicity of carbamate-bran pellets to *Limax flavus* L. Experimental period 4 days. (After GODAN 1965)

Active ingredient	Carbamate-bran mixture			
	2%		10%	
	Attractiveness (% slugs on bait)	Toxicity (mortality as %)	Attractiveness (% slugs on bait)	Toxicity (mortality as %)
Carbaryl	90	0	90	0
Isolan	90	13	99	100
Mexacarbate	95	100	10	25
Promecarb	100	10	95	45
Metaldehyde	80	50	–	–

which is attractive to Cellar Slugs which feed on starch-rich food. The high dose mexacarbate bait was an exception. This had a repellent effect, which resulted in only a few slugs feeding on it, and then consuming such small quantities that the effect was non-lethal. The 2% mexacarbate-bran mixture (with 0.5% active ingredient) caused high mortality, because the low concentration was not repellent and lethal amounts of bait were consumed. This mexacarbate mixture was, indeed, more attractive and more toxic than the metaldehyde pellets. High mortality was achieved with isolan only at a concentration of at least 10%, while promecarb did not yet prove satisfactory and carbaryl was without effect, even though the bait contained 5% active ingredient in each case.

Mexacarbate in a 3% mixture with bran and isolan in a 10% bait preparation achieved equally high catches and were equally toxic. The mexacarbate used in the bait mixture was a 25% emulsion, while the isolan was a 6% spray preparation; the final concentration in the baits used in the experiments were 0.5% and 0.6% respectively.

A mexacarbate concentration of 0.5% is optimal both in attractiveness and its molluscicidal effects. A higher dose (for example, 2.5% active ingredient) was found to be repellent, the number of Cellar Slugs attracted being low (only 10%) and the mortality correspondingly low (25%).

The molluscicidal effects of carbaryl, isolan and mexacarbate were also tested in contact and feeding experiments on the snails *Cepaea nemoralis* (L.), *Helicigona arbustorum* (L.) and *Helix pomatia* L. (GODAN 1967). Carbaryl was again found to be unsuitable for practical application, since 100% mortality was achieved only with the most sensitive of the three species (*Cepaea*), and then only at a concentration of 1% which was high compared with the lethal dose of other carbamates; isolan and mexacarbate were toxic for *Cepaea* at concentrations of only 0.12% and 0.05% respectively.

Experiments by DAXL (1971) confirmed the excellent molluscicidal action of isolan, which produced better results in the field than metaldehyde (Table 30).

DAXL also found that while damp cool nights reduce the effectiveness of metaldehyde, they increase the effectiveness of isolan.

Table 30. Mortality of *Deroceras reticulatum* treated with isolan and metaldehyde bait. (Modified from DAXL 1971)

Bait		Affected	Dead
Metaldehyde	5 %	20	1
Metaldehyde	1 %	34	–
Isolan	0.6%	13	46
Isolan	0.3%	20	33
Mixture (Isolan 0,3% + Metaldehyde 1%)		59	23

In tests on the effects of 32 substances on *Helix aspersa* Müller in citrus plantations in California, the best results were obtained with **isolan** (PAPPAS and CARMAN 1955) (bran pellets with 0.5%, 2% and 8%). The substances tested can be ranked in effectiveness as follows:

a) Toxicity either nil or almost nil: DDT, heptachlor, dieldrin, chlordan, prolan, aldrin, isodrin, endrin, perthan, TDE;
b) Toxicity low to moderate: potasan, toxaphen, metoxychlor, diazinon, malathion;
c) Toxicity average: stroban, HCH, parathion, chlorthion;
d) Toxicity excellent: isolan.

Bran baits containing 8% of the active ingredient were slightly repellent in the case of isolan, parathion, chlordan and chlorthion.

Spraying with isolan at concentrations of 30 ml, 60 ml or 120 ml or 100 l water (per orange tree) produced excellent mortality rates of 93.2%, 95.6% and 95.9% respectively with *Helix aspersa* Müller.

Isolan has also been used against the very damaging *Theba pisana* (Müller) in citrus plantations in Israel (HARPAZ and OSERI 1961). 24 h after spraying with a concentration of 0.5%, the mortality was 93.0%, while with a moistened bait containing 1% isolan, it was only 28.2%. This shows how difficult it is to use bait in competition with the preferred food, in this case oranges. Bait containing 0.75% isolan achieved 92.4% mortality after 72 h. *Helix aspersa* Müller in California seems to be more sensitive than *Theba pisana* (Müller) in Israel, for a 0.07% spray preparation of isolan caused mortality of 95% when used against *Helix aspersa*, while a seven times higher concentration was required to achieve the same mortality for *Theba pisana*. Further, broadcasting isolan bran bait containing 1% of the molluscicide caused a high mortality of *Helix* (88.5%), but this was low for *Theba* (28.2%).

Mortality of the slug *Prophysaon andersoni* (Cooper) three days after contact of the sole with **mexacarbate** (Zectran*), applied at the rate of 50 μg/cm², was as high as that achieved with „Schneckenkorn Mesurol“. Dermal application is, however, not effective, since the substance is removed by mucus secretion before its toxic effect has developed (GETZIN and COLE 1964). *Helix aspersa* Müller was kept

* Not commercially available in the Federal Republic of Germany

under control even at low temperatures, using an emulsion with 84 g active ingredient per 100 litres, or by spraying with a 25% wettable powder (60% to 80% RH). As early as the third week of application, the emulsion caused mortality of 96%; even with a low dose of 10 g/100 l water, the mortality was, at 80%, still quite considerable (ARNOLD 1962). According to the manufacturers, mexacarbate as a 0.2% wettable powder should be applied at the rate of one to 5 l (or kg) per ha; if in the form of pellets, at the rate of 20 kg/ha. It is non-toxic to plants and is therefore very useful in horticulture. DAVIDSON (1962) used it in orchid *(Cattleya)* cultivations against the snail *Zonitoides arboreus* (Say), which attacks the roots. The orchid roots were drenched with a suspension of about 600 g of a Zectran wettable powder in 100 l of water; completely successful control was achieved even with only a single application and no further damage to the orchids could be detected. Spraying with mexacarbate kept field-grown strawberries free from attack by Field Slugs for 4 to 6 weeks, provided both soil and straw (hiding places for slugs) under the infested plants were saturated (HAMMER 1962).

CROWELL (1967, 1977) found in tests with various insecticidal methylcarbamates and acaricides on *Helix aspersa* Müller and on *Arion ater* (L.) that **mercaptodimethur** (methiocarb) is the most suitable for control. It has been commercially available for several years as „Schneckenkorn Mesurol“. A 4% concentration of the active ingredient achieved a mortality of 79.3% for *Helix aspersa* Müller and of 84.6% for *Arion ater* (L.) as compared with mortalities of 17.8% and 20.0% respectively with a comparable dose of metaldehyde pellets. In addition to mercaptodimethur, the following substances were found to be effective against terrestrial gastropods: 4% tranid – 78.9% mortality of *Helix;* 2% formetanat – 100% mortality of *Arion;* 2% methomyl – 86.7% mortality of *Arion* and at 4% – 77.1% mortality of *Helix;* the gastropods usually died after one week. Mercaptodimethur is thus very much more effective than metaldehyde (Table 31).

In a damp environment mercaptodimethur is far more effective. Mortality of slugs poisoned with metaldehyde and thereafter kept in 100% relative humidity for eight days was only 50%, while the other half recovered by adjusting their body water content; by contrast, 95% of the experimental animals which had eaten mercaptodimethur bait died within four days. In comparing the practical results of using mercaptodimethur bait with those obtained using metaldehyde bait (10 g per trap in each case), MARTIN and FORREST (1969) have found that mercaptodimethur is far more effective than metaldehyde: mortality of *Deroceras reticulatum* (Müller)

Table 31. Comparison of mortality rates (%) with mercaptodimethur and metaldehyde. (After CROWELL 1967)

Molluscicide	Concentration	Mortality in %	
	%	*Arion ater*	*Helix aspersa*
Schneckenkorn-Mesurol	2	68.9	82.2
	4	84.6	79.3
Metaldehyde pellets	2	24.4	24.4
	4	20.0	17.8

was 94% and 58% respectively; of *Arion subfuscus* (Drap.) 100% and 46% respectively; *Arion hortensis* Fér. 100% and 11.1%; *Limax maximus* L. 100% and 33.3%, and *Milax budapestensis* (Hazay) 92.3% and 14.3%.

The carbamate preparation was also more effective than metaldehyde against Field Slugs, when applied after sowing, on spring greens, winter wheat and summer barley. Thus MARTIN and FORREST, in experiments on winter wheat, estimated that with a mercaptodimethur bait „Schneckenkorn Mesurol" applied at the rate of 5.6 kg/ha, mortality was 81.8% and at the rate of 11 kg/ha it was 82.9%, in comparison, a far heavier dose of metaldehyde of 31.0 kg/ha mortality was only 57.3%. The percentage of plants damaged was 7.83% and 6.93% respectively for the two doses used; with metaldehyde 17.30%; this figure was of course much higher for untreated areas, where it was 40.45%. In this connection an experiment on inhibition of feeding on drilled wheat seeds is interesting. Molluscicidal pellets were either worked into the upper layer of soil or spread over the surface at the rate of 11 kg/ha. The results were compared with those obtained using the normal dose of metaldehyde bait (Fig. 88). Mercaptodimethur was better than the metaldehyde even 23 days after application. A thorough investigation of the most suitable time of application is, however, desirable, to establish whether application before sowing or planting is necessary to prevent gastropod damage to drilled cereals or to tubers and roots (potatoes and carrots).

Applied at the same concentration of molluscicide (4%), mercaptodimethur bait is more effective against slugs than metaldehyde bait (MARTIN et al. 1969); according to WEBLEY (1969), however, this is not true for all species: the very resistant large *Arion ater* (L.) is affected only by the carbamate, while the highly sensitive, smaller *Deroceras reticulatum* (Müller) is equally susceptible to both poisons.

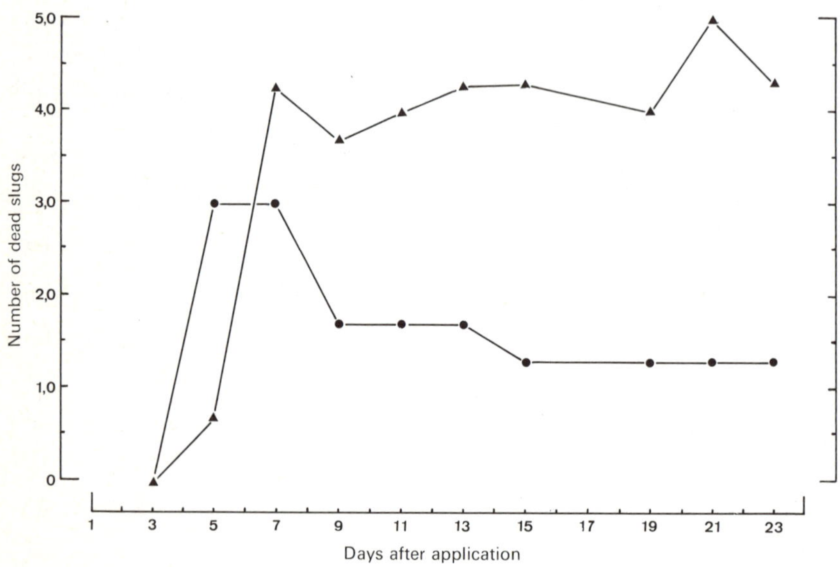

Fig. 88. Number of dead slugs found on the ground after application of mercaptodimethur and metaldehyde. ▲——▲ = mercaptodimethur (11 kg/ha), ●——● = metaldehyde (31 kg/ha)

Mortality of Field Slugs was over 90% with a 4% carbamate bait at a dose of 100 kg/ha on strawberries, rhubarb and grassland, where metaldehyde frequently fails. Damage to strawberries was reduced to between 3% and 4% in this way, the loss of fruit on untreated areas being 13% (GETZIN and COLE 1964; GETZIN 1965). With an initial mortality rate among Field Slugs of 99%, mercaptodimethur was still significantly toxic 25 days later. MOENS (1970) found a low dose of mercaptodimethur-bran bait (4% active ingredient) to be more effective than a larger dose of metaldehyde bait (5% active ingredient). Control measures were also successful in the field against *Deroceras laeve* (Müller) and *Arion circumscriptus* Johnston, in greenhouses cultivations of *Gerbera jamesonii* hort, against *Aegopinella nitidula* (Drap.), *Discus rotundatus* (Müller) and *Deroceras laeve* (Müller) in cold frame cultivations of Chinese Cabbage *(Brassica pekinensis)* against *Arion circumscriptus* Johnston and *Deroceras laeve* (Müller). Spraying with a 4% suspension have successfully protected cultivations of *Anthurium scherzerianum* hort. from damage by the Rounded Snail, *Discus rotundatus* (Müller) (DIRLBECK 1969).

The following results were obtained in field experiments in which different molluscicidal preparations were applied after the sowing of spring greens: with Schnekkenkorn Mesurol at a dose of 5.6 kg/ha, there were 16 dead and moribund Field Slugs and only 7 survivors; at a dose of 11 kg/ha there were 38 dead and 5 survivors; with metaldehyde pellets at a dose of 21 kg/ha there were only 2 dead and damaged and 38 survivors; this demonstrates impressively the superior molluscicidal properties of the carbamate preparation. In the damp climate of Great Britain Schneckenkorn Mesurol was used with success against slugs even at the low application rate of 5.6 kg/ha (MARTIN and FORREST 1969).

The superiority of mercaptodimethur, particularly when difficult conditions make aerial spraying necessary, has been demonstrated in its use on slug infested rape (rate of application of 4.5 kg/ha of mercaptodimethur preparation, compared with 9.7 kg/ha of metaldehyde preparation), experimental area 2.25 ha (250 m long), 5 aerially treated sectors, (each 18 m long) (HERMANN 1971). Table 32 shows the result of this treatment based on counts of slugs found in ten squares each 0.25 m^2.

Table 32. Control of slugs in stubble rape by aerial application of bait pellets. (Modified from HERMANN 1971)

Molluscicide application rate	Result	Test square number										Total
		1	2	3	4	5	6	7	8	9	10	
Mercapto-dimethur 4.5 kg/ha	Dead	30	12	10	27	37	9	23	35	1	12	196
	Affected	8	12	3	12	11	5	6	18	10	11	96
	Healthy	2	17	1	2	2	2	4	12	13	12	67
Metaldehyde 9.7 kg/ha	Dead	2	1	4	8	7	1	10	7	4	4	58
	Affected	6	6	3	9	5	5	10	8	4	7	63
	Healthy	8	9	3	7	4	24	9	12	11	15	102
Untreated	Dead	0	0	0	0	0	0	0	0	0	0	0
	Affected	0	0	0	0	0	0	0	0	0	0	0
	Healthy	47	12	9	25	16	16	18	31	17	54	245

With mercaptodimethur, 54.5% of the slugs found were dead, 26.7% showed adverse effects of the poison and 18.8% were unharmed: with metaldehyde the proportions were reversed: 26% of the animals found were killed, 28.2% damaged and 45.8% were healthy.

Schneckenkorn Mesurol has also been successfully used against *Helix aspersa* Müller, *Cepaea nemoralis* (L.) and *Cepaea hortensis* (Müller) which attack blackcurrants, against *Zonitoides arboreus* (Say) in anthurien cultivations in the Netherlands, and against *Helix aspersa* Müller, and *Theba pisana* (Müller) in citrus plantations in Spain, where bait containing 1% or 2% active ingredient achieved mortalities of 92% and 100% (DEN BOER 1968 cited by HERMANN 1971; DEL RIVERO and ROCA 1969; STRINGER and MORGAN 1969). CRAWFORD-SIDEBOTHAM (1970) found that mercaptodimethur bait was better than metaldehyde bait not only in the mortality rate achieved immediately after application, but also with respect to the rate of recovery of slugs which appeared to be heavily poisoned but were not dead; most of those which had eaten metaldehyde bait recovered after 48 h, whereas those which had eaten mercaptodimethur bait died; this applies to the slugs *Deroceras reticulatum* (Müller), *Deroceras caruanae* (Pollonera), *Milax budapestensis* (Hazay), *Milax sowerbyi* (Fér.), *Arion ater* (L.), *Arion fasciatus* (Nilsson), *Arion hortensis* Fér., *Arion subfuscus* (Drap.) and *Arion lusitanicus* Mabille.

The critical importance of **humidity** is clearly shown by the investigations of MARTIN and FORREST (1969). When kept in a relative humidity of 100% for eight days after treatment with bait, 50% of the "metaldehyde slugs" recovered, whereas 95% of the "carbamate slugs" had died within only 4 days.

The influence of **temperature** has been investigated by MALLET and BOUGARAN (1971) (Table 33). In contrast to metaldehyde, mercaptodimethur achieves a high mortality even at low temperatures such as 6 °C – this applies at least to Field Slugs, which are still active at such temperatures. Metaldehyde, on the other hand, is effective only at temperatures of 20° C and over, when the air and soil are sufficiently dry. The other two species are less active at low temperatures (Chap. 2), so that the bait is ignored. The experiment also provides further evidence for the

Table 33. Influence of temperature on the toxicity, expressed as a percentage of mortality, of metaldehyde and mercaptodimethur pellet baits. (After MALLET and BOUGARAN 1971)

Species	Molluscicide	Temperature						
		6 °C %	8 °C %	10 °C %	16 °C %	20 °C %	22 °C %	25 °C %
D. reticulatum	Untreated	0	0	0	0	0	0	0
	Metaldehyde (5%)	0	15	15	40	80	100	100
	Mercaptodimethur (4%)	90	95	90	85	90	95	85
A. hortensis	Untreated	0	0	0	0	0	0	0
	Metaldehyde (5%)	0	10	20	50	50		
	Mercaptodimethur (4%)	0	20	20	40	40		
M. sowerbyi	Untreated	0	0	0	0	0	0	0
	Metaldehyde (5%)	0	10	20	20	20		
	Mercaptodimethur (4%)	20	40	40	80	100		

different effectiveness of molluscicides on different species, according to activity and behaviour, a point which has already been discussed several times.

Mercaptodimethur is an odourless, white, crystaline powder, which is insoluble in water. According to UNTERSTENHÖFER (1962) its toxicity to mammals, including man, after inhalation, cutaneous or oral intake in very small amounts can be regarded as negligible.

The carbamate **carbofuran** (2.2-dimethyl-2,3-dihydrobenzofuranyl-7-N-methylcarbamate) is extremely toxic to gastropods (CING-CHIEH YU et al. 1974). **Promecarb** is effective against *Limax flavus* L. as a 2% spray, but not as bait (GODAN 1965).

A further advantage of the carbamates, in particular of mexacarbate (Zectran) lies in their long **persistent toxicity** (GODAN 1965, 1967). Even after 60 days, 70% mortality was achieved with *Limax flavus* L., one of the most resistant species. The base, sides and covers of glass dishes were sprayed with a 2% solution in acetone, and Cellar Slugs placed in the treated dishes 1, 3, 7, 11, 15, etc., days later; in each case 48 h sole contact with the treated surfaces was allowed before the slugs were removed and kept for a week at 85% RH for further observation (Fig. 89). From the time of preparation until they were used, the dishes were kept open and in the dark, because of the light sensitivity of mexacarbate (KENAGA et al. 1962). Mexacarbate was considerably more effective than metaldehyde and after more than 45 days mortality rates from 95% to 100% were observed. The other carbamates (isolan, carbaryl) were not as good as metaldehyde.

The extreme persistant toxicity of Zectran was also demonstrated with *Cepaea nemoralis* (L.) (Fig. 90) and *Helix pomatia* L. (Fig. 91). Zectran was effective even

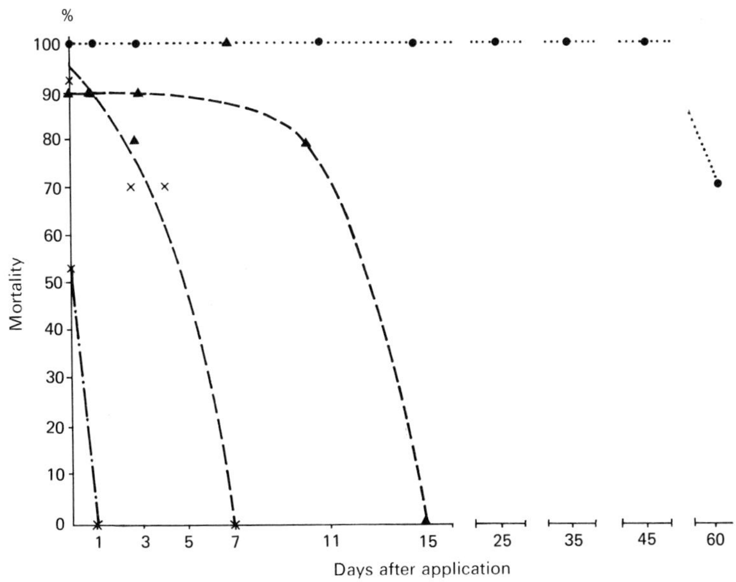

Fig. 89. Persistance of effectiveness of some molluscicides as shown by percentage mortality of *Limax flavus*, following 48 h contact, at intervals after initial application; carbamate formulations each 2%, metaldehyde wettable powder 0.2%. ×–·–·–·–× = carbaryl (50% a.i.); ×– – – – –× = isolan (6% a.i.); ●·········● = mexacarbate (25% a.i.); ▲– – – – –▲ = metaldehyde

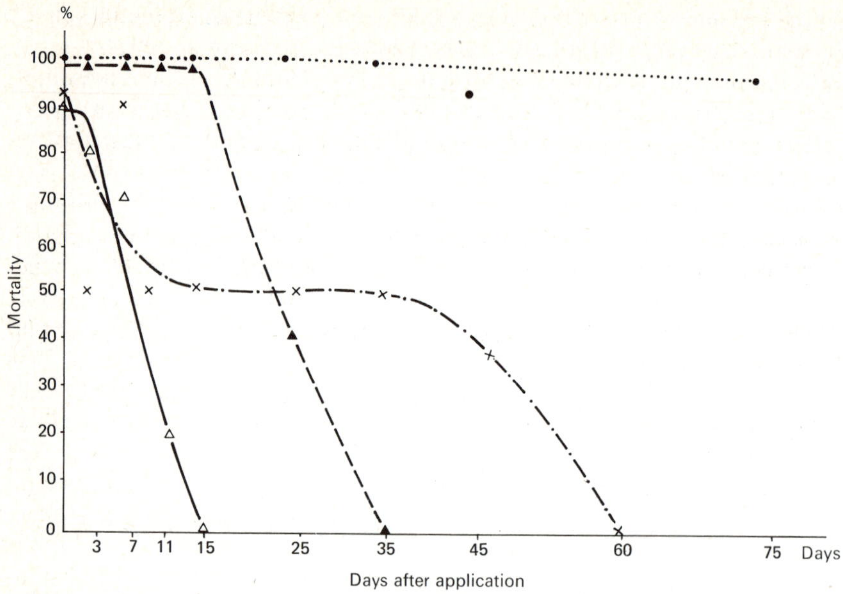

Fig. 90. Persistance of effectiveness of some molluscicides as shown by percentage mortality of *Cepaea nemoralis*, following 48 h contact, at intervals after initial application; carbamate formulations each 2%, metaldehyde wettable powder 0.2%. ×–·–·–·–× = carbaryl (50% a.i.); ▲- - - - -▲ = isolan (6% a.i.); ●·········● = mexacarbate (25% a.i.); △———△ = metaldehyde

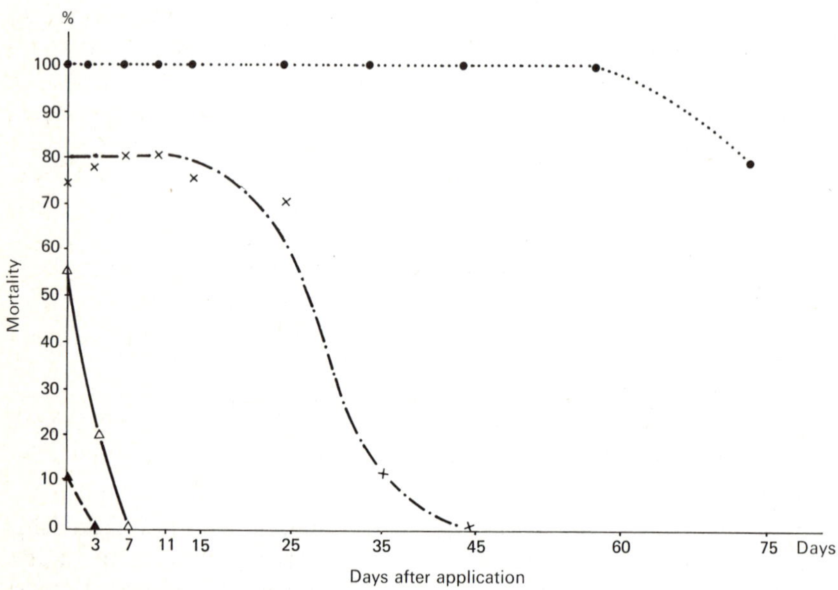

Fig. 91. Persistance of effectiveness of some molluscicides as shown by percentage mortality of *Helix pomatia*, following 48 h contact, at intervals after initial application; carbamate formulations each 2%, metaldehyde wettable powder 0.2%. ×–·–·–·–× = carbaryl (50% a.i.); ▲- - - - -▲ = isolan (6% a.i.); ●·········● = mexacarbate (25% a.i.); △———△ = metaldehyde

against the very resistant *Helix pomatis* L. and even more than 65 days after application, mortalities of 95% and 100% were recorded for *Cepaea* and *Helicigona*. With *Helix pomatia* L. toxicity of mexacarbate lasted for 60 days, but by that time the value for metaldehyde was practically zero. Again laboratory investigations by MORETON (1953) have demonstrated the rapid loss of toxicity of metaldehyde. At an application rate of approximately 5 kg/ha, metaldehyde retained its full toxicity to slugs for only 5 days. *Helicigona arbustorum* (L.) proved to be less sensitive than *Cepaea nemoralis* (L.) in these tests; it was not affected at all by isolan and carbaryl; even immediately after application, these two carbamates caused only 60% mortality. By contrast, exposure to isolan results in 100% mortality of *Cepaea* for as long as 15 days after first application. Carbaryl, on the other hand, was completely ineffective 24 h after application.

The effectiveness of the carbamates is dependant on body size and surface area of the sole, age and species of gastropod (GODAN 1961, 1965, 1966). They have **ovicidal** properties, as DAXL (1970) has demonstrated for isolan. This substance has a strong toxic action on the 8-day-old embryo, giving rise, amongst other symptoms, to a slowing of the heart beat. More details have already been given in Chap. 3.2.1.1.

The *systemic* properties of many carbamates such as carbaryl, isolan, oxamyl and mexacarbate (DAVID et al. 1960; METCALF 1961; MOENS and FRASELLE 1980) can be regarded as advantageous; this is not a characteristic of metaldehyde. It could be important in control of those gastropods which feed on the green parts of plants, but of course phytotoxic damage must not occur. Experiments by DAXL (1968) have demonstrated the systemic effect of isolan on Field Slugs; however, the dosages used (3 ml, 7 ml and 12 ml of a 0.5% isolan preparation) were toxic for the plants, *Datura stramonium* (thorn-apple), a favourite food of these slugs. After watering the soil in the plant pots, leaves were removed from the plants after 1, 2, 3, 4, 5 or 6 days and presented to the Field Slugs (10 animals per experiment) as food. The mortality of the slugs rose progressively with the length of time the leaves had remained attached to a treated *Datura* plant and also with the amount of molluscicide applied. After only 1 day, dead slugs were found, even with the relatively low application rate of 3 ml per pot. As a result of the systemic effect, the quantity of isolan contained in the leaves steadily increased, with the result that the slugs died more rapidly after eating leaves which had remained on the plant for 3 days than they did after eating leaves which had been removed after only one day. With plants treated with 12 ml of the solution, the following mortalities were obtained: with leaves removed from the plant after one day, 9 out of the 10 slugs died, but only after feeding on these leaves for 6 days; with leaves removed after 2 days, all the slugs died after 4 days feeding on the leaves; with leaves removed 3 days after treatment, all the slugs died after 3 days feeding.

The carbamates are undoubtedly more effective molluscicides than metaldehyde. Experience shows that gastropod plagues usually occur in moist regions or during long periods of damp weather, so that control with metaldehyde nearly always takes place in weather conditions which are quite unsuitable for this poison.

Apart from the possible side effects on the fauna and flora, which are still to be discussed, the carbamates are superior to metaldehyde in the following respects:

1. Particularly effective in damp weather, in permanently moist areas and at low temperatures.
2. Some carbamates, such as mexacarbate, have a long-term effect; this is not shown to the same degree by metaldehyde.
3. Molluscicidal carbamates are also effective against snails, which are relatively difficult to control.
4. The marked systemic properties of isolan and mexacarbate could be advantageous in control of gastropods which live on the green parts of plants. There is no systemic effect with metaldehyde.

3.2.1.3 Differences in the Symptoms of Poisoning by Metaldehyde and Carbamates

The symptoms shown by gastropods poisoned with metaldehyde differ markedly from those following carbamate poisoning (Daxl 1968, 1970; Godan 1965, 1966). A knowledge of these differences might be useful, since one carbamate-containing preparation is already commercially produced and others are likely to become available in the future.

As a result of over-production of mucus following metaldehyde uptake, gastropods are soon immobilised and are to be found widely scattered over the treated area, since some manage to creep a short distance before being completely inactivated. Later their shape becomes very compact; the foot muscle is contracted and hard and only the anterior part of the body continues to move convulsively, being lifted frequently from the soil. Small, rapid waves of contraction run through the skin musculature. Finally, the head remains permanently withdrawn under the mantle shield and muscular contractions occur in the front part of the foot, while the posterior part is motionless. In the final stage of poisoning, the oesophagus is usually everted from the mouth, rather like a prolapse (Fig. 92).

Snails, such as *Helix pomatia* L. show the following symptoms after oral ingestion of metaldehyde powder with the food: after 10 to 15 min, withdrawal of tentacles and head and raising of the anterior part of the body; this is soon followed by complete withdrawal of the body into the shell, while large quantities of foamy mucus are secreted.

In contrast to these symptoms which are typical of metaldehyde poisoning, gastropods poisoned with carbamates continue to crawl about for a considerable length of time, since mucus secretion and loss of body water remain almost normal. They remain active for a while, are excited and move with the body longitudinally extended and optic tentacles protruded. Metaldehyde-poisoned slugs, however, re-

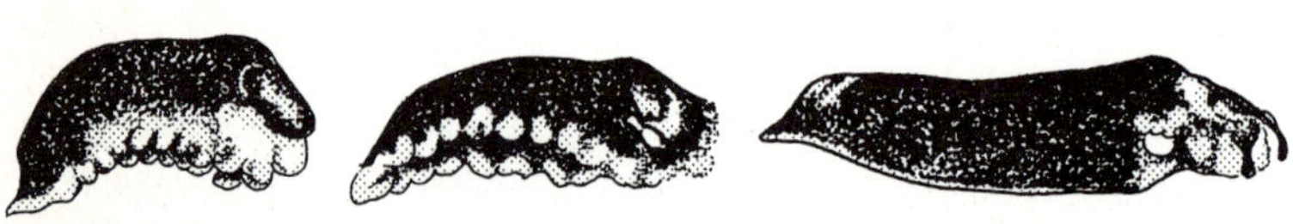

Fig. 92. Comparison of the effects of metaldehyde and carbamate on *Limax flavus*. Metaldehyde: contraction of musculature, especially strong in foot fringe and sole (*left*), start of "mouth rupture" (*centre*). Isolan: loss of muscular tone, here the optic tentacles (*right*)

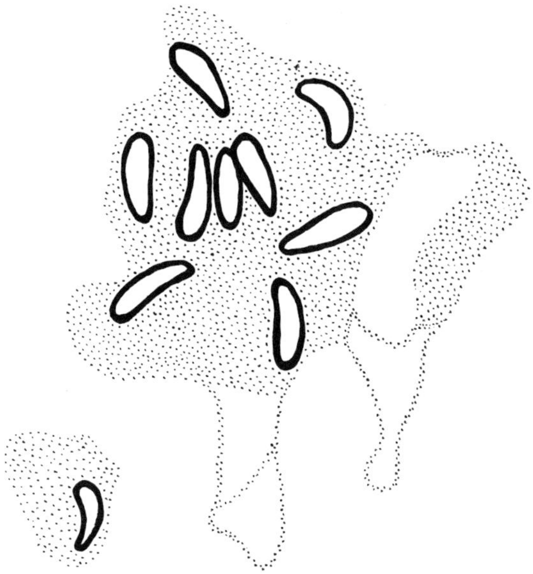

Fig. 93. Diagram showing the effect on slugs of 10 to 15 min sole contact with metaldehyde (0.2% wettable powder): immobile contracted animals. *Dotted areas* and *lines* area covered with mucus and mucus trails

main motionless, in a contracted state, surrounded by a large patch of mucus (Figs. 93, 94).

About 10 to 15 min (depending on concentration) after the onset of carbamate poisoning **flight reactions** are observed. After initial excitation, muscle tonus is lost, so that the body becomes soft and limp. In a slug the body becomes spirally twisted, because the animal attempts to move, but is unable to do so because of the lack of tone of the musculature. This stage of poisoning takes a different form in snails (Figs. 95, 96). Since tone is lost both by the foot musculature and the columella muscle, which is firmly anchored in the visceral hump, the body can no longer be withdrawn into the shell and thus protrudes from it. Since the visceral hump can no longer be held firmly in the shell, the edge of the mantle protrudes as a thick swelling and obstructs the rhythmic opening and closing of the pneumostome, making respiration difficult. At this stage the mantle edge is blown up into a voluminous structure which contains a colourless, transparent, foamy fluid. According to GEBHARDT-DUNKEL (1953), the glandular mantle edge has a water-regulatory function, mainly serving to protect the snail from too great water loss, and is in fact the largest reservoir of water in comparison with other organs of the body. In slugs, this role is performed by blood and skin. The heavy collection of fluid in the mantle edge is observed only after poisoning with carbamates, not after metaldehyde poisoning. It is clearly a defensive reaction of snails, to prevent an excessive loss of water. This succeeds at first, until, as poisoning progresses, muscle tone falls further and any protective function of the mantle edge ceases. The snail dies as a result of relaxation of the muscles and loss of blood.

If contact with the carbamate is discontinued at the time of spiralling of the body, recovery is possible, provided the slug does not come into contact with water. If the side-position typical of carbamate poisoning has already been taken up, death always follows. The **side position** is in slugs an infallible sign of lethal poison-

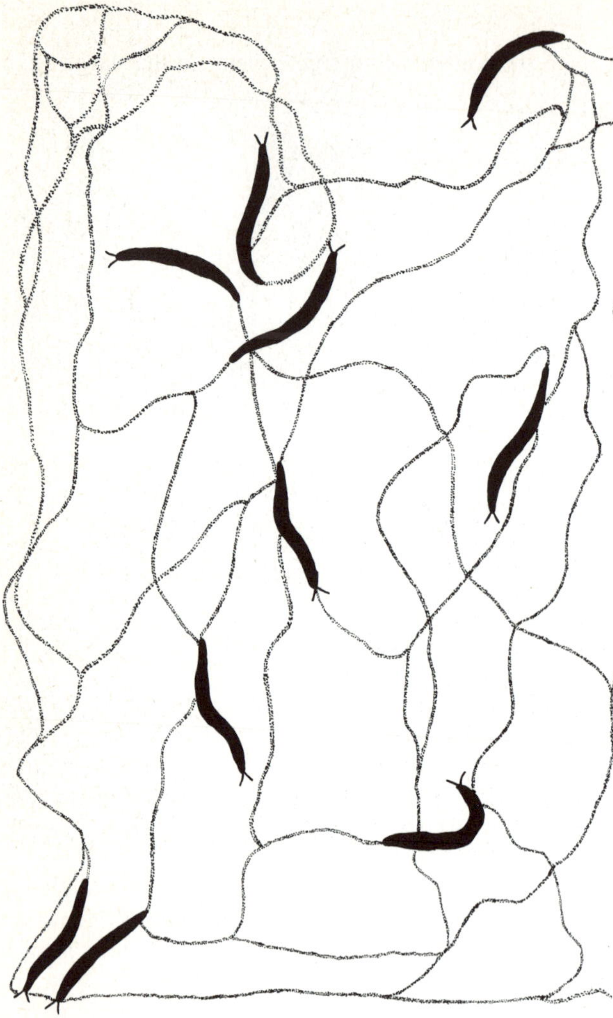

Fig. 94. Diagram showing the effect on slugs of 10–15 min sole contact with isolan (0.2%): animals still active. *Dotted lines* mucus-covered trails

ing. The loss of muscle tonus is very greatly increased by contact with water. Animals which appeared little affected by contact with carbamates died after being wetted, while others with the same degree of poisoning, but which had not been wetted, survived. With metaldehyde, however, provided the poisoning has not yet reached an irreversible stage, gastropods recover after being rinsed with water or in high humidity (about 95% RH) (STRINGER 1946 and others).

Their increased effectiveness following contact with water is of great practical importance and makes carbamates very suitable for use in gastropod control. In experiments conducted by GODAN (1966), two groups of slugs were similarly treated with isolan (2%) but thereafter kept for 8 days under quite different conditions, both, however, receiving fresh food daily. After treatment with carbamate (5, 6 or 2 h exposure to the molluscicide, depending on species) group I came into constant contact with water, while group II was kept for the same length of time at a rela-

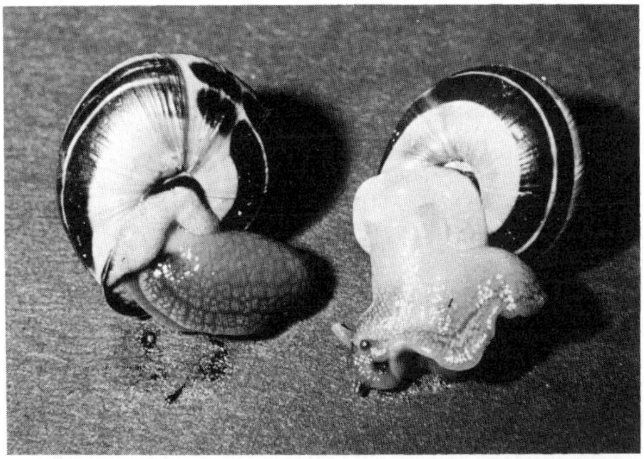

Fig. 95. Loss of muscular tone in foot and optic tentacles in snails (*Cepaea nemoralis*), isolan preparation (2%)

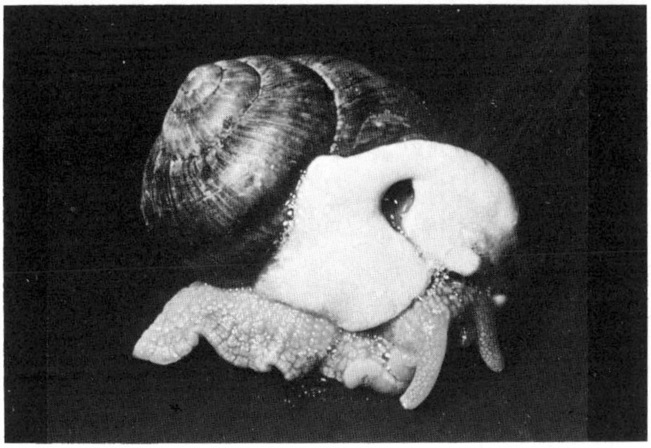

Fig. 96. Prolapse of mantle edge following loss of tone of columellar muscle, *Helix pomatia*, isolan preparation (2%)

tively low humidity (maximum 70% RH). It was not possible to keep slugs for a long period at humidities below 70% RH because of the danger of desiccation. The specimens in the two groups had the same average age, body weight and sole surface area (Table 34).

The toxic effect on species with a low body water content, like *Limax maximus* L. and *Limax flavus* L., was very greatly increased after wetting the animals. Species with high body water content and also a small species, like *Lehmannia marginata* (Müller) are already so highly sensitive that it is impossible to increase their susceptibility by treatment with water. *Deroceras laeve* (Müller) from group I and group II show the same degree of poisoning. Injection of isolan (1%) into the sole of adults and juveniles had no effect on the water content of *Limax flavus* L.; metaldehyde, however, caused increased water loss by stimulated secretion of mucus (Daxl 1968, 1970). The doses used for adult slugs were 0.010 ml/g body weight of metaldehyde, 0.009 ml/g of isolan (temperature 20 °C) and for juveniles 0.01 ml/g body weight for both molluscicides (temperature 10 °C) (Figs. 97, 98). Water loss was estimated by measuring loss of body weight within a given period of time.

The difference in amounts of water lost, which is attributable to the different action of the two molluscicides, is confirmed by the practical observation that, in

Table 34. Influence of treatment with water on carbamate-affected slugs. (After GODAN 1966)

Species	Average age (months)	Average weight (g)	Average sole (mm^2)	Group	Degree of poisoning side position 1 h after treatment %	P	Total mortality %	P
Limax maximus	7–8	6.7	676.1	I	90.0	<0.02 sign.	83.3	<0.02 sign.
				II	46.7		46.7	
Limax flavus	4	2.6	358.8	I	86.7	<0.01 sign.	58.5	<0.02 sign.
				II	45.3		26.9	
Lehmannia marginata	6–7	1.4	191.0	I	100.0	>0.05 not sign.	100.0	>0.05 not sig
				II	93.5		96.8	

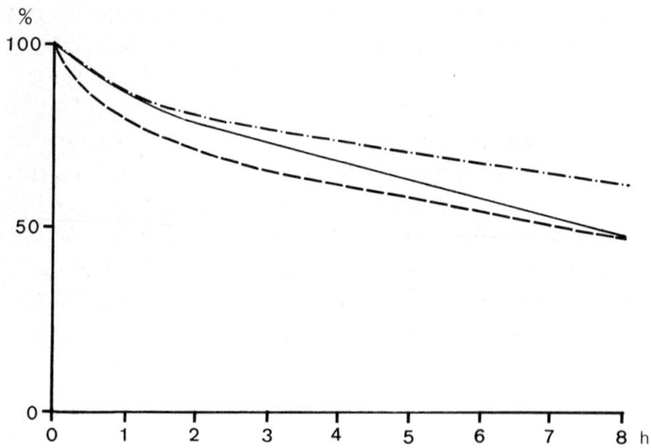

Fig. 97. Weight loss in juvenile *Limax flavus* following treatment with molluscicides, 10 °C, 75% RH, hourly weightings. – – – – = metaldehyde; – · – · – = isolan; ——— = untreated

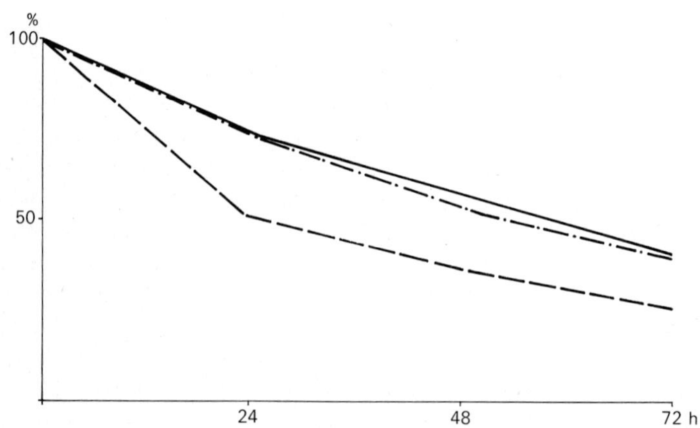

Fig. 98. Weight loss in adult *Limax flavus* following treatment with molluscicides, 20 °C, 75% RH, – – – – = metaldehyde; – · – · – = isolan; ——— = untreated

contrast to metaldehyde-treated animals, mucus production of carbamate-poisoned specimens is more or less normal.

3.2.1.4 Combination of Metaldehyde with Other Active Ingredients

Since it is especially in damp areas that gastropod control is necessary, the marked influence of weather conditions on the efficacy of metaldehyde is a disadvantage which must be counteracted in some way. Since, however, environmental pollution by metaldehyde is minimal, it will be abandoned as a molluscicide with reluctance. Attempts have therefore been made to combine it with other poisons, in the hope that the problems associated with the use of metaldehyde on its own could be reduced; at the same time the highly toxic effects on the environment of other, less weather-dependant molluscicides could be avoided, if low doses of both compounds in combination were effective. **Calcium arsenate** was the first substance to be tried, since arsenic compounds had, at earlier times, been frequently used for gastropod control. LANGE and MACLEOD (1941) recommended bran or mollasses bait containing 1.5% to 2.5% metaldehyde powder and 5% calcium arsenate. Calcium arsenate alone had proved more effective in bait formulations than metaldehyde for the control of *Helix aspersa* Müller (LEWIS and LAFOLLETTE 1942). Mixtures of isolan and calcium arsenate were also tested; they achieved a mortality of 91.1% (PAPPAS and CARMAN 1955). Arsenic containing preparations may, however, no longer be used in the Federal Republic of Germany.

BÖHM (1957) tested the effect of adding sodium silicofluoride, a compound which absorbs water from the body of gastropods, at a concentration of 28.5 g to 9 kg metaldehyde-bran mixture; he also tested calcium arsenate (74 g with 70% active ingredient). Neither additive had a favourable effect on the control of either *Deroceras laeve* (Müller) or the snail *Zonitoides arboreus* (Say) in a heated greenhouse.

The combination of metaldehyde with **dimethylglyoxim** has been found to be satisfactory (RAMMOS 1962). Although the molluscicidal effect of this substance alone is negligible, the mixture of 8% dimethylglyoxim and 2% metaldehyde caused 100% mortality within 12 h.

Attention was then turned to **insecticides.** A combination of 2.5% **malathion** with 6% metaldehyde and 91.5% bran was, however, no more effective against slugs and snails than commercially available metaldehyde pellets. MOENS (1971) tested bait containing 3% metaldehyde and 2% **methomyl**, at that time a newly developed insecticide, on terricolous slugs; this was found to be almost as effective as a mercaptodimethur bait (5%), whereas the insecticide alone led only to a 51% mortality and was thus not satisfactory. By using **methylcarbamates** in mixtures with metaldehyde, it was hoped that the high toxicity of carbamates in damp conditions would, by preventing the recovery of gastropods in a damp environment, eliminate the problems associated with using metaldehyde alone. The combination *metaldehyde-carbaryl* indeed yielded better results than either of the substances used alone (RUPPEL 1959; WEBLEY 1962, 1965).

It is thus possible to increase the effectiveness of molluscicides by using them in combination. If, as seems to be the case, metaldehyde continues to be regarded as the standard molluscicide against terrestrial pulmonates, many more investigations will be necessary to establish an optimal combination.

3.2.1.5 Methods of Application

Spraying, Dusting and the Use of Pellet Bait. Today the use of bait is the most common method of gastropod control in both agriculture and horticulture, whereas spraying and dusting are less frequently used.

Initially metaldehyde spraying and dusting formulations were regarded as better than bait, particularly for ornamental plants such as orchids (ALICATA 1950; JEFFERSON 1953; LANGE and SCIARONI 1952 and others). They effectively reduced gastropod populations in bean, maize and strawberry crops, but were ineffective when the leaves alone were sprayed (HOWITT and COLE 1962). HOWITT (1961) applied a metaldehyde suspension through an irrigation system at the rate of 1 or 2 kg/ha, and achieved a considerable reduction in gastropod infestation of the clover cover of a fruit orchard. Metaldehyde as a spray was also successfully used on seedbeds with young plants and as a dust on maize. This led to the discovery by HOWITT and COLE that a low application rate of powder achieved greater mortality of slugs than did a higher rate; the reason for this lay in the repellent effect of the higher concentration (GODAN 1961; LANGE and SCIARONI 1952). According to BÖHM (1961), a metaldehyde wettable powder used as a spray (dose 30 ml/5 l water/10 mm^2) afforded excellent protection for up to four days from attack by *Deroceras laeve* (Müller), *Arion subfuscus* (Drap.), *Helicigona arbustorum* (L.), *Bradybaena* fruticum (Müller) and *Helix pomatia* L.

The use of spray and dusting preparations is, however, justifiable only when drastic action has become necessary to prevent sudden and unusually heavy losses through feeding damage. According to WEBLEY (1965), molluscicide sprays should be used only for sudden *gastropod plagues*, while bait should be applied continually, to keep damage down to a tolerably low level. In cool weather, however, because of the low activity of gastropods and the relatively uncertain effect of metaldehyde, spraying is to be preferred to the use of bait.

From the outset, opinion was divided on the advantages of using **bait,** and its effectiveness in strawberry cultivations in particular was questioned (HOWITT and COLE 1962 and others). Nevertheless, the use of molluscicidal baits has come to be accepted in practice. Bait pellets usually contain 3% to 8% of the active ingredient (metaldehyde or mercaptodimethur), an attractant and a fungicide. Their toxicity depends on the amount of poison taken up, the dispersal of the slugs and snails and the intensity of feeding activity, as well as the attractiveness of the bait to the animals which must feed for some time in order to ingest a lethal dose (HUNTER and SYMONDS 1970). VAN DINTHER (1973) recommends metaldehyde bait comprising 200 kg molluscicide powder and 12 kg wheat or rice bran as attractant, spread at a rate of 30 kg/ha; this should be moistened with 7 l water because gastropods normally eat only moist food and reject it when it is dry. Thus commercial products are eaten only after they have become swollen through uptake of moisture from the soil.

Broadcasting of bait pellets should be avoided with artichokes and cabbages; here they should be placed in little clumps around the individual plants. Strawberries and lettuce, both very attractive to slugs, can be protected by barrier poisoning, that is by spreading pellets in a belt around the cultivated area (MAYER 1957). This method of application (2% metaldehyde) protected narcissus bulbs for up to 12 days from attack by *Arion ater* (L.) while dusting with 15% metaldehyde was not so effective (DOUCETTE 1954); experience points to the likelihood of this being due to a repellent effect in the latter case. Broadcasting of bait pellets was successful in various strains of rye and wheat, barley and rape (KOLTERMANN 1955).

Several decades ago bait was usually applied in **clumps** but not always with success, because insufficient attention was paid to behaviour, learning capacity, habituation and activity rhythm of the slugs and snails or these were ignored altogether (Chap. 2).

The effectiveness of this procedure is partly determined by the *distance* between the pellets clumps. BÖHM (1957) found that in the field the attractive effect is limited to a radius of a 0.5 to 1.0 m around each clump; according to THOMAS (1948), *Milax budapestensis* (Hazay) were lured to the bait up to a distance of 91.5 cm from a clump, while this distance was 45.7 cm for *Arion hortensis* Fér. These distances appear, however, to be too great for practical application, since GODAN (1961) found in laboratory experiments on the olfactory sense that, in still air, 12 cm was the upper limit for detection. As there are usually only slight or even no air movements between densely growing plants or below them on the surface of the soil where the slugs live, only experimental results obtained in still air are meaningful. Clumps of bran or metaldehyde-bran bait, placed in the middle of 25 cm^2 dishes, were detected by *Limax maximus* L. (50 animals) up to distances of 12 cm. At distances of more than 12 cm only 14% of the slugs found the bait. In moving air, by contrast, detection over far greater distances had been observed. Thus *Helix pomatia* L. was able, in strong air currents, to detect cabbage leaves up to 40 cm away, but only to 6 cm in still air (FISCHER 1950).

It is interesting, in this connection, that there are species-specific differences in the degree of development of the olfactory sense. In still air, the upper limit of the range of the olfactory sense of *Helix pomatia* L. was 6 cm, while for *Limax maximus* L. it was 12 cm. For practical purposes, therefore, a distance of at the most 10 to 12 cm between the piles of bait can be recommended. Otherwise it is a matter of chance whether the slug or snail finds the bait or not. HUNTER and SYMONDS (1970) recommend 10 to 20 cm as the optimal distance between individual pellets of metaldehyde-barley meal or mercaptodimethur-wheat meal bait for successful control of field slugs; this distance is appropriate to the olfactory capability of the slugs as established experimentally by Godan. The olfactory sense of the gastropods is of critical importance particularly when a large area is to be controlled and the bait must be attractive at a considerable distance. Here the substance used to make the bait enticing is of great importance. If the bait contains a "carrier" base that is not very attractive to most of the species infesting an area, control is bound to fail, unless the animals happened to be so active that they encountered the bait purely by chance as they crawled about.

The largest catch after treatment with bait occurs during the first night, as was first noticed by BARNES and WEIL (1942) for metaldehyde. Over a period of 7 nights, the number of slugs caught (total number 4,684) during the first night was 55%, or more than half, during the second and third nights 18% and 10% respec-

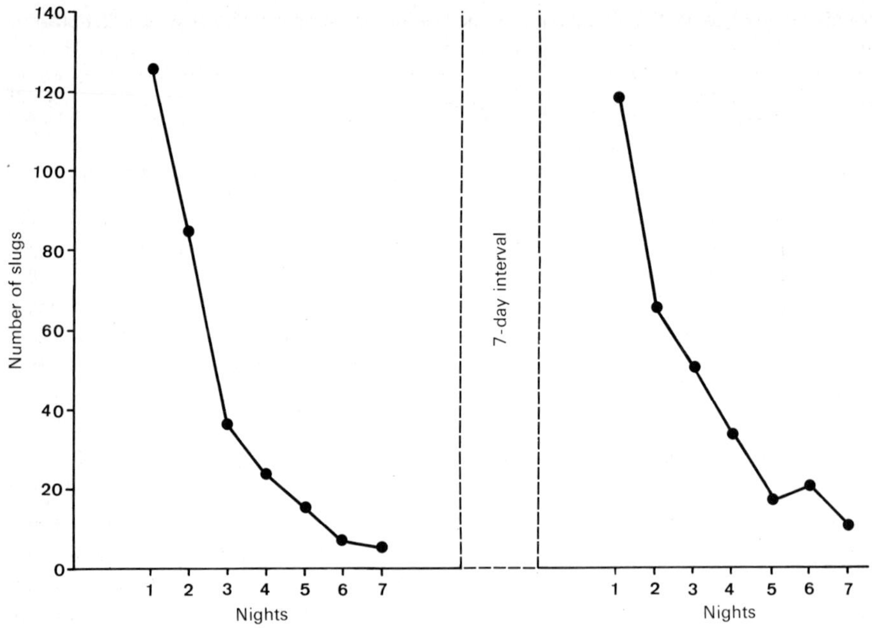

Fig. 99. Daily catches of slugs on 10 metaldehyde bait "biscuits," before and after a 7-day feeding interval (removal of "biscuits")

tively. WEBLEY (1965) also noted a diminution of the catch on succeeding nights (Fig. 99). It is interesting that after an interruption of treatment for seven days, followed by renewed application of bait, the catch of *Limax maximus* L. at first rose again and then declined from the second day. With mercaptodimethur, too, attractiveness was lost after 3 or 4 days; this returned on the first day after changing the position of the traps, and then again disappeared (MARTIN and FORREST 1969). It is possible that the explanation lies in the habituation of gastropods to the active ingredient or "attractant", which then is lost again during an interval in the treatment, when the bait is no longer available or when its position is changed.

Broadcasting of bait pellets is preferable to setting them out in clumps, since the distance between the individual pellets is of necessity only a few centimeters. The likelihood of gastropods encountering the pellets as they creep about is far greater when the pellets are scattered than if they are put out in clumps. Confirmation of this is provided by greenhouse experiments [15 °C to 16 °C, 300 *Deroceras reticulatum* (Müller)]: mortality following broadcasting was three times that achieved with clumps of pellets (GODAN 1961). The baits used were (I) a trade preparation of metaldehyde and (II) a specially prepared metaldehyde-bran bait; each clump contained 2.5 g; the distance between them was 40 cm, a distance which was still within the limit recommended by BÖHM (Table 35).

The increase in mortality by broadcasting is clear; in the experiment with pellet clumps a significant but by no means high mortality was observed on only the first day after treatment, and thereafter not at all. Further, the length of time during which the bait remained effective was increased.

Table 35. Comparison of the effectiveness of application broadcasting with that of pellet clumps. (After GODAN 1961)

Metal-dehyde-bran mixture	Application method	Days after application									Total catch
		1			3			7			
		0	∅	+	0	∅	+	0	∅	+	
I	Broadcasting	0	1	61	5	3	3	8	1	21	103
	Pellet clumps	0	6	24	4	1	2	1	0	0	38
II	Broadcasting	12	2	67	0	3	17	7	1	29	138
	Pellet clumps	19	0	11	7	0	0	3	0	0	40

0 = healthy; ∅ = affected; + = dead

It can be concluded that for practical purposes the distribution of pellets by broadcasting is to be preferred.

When a molluscicidal preparation is evenly distributed over an area, for instance by broadcasting, its attractiveness to gastropods plays only a secondary role: its effectiveness is mainly dependant on the quantity of molluscicide taken up by each animal. For in addition to the amount actually ingested, some is absorbed by direct contact as the animals creep over the soil which is densely strewn with moist and thus swollen pellets. This is of considerable advantage because bran, the usual additive of pellets, is by no means attractive to all species of gastropods.

The **size** of the **pellets,** too, is not without importance. According to observations by RAMMOS (1962), gastropods feed on metaldehyde-bran bait for about 1 $^1/_2$ to 3 min and then suddenly stop. Thus only a part of the pellet is eaten: production of large pellets is therefore wasteful and uneconomic. Small, so-called minipellets, may lead to failure of control, because as a result of their small size, they fall into cracks in the soil. They are encountered by the terricolous species, such as *Arion hortensis* Fér., but less frequently by *Deroceras* sp. and other gastropods which live on the soil or on the plants and practically never those snails which are found in bushes well above the ground. MALLET (1968) consequently regards the most suitable size to be 40 pellets to 1 g. Small granules are produced commercially on the grounds that such a formulation (metaldehyde) has a greater surface area than the larger pellets and is thus more effective by offering a greater chance of contact; further, the quantity used is considerably reduced (MERCK 1968).

In control of *Arion hortensis* Fér. by broadcasting of pellets, a distribution rate of 25 metaldehyde-bran pellets to 1 m^2 has been found to suffice; and for Field Slugs 28 pellets to 1 m^2 (MALLET 1968). HUNTER and SYMONDS (1970) consider at least 25 pellets/m^2 or a distance of 10 cm between individual pellets to be necessary. They can be spread evenly by using a fertilizer spreader adapted for sowing of grass seed or the distribution of insecticidal pellets. Aerial distribution has proved very useful for treatment of large areas. According to WEBLEY (1970) the best results in controlling of *Deroceras reticulatum* (Müller) were obtained with an application rate of 10 pellets/0.75 m^2; for *Arion hortensis* Fér. and *Milax budapestensis* (Hazay) a rate of 10 pellets/0.75 m^2 was no more effective than one of 5 pellets/0.75 m^2; this may be because neither of these two species is attracted to the bait to the same extent as *Deroceras*. PAUCK (1955) regards an application rate of 1 kg/1,000 m^2 as necessary.

Seed Dressing with Molluscicides. Three species of slugs can cause considerable damage to drilled seeds: *Deroceras reticulatum* (Müller), *Arion hortensis* Fér. and *Milax* sp., mainly *M. gagates* (Drap.). In particular Field Slugs often cause total destruction of winter wheat in marshy areas, when this is sown after a very leafy crop. Their control with bait is, however, unsatisfactory.

Treatment of the seeds or tubers with molluscicide prior to planting may be a more effective way of preventing damage to cereal seeds or to underground parts of plants, such as potato tubers or carrots. MOENS and VAN DEN BRUEL (1958) succeeded in preventing damage to young rye plants through **treatment of the soil before planting.** Such measures are, according to GOULD (1962), more effective before than after ploughing in controlling *Deroceras reticulatum* (Müller); this is because most of the slugs move into the soil after ploughing to feed on the plants which have been ploughed in, and are thus unaffected by treatment of the soil surface.

Since dressing cereal seeds with insecticides has given such excellent results, similar experiments were tried with molluscicides (GOULD) (Table 36).

The substances used were: copper (II) oxychloride 45% copper (equivalent 76% copper (II) oxychloride); Paris Green (copper acetate arsenite) with 25% copper, 57% As_2O_3 as well as 1% soluble As_2O_3 and Dieldrin with 60% active ingredient in powder form. The germination of wheat seeds was practically unaffected by the treatment; the percentage of seeds germinating was 70% to 91% after treatment with copper (II) oxychloride, and 71% to 80% with Paris Green, 70% to 87% with copper carbonate, and between 69% and 87% with metal-

Table 36. Effectiveness of molluscicides applied as seed dressing in control of Field Slugs. (Modified from GOULD 1962)

Compound	Amount of active ingredient per 35.24 l seed	Stay of seed grains in soil (days)	Dead slugs		Destroyed grains per box		Reduction of loss of seed as %
			Treated	Untreated	Treated	Untreated	
Copper (II) oxychloride	56.7	10	2	0	32	31	–
	141.75	14	0	0	43	82	47.5
	283.5	14	2	0	39	81	51.8
	283.5	20	0	0	38	114	66.6
Paris Green	283.5	12	8	2	6	53	88.6
	283.5	14	11	4	8	38	79.0
Copper carbonate	283.5	14	0	0	33	62	46.7
Metaldehyde	283.5	10	16	0	4	48	91.6
Dieldrin (commercial product)	56.7	8	2	0	19	20	5.0
Carborundum	283.5	18	4	5	95	74	–

dehyde. The value of 68% to 87% germination following treatment with carborundum as a control shows that there were no significant looses.

In the conditions of this experiment, the protection afforded seeds in the soil was considerable, particularly after seed dressing with metaldehyde and Paris Green. Copper oxychloride and copper carbonate appear to be mild deterrents, since only a few slugs were killed, but nevertheless reduction of feeding damage to the germinating seeds was satisfactory.

In another series of experiments, in which carbaryl and organic mercury compounds were tested in addition to the above list of substances, copper (II) oxychloride and metaldehyde proved the most effective in protecting seeds from attack by *Deroceras reticulatum* (Müller), *Arion hortensis* Fér. and *Milax* sp., while carbaryl must be rejected because of its high toxicity to plants. Paris Green applied at the rate of 0.6 g/l seed, was also slightly phytotoxic to the seedlings. In laboratory conditions, metaldehyde and Paris Green (about 1.2 g/l seed) gave the best results, but in the field copper (II) oxychloride applied at the same rate is preferable. Treatment with mercaptodemethur has the advantage that, in addition to protection against gastropods, the seeds, such as those of maize, are given a bitter taste which protects them against birds (HERMANN and KOLBE 1971). In 1968 this substance was, in fact, introduced as a seed dressing to protect maize from attack by birds; it is possible that the germinating maize seeds are also protected from gastropod damage. Observations by the Plant Protection Service of the FRG suggest that dressing treatment wheat seeds with mercaptodimethur is not effective against Field Slugs; furthermore, birds have died after eating treated maize.

STEPHENSON (1961) considered the possibility of seed dressing with dimethylphthalate, dibutylphthalate, benzylbenzoate and salicylaldehyde, all of which are repellent to gastropods. The first two substances are also insecticides, while the third is an acaricide. These compounds are, however, phytotoxic at the required dose, and at lower concentrations do not protect seeds from attack by Field Slugs.

These investigations have shown that cereal seed dressing can protect them from slug damage, or at least reduce this to a tolerable level. According to GOULD, however, seed dressing with molluscicides affords only limited protection, because it is effective only immediately after drilling. Paris Green and copper sulphate, while mild deterrents, are not lethal to slugs; they merely repell the pests, thus protecting seeds from feeding damage, but the young plants are then eaten more voraciously. There is also the danger of a toxic effect on the seedlings. To counteract this disadvantage, a larger quantity of seed than normal has to be sown, which makes drilling more difficult.

Application of the molluscicide together with the seeds, instead of dressing them with it, has also been tried. WEBLEY (1966) advocated drilling pellets and wheat seeds at the same time. Commercially available slug pellets are, however, not suitable for this purpose, because they break up easily during drilling and because of their relatively shortlived toxicity in the soil. To remain effective they would have to be protected against destruction by dampness and fungi in the soil, since it is important that they should still be toxic when slugs become active. Pellets could be adequately protected by dressing them with siloxane, alkoxysilane or silicon resin. STEPHENSON (1972) recommends impregnating the pellets with formaldehyde-hardened gelatine to improve their durability in damp soil.

3.2.2 Dehydrating and Caustic Compounds

Many earlier control substances depended for their lethal effect on the stimulation of excessive secretion of mucus, as is the case in the initial stages of the action of metaldehyde, a substance which later came to the fore. The result is a pronounced desiccation of the slug which finally leads to death; however, no real poisoning occurs. In this category are dehydrating substances such as Hederich-Kainite and calcium cyanamide which are used as herbicides or inorganic fertilizers. Other substances have a corrosive effect on the skin of gastropods. For example, quick lime (calcium oxide) withdraws water from the body of a snail or slug as it turns to slaked lime, and has, as well, a caustic effect on the skin. Calcium cyanamide and copper sulphate are also dehydrating, but are toxic as well, as a result of the cyanamide or the copper ions.

Dehydrating and caustic compounds are suitable for keeping gastropods away from small areas or valuable single plants in greenhouses. The substances used are caustic soda, quicklime, Kainite, calcium cyanamide, Chili saltpetre, copper sulphate and others, and they are applied *before* sowing or planting. In fields where large gastropod populations appear during ploughing and cultivation of the soil before either autumn or spring planting, treatment with waste lime is recommended. During dry weather the area to be treated should first be gone over with a skimmer and a disc harrow and, if need be, a roller, to achieve a soil surface of horticultural standard (Tomaszewski 1949). Field Slugs shelter in the soil in winter, but come to the surface if there is a long period of mild weather. In endangered areas newly sown cereals should therefore be carefully watched and at the first sign of infestation caustic soda should be applied immediately (Koltermann 1955). In cabbage fields, gastropod damage has been kept within tolerable limits by spreading copper sulphate 1 to 3 weeks before planting. According to Schreier (1969), Field Slugs in sugar beet can be controlled successfully by treatment with freshly slaked lime (300 to 400 kg/ha) or with unoiled calcium cyanamide (100 to 150 kg/ha). In store rooms and cellars, spreading lime is effective. The areas must, however, be clean, free of rubbish, frequently aired and above all dry.

Copper and iron sulphate alone can cause plant damage; the following combination, which is harmless to plants, is said to be preferable: 2.5 kg copper sulphate to 50 kg Kainite, applied at the rate of 300 to 400 kg/ha.

Treatment with dehydrating and caustic compounds must always be carried out twice, with the result that large amounts are used. They are for:

Freshly slaked lime	3	to	6 dt/ha
Kainite	4.5	to	6 dt/ha
Unoiled calcium cyanamide	1.5	to	2 dt/ha
Calcium cyanamide + Cainite (1 part + 4 to 6 parts)	3	to	4 dt/ha
Caustic soda			3 dt/ha
Chili saltpetre	6	to	9 dt/ha
Copper sulphate	125	to	250 kg/ha
Iron sulphate	125	to	250 kg/ha
Calcium cyanamide	300	to	500 kg/ha

In each case, half the amount is distributed in the late evening and the other half in the morning, if the gastropods are active. It must also be remembered that these substances are successful only in dry or not very moist conditions. A third important condition is that they be distributed twice, in rapid succession, within 15 to 30 min, since the first application leads only to excessive, but not lethal, secretion of mucus. It is only the second application which is lethal, provided the slug is unable to correct its water content. If the time lapse between the two applications is longer than half an hour, and if, in addition, conditions are very humid, the slugs still recover. Ideally, the substance being distributed should land directly on the slug or snail. The gastropod will normally shed a mucus sheath laden with the powder; but if this sheath cannot be renewed rapidly, because of the difficulty of replenishing its water content, then, on the second application the caustic substance is effective as it comes into direct contact with the sensitive skin. This only applies, however, in dry environments, for with rain, watering or in an already humid atmosphere the water deficit is rapidly made good through the skin and a new mucus sheath is secreted. Various conditions must be fulfilled if treatment with dehydrating or caustic substances is to be successful: direct application to the slug's body, absence of hiding places such as stones, planks, earth clods, grass clumps and dense vegetation lying close to the soil surface, continuous dry weather and the absence of heavy dew. Since these favourable conditions seldom all prevail at the same time, often only weaker and very young gastropods are killed; the strong ones recover. It is essential to make applications of the substance, with strict adherence to the time lapse – at best 15 min – between the two applications.

Since calcium cyanamide and copper salts were formerly very important in gastropod control, they will now be considered in more detail.

3.2.2.1 Calcium Cyanamide

This substance is a very effective molluscicide. It is the chief synthetic nitrogen fertilizer with organically bound nitrogen; this is in the form of cyanamide. As a result of biological processes in warm, moist conditions in the soil, it is converted by way of urea to ammonia and calcium nitrate, during which process cyanamide is temporarily released. Calcium cyanamide is used as a source of nitrates for cereals, maize, root vegetables, fodder and also in vegetable, fruit, wine and hops growing. These are all crops particularly plagued by infestations of Field Slugs.

To be effective against Field Slugs, calcium cyanamide must be spread as a rate of at least 300 kg/ha; Getzin and Cole (1964) recommend about 400 kg/ha. It remains effective even at temperatures as low as 0 °C, thus making control measures possible during cool weather and in the autumn and winter; this is not possible with metaldehyde. The toxicity of calcium cyanamide is of longer duration than that of metaldehyde, lasting up to 7 days in dry weather and 5 in rainy conditions when metaldehyde would be virtually ineffective. Calcium cyanamide is not as rapidly affected by brief, intermittent rain as are other dehydrating and caustic molluscicides, but is rendered useless by heavy, continuous rain. It is thus relatively independant of weather conditions. One application usually suffices instead of the two which are normally recommended, thus saving time and labour if large areas are to be treated.

Van den Bruel and Moens (1956, 1958, 1960, 1964) tested calcium cyanamide in comparison with other substances and have concluded that it is very effective, since at even 5.5 °C (300 kg/ha) a mortality of 97% of Field Slugs was observed. At as low an application rate of 100 kg/ha this substance was better than metaldehyde. The low mortality at 0.5 °C is not to be attributed to a fall in toxicity of calcium cyanamide, but to the limited activity of Field Slugs as such a low temperature so that they do not come into contact with the substance.

In the literature calcium cyanamide is recommended for use in cereal cultivation, first, to protect the seeds from slug damage because at high concentrations it has a repellent effect, secondly, for the formation of barriers around the edge of fields and thirdly, to protect seedlings.

It is not, however, used for control of terrestrial gastropods in agriculture today for it can cause permanent damage to young cereal plants, mainly to the root system down to a depth of 2 cm below the soil surface, the extent of penetration of the free cyanamide. It is therefore inadvisable to use it either for very young plants at the 4- to 6-leaf stage, in a thinly sown cultivation which is susceptible because of badly rooted plants or for plants which, following damage by frost, may have their root system partially destroyed. As a protective measure against gastropod attack, calcium cyanamide should normally be applied only *before* drilling of the seed; it can be applied after sowing if the seeds are covered with at least 3 cm of soil, and also after the plants have passed the 4- to 6-leaf stage.

The strong repellent effect of calcium cyanamide drives Field Slugs into shelter where they remain for several days; during this time no feeding damage is to be feared, and plants may grow sufficiently to be no longer susceptible to attacks by slugs.

Van den Bruel and Moens (1958) recommend the following rates of application for different crops:

300 to 500 kg/ha – winter wheat and rye
300 to 400 kg/ha – pastures, meadows, winter rape (December to February for Belgian conditions)
250 to 325 kg/ha – (22% active ingredient) – cereals (for plants 7 to 10 cm in height)
200 kg/ha – (22% active ingredient) – cereals (for plants 20 to 25 cm in height), in spring very susceptible
200 kg/ha – irises, gladioli (plants about 10 cm in height), both very resistant
530 kg/ha – (18% active ingredient) – bulbs (initially very sensitive; 3 weeks after appearance of the leaves relatively resistant)
150 to 250 kg/ha – (18% active ingredient) – cereals (April to May for Belgian conditions).

According to Böhm (1957), protective barriers of calcium cyanamide around valuable cultivations have proved successful. This substance is also used against amphibious snails on pastures and meadows. It is possible to clear an area completely of a snail such as *Galba truncatula* (Müller) by repeated application at monthly intervals. Total elimination is, however, made difficult in very damp fields or near ponds, in places where water may collect in the hollows left by footprints of grazing stock, where difficult road conditions reduce the efficiency of the spreading machines or, finally, where a large number of ditches, like the polders of Belgium, prevent speedy treatment and thus reduce the efficacy of the molluscicide. In these conditions, individual snails can escape contact with the material; these will form reservoirs from which a new pest population can arise the following spring.

Despite its merits, calcium cyanamide may not be used for fruit and vegetables because of the possibility of caustic damage to the treated plants and because of its toxicity to man and his domestic animals. If the mowings are not used as fodder, calcium cyanamide can be used to treat slopes, the edge of ditches, fallow fields, weedy borders of cultivated land and around compost heaps. With one application in early winter and a second in late winter, when there is no frost or snow, older gastropods can be destroyed and, in mild winters, even the newly hatched young.

3.2.2.2 Copper Salts, Compounds of Heavy Metals

Copper compounds are extensively used in agriculture as fungicides. Copper sulphate ($CuS0_4$) is the *classical molluscicide* (STRUFE 1968), in particular for parasite vectors which inhabit irrigation ditches (SHARAF EL DIN and EL NAGAR 1955; CHENG and SULLIVAN 1977; CHRISTIE et al. 1978 and others) and, together with calcium cyanamide, used to be the most commonly used substance.

Copper sulphate has also been used against terrestrial pulmonates and is regarded as a favourite compound for the control of *Achatina fulica* Bowdich. BORG (1953) recorded 94.3% mortality of Field Slugs with a dose of 800 l water/ha of 3% copper sulphate spray. An equally high mortality was obtained with calcium hydroxide applied at 400 kg/ha in two equal treatments each at half this rate of application. Superphosphate caused 66.7% mortality at 300 kg/ha while with calcium nitrate mortality was 73.3% at the same application rate. Finally, copper sulphate can be used as a spray (12 kg/ha, dissolved in 400 l water). Mixtures of copper sulphate (10 to 20 kg/ha) with Kainite (400 g) and magnesium sulphate or calcium carbonate (400 g); or a mixture of one part of dehydrated copper sulphate with 10 parts of calcium hydroxide; or 1 part of copper sulphate with 20 parts of Kainite; these are spread over the soil (CAIRNCROSS 1957; GLENDENNING 1952; VAN DINTHER 1973).

According to HENDERSON (1965, 1968, 1969), the lethal dose LD_{50} for Field Slugs after 1 h contact at 10 °C or 20 °C is 71.9 ± 1.9 µg per slug, and if ingested 129.2 ± 5.9 µg per slug. In comparison with other poisons (LD_{50} for sodium pentachlorophenolate (Na PCP) is 22.9 ± 2.5 µg/slug and metaldehyde is 85.2 ± 4.0 µg/slug). Copper sulphate is not very toxic if ingested. With Field Slugs copper sulphate is more toxic if injected into the haemocoel (LD_{50} 18.4 ± 1.1 µg/slug) while its effect is less pronounced if ingested.

Since copper sulphate is a contact molluscicide absorbed through the foot epithelium (RYDER and BOWEN 1977), its application should be carried out at times when gastropods are most active, that is on warm, damp evenings and in the early morning. Control should not be carried out on wet, rainy nights, because the molluscs do not then leave their shelters. According to WINKLER and CHI (1964), the substance has a certain repellent action at high concentrations.

Copper sulphate has also been used in rice fields in Surinam for the control of the snail *Pomacea lineata* (Spix) and *P. glauca* (L.) at an application rate of 2 to 5 ppm or 2 to 5 kg/ha (VAN DINTHER 1956, 1957). It was applied four days after flooding of the fields and one day before planting of the already germinating seeds; this is probably the time of greatest activity of the snails. The efficacy of copper sulphate, like that of both NaPCP and niclosamide, which are still to be discussed, is very greatly reduced by the mud of the fields (HOLZ 1962); however, copper sulphate is today again recommended for use in fields against Helicidae (ANON. 1977).

Copper sulphate is effective also against **snails' eggs:** it was lethal to the eggs of the terrestrial *Helix aspersa* Müller with an LD_{50} of 4 to 5 mg/egg (STRINGER and MORGAN 1970); at 10 ppm it caused 99.5% mortality of the eggs of the freshwater snail *Oncomelania formosana* (PILSBRY and HIRASE), while at 1 ppm mortality was 79.6%, compared with 93.8% caused by NaPCP (CHI and WINKLER 1962; WINKLER and CHI 1964). In general, the eggs of *Oncomelania* are more difficult to kill than those of the freshwater snail *Biomphalaria* because of their mud-cocoon.

Opinion is divided about the toxicity of copper sulphate to plants. GLENDENNING (1952) regards it as harmless in this respect, while VAN DINTHER (1973) has recommended a thorough investigation of this matter. Until conclusive evidence is available it should invariably be applied *before* sowing or planting and not later. On the other hand, a herbicidal effect on water plants in irrigation ditches is very much to be welcomed (SHARAF EL DIN and WILSON JONES 1954).

Like other molluscicides which consist of heavy metals and which are applied to water, copper sulphate is adsorbed by plants, minerals and mud, processes which depend on the pH of the water and the amount of organic matter it contains. It therefore accumulates in the substratum of treated waters; 32% to 49% of applied copper sulphate is absorbed by the mud within the first five minutes and 52% to 67% within 24 h (PAULINI 1963, cited by STRUFE 1968); mud can reduce its toxicity to gastropods by as much as 86%.

Its molluscicidal action is very greatly reduced by alkaline or very hard water and by vegetation; this is not true of niclosamid. MÜLLER (1962) observed for this reason a considerable discrepancy between the results of applying copper sulphate in the laboratory and its effect in the field. In the laboratory 4 ppm killed all the freshwater snails, but was commonly completely without effect in the field, particularly in very alkaline water and in the presence of a large quantity of organic matter. TARCZYŃSKI et al. (1965) used a 5% copper sulphate solution at the rate of 5 kg/ha with great success against *Galba truncatula* (Müller) on pastures in Poland.

Just as copper sulphate can be applied in protective strips to control terrestrial gastropods, so too, according to TEESDALE et al. (1961), it can be used as a chemical barrier in water to repell aquatic and amphibious intermediate hosts. To this end it should temporarily be applied at a rate of 0.125 ppm to 0.25 ppm at the point where irrigation canals receive their charge of water from a main channel or reservoir.

More recently two other copper compounds have been found to have useful molluscicidal properties, namely **copper (I) oxide** (Cu_2O) and **copper (II) oxychloride.** Copper (I) oxide is ineffective against snails' eggs, but proved to be extremely toxic to newly hatched *Biomphalaria* sp., on the other hand the fungicide copper (II) oxychloride delayed hatching and in some cases prevented it altogether. According to STRINGER and MORGAN (1969, 1972) eggs in treated soil were damaged to such a degree that only 30% of them hatched.

The LD_{90} of Cu_2O was 0.74 ppm for juvenile *Biomphalaria glabrata* (Say) (3 to 5 mm) and 2.7 ppm (average value) for adults (13 to 15 mm). Thus for adult snails the lethal dose was three times that for young snails. Because of its highly lethal effect on newly hatched and juvenile snails, Cu_2O is more effective in the field than copper (II) sulphate. Like copper (II) sulphate this compound is also absorbed by mud; it becomes completely ineffective in very muddy water (CHU et al. 1968).

In experiments with the terrestrial *Helix aspersa* Müller, *Cepaea hortensis* (Müller) and *C. nemoralis* (L.) spraying with copper (II) oxychloride (2.4 kg Cu/ha in 1,000 l water/ha) was found to be inadequate. This substance is nevertheless useful since in feeding experiments, it was found to have a repellent effect. Thus wet treated lettuces were kept free of gastropod attack for three weeks (STRINGER and MORGAN 1969).

Heavy metal compounds such as one of tin (Brestan) and another of zinc (Ziram), are used in some countries to destroy water weeds, and in the case of Ziram, as a larvicide and fungicide as well. These have been found to be good molluscicides for control of freshwater pulmonates (GRÉTILLAT 1961; MOENS and PECHEUR 1964). Ziram is preferable to copper sulphate in that it is harmless to man and domestic animals, remains stable in alkaline waters, and does not lose its molluscicidal effect in water containing a large amount of organic matter (ANON. 1962). Ziram is lethal to snails at 10 ppm, to mosquito larvae (*Culex fatigans* Wiedemann, *Anopheles* sp.) at 1 to 2 ppm and also kills water weeds of the genus *Pista*, the roots of which are a habitat of certain mosquito larvae. "A preparation which is at the same time a fungicide, a molluscicide, a larvicide and a herbicide can be of great use in areas where bilharzia, malaria and filariosis are all common." Because of its phytotoxicity, Ziram is not, however, entirely suitable for use on inundated fields and the same applies to Brestan which is very toxic to young rice plants, so that the application must be made as much as seven days before planting (VAN DINTHER 1973).

The heavy metals copper, chromium, cadmium and lead have the following effects on the freshwater snail *Biomphalaria glabrata* (Say) (RAVERA 1977): survival of adults in solutions of 1 ppm: cadmium chloride – 15 h; copper (II) sulphate – 26 h. Low concentration of copper, chromium and cadmium reduced the fertility of adult snails. The embryos survived half a day in 1 ppm Cd, 1 day in 1 ppm Cu and 4 days in 1 ppm Pb [$Pb(NO_3)_2$]. Chromium was less effective than the other heavy metals, but delayed embryonic development at concentrations lower than 0.7 ppm.

In both the review article by STRUFE (1968) and in the Bulletin of the World Health Organization, many references are to be found to the literature dealing with the molluscicidal properties and application of copper sulphate, other copper derivatives and heavy metal compounds for the control of trematode vectors.

There are, in addition to these molluscicides which are effective against both terrestrial *and* freshwater gastropods, other compounds used exclusively for aquatic snails; these are discussed in the next section.

3.2.3 Molluscicides for Control of Aquatic and Amphibious Snails

Upon application to rivers, ponds, canals, etc., a molluscicide may precipitate or be adsorbed; the concentration in the water may thus fall below the lethal dose for gastropods and control fails, despite excellent toxicity of the substance used. The pH of the water may also be important. In laboratory experiments 100% mortality of *Biomphalaria glabrata* (Say) was obtained after 24 h exposure to 0.2 ppm niclosamide in tap water with a pH of 7.0 to 8.2. At both a higher acidity (pH 4 or pH 5) and a higher alkalinity (pH 9.0 or pH 9.9), four to five times this concentration of the molluscicide was required to achieve the same result. FOX et al. (1963) and MEYLING and PITCHFORD (1966) therefore recommend first establishing the pH and also the degree of turbidity of the water before deciding on the application rate. Organic matter present in the water can also interfere with the control measures. Interesting in this connection is an investigation of DE SOUZA and PAULINI (1968) on the effect of the presence of living yeast cells on lethality; the LD_{50} of both niclosamide and pentachlorophenol rose when the snails were tested in the presence of yeast cells: this was because more than 50% of the molluscicide was absorbed by the yeast.

As has already been described for terrestrial gastropods, the behaviour of freshwater snails may limit the success of control, even with a highly toxic substance. This occurred in the "Wageningen Rice Scheme" in Surinam (1954 to 1960) during attempts to destroy the snails *Pomacea glauca* (L.) and *P. dolioides* Reeve which inhabit the fields (VAN DINTHER 1973). Niclosamide had been found to have an excellent toxic effect at 0.4 ppm in the laboratory, but in field conditions this was not always satisfactory. Here Bayluscid was used at an application rate of 0.5 to 1 kg/ha of a 70% wettable powder in 40 l water. After the surface of the water in flooded rice fields had been sprayed, many snails left the water and crawled on to projecting clumps of earth, weeds and rice stubble, thus escaping the action of the molluscicide. Control was successful only after introduction of aerial spraying when objects above the water surface were also thoroughly sprayed. Since it brought better results than application from the ground, aerial spraying is today widely used for extensive treatment in this area.

As with land gastropods, destruction of amphibious and freshwater snails requires an adjustment of chemical control measures to the ecological situation (LEVINE 1970; CHU 1978). HAIRSTON and SANTOS (1961) investigated the possibility of ecological control of *Oncomelania quadrasi* (Möllendorff), the intermediate host of the trematode *Schistosoma japonicum* Katsurada on the Phillipines.

It is worth mentioning a project aimed at keeping intermediate hosts from entering snail-free irrigation canals by the repellent effect of a molluscicide. To this end ETGES and GILBERTSON (1966) tested compounds which are also used for plant protection, such as copper sulphate, niclosamide, trifenmorph, NaPCP. As bait, both for the intermediate hosts and their predator *Marisa cornuarietis* (L.), germinating wheat seeds were mixed with the molluscicide and placed at the openings of the irrigation canals. The maximum concentration of molluscicide released into the water in this way was 200 ppm. After 48 h it was found that since both copper sulphate and NaPCP are avoided by snails, mortality was low (0.6%); on the other hand, trifenmorph and niclosamide were only slightly repellent and were lethal even at very low concentrations. An experimental strain of *Biomphalaria glabrata* (Say) which was amphibious instead of aquatic, proved to be little affected; this was because the snails could leave the treated water and thus escape death. Because of this behaviour even nonrepellent substances were without effect, as the snails did not come into contact with them. The predatory snail *Marisa cornuarietis* (L.) was the most resistant and since it preys upon the intermediate host of trematodes, *Biomphalaria glabrata* (Say), it is highly suitable for combined chemical-biological control (Chap. 3.6).

In the following section the most important chemical substances used to control aquatic and amphibious snails will be discussed: their use as plant protection agents for the control of Lymnaeidae in cress cultivations (niclosamide) and *Pomacea* spp. in rice fields (trifenmorph, NaPCP) has been considered or has already found practical application.

3.2.3.1 Niclosamide

Niclosamide, or 5,2′-dichlor-4′-nitro-salicylanilide is a phenolic derivative of salicylic acid (DUHM et al. 1961). It is the active ingredient in the preparation Bayluscid, and is used in flooded rice fields and in plantations with ditches and canals, as well as in streams, ponds and other water bodies; it has been in use since 1959. Since that time much has been published on niclosamide, mainly in the plant protection review "Bayer" and in the Bulletin World Health Organization (CROSSLAND 1963; DEGRÉMONT 1972; GÖNNERT 1962; GÖNNERT and WEBBE 1966; SHIFF et al. 1973; STURROCK 1974; TECHNAU 1974; JORDAN et al. 1978 and others).

Niclosamide is lethal to freshwater snails and their eggs at a concentration of 1 to 8 ppm for 1 h exposure; it is lethal to the larval stages of *Schistosoma* sp. (Miracidiae, Cercariae). A niclosamide preparation with sand has been used with success against dense populations of *Biomphalaria choanomphala* Martens on the floor of Lake Victoria (MAGENDANTZ 1974).

At the concentrations which are lethal for gastropods, the molluscicide is not toxic to economically important crops such as oats, rice, sugar cane and mustard (STRUFE 1968). No harmful consequences to fruit or leaves could be detected following niclosamide application to 3- to 5-year-old orange trees, even when the concentration exceeded the lethal dose for snails (MEDENWALD and SHIFF 1965).

Compared with NaPCP and copper sulphate, niclosamide has proved to be the most promising chemical compound for the control of the snail *Biomphalaria glabrata* (Say) and its eggs (FLETCHER 1965).

It has also been considered for control of Lymnaean snails which present a particular problem in the ditches and water reservoirs in which cress is cultivated. Before Bayluscid was available, VAN DEN BRUEL and MOENS (1957, 1958) tested **isochlorthion** but regarded it as unsatisfactory because its efficacy was strongly influenced by temperature; at 0 °C two or three times the normal dose became necessary. Furthermore, young animals survived more often than adults. In addition, reinfestation of a cleared area within 15 days by gastropods from the neighbourhood is likely. Isochlorthion was, furthermore, toxic to fishes and crayfishes and reduced their populations in treated ditches (RICOU and MARRE 1960).

Moens (1965) described two methods of applying niclosamide against lymnaeans; these can be used in combination. In the first method, the molluscicide should be spread over the surface of the water (application rate 0.5 g/m^2 for older plants and 0.25 g/m^2 for young plants) and this concentration maintained for 24 h, optimally at a temperature between 10 °C and 18 °C. Circulation of water in the ditch, which is normally essential, should be stopped for 1 or 2 h, so that the water becomes still; in this way an 85% mortality of the snails can be achieved. The second method involves application of the preparation to flowing water, the entrance to the ditch being recommended as the place of application, this time without stopping circulation of the water. The snails collect mainly at the entrance to the ditch because the cress grows especially well there and affords a better feeding chance and shelter than the other more sparsely covered parts of the ditch. At this point almost the entire snail population can thus be destroyed. It is advantageous to combine the two methods, and by so doing ditches have been rid of *Radix peregra* (Müller), *Galba palustris* (Müller), and *Galba truncatula* (Müller); the harmless *Hydrobia* sp. was, however, killed as well – an unintentional side effect.

In using this molluscicide, however, burns to the leaves and stalks, particularly of young cress plants, have to be taken into account; older and well rooted plants are not affected. For the protection of young plants, lymnaean snails and their eggs can be destroyed by an application of 0.25 g/m^2 of the commercial formulation, provided this dose is maintained for 24 h. Older cress plants can withstand a dose of 0.5 g/m^2.

By addition of a phenyl group to niclosamide, a compound known as **PSA** is formed (2'.5-dichlor-4'-nitro-3-phenylsalicylanilide) a molluscicide with a new and unique effect on freshwater snails. The LD_{90} for 1 h exposure of the experimental animals *Biomphalaria glabrata* (Say) and *Bulinus (Physopsis) globosus* (Morelet) was greater than that for 3, 6 or 24 h exposure. Expressed as ppm, the LC_{90} values after exposure of the snails for 1, 3, 6 and 24 h were 0.283 ppm, 0.23 ppm, 0.125 ppm and 0.048 ppm. *Biomphalaria glabrata* (Say) proved to be more sensitive than the other species after 1 h exposure to PSA, while *Bulinus globosus* (Morelet) was more resistant than the other species after 24 h of treatment. A rapid action of the molluscicide followed by chemical decomposition within a short time is of particular advantage for treatment of stagnant water, but not for flowing water, as then the concentration falls too rapidly (Ritchie and Fox 1968).

In order to avoid this undesirable fall in concentration, Fenwick (1969) tested the suitability of carrier materials and found that towelling or linen cloths impregnated with tri-n-butyl tin oxide (3%) as well as niclosamide (8%) in a neopren (synthetic rubber) base ("rubber sheeting") could be used to release a lethal dose into the water over several weeks. Fenwick regards this procedure as particularly suitable for creating a **chemical barrier** to prevent reinfestation of cleared ditches through migration of snails from untreated lengths of the canals. This procedure derives from the common use of "toxic rubber" to prevent fouling by marine organisms of surfaces, particularly of boats, exposed to sea water. Linen cloths, first thoroughly soaked in molluscicide and then cut into pieces of 6 or 12 cm^2 were first used by Berrios-Duran and Ritchie (1968); these caused almost 100% mortality of *Biomphalaria glabrata* (Say) and *Bulinus globosus* (Morelet); the LD_{50} for these snails, using a 24-h exposure, was found to be 0.1 to 0.2 ppm. Gelatine granules are also recommended as carriers for Bayluscid (UPATHAM and STURROCK 1977).

The molluscicide is toxic for eggs, early embryonic stages and newly hatched snails and even for *Schistosoma* cercariae; it has also insecticidal properties and an excellent long term effect.

3.2.3.2 Trifenmorph

Trifenmorph or N-tritylmorpholin is just as strongly toxic to aquatic and amphibious snails. Since 1966 it has been used successfully as a molluscicide (Frescon) against the intermediate hosts of *Schistosoma* sp. and is still used today in the rice and cotton plantations of Madagascar (DEGRÉMONT et al. 1972). It has a neurophysiological effect (MORETON and GARDNER 1976; GARDNER and MORETON 1978). Amongst the first to draw attention to this substance were BEYNON and WRIGHT (1967), BOYCE et al. (1966), CROSSLAND (1967, 1976) and SHIFF (1966). It is also lethal to the snails which act as intermediate hosts for *Fasciola gigantica* Cobbold (WARLEY and WRIGHT 1977).

Trifenmorph is very toxic to *Bulinus truncatus* (Audouin), *Biomphalaria glabrata* (Say) and *Biomphalaria pfeifferi* (Krauss); the LD_{50} is 0.1 tp 0.5 ppm with 1-h exposure, and 0.01 to 0.05 ppm with 24-h exposure. In field experiments, a concentration of 0.1 ppm over 6 h was lethal to *Biomphalaria pfeifferi* while 0.25 ppm was lethal to *Bulinus truncatus* and *Biomphalaria glabrata.* Since trifenmorph is not ovicidal, the heavily reduced snail population can recover; if the lethal dose is, however, maintained for 16 to 30 days the whole population can be destroyed because newly hatched young are also killed. In this way treating irrigation canals led to total elimination of *Biomphalaria pfeifferi* (Krauss) at a concentration of 0.02 ppm and of *Bulinus truncatus* (Audouin) and *Biomphalaria glabrata* (Say) of 0.025 ppm. While 5 h exposure to 0.25 ppm caused 83% mortality of *Bulinus* and 75% of *Biomphalaria*, 0.05 ppm sufficed to cause a 75% mortality of the very sensitive *Lymnaea natalensis* (Krauss); 15 min exposure to a 1.0 ppm concentration was lethal for this snail. If necessary, treatment can be repeated several times at 14- to 21-day intervals, so that young animals which have hatched in the meantime are also killed.

In the field, when a dose of 1.0 ppm was applied in water reservoirs, canals and natural waters, the molluscicide formed a milky emulsion and slowly sank to the bottom. Even the brief period before sedimentation was complete sufficed to kill the snails. The speed at which the emulsion settled was influenced by the density of the plants in the water and also the temperature; it was quickest – 24 cm in 10 min – in water which was free of plants.

Furthermore, snails buried in mud in the substratum remained unaffected. They are only affected by the molluscicide when they emerge from the mud. Its settling rate combined with a strong molluscicidal action makes trifenmorph an excellent control substance for use in *stagnant* and slowly flowing waters. In these conditions a non-ovicidal but highly toxic molluscide can be very useful if it is effective over a long period at low concentrations (YEO 1965).

There are species-specific differences in the effect of trifenmorph on freshwater snails. *Physa acuta* Draparnaud, for example, was 260 times more resistant than

Bulinus truncatus sericinus (Jickeli) after 6 h exposure and 60 times more resistant with 24 h exposure.

FENWICK (1970) recommends a method for the elimination of *Biomphalaria pfeifferi ruppelli* (Dunker), the intermediate host of *Schistosoma mansoni* Sambon, from sugar cane plantations. This involves repeated and lengthy treatment of the irrigation canals, with 5-day application of 0.025 ppm repeated at intervals of 7 weeks; in this way the young snails which hatch from resistant eggs are also killed. Repeated application is also necessary for the elimination of *Lymnaea natalensis* (Krauss) when a concentration of 0.25 ppm must be maintained (FENWICK and LIDGATE 1970). Despite 100% mortality of the snails, young animals were found 9 weeks later, so that in practice treatment at 5-week intervals is recommended. Comparative investigations of SHUKLA and ROY (1975) with trifenmorph (0.025 ppm) and copper sulphate (10 ppm) in stagnant water have likewise demonstrated the greater efficacy of Frescon against various *Lymnaea* species as well as the fact that it is harmless to rice plants.

The lack of ovicidal effect of trifenmorph is a particular disadvantage in running water because frequent and repeated applications which are necessary are usually costly. On the other hand it has the advantage that it is largely harmless to crops such as rice and cotton (STRUFE 1968). PERRET et al. (1972) attempted to improve the effect of Frescon by the addition of **urea**. The urea leads in a vast increase in the biomass of phytoplankton, which in turn results in a great reduction in the oxygen concentration in the water. The endothermic reaction which occurs as the urea dissolves in water causes a fall in temperature which results in improved distribution of the molluscicide in the water. The toxic effect was so strong that the treated reservoir remained free of snails for 5 months, whereas with the conventional method, that is without urea, renewed treatment with triphenmorph was necessary every 2 months.

In Europe, MOENS (1970) tested this molluscicide on *Galba truncatula* (Müller), which had already been shown in laboratory investigations to be poisoned by 48 h contact with a concentration of only 0.03 ppm. In the field, dusting at an application rate of 300 g active ingredient/ha successfully controlled these snails in pastures and meadows. At 500 g/ha ($\approx$ 3 l/ha of an emulsion with 16% active ingredient), a single treatment in spring was sufficient to clear the periphery of meadows, while two or three applications at intervals of 1 or 2 months were necessary for the very damp central areas.

Frescon was also considered as a means of controlling freshwater snails in fish ponds. CROSSLAND et al. (1971) tested its suitability for the control of *Radix peregra* (Müller), the intermediate host of the trematode *Diplostomum spathaceum* (Rudolphi), which causes blindness in Rainbow Trout in rearing ponds. The snails were successfully eliminated by trifenmorph and the fish withstood the treatment well. While concentrations ranging from 0.01 to 0.1 ppm was critical for both snails and fishes, one of 0.05 ppm resulted in 100% mortality of the snails but only 10% of the fishes. Although 5 months later the pond was found to be reinfested with a new population of snails, which had arisen from the young animals which had hatched in the intervening period, treatment of the edges of the pond was considered adequate, as the gastropods primarily or exclusively inhabit this zone.

Meier-Brook and Kuang Yung Tan Tjheng (1977) tested the sensitivity of the predatory snail *Marisa cornuarietis* (Linnaeus) to trifenmorph.

3.2.3.3 NaPCP

The sodium salt of pentachlorophenol is, when manufactured industrially, a grey powder which smells of phenol. It has many uses: for the preservation of wood against boring insects and against fouling, as a fungicide, a herbicide, a strong defoliant, a termite deterrent and, finally, also as a molluscicide. As such it was first utilized in 1948 in Egypt and in Japan (Müller 1962). More recently pentachlorophenol has also been found to have an algicidal effect (Mukherji 1972). An extensive literature is available in the Bulletin of the World Health Organization from about 1955.

NaPCP is used on a large scale for the control of snails which act as intermediate hosts of trematodes on farms in the tropics and subtropics and also in Japan (Komiya 1961); this is done by placing *briquets* containing the chemical in the irrigation canals; these dissolve within 6 to 8 h. Ritchie et al. (1969) have tested the material formulated as a hard bar which is laid under vegetation on the bank of an irrigation canal and which gives off the molluscicide over a very much longer period (2 days and more), so that a constant dose of 1 ppm in the water can be maintained. The effect is considerable, even with swiftly flowing water.

In tests on *Bulinus truncatus* (Audouin) comparing different molluscicides, a mortality of 90% to 100% resulted with doses of 0.7 ppm niclosamide (70% wettable powder), of 7 ppm NaPCP (briquet-formulation) and of 20 ppm copper sulphate (80% wettable powder) after 24 h exposure (Massoud et al. 1969). In field conditions, higher concentrations were found to be required: 1.4 ppm for niclosamide, 14 ppm for NaPCP and 40 to 50 ppm for copper sulphate. For control of *Galba truncatula* (Müller) in pastures and meadows a dose of 20 kg/ha or 1 ppm in water is necessary, but according to Moens (1970) this substance is markedly toxic to fishes.

Because of its caustic action on the skin and mucous membranes of man, handling this molluscicide is not without danger (Blair 1961), so that Dawood et al. (1966) give the preference to niclosamide, although both molluscicides – NaPCP and niclosamide – caused in experiments a mortality of over 98% of *Bulinus* and *Biomphalaria*. Van Dinther (1973) nevertheless considers NaPCP as the most suitable molluscicide for control of *Pomacea* species in the rice fields of Surinam, where it is applied over large areas by means of aerial application at a rate of 3 to 4 kg of pellets (80% active ingredient) dissolved in 40 l water; this is sprayed over the flooded fields 3 to 5 days before planting. Snails can survive, however, in the deeper water layers, but renewed application is not advisable because of the high toxicity of NaPCP to the young rice plants; in this case niclosamide (70% wettable powder) must be applied at a rate of 0.5 kg/ha.

Despite the excellent molluscicidal properties of NaPCP (Batte and Swanson 1952; Strufe 1968), it has one disadvantage compared with niclosamide, namely that at a concentration of 10 ppm it is an irritant to snails (Fletcher 1965). A repellent action is undesirable in dealing with aquatic and amphibious gastropods, because it releases flight reactions and the treated areas are later reinfested when the snails return.

Although not lethal, very low concentrations of NaPCP can be of practical importance in that they greatly reduce the fertility of snails and the rate of hatching of eggs; this occurs at concentrations of 0.05 ppm to 0.1 ppm (OLIVIER and HASKINS 1970).

NaPCP is ovicidal and is lethal to eggs at a concentration of 10 ppm in the water (FLETCHER 1965; STRUFE 1968). Its effect on the eggs of *Galba truncatula* (Müller) at 20 °C is 68 times and at 30 °C 82 times greater than that of copper sulphate (ENIGK and DÜWEL 1960). BATTE and SWANSON (1952) describe an ovicidal effect with 2.5 ppm of NaPCP and with 5 ppm of DNOC, a substance still to be discussed.

NaPCP has a very much greater effect on the intermediate hosts of Bilharzia than has copper sulphate, but up to 80% of it is absorbed within 7 days by the mud and water plants in pools and ponds.

For practical purposes it is desirable to know how long NaPCP remains on or in the grasses. Investigations of this are important in that they allow an estimate to be made of the length of time grazing must be stopped on treated pastures, so as to protect cattle from being harmed by the caustic substance. According to ENIGK and DÜWEL (1960), NaPCP can be detected on grasses for only 9 days after application and is washed off within only 48 h by heavy rain. It was not, however, possible to establish whether this substance is absorbed from the soil by the roots of the meadow grass. NaPCP application should be made in late afternoon and when the sky is overcast since this compound is rapidly decomposed by UV light, thereby losing its molluscicidal properties, which can persist for several days in diffuse light.

3.2.3.4 Other Chemical Substances

The search for satisfactory molluscicides continues, and several substances have been found, but only their most important properties can be discussed here; included amongst these are plant derivatives.

Yurimin (3,5-dibromo-4-hydroxy-4′-nitroazobenzene) was first used in Japan for control of the annelid *Limnodrilus* sp. (Oligochaeta) and subsequently for control of the terrestrial snail *Bradybaena (Acusta) despecta sieboldiana* (L. Pfeiffer) (Fam. Bradybaenidae); it was also tried against the aquatic *Biomphalaria glabrata* (Say) in Brazil (PAULINI and DE SOUZA 1968; YASURAOKA et al. 1968). It has been found to be toxic to the amphibious *Oncomelania nosophora* (Robson) at a dose of 5 g active ingredients/m^2 of ditch. The LD_{50} was 6.3 ppm (4.3 ppm to 9.1 ppm) at 25 °C and 24 h exposure, or 5.2 ppm (between 4.2 ppm and 6.5 ppm) at the same temperature and 48 h exposure. Acidity of the water between pH 4.5 and pH 5.2 lowers its toxicity, while alkaline water with a pH between 7.2 and 8.7 increases it. It is hardly affected at all by 4 h of direct sunlight on the water. Even after 18 h of sunlight a concentration of 7 ppm Yurimin in ethanol was reduced only to 3.7 ppm, which still has a molluscicidal effect, while NaPCP is completely destroyed by ultraviolet radiation. This stability of Yurimin in the direct sunlight makes it highly suitable for use in areas where there is strong sunshine. VILLIERS and ROSSOUW (1967) found in a series of tests on amides that **dodecanoyl-** and **tetradecanoyl-morpholide** have a very marked molluscicidal effect on *Bulinus tropicus* (Krauss), while being completely harmless to man; unsubstituted amides, and also those with only one substituted group, are ineffective as molluscicides.

Another substance which should be mentioned is 4′-chloronicotinanilide (DUNCAN et al. 1977).

One of the major problems in control of freshwater snails is the reinfestation of the treated irrigation canals, ponds and other water bodies; this usually becomes apparent 2 or 3 months after clearance and is because many snails survive in the mud. It is possible to deal with this problem by using TBTO (GILBERT et al. 1973), a bactericide and fungicide which is used in wood preservation against antibiotic-resistant staphylococci, and has proved to be effective against various *Biomphalaria* species. Tests were conducted with a pasty product (a sort of India rubber) which was very durable in water; this was shaped into "stones" 5 to 15 kg in weight, with about 30 g active ingredient per "stone"; after drying, these were placed on the substratum in the water. The material cannot be used in granular form. The molluscicide is slowly released into the water from the "stones"; these remain effective for several months, for sunlight and mud in no way affect this preparation. Stagnant or slow-flowing water was kept snail-free for two years with a total application of 300 g/m^2 of ditch. TBTO probably accumulates in the body of a snail. It is lethal to aquatic insects and fishes at the start of the treatment, but the stock soon recovers, while the waters remain free of gastropods for a long time.

Other substances found to have molluscicidal properties are **phenacile halide** (VILLIER and MACKENZIE 1963), **thiol** reagents (ZSOLNAI 1970), **acrolein** (aqualin), which is twice as effective as NaPCP and at the same time destroys water weeds (FERGUSON et al. 1971; VAN OVERBECK et al. 1959) and also **tetralin** (METRI 1973).

Acetylcholine and substances with similar actions to those of **curare** reduce the respiratory rate and cause tetanus-like convulsions, as CHICHERY (1971) found by injecting the dorsal region of the visceral nerve of the cephalopod *Eledone cirrhosa* Lamarck. JUORIO and KILLICK (1972) have investigated the effect of the drug **reserpine** on the synthesis and storage of monoamines in the nervous tissue of *Octopus vulgaris* Lam. and of *Helix pomatia* L.

3.2.3.5 Inhibitors

This is a group of substances which, although not lethal to gastropods, reduce the density of a population and keep it below the economic threshold value by inhibiting **growth** and **oviposition.** WARREN and WEISBERGER (1966, 1967) discovered that **chloramphenicol** not only inhibits protein synthesis in *Schistosoma* larvae, which develop as parasites in snails, but also has an effect on the snails themselves. The growth rate of *Biomphalaria glabrata* (Say) was reduced by a concentration of 156 ppm chloramphenicol and ceased altogether at 625 ppm. Oviposition was noticeably reduced by 156 ppm and dropped by 80% at 312 ppm. This substance is, however, lethal only at the high concentration of 1,294 ppm (LD_{50}) with 72 h exposure.

Normally about 700 *Schistosoma* cercariae are released daily into the water from a single snail, until its death. If, however, a snail which has been infected with miracidia from the trematode is kept for 24 h in water containing chloramphenicol, release of cercariae ceases for the duration of the treatment. This effect was observed at concentrations of 5 ppm 4 l water and above and at 20 ppm it lasted for 10 and more weeks. It can be seen that this inhibitor has what might be described as a parasitostatic effect; although the parasite in the liver of the snail is not destroyed, its development is inhibited. Exposure to the chemical for 16 weeks leads to complete elimination of the parasite, although neither the miracidia nor the cercariae in the snail were directly affected.

RAFF (1972) investigated the inhibiting effect of **cytochalasin B, colchicine** and **vinblastine** on embryos of the marine snail *Ilyanassa obsoleta* Say.

Compounds of heavy metals can also reduce the fertility of snails. Further sublethal doses of **triphenyl-lead-acetate** (0.01 ppm) disturbed the development of sporocysts in the intermediate host, but the age of the snail at the time of treatment is also decisive (HIRA and WEBBE 1972). Niclosamide, NaPCP and copper sulphate inhibit the oxidation of succinate in homogenized tissue of *Biomphalaria alexandrina* (Ehrenberg) (ISHAK et al. 1972).

3.2.3.6 Naturally Occurring Molluscicidal Substances

There are naturally occuring substances which kill gastropods or which reduce their fecundity and which could be important from the viewpoint of environmental protection. These are derivatives of plants and of fungi, of which antibiotics are an example.

Compounds of Plant Origin. Many plant extracts have molluscicidal properties, which depend on the presence of saponines. The most important of these are *Balanites aegytiaca* Del. (Zygophyllaceae), *Sapindus saponaria* L. (Sapindaceae), which is called damsissa, and finally *Swartzia madagascariensis* Desv. (Leguminosae). The insecticide **pyrethrum,** which is derived from plants, has, according to THOMAS (1948), no lethal effect on terrestrial molluscs, but ADLUNG and KAUTH (1956) found that a 5% oil emulsion was markedly toxic to freshwater snails *Radix auricularia* (L.), *Lymnaea stagnalis* (L.) and *Physa fontinalis* (L.), while even 3 ppm was lethal after 20 h exposure.

Dried and pulverised berries (soap berries) from the plant known as **Endod** *Phytolacca dodecandra* l'Hérit (synonym *Ph. abyssinica* Hoffm. and *Pircunia abyssinica* Moq., Family Phytolaccaceae) found in tropical and southern Africa and Madagascar, are used by the local population for washing clothes in rivers and lakes in Ethiopia, and the snail populations present in these waters die off. LEMMA (1970) confirmed that the berries have a toxic effect on *Bulinus truncatus sericinus* (Jickeli), *Biomphalaria pfeifferi ruppelli* (Dunker), *Lymnaea natalensis* (Krauss) and *Physa acuta* Drap. with a LC_{90} of 20 ppm. In field conditions rivers and irrigation canals treated with 50 to 100 ppm were free of snails after 3 to 6 h. Following butanol extraction of the dried berries, the lethal effect on *Biomphalaria* of a 24-h exposure was increased seven to ten fold as compared with the effect of 15 ppm to 30 ppm suspension. The butanol extract of soapberries caused, at a concentration of 6 to 7 ppm active ingredient, 100% mortality of snails within 24 h and after only 6 h at a concentration between 19 ppm and 25 ppm (BAALAWY 1972).

Glycoside or glucuronide derivatives of oleic acid are the molluscicidal ingredients of soapberries (HORTON 1968). JOHNSON and KINGSTON (1974) found that the active ingredients in a butanol extract of the berries of *Phytolacca americana* L., which are also toxic to snails, were the glycoside glucosylxylosyl, phytolaccagenine and also a substance which can be extracted by hydrolysis and emulsification from this saponin, namely xylosylphytolaccagenine. PARRHURST et al. (1973) discovered oleanolglycotoxin-A in the berries of *Phytolacca dodecandra* L'Hérit, one of the most important molluscicidal saponins.

Endod retains its toxicity with UV radiation of the water, whereas NaPCP becomes ineffective very rapidly, after only four hours irradiation. The molluscicidal effect is maintained in water within a wide range of both temperature and pH. Endod is not, however, ovicidal, so several applications are necessary to kill newly hatched young. The rapidly growing climbing plant which flowers twice a year, is completely harmless to mammals, birds and water plants, but is just as toxic to fishes as are the other standard molluscicides. Endod is lethal not only for snails but also for leeches (*Lymnotis niloticus* Savigny) as well as miracidia and cercariae of *Schistosoma* sp., a very desirable effect in the control of bilharzia in man.

Endod is neither phytotoxic to crops (barley, maize, wheat, millet, beans, peas, amongst others), nor does it interfere with their germination or growth, even at concentrations as high as 500 ppm and 1,000 ppm; thus in this respect, too, it is ideally suited for control of gastropods on agricultural land.

Its good molluscicidal properties combined with the absence of ecologically harmful side effects, make Endod a molluscicide which meets todays demands for environmental protection.

Extracts from *Polygonum senegalense* Meissner have proved to be a potent agent against *Lymnaea natalensis* (Krauss) and *Biomphalaria pfeifferi* (Krauss) (DOSSAJI et al. 1977; MARADUFU and OUMA 1978).

Finally, the **algal complex** *Microcystis farlowiana* + *Pseudanabaena franquettii* (Cyanophyceae) has been found to be toxic for *Radix auricularia* (L.) and *Galba truncatula* (Müller); the lysed cells kill the snails within 48 h (GEVREY et al. 1972). The toxic factor is heat-resistant (100 °C) and water-soluble.

Antibiotics. Besides chemical inhibitors (for example, chloramphenicol), there are also naturally occurring molluscicidal substances (antibiotics) which are derivatives of fungi and which interfere with the growth and reproduction of molluscs. Investigations have so far dealt only with mussels (LE PENNEC and PRIEUR 1972; LE PENNEC et al.1973) and freshwater snails, but not terrestrial species.

CHERNIN (1957) investigated the effect of streptomycinsulphate, dihydrostreptomycinsulphate and penicillin G in pure water on *Biomphalaria glabrata* (Say) with exposure times of 5, 12 or 18 days (Fig. 100). Low concentrations of streptomycin, such as 100 µg/ml water inhibited growth and egg-laying of the snails. Penicillin G alone had no effect, but only when combined with streptomycin; the inhibitory effect was, moreover, entirely due to this antibiotic. The effect is not lethal. Streptomycinsulphate and dihydrostreptomycinsulphate had similar effects at concentrations of 50 µg/ml and 100 µg/ml water respectively. Not only growth, but also reproduction of the snails is inhibited, while oviposition ceases (Table 37).

Upon being introduced to streptomycin-containing water, the snails sank to the bottom and began to turn and twist in an uncontrolled and convulsive manner; this continued with occasional pauses, for 2 to 4 days. At the end of this period the

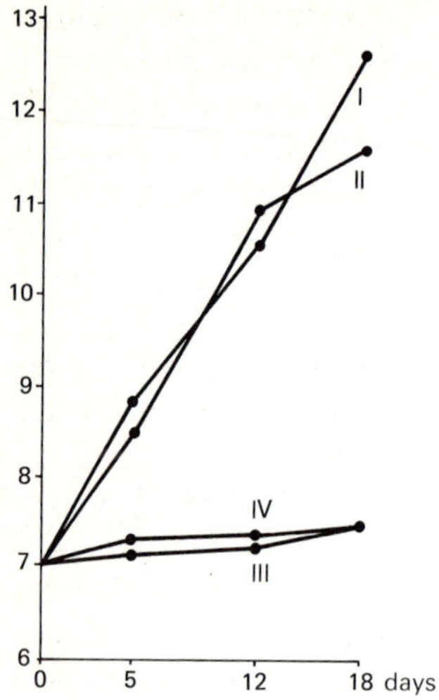

Fig. 100. The effects of streptomycin on growth of groups of 20 *B. glabrata.* Group *I* controls; group *II* 100 units of penicillin per ml; group *III* 100 micrograms of streptomycin per ml; group *IV* 100 units of penicillin plus 100 micrograms streptomycin per ml

snails again behaved normally, except that their movements were slower and they were less active than completely healthy animals. The convulsive movements did not occur in the presence of Ca^{++} ions, but this depended on their concentration and on that of the streptomycin. Presumably certain inorganic ions or mixtures of them protect the snails from antibiotics. Addition of 2.1 g $CaCl_2$/l water was very effective in preventing the inhibition of shell growth by 50 µg/ml streptomycin. At a high dose of 100 µg/ml of the antibiotic the preventive effect of $CaCl_2$ was much less significant and at low concentrations of $CaCl_2$ no longer affective.

The mechanism by which streptomycin inhibits growth and reproduction in *Biomphalaria glabrata* (Say) is not known. CHERNIN (1957) believes it is possible that the antibiotic may have a direct effect on the snails, not through various intermediate processes. The reasons for this view are first, that the reactions of the snail to the antibiotic, described above, occur immediately upon exposure and resemble the rapid action of drugs which attack the respiratory surfaces or the nervous system; secondly, streptomycin strongly inhibits the growth of the snails even on a diet free of microorganisms; thirdly, normal growth and reproductive capacity can be resumed when the snails are placed in fresh water free of antibiotic, but this capacity is limited and depends on the length of exposure. The investigations of MORALES and CHALAZONITIS (1970/71) on the effects of a substance isolated from

Table 37. The number of egg masses produced by two groups of *B. glabrata* after the addition of 100 µg streptomycin per ml to one of the aquaria. (Modified from CHERNIN 1957)

Days after streptomycin	Streptomycin added	Control (without streptomycin)
0– 7	0	45
8–14	1	102
15–21	0	74
Totals	1	221

Streptomyces griseus (Krainsky) Waksman and Henrici on the marine snail *Aplysia californica* (Cooper) have shown that antibiotics can affect the nervous system.

Substances with an effect like that of streptomycin could become more important in the control of freshwater snails. Even if not lethal at the low concentrations which are permissable, reduction of the size of a snail population might nevertheless justify their use. Investigations on possible side effects on vertebrates, including man, would, however, be necessary, since there is opposition to the use of antibiotics for non-medical purposes.

Antibiotics can, however, also be toxic, causing the body of the snail to shrivel, and they may be repellent, so that the animals attempt to escape from the treated water.

Seneca and Bergendahl (1955) have divided antibiotics into the following three groups, according to their toxicity to freshwater snails with an exposure of 24 h:

I. strongly toxic: thiolutin, rimocidin, anisomycin, and nystatin;
II: very weakly toxic, with almost no effect: penicillin, streptomycin, erythromycin, carbomycin, viomycin and bacitracin;
III: moderately toxic: terramycin (oxytetracycline), polymyxin B and neomycin.

Thiolutin, in particular, is considered to be excellent, for it is lethal at a dose of 0.25 μg/ml over a 12-h period, and at 0.06 μg/ml over 24 h, while at a concentration of only 0.003 μg/ml maintained for a week it is still toxic to snails.

3.2.4 Molluscicidal Effects of Herbicides, Insecticides and Repellents

It has often been suggested, that it would be desirable to have a single substance which would control not just one type of plant pest, but be toxic to many – insects, nematodes, gastropods – and even be a weedkiller at the same time. In the case of herbicides this would mean that such a substance would not only kill weeds but also molluscs, which in many areas are the most difficult of all pests to combat. In this way the present high rate of utilization of chemical substances for control of weeds *and* gastropods – namely, a herbicide *and* a molluscicide, – would be reduced, since instead of two, only *one* preparation would be necessary. In this way costs, time and labour could all be saved.

A toxic effect is not necessarily of primary importance in such a preparation; even inhibition of feeding could be useful, if it afforded sufficient protection to the cultivated plants which are threatened from all sides by pests. It might also be possible with the use of herbicides or insecticides to reduce the activity and resistance of pest gastropods so as to enhance the action of the very weather-dependent metaldehyde, the use of which would be abandoned with reluctance by those involved in gastropod control. A reduction in fecundity of slugs and snails would also suffice to keep the population density within tolerable limits.

Many investigations of this sort have been carried out and in the course of time substances have been found amongst the herbicides, insecticides, fungicides and others, which have an effect on gastropods. None of these substances is, however, used today as a molluscicide.

3.2.4.1 Herbicides

In terms of quantity, herbicides can claim first place amongst the substances used for plant protection in agriculture and horticulture; thus in 1972, for example, they comprised 63% of all the preparations used (DREES 1974). They are not without influence on insects and the soil fauna (FOX 1964; GODAN 1969, 1970 and others). They might be expected also to have a controlling effect on gastropods. Certainly one primary requirement is fulfilled, namely that the molluscs should come into contact with them over long periods of time, since herbicides may persist in the soil or in the plant for a considerable time, and may even accumulate. Gastropods will thus come into frequent or constant contact with them, whether this be through the skin or by ingestion and through the intestine. Many herbicides are retained in the soil as a result of strong adsorption, and this facilitates contact with terricolous species. Heavy soils with a high humus content create optimal conditions for slugs, and such soils absorb more herbicide than do lighter soils. Triazine and urea derivatives are mainly bound to humus.

GODAN (1973) tested the effects on *Deroceras reticulatum* (Müller), *Lehmannia marginata* (Müller), *Limax maximus* L., *Limax flavus* L., and *Arion rufus* (L.) of those herbicides generally used on plant cultivations likely to be infested with Field Slugs, namely strawberries, vegetables, lettuce, winter cereals, ornamental plants, clover, lucern and others. These are also herbicides which, according to information provided in „Leitfaden für Pflanzenschutzmittel" (Gesunde Pflanzen 1971), decompose slowly: this may be for simazin (triazin) several months to 1 year; for chlorpropham (CJPC) 3 or more months; for dalapon (2.2-dichloropropionic acid) 4 to 8 months; for linuron (urea derivative) 3 to 4 months; for lenacil 5 to 6 months and aminotriazol 4 to 5 months.

Slugs were kept (and fed) at high humidity (80% to 90% RH) for 10 days' observation after treatment with each herbicide. After contact with the sole, mortality was either zero or very low, even at high concentrations and even for the sensitive small species; the effect was different when the substance reached the intestine as a result of being eaten (Table 38).

In the feeding experiments all the herbicides tested with the exception of Simazin and Dalapon caused high or significant mortality rates only of the small species, *Deroceras reticulatum* (Müller) and *Lehmannia marginata* (Müller). A species-specific effect of these chemical substances was observed, an effect already mentioned several times in this monograph.

This result indicates that it is perfectly possible for chlorpropham, linuron, lenacil and aminotriazol to reduce populations of the small species if the substance is ingested.

Apart from toxicity, it should also be established whether herbicides might have an inhibitory effect on feeding. In Figure 101, the average quantities of treated and of untreated potato slices consumed are compared. Feeding behaviour differed not only from slug to slug but was also influenced by the herbicide used and its concentration. Those with practically no effect on feeding behaviour were simazin, linuron, lenacil and, in the case of Cellar Slugs alone, aminotriazol. Potato slices treated with a low dose of chlorpropham (1 kg/ha) were preferred to untreated slices, particularly by *Limax flavus* L.; aminotriazol was attractive to *Leh-*

Table 38. The molluscicidal effect of some herbicides on various species of slugs. (After GODAN 1973)

Herbicide	Application rate µg/cm²	Slug	Mortality (%) after	
			contact	ingestion
Simazin	7.5	*D. reticulatum*	0	–
		A. rufus	0	–
	50	*D. reticulatum*	0	55
		L. marginata	0	5
		A. rufus	10	25
	100	*D. reticulatum*	0	–
		L. marginata	0	–
		A. rufus (young)	0	–
		(old)	10	–
Chlorpropham	5	*D. reticulatum*	–	0
		L. flavus	–	0
	30	*D. reticulatum*	0	100
		L. marginata	0	96
		L. flavus	30	30
Dalapon	127.5	*L. marginata*	0	55
		L. flavus	0	0
Linuron	30	*L. marginata*	0	85
		L. maximus	10	–
		L. flavus	0	0
Lenacil	16	*L. marginata*	5	78
		L. flavus	5	0
		A. rufus	10	–
Aminotriazol	150	*L. marginata*	2	75
		L. flavus	0	5

mannia marginata (Müller) (Fig. 102); chlorpropham at higher concentrations and also dalapon were repellent; MALLET and BOUGARAN (1971) also found that a 5% concentration of chlorpropham was repellent to slugs.

The effects of the most important herbicides on gastropods can be summed up as follows:

1. A lethal effect depends on the manner in which the substance is taken up by the slug.
2. Contact of the sole with simazin, chlorpropham, dalapon, linuron, lenacil and aminotriazol has no effect but, apart from simazin and dalapon, these substances are significantly toxic to the small species [*Deroceras reticulatum* (Müller), *Lehmannia marginata* (Müller)] if ingested with treated food.
3. Chlorpropham at a high concentration (6 kg/ha) and dalapon have a repellent effect (inhibition of feeding).
4. Chlorpropham at a low concentration (1 kg/ha) and aminotriazol are attractive to the small *Lehmannia marginata* (Müller).

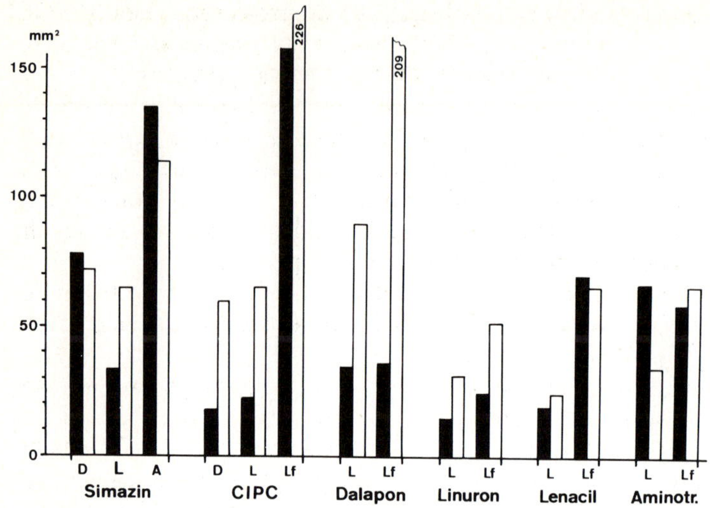

Fig. 101. Average consumption (mm^2) of potato slices treated with herbicides (*black columns*) compared with untreated potato slices (*white columns*). *A Arion rufus; D Deroceras reticulatum; L Lehmannia marginata; Lf Limax flavus*

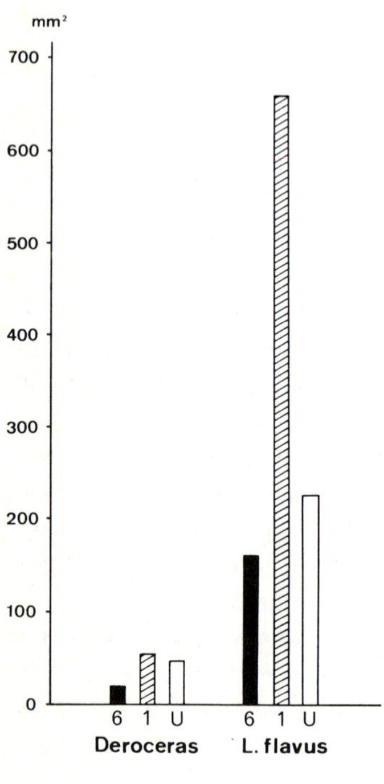

Fig. 102. Average consumption (mm^2) of potato slices treated with chlorpropham compared with untreated potato slices. Application rate of herbicide: *black columns* 6 kg/ha; *striated* 1 kg/ha; *white* untreated

DNOC: Of the group of dinitro-ortho-phenols 4,6 dinitro-ortho-cresol (DNOC) and 4,6-dinitro-2-cyclohexylphenol (DNCP) are mainly used in control of the intermediate hosts of the bilharzia parasite (*Schistosoma* sp.) (STRUFE 1968).

DNOC consists of yellowish crystals, it persists for 6 to 14 weeks in the soil, for 2 to 4 weeks in water, and in some cases up to 2 months in water (DREES 1974; STRUFE 1968). A residue of less than 0.01 ppm active ingredients in soil as a result of repeated application as a weedkiller was found to persist for many years (HURLE and RADEMACHER 1970). At a pH of 5.5 DNOC is rapidly absorbed by mud in the waters to which it is applied, but almost not at all at pH 8.4.

DNOC is highly toxic to aquatic and amphibious snails, a concentration of 2 ppm to 3 ppm causing 100% mortality after 20 to 24 days (KUNTZ and WELLS 1951).

DNOC and DNCP are also toxic to terrestrial gastropods, but are not registered for use as molluscicides in the Federal Republic of Germany. VAN DEN BRUEL and MOENS (1960) and also MOENS (1962) recommend use of DNOC for large areas and to protect old cultivations of leguminous plants before cereal crops are sown. Treatment in autumn after the harvest or before a new sowing or planting considerably reduces damage caused by the terricolous *Milax* spp. Further, DNOC is suitable for use as a barrier to protect market gardens when the areas are cleared after harvesting; this should not be done during rainy or cold weather, because persistant dampness largely abolishes the molluscicidal effect. This substance is still effective at low temperatures. A disadvantage is its toxicity for cultivated plants at the high application rate required, which is many times that required for its use simply as a weedkiller.

The ammonium salt has proved to be an excellent molluscicide for use against Field Slugs on cereals; it caused a mortality of 70% to 80% at an application rate of 10 kg/ha, 85% with 15 to 20 kg/ha and 97% at 25 kg/ha. The toxicity of this very high dose is the same as that of calcium cyanamide applied at the rate of 500 kg/ha. Even a 0.3% concentration had a significant effect on *Deroceras reticulatum* (Müller) (MALLET 1968; MOENS 1962; VAN DEN BRUEL and MOENS 1958, 1960). The usual rate of application as a herbicide of 5 kg/ha is, however, not satisfactorily effective against slugs. DNOC is ovicidal; after contact with this substance the eggs of *Deroceras* turn yellow and the embryos subsequently die (GOULD 1962).

Ioxynil (4-hydroxy-3,5-di-iodobenzonitrile) belongs with bromoxynil (3,5-dibromeanaloges) to the group of 3,5-dihalogen-4-hydroxybenzonitrile, the molluscicidal properties of which were first mentioned by WAIN (1963, 1964). Ioxynil is ten times more effective than metaldehyde. The LD_{50} for *Biomphalaria glabrata* (Say) is 0.5 ppm for ioxynil and 2 ppm for bromoxynil; for the less sensitive *Biomphalaria sudanica tanganyicensis* (Smith) these values are 1 ppm and 3 ppm respectively. The herbicidal effect of ioxynil can be exploited to advantage, since it also destroys the water hyacinth *Eichhornia crassipes* (Mart.), which is one of the major scourges of irrigation canals in Egypt.

Ioxynil is also strongly toxic to terrestrial gastropods and their eggs, but its rapid decomposition in soil is a serious hindrance to its practical application (DAXL 1968, 1970, 1971). Furthermore, molluscicidal concentrations of this substance are very much greater than those which are adequate for its use as a weedkiller, and exceed the maximal tolerance of the plants; its use has therefore to be limited to slopes, headlands, fallow land and ditches. DAXL considers that it might be used in greenhouses and gardens to form repellent barriers around valuable plants. Ioxynil has indeed a strong repellent action and could be of value as an inhibitor of feeding, since Field Slugs avoid surfaces treated with a concentration of 0.05% ($\approx$ 110 mg active ingredient/m^2).

3.2.4.2 Insecticides

Over the years many insecticides have been examined for possible molluscicidal effects, and this work continues today (ADLUNG and KNAUTH 1956; CROWELL 1967; GETZIN and COLE 1964; HALAWANI 1947; JONES 1965; JUDGE 1969; JUDGE and KUHR 1972; MALLET 1968; MEAD 1961; PAPPAS and CARMAN 1955, 1961; RIEMSCHNEIDER and FRÖMMING 1951 and others). Amongst the methylcarbamates mercaptodimethur (methiocarb) is now produced commercially as a molluscicide, but in general the usual insecticides are not suitable for practical application in gastropod control. They are either non-toxic or only moderately toxic, or required at such high concentrations that the problem of residues is exacerbated. In the Federal Republic of Germany the use of chlorinated hydrocarbons is largely prohibited or their use limited by law.

For terrestrial gastropods the following substances have been found to be non-toxic to moderately toxic: DDT, heptachlor, aldrin, dieldrin, isodrin, endrin, perthan, chlordan, parathion, toxaphen, malathion and metasystox. Allethrin was found to be very toxic to freshwater snails (*Lymnaea* sp., *Physa* sp.): the snails were affected by a two hour exposure to 3 ppm in the water and died after 15 to 20 h; a concentration of 10 ppm was lethal within only 2 h. In experiments using a 24-h exposure, after which the snails were kept in fresh water, the following concentrations were found to be lethal: HCH – 5 ppm; chlordan – 7 ppm; aldrin – 10 ppm; dieldrin – 8 ppm; parathion – 3 ppm. At 10 ppm metasystox caused only a light-reversible paralysis.

The δ-isomer of **hexachlorcyclohexane** has some molluscicidal effect on freshwater snails (Lymnaeidae, Planorbidae, Physidae), while the insecticidal γ-isomer has no such effect. A dusting powder (15%) of HCH with a high percentage of the δ-isomer was used against the amphibious *Pomacea* spp. in the rice fields of Surinam and found to be *very effective* when the snails came into contact with it; it was, however, regarded as unsatisfactory as the powder remained on the surface of the water, so that only a limited number of snails came into contact with it and the overwhelming majority which remain below the surface survived (VAN DINTHER and STUBBS 1963).

Aldrin is lethal to freshwater snails like *Lymnaea* sp. and *Physa* sp. after 15 to 20 h exposure to a concentration of 10 ppm and according to STEPHENSON (1959) it is also effective against terrestrial gastropods, although BÖHM (1957) observed no such effect. It is interesting in this connection that JONES (1965) reported an increase in feeding damage in a field treated with aldrin against the Carrot Fly *Psila rosae* Fab.; a possible explanation is that aldrin is to some extent attractive to *Arion hortensis* Fér. and *Milax budapestensis* (Hazay). When harvested, the crops from those fields which had been most frequently treated with aldrin against Carrot Fly showed the greatest gastropod damage to the carrots. Thus while only 8% of the carrots from untreated fields showed evidence of gastropod attack, this value was 17.3% for fields treated with aldrin at a rate of 110 kg/ha and as much as 24% with 220 kg/ha.

BARRY (1969) recommends a phosphoric acid ester, **phorat** (Thimet, a systemic insecticide) in the form of an emulsion and at an application rate of 1.0 to 1.5 kg/ha against Fields Slugs, the major pests to maize. Application should be made 4 to 6 h after dusk, when the activity of the slugs is at its peak. Phorat was in fact better than Zectran and reduced slug populations by 95% after 24 h compared with 85% after 48 h. Although zinophos granules resulted in a similar fall in the slug population after 5 days, only phorat reduced feeding damage to the plants. This substance has proved particularly effective in treating young maize plants, even a single application affording protection for 20 days, where with other molluscicides two applications at an interval of 8 days had been necessary. According to MUSICK (1972), the single treatment is useful as it suppresses the gastropods during the critical initial growth phase of the young maize plants. This does not, however, prevent damage at a later stage of development of the plants and the tap root is often studded with holes made by slugs. This later type of damage can assume considerable proportions as a result of modern methods of field cultivation involving either minimum tillage, or no tillage at all, so that the surface of the field becomes covered with a dense layer of mulch. The first application should be made at the time of peak activity of the slugs, and the second 15 to 20 days later, or when the gastropod population threatens serious economic damage.

Zinophos has been used with success in the control of the slug *Prophysaon andersoni* (Cooper); the lethal concentrations were 50 $\mu g/m^2$ for phorat and 10 $\mu g/m^2$ for zinophos, which caused 100% mortality after only 1 or 2 days contact with treated filter paper. Even so the efficacy of these two compounds did not approach that of mercaptodimethur. A mixture of equal parts of phorat and zinophos at a concentration of 7.5% active ingredient in the formulation, proved effective against Field Slugs in peas.

Parathion has been found to be toxic to the freshwater snails *Lymnaea stagnalis* (L.) and *Radix auricularia* (L.) at concentrations of 3 ppm in the water. Further, application of diazinon at 1 ppm causes paralysis and death.

Gusathion (Guthion) has proved to be a molluscicide with a strong toxic effect on *Helix aspersa* Müller in citrus plantations in California, an area where metaldehyde frequently fails; unfortunately gusathion seems to be less effective against slugs. DEL RIVERO and ROCA (1969) confirm that both gusathion and ioxynil can be used for the control of *Helix aspersa* Müller and *Theba pisana* (Müller), but emphasize that the high concentration at which ioxynil must be used is damaging to plants. The studies of CHEYMOL et al. (1975) and FITZPATRICK (1975) must also be mentioned here.

Molluscicide properties have also been discovered in **metacil** and **ultracil;** these substances in a 2% bait formulation caused a mortality of 100% with *Theba pisana* (Müller) and of 90% with *Helix aspersa* Müller. *Theba* is more sensitive than *Helix*, suffered a 98% mortality at a concentration of only 1% of the active ingredient. In comparative studies promecarb (GODAN 1965, 1966), which has already been discussed under carbamates, caused 82% mortality of *Theba* with a 1% concentration, while 2% was necessary to cause 96% mortality of *Helix*.

3.2.4.3 Repellents

These are substances which are not lethal, however high the concentration, but are always repellent, in contrast to the other substances discussed so far, which are repellent only at such relatively high concentrations that their molluscicidal effect is lost, because the pests avoid them. Gastropod repellents have not been as fully investigated as have insect repellents. Substances have been found, however, which are repellent both to gastropods and insects. STEPHENSON (1962) tested the three fly repellents dimethylphthalate, dibutylphthalate and benzylbenzoate and found that the first two are also repellent to adult *Deroceras reticulatum* (Müller).

If the response to the substance is elicited from a distance, it is regarded as a *repellent*, while if it requires contact with a bait formulation which is then rejected, it can be regarded as a *deterrent*. As far as terrestrial pulmonates are concerned, such substances cannot necessarily be recommended, since the slugs and snails which are driven away move into adjacent areas in search of food and these suffer more than usual damage as a result. Only if a campaign to repel and inhibit feeding is conducted over a very large area, can it be successful, since, lacking an escape route, the gastropods are forced to remain in the area being treated. As a result of the prolonged starvation, their resistance to disease and parasites is lowered, and these agents finally kill them. Repellents might thus be of use in the future, preceeding biological control measures, serving to weaken the pest population as a result of inhibition of feeding.

3.2.5 Molluscicides

3.2.5.1 Mode of Action and Detoxication

In pulmonates, the most important organ metabolising foreign substances is the mid-gut gland (liver, hepatopancreas). In vitro experiments by CARLSON (1972) have shown that the liver of the mussel *Mercenaria mercenaria* L. is able to reduce p-nitrobenzoic acid to p-aminobenzoic acid. The same metabolic pathway has been demonstrated in tissues of the mantle, gills, gonad and foot.

In higher animals **carbamate** metabolism, which occurs mainly in the liver, has been thoroughly investigated. These compounds are normally rapidly degraded and thus do not accumulate. In vertebrates it has been found that usually 70% to 80% of the applied dose was excreted with the urine within 24 h and 0.5% to 15% eliminated with the faeces (DOROUGH 1970; KUHR 1970). In insects, however, the fat body is the most important carbaryl-metabolising tissue. Plants, too, break down methyl carbamate in a relatively short time, but little is known about the enzymes of carbamate metabolism in plants. In insects and mammals the carbamates cause a very rapid, short-lived *inhibition* of cholinesterase by blocking the esterase and anionic groups on the enzyme; this leads to endogenous acetylcholine poisoning leading to combined muscarine- and nicotine-like symptoms, which correspond closely to the effect of organo-phosphorous compounds (KLIMMER 1971).

It is only fairly recently that acetylcholinesterase action has been studied in molluscs, and this in the nervous systems of marine and freshwater molluscs (GABBOT and HOLLAND 1973; GRIGOR'EVA 1973; HEILBRONN et al. 1971; JURCHENKO et al. 1973; KATO 1972; KATO et al. 1972; KEYL et al. 1957; KREMZNER and ROSENBERG 1971; VULFIUS and JURCHENKO 1972). Amongst pest pulmonates only *Helix pomatia* L. has so far been investigated (BOGUSCH 1972; DUNNING and MACHINE 1972; OSBORNE and NEUHOFF 1973 with citations of the literature up to 1972).

Inhibition of cholinesterase results in over-excitation of the cholinergic nervous system and later in a central nervous paralysis and loss of muscular tone. This is the main characteristic of carbamate poisoning in gastropods. Their muscles are regarded as being cholinergic (HOYLE 1964).

The metabolic disturbance caused by carbamates can last for a considerable time and may finally lead to death, even when the appearance of the animals does not suggest that this is likely. This is very clear with mexacarbate, when the pulmonates died six days after treatment, although the symptoms appeared to be slight. CASIDA (1963) found this substance to be far more effective than carbaryl as a cholinesterase inhibitor in insects. Experiments by GODAN (1965, 1966) showed that in terrestrial gastropods, too, the effect of mexacarbate is much stronger.

Methylcarbamates persist in bait pellets for a relatively short time only (STRINGER et al. 1970). There was a steady loss of the anticholinesterase action of a 4% mercaptodimethur bait formulation on damp soil at 15 °C, with the result that at the end of a 15 day observation period, only 60% of the active ingredient was left.

Like the carbamates, **phosphoric acid esters** inhibit cholinesterase activity in molluscs; this has been demonstrated in mussels by KULIEVA et al. (1971) and in gastropods by SINGH and AGARWAL (1978). RORKE et al. (1974) found in vitro more than 90% inhibition of blood cholinesterase of *Helix aspersa* Müller by fenitro-oxon and physostigmine (for comparison) at a dose of 10^{-4} g/ml, while fenitrothion and diethylphenylphosphate at 10^{-3} g/ml had no such effect. Fenitro-oxon and fenitrothion applied locally to the foot caused varying rates of mortality, which cannot, however, be related to an inhibition of cholinesterase activity. The behavioural disturbances observed in *Helix aspersa* Müller after contact with the active ingredient may take the form of withdrawal of the body of the snail into the shell; thereafter the animal either dies or recovers within 6 to 8 h; alternatively, the animals may crawl about excitedly with drooping, flaccid optic tentacles and/or lowered co-ordination of the musculature; thirdly, the snails may become motionless, with the tentacles completely relaxed and lying against the sides of the body. DEPP showed no sign of causing cholinesterase inhibition, but did affect the behaviour of this snail; on the other hand physostigmine, the effects of which were studied for comparison, caused both cholinesterase inhibition and a change in behaviour. The two effects, namely on mortality and on behaviour of the snail, show that toxic organo-phosphorous insecticides have more than a simple anticholinesterase effect on the snail *Helix aspersa* Müller; the effects of these compounds on the membrane potentials of the neurones are being studied. Detoxication of simple organo-phosphorous compounds is effected by an enzyme which hydrolyses them; this enzyme has been extracted in a pure form from the cerebral ganglia of several species of cuttlefish and also the snail *Aplysia californica* (Cooper) (HOSKIN and LONG 1972).

The actions of **niclosamide** (Bayluscid) have been studied using *Biomphalaria glabrata* (Say); the substance affects the activity of various enzyme systems, primarily those of the respiratory chain and of carbohydrate metabolism. Oxygen uptake of the snail is inhibited by concentrations greater than 0.2 ppm. Pentachlorophenol is assumed to work in the same way, the main difference between the two substances being that to achieve a particular metabolic effect, such as the inhibition of oxidative phosphorylation in liver mitochondria, the required concentration of PCP is at least 10 times that for niclosamide. GÖNNERT (1962) reports that a concentration of Bayluscid one fiftieth that of PCP was twice as effective in inhibiting the respiration of the snails, while with solutions of equal toxicity, respiration was only reversibly blocked by PCP, but irreversibly by Bayluscid.

Niclosamide persists in the mud of the substratum (STURROCK 1974) which, according to HOLZ (1962) absorbs ten times more than the amount dissolved in the water above it.

On the basis of studies on the effect of **pentachlorophenol** (PCP) on the metabolism of *Biomphalaria glabrata* (Say), WEINBACH and NOLAN (1956) suggested that at least part of the toxic effect of PCP results from uncoupling of oxidative phosphorylation. At a concentration of 2.10^{-6} M there is, within 5 h, a 28% increase in respiratory rate. A concentration of 10^{-5} M causes a 28% increase with tissue homogenates, but at a higher concentration (10^{-2} M) respiration of the homogenate is strongly inhibited (62%).

After application of **copper sulphate,** adults of the amphibious snail *Oncomelania formosana* (Pilsbury and Hirase) were found to store Cu^{++} ions in special cells outside the wall of the intestine. Even at very high concentrations in the water, the snails ingest very little as a result of the salt's repellent effect, but in this case it is absorbed through the outer surface of the body and mainly enters the blood. Detoxication takes place in protective cells in the haemocoel and liver. The Cu^{++} ions also penetrate the shell of the egg and are taken up, together with food substances, by the embryos which become so lethargic and weak that they are unable to rupture and emerge from the mud cocoon: as a result they die. Histochemical studies of WINKLER and CHI (1964) have shown that Cu^{++} ions are bound to the wall of the mud capsule which encloses the egg of *Oncomelania* and also by proteins in the fluid which surrounds the embryo in the egg. Since the embryo is nourished by these proteins during the course of its development, Cu^{++} ions enter the body of the embryo and are stored in special cells in tissue around the stomach and the first part of the intestine. The copper ions are, however, rendered non-toxic in adult snails by large cells in the liver, similar to those in the wall of the stomach, which lie mainly along the inner edge of the haemocoel, and which now clump together and shrivel up. CHENG and SULLIVAN (1973) compared the toxic effect of two organo-copper compounds on the freshwater snail *Biomphalaria glabrata* (Say) and concluded, in relation to the "design" of synthetic copper molluscicides, that copper in complex binding seems to be able to activate biological processes as well as to be protected from absorbency by other molecules, colloids and ions from the snail's environment.

According to BEYNON et al. (1972), half the total amount of **trifenmorph** (Frescon) applied to pastures disappears from the grass within 3 weeks, and from the soil within 1 to 4 weeks. N-tritylmorpholin is rapidly hydrolyzed to triphenylcarbinol in suspensions at pH 7. The concentration measured at the outset of the experiments was between 12 and 71 mg/kg of freshly cut grass, which corresponded to an application rate of 0.45 and 0.5 kg/ha. After 8 weeks the maximum residues of grass samples was 0.3 mg/kg and after 12 weeks less than 0.05 mg/kg.

A series of molluscicidal **vinylsulphones** showed a direct relationship between the sulphhydryl-reaction and their molluscicidal effect. No inhibition of tissue respiration could be established in in vitro experiments with *Biomphalaria glabrata* (Say); this also applies to rats, and so can be ruled out as the basis of the toxicity of vinylsulphones to gastropods. The sublethal toxic effect is probably the result of disturbance of normal kidney function.

Like mussels (VILPO et al. 1972; READE and READE 1972), terrestrial snails are able to rid themselves of non-metabolizable foreign particles by phagocytosis; this has been demonstrated in *Helix aspersa* Müller by BROWN (1967) and in *Cepaea hortensis* (Müller) by DINDAL and WURZINGER (1971). A colloidal suspension (0.35 ml)of thorium dioxide was injected into the pedal sinus and the process of its elimination followed. Amoebocytes transport the particles from the body tissues to the reproductive organs and the epithelium of the mantle, where they are rendered innocuous by the cells in these organs. Although the mantle epithelium is involved in the process of elimination of non-metabolized particles through phagocytosis by macrophagous haemocytes, in *Helix aspersa* Müller the reproductive organs seem to be more important in this respect that they are on the aquatic *Biomphalaria* sp. Small quantities of chemical substances are also eliminated in the mucus.

3.2.5.2 Resistance

We are not concerned here with species-specific or age-dependent resistance, but with the resistance to chemical substances which can develop in time as a result of exposure to sublethal doses.

In insects the development of resistance is a well-known phenomenon. More recently it has also come to the fore amongst gastropods and more attention should be given to this in future, when repeated applications of a molluscicide are made.

It may be that freshwater snails are particularly predisposed to the development of resistance, since before 1961 it had been found that *Oncomelania* sp. had become resistant to the frequently used NaPCP (KOMIYA 1961; KOMIYA et al. 1961; GANCARZ 1958; WALTON et al. 1958). Investigations by NEWTON (1963) showed that even within the relatively short apace of ten generations, a resistant strain can be selected. Two strains of *Biomphalaria glabrata* (Say) ("Brazilia" and "Venezuela") were kept for 24 h in containers with the LD_{90} dose of NaPCP (1.0 and 1.5 ppm). After a recovery period, on each occasion of 24 h, the surviving snails were allowed to reproduce, and each succeeding generation subjected to a dose of 0.25 ppm more. After ten generations, the selected strain of each of the original stocks showed resistance to NaPCP. Thus mortality of the normal "Brazilia" stock was $85\% \pm 3.2$ at a concentration of 1.0 ppm, but only $21.4\% \pm 3.9$ in the selected strain. Mortality of untreated "Venezuela" with the same concentration was $92.5\% \pm 2.4$, and for the selected strain of this stock it was $34.2\% \pm 4.3$.

In view of these results, NEWTON regards it as perfectly possible that in suitable conditions in the field *Biomphalaria* can develop resistance to pentachlorophenol (PCP). YASURAOKA and HOSAKA (1971), however, consider this to be unlikely under field conditions.

In 1975 a whole series of cases of resistance of terrestrial pulmonates to metaldehyde were reported for the first time. In California FISHER and ORTH (1975) found very clear lack of sensitivity of *Helix aspersa* Müller to metaldehyde pellets with 4% active ingredient.

3.2.5.3 Residues

There are only a few reports of investigation of residues of **metaldehyde** in the literature, VERSCHUUREN et al. (1975) and VASIAGINA et al. (1976) have, however, found that the residues of this substance disappear relatively quickly. Although the initial deposits on leaf surfaces may amount to more than several hundred ppm, they are reduced after 3 to 13 days to a few ppm only. Analytical methods for the estimation of metaldehyde in plant material are described by GIANG and SMITH (1956) while BECKER (1969) describes a modified, more sensitive method. In greenhouse experiments in which a 15% wettable powder was applied at the rate of 1 kg/100 m^2 to potted plants, the residues were determined at intervals; the results are presented in Table 39.

Metaldehyde was found to disappear after several days. Only on chrysanthemum leaves was there still a relatively high level 20 days after application. The degradation of metaldehyde takes place at about the same rate, on lima beans and geraniums.

These variations in the residues and their lower value in lima beans and geraniums as compared with *Chrysanthemum* leaves can possibly be explained on the basis of studies (1961, 1962) by BECKER, MÜLLER and GODAN on metaldehyde residues in vegetables. The results, which have not been published, are interesting in that they show that the amount of residual metaldehyde is related both to the nature of the leaf surface and to the shape of the plant.

Field allotments with different vegetables such as lettuce cv. "Attraktion", kohlrabi cv. „Weisse Delikatesse“, Savoy cabbage cv. "Vorbote", red cabbage cv. "Dauerrot" and carrots cv. „Rote lange stumpfe ohne Herz“ were treated in the normal manner with commercial metaldehyde wettable powder (application 1 l/m^2 of 0.2% Slugit) and samples for testing the residue were taken one hour later and

Table 39. Metaldehyde residues in plants. (After GIANG and SMITH 1956)

Days after treatment	Lima beans (plants) ppm		Chrysanthemums (leaves) ppm	Geraniums (leaves) ppm
0	292.5		–	378.2
1	232.2		657.1	297.2
3	201.3		503.6	208.3
8	17.8	(9 days)	137.3	5.3
10	0.3	(13 days)	38.9	2.0
20	–		21.5	–

then daily, as well as at other times as required. The first test was made when the plants were at the following stages: lettuce – 8 days after removal from the cold frame; kohlrabi – when the tuber had reached a diameter of 2–3 mm; Savoy cabbage – when the head had started to form; red cabbage – when the head was 9–10 cm in diameter; carrots – when the root was 4–5 cm long. The month of the year (1961) and temperatures were as follows: lettuce – end of May, 18 °C; kohlrabi – end of June, 30 °C; savoy cabbage – beginning of July (7.7), 16 °C; red cabbage – mid-July (11.7), 20.5 °C and carrots – end of July (18.7) 15 °C.

The behaviour of the spray fluid on the plants differed according to the texture of the leaves and the shape of the particular type of vegetable; lettuce – leaves used as test material; these were tender, without a covering layer of wax; the spray fluid normally ran off the leaves and accumulated at the base of the rosette-like arrangement of leaf stalks in the heart of the plant; red cabbage – leaves used as test material; these had a covering layer of wax; the spray fluid did not wet the leaves, but formed droplets on the leaves; savoy cabbage – leaves used as test material, these had a thick wax covering; spray fluid remained in the many indentations of the very wrinkled leaves, and penetrated from here into the tissue of the leaf blade; kohlrabi – tuber as test material; this may contain a large or small residue according to the density of the leaf tuft, for the fluid collected between the young leaves in the heart of the plant and also in cracks and indentations in the tuber; carrots – root as test material, this not directly affected, fluid flows along the shaft of the very feathered leaves, weted the soil and the upper part of the root. In the tests with gastropods, the aim was to establish whether the various plant parts (leaves, tuber, root) contained adequately toxic amounts of metaldehyde.

To test whether it is, in fact, possible to obtain an accurate estimate of a metaldehyde residue with the method used by K.F. Becker and Müller, 6 mg active ingredient was added to 100 g samples from untreated vegetables. After appropriate treatment of the samples, the quantities recovered were 95% for lettuce, 86% for the leaves of kohlrabi and 94.7% for its tubers, and 90.2% for red cabbage.

The results of the residue estimates by K.F. Becker and Müller are set out in Table 40.

After spraying the plants individually, the maximum observed values were 90 ppm for lettuce and 71.4 for Savoy cabbage; both these values were strikingly high. The tender leaves of lettuce retained a larger metaldehyde residue than Savoy cabbage leaves which are covered with a thick wax layer. The relatively rapid reduction in metaldehyde content of plants previously observed by Giang and Smith and shown in Table 39, was found also in these experiments.

In a second series of experiments, analyses were made of the residues after dusting with metaldehyde. The application rate was 2 g/m^2 of a 25% metaldehyde powder, which resulted in a dose of 0.5 g active ingredient per m^2. The results are summarized in Table 41.

The conclusions of Becker and Müller (pers. comm.) on the results of their analyses of metaldehyde residues can be summarized as follows:
1. Application as a **spray formulation** (wettable powder): There was little difference between the residues resulting from prescribed and non-prescribed application. Only a little of the spray remained on plants with a thick layer of wax; the greater part ran off. In warm weather metaldehyde evaporates rapidly, and it is washed

Table 40. Metaldehyde residues in ppm (average values) after spraying with wettable powder (Sluggit). (After BECKER and MÜLLER)

Sample		1 h after treatment	Days after treatment 1	2	3	4	5	6	7
Lettuce	1. Experimental series	E 73.0	51.1						nd
		B 83.0	43.7						nd
	2. Experimental series	E		11.1			2.1		nd
		B		12.1			3.9		nd
Kohlrabi	Tuber	E 1.3	0.7		0.4				nd
		B 1.2	0.8		0.2				nd
	Leaves	E 1.8	nd						
		B 3.4	nd						
Savoy cabbage	Heart	E 20.7	7.6			0.4			nd
		B 62.5	4.7			1.4			nd
	Outer leaves	E not applicable							
		B 11.3	4.5			2.3			nd
Red cabbage	Heart	E 8.5	4.8		0.7				nd
		B 5.0	4.7		0.2				nd
	Outer leaves	E 7.0	4.4		1.1				nd
		B 7.0	6.7		2.1				nd
Carrots	Root	E 2.8	0.2		nd		nd		
		B 0.7	1.6		nd		nd		
	Leaves	E 24.2			nd		nd		
		B 51.4			nd		nd		

E = individual treatment (90–100 ml/plant); this series was conducted to establish the size of the metaldehyde residue following application which was not carried out in accordance with instructions

B = treatment of area as a whole, in accordance with instructions

nd = active ingredient not detectable

off the plants by rain. Washing of the harvested plants greatly reduced the metaldehyde residues. Thus washing as thoroughly as in food preparation (three rinses in still water or thorough rinsing under running water) reduced sample residues of 2 to 40 ppm as measured in the plant samples to non-detectable levels. No metaldehyde could be demonstrated in any of the samples taken 5 to 8 days after treatment.

2. Application as a **dusting powder:** After prescribed treatment of the area as a whole, the samples were taken rather late, because the plants were still small. Dusting of individual plants was undertaken shortly before harvesting. In this case there were no demonstrable residues on lettuce, Savoy cabbage or red cabbage after 7 days and on kohlrabi after 14 days. The increase in the amounts of residues in carrots, which reached a maximum after seven days, is remarkable; only after 22 days could no metaldehyde be detected. The values obtained for the individual samples fluctuated greatly, since an even distribution of dusting product over the individual plants is not possible. In all cases, the residues had fallen below the detectable level after an interval of 21 days even with non-prescribed treatment of the individual plants.

Table 41. Metaldehyde residues in ppm (average values) after dusting. (After BECKER and MÜLLER)

Sample		1 h after treatment	Days after treatment									
			1	3	6	7	9	14	19	22	24	28
Lettuce	E	10	12	4	nd							
	B							nd				
Kohlrabi tuber	E	33	15	1		3		<1				
	B					nd						
leaves	E			2		nd						
Savoy cabbage	E	11	1	<1			nd					
	B					4			7		1	nd
Red cabbage	E	5	2	<1		nd						
	B					2			6			nd
Carrot	E	nd	nd	3		5		2		nd		
Turnip	B									nd		

E, B, and nd as in Table 40
< = less than

The **gastropod tests** conducted by GODAN have, like the residue estimation, shown that far more of the molluscicide is retained on the tender leaves of lettuce than on the other vegetables. For when lettuce leaves were fed to slugs, 1 or 72 h after being sprayed or dusted they caused 100% mortality of both *Deroceras reticulatum* (Müller) and *Deroceras laeve* (Müller) and as high a mortality rate was found with *Limax flavus* L. using material 2 h after treatment. Of the other vegetables, only carrots were lethal to the small species 2 h after treatment, and the mortality rate was low, at 20%.

Other investigations have been made on metaldehyde residues in cultivated plants. SELIM (1974) found that 14 days after treatment, a residue of only 0.1 ppm could be detected on strawberries treated with metaldehyde wettable powder (application rate 3 kg/ha) or metaldehyde dust (1.3 to 2.6 kg/ha) and on artichokes treated with wettable powder (1 to 2 kg/ha). After application of metaldehyde as bait pellets (2 kg/ha) to strawberries, the residue was 0.3 ppm after the same length of time. SELIM considers that contamination of the strawberries was brought about by the pickers or by bait pellets which fell near the fruits.

In many cases it is desirable to know whether there is a metaldehyde residue in the soil from previous treatments. Since several applications are usually made, the molluscicide is frequently worked into the soil and not simply applied to the surface and sprays often penetrate to lower layers of the soil, an accumulation of the poison cannot be excluded. Even if the following experimental findings of GODAN on the presence of metaldehyde in soil samples must be regarded as preliminary, they may nevertheless form a basis for further investigations.

It is possible to concentrate the metaldehyde residues from soil samples and to demonstrate its presence by its effect on slugs. For this purpose, very young specimens of the small, sensitive species *Deroceras laeve* (Müller) and *Lehmannia marginata* (Müller) are most suitable. Initially, experiments were conducted to establish the presence of metaldehyde in soil samples which had been treated with known concentrations of the substance. Since metaldehyde is only very slightly soluble in water, this was not suitable for use as a solvent to extract metaldehyde from the treated soil. It does dissolve, however, in hot alcohol, and crystalizes out on cooling (KLIMMER 1971). If a soil sample is washed with hot alcohol, metaldehyde dissolves and after cooling this solution, it settles out as flakes and can be tested.

Wet garden soil (more that 90% RH) was treated over an area of 0.25 m^2 and to a depth of 2 to 3 cm with 5 ml of a 2.0% or a 0.2% Slugit concentration. The effect of this soil on juvenile slugs was tested 2 h after spraying and again several days later, to check the persistance of the molluscicide. For this purpose a 500 g (10×10 cm surface area) sample of treated soil was removed and washed several times with 300 ml hot alcohol in a waterbath, always through the same filter, to wash the compound and to concentrate it in the filtrate. The hot solvent is clear because the metaldehyde dissolves in it, but on cooling becomes cloudy as the metaldehyde precipitates. This takes place so slowly, particularly at low Slugit concentrations, that it is only completed after many hours or overnight. After long drying on the filter paper, so that all the remains of the alcohol evaporated, the residue was moistened with 5 to 10 ml water and 8-day-old specimens of *Deroceras laeve* (Müller) or *Lehmannia marginata* (Müller) placed on it. At the high dose employed in the experiment 2 h contact sufficed to demonstrate the presence of metaldehyde; all the slugs were dead within 2 days. To demonstrate the presence of a ten times weaker concentration, a soil sample of 1,000 g is necessary. The slugs tested with this residue remained alive, but showed the symptoms typical of metaldehyde poisoning: over-production of mucus, convulsive lifting of the head region, and thereafter a contracted posture of the body. The experiments were not carried further; they are mentioned here only to stimulate interest in the problem and encourage the further search for a less expensive solvent than alcohol. It may be that chloroform, which has been used by GIANG and SMITH (1956) and BECKER (1969) for the extraction of metaldehyde from plant material, would be more suitable.

The intensity of the symptoms of poisoning in the test slugs depended on the concentration of metaldehyde in the soil sample and the age of the residue, if this had been kept for several days before testing. In a prolonged experiment extending over 7 days, the soil was moistened daily, and maintained at a relative humidity of 85%. The test slugs reacted only weakly on the last day, showing almost no symptoms of poisoning and appearing healthy. On the other hand they showed marked signs of poisoning after contact with soil samples taken 2 h after application of the molluscicide. One dried residue of a high concentration extracted from 500 g soil weighed, after subtracting the weight of the filter paper, 72.3 mg; another weighed 60.3 mg and a third 118.3 mg. A soil sample treated with 0.2% Slugit solution gave a dried residue of only 49.5 mg.

Methylcarbamates are characterized by the short time they persist in plants and in soil, and are therefore not stored (KLIMMER 1971; KUHR 1970): in this case problems of residue are of little significance. Nevertheless, mention should be made of experiments by PIELOU et al. (1962) on the toxicant film on plants after spraying with carbaryl. These show that plant protection substances have a tendency to accumulate in larger amounts on the lower surface than on the upper surface of a leaf, quite independently of the action of rain. Cherry leaves were sprayed with a

50% carbaryl wettable powder at a concentration of 1 g/l water and 100 leaves were analyzed immediately after excess fluid had been drained off them. In 75 of the leaves the amount of carbaryl on the lower surface (2.38 $\mu g/cm^2$) was greater than on the upper surface (1.65 $\mu g/cm^2$).

Since **niclosamide** is used against freshwater snails on meadows and rice fields as well as in irrigation ditches, it is important to know the size of the residue in agricultural crops. MEDENWALD and SHIFF (1965) found the following residues in young citrus trees watered during the course of their growth with a solution of niclosamide labelled with ^{14}C (concentration 0.05 to 1.0 μC/ml). After 11 days growth and a water consumption of 15.5 ml, the residue was 0.7×10^{-3} μC/g of fresh plant material; with a water consumption of 10.5 ml it was 1.0×10^{-3} μC/g; after 14 days growth and a water consumption of 22 ml the residue was 0.7×10^{-3} μC/g. In young banana trees phytotoxic symptoms were not observed, nor could residues of either niclosamide or trifenmorph be found in the fruits; these two substances are thus suitable for use in irrigation canals of banana plantations (STURROCK et al. 1974).

The nematicide Temik 10 G **(aldicarb)** leaves residues in tomato fruits of 0.025 ppm, 0.014 ppm and 0.010 ppm after 78, 84, and 98 days respectively following application of 0.25 g/plant pot (LINDQUIST et al. 1972). Ripe tomatoes are, however, eagerly eaten by slugs, and this would mean that the animals take up the residue as they feed. This result shows that gastropods may incidentally take up residues of other plant protection substances which have been applied in their habitat, and which are either not toxic to them or taken up in sub-lethal doses. These may be ingested by feeding on treated plants or by contact with them or with treated soil, and in this way enter the body of the slug or snail.

3.2.5.4 Side Effects

WEINMANN (1968) has discussed the implications and causes of contamination of the environment with pesticides and on the threat this represents. In an analysis of the problems involved and the results of residue estimation of molluscicides, STURFE (1968) dealt with, amongst others, the following substances: copper sulphate, organic heavy metal compounds, phenol compounds, triphenyl-methyl-amine derivatives, brestan, ziram, pentachlorophenol (PCP), DNCP, niclosamide, molucid and trifenmorph. The side effects on mammals, fishes, plants (including vegetables) as well as on the aquatic fauna and flora are described. When agricultural crops are sprayed with water containing a molluscicide, as part of a control campaign, a certain amount of the active ingredient enters the soil and is then taken up by the plants (cereals, citrus, rice and sugar cane). Molluscicides like metaldehyde and mercaptodimethur, which are used exclusively against terrestrial pulmonates, are not listed by STRUFE, but two others which are effective against land gastropods are dealt with, namely, copper sulphate and PCP.

Information is available on the undesirable side-effects of **metaldehyde.** The main effects are on hedgehogs, which eat slugs eagerly and may consume large numbers of poisoned slugs. A thorough investigation of this and related problems is necessary. BARNES and WEIL (1942) have observed digestive disturbances in hedgehogs. Metaldehyde is also toxic to birds (MEAD 1961; OLDCORN 1938). Slug pellets are said to have caused the death of hares and rabbits in the vicinity of the Plant Protection Office in Oldenburg.

Dogs and cats should be kept away from slug pellets containing metaldehyde, for there has been an ever increasing number of reports of sickness and death resulting from ingestion of bait pellets. HABERKERN (1977) recommends the use of chlorprothixen as an antidote. It would be desirable to make metaldehyde preparations unattractive to domestic animals by altering the present formulation. PLATE and FRÖMMING (1952) found diplopods of the species *Cylindroiulus britannicus* Verh. on bait pellets with signs of having been affected or even killed by them.

The undesirable side effects of **carbamates** are much more marked than those of metaldehyde; damage to earthworms and beneficial Coleoptera may occur, as the experiments of DAXL (1971) with isolan sprays (application 10 ml/m^2 ≙ 600 mg active ingredient/m^2) have demonstrated. The side effects of isolan applied as bait pellets are, however, found to be less marked than those resulting from spraying. Since 1981 a slug pellet formulation has been available, which has a bitter taste, to prevent its being eaten by cats and dogs.

The effect of **carbaryl** on the soil fauna is well known as a result of the studies of ASPÖCK and AN DER LAAN (1963) and AN DER LAAN and ASPÖCK (1962). It is stable in light and not affected by heat up to 70 °C, and at a pH between 4.5 and 9.5, its effectiveness may even persist for years. In forest soil without plant cover, this pesticide shows remarkable stability for at least one year, whereas in soil which is overgrown, the side effects disappear very rapidly.

Owing to its extreme toxicity, earthworm populations in particular are reduced by carbaryl. Since earthworms contribute greatly to soil fertility and thus to crop yield, this is a serious matter. High concentrations of carbaryl can result in the destruction of the entire population of the small enchytraeid worms; the enchytraeid fauna remained either very reduced in numbers or were totally absent even a year after treatment with carbaryl.

In these experiments carbaryl suspensions and powder were tested on the soil of pine woods, beech woods, mixed woodland, with and without ground cover. A 0.08% suspension was found to have considerable side effects on collembolans. Other insects which proved to be sensitive were the other orders of Apterygota, some Coleoptera *(Staphilinidae*, *Pselapidae)* and Psocoptera. Relatively resistant were dipteran larvae, Diplopoda, Chilopoda, Hemiptera, and amongst the Protozoa, the coccidian *Orthezia.*

Earthworms show the following symptoms when poisoned with carbaryl: a few minutes after contact with the substance the two extremities of the body are thrashed to and fro, while the animal writhes in coils. Mucus secretion increases, capacity for movement is reduced and finally ceases altogether as paralysis develops. Several adjacent segments start to swell up on one side of the body, while segments further away become flaccid or constricted in appearance. Later, small epithelial swellings develop, scattered irregularly over the surface of the body. They grow rapidly and become hard, while the rest of the body remains slack. Finally, the largest swellings burst, revealing the musculature and releasing coelomic fluid and blood; the body tissues disintegrate and the animal dies. Large earthworms are more sensitive than younger animals; the swellings increase in size more rapidly in older animals and burst sooner. Epithelial swellings developed within 2 days after contact of worms with a 0.1% Sevin suspension, and the treated animals died within 6 days. With a 0.3% suspension, the swellings developed on the first day after contact and death occurred within 5 days.

Treatment of soil surface of fields with carbaryl powder applied at the usual rates of 0.2 g, 0.5 g or 1 g/m^2 resulted in a reduction in locomotory capacity of earthworms so that they were no longer able to burrow into the soil; indeed, as a result of their writhing movements, they came into contact with more of the carbaryl. At low humidity and a relatively dry upper soil surface, death occurred as a result of desiccation and thus as an indirect effect of poisoning. If the humidity of the air is sufficiently high and the soil adequately moist as a result of rain or irrigation, the animals die when the epithelial swellings burst. Earthworms which happen to be in the soil at the time of application of the molluscicide nevertheless come into contact with it when they emerge from the soil at night or after rain.

The symptoms shown by earthworms at the onset of poisoning (thrashing of extremities of the body, uncoordinated movements) suggest that the nervous system is affected (Aspöck and an der Laan 1963). In the Lumbricidae, fishes and Turbellaria there is first a phase of excitation, which then changes to one of progressive paralysis. Godan has observed the same sequence in poisoned gastropods. The effects of carbaryl on Lumbricidae are irreversible. The rate of application of the substance is of little importance, for sooner or later the animals come into contact with it and even the smallest amounts suffice to harm a lumbricid population.

The effects of poisoning in *Tubifex* sp. are shown in the loss of ability of the worm to roll into a ball, a behaviour it normally displays in clear water in the absence of mud; signs of paralysis follow and finally a striking change in the colour of the blood to dark red. As a result of swelling of single segments, several adjacent segments become restricted. These become white in appearance and finally break away, while the posterior end of the still living animal starts to disintegrate. Unlike those formed in the Lumbricidae, these swellings do not burst. Carbaryl is markedly toxic to flatworms such as *Crenobia alpina* (Dana).

Carbaryl is also toxic to fishes. A concentration of 0.05 ppm in the water caused 60% mortality of young specimens of *Lebistes reticulatus* Peters, while concentrations between 2.5 ppm and 7.5 ppm caused the same mortality of adults. This fish, which is introduced to rice fields to control the larvae of malaria mosquitos and, as a member of the Poeciliidae, is in no way sensitive, shows the following symptoms of poisoning at concentrations of 10 ppm or 1 ppm of carbaryl: uncoordinated jerky and twisting swimming movements and finally, while the fish is not yet dead, passive floating on the surface of the water for many hours with the body on its side or upside down. Eventually the animal lies in a coma-like condition on the bottom until it dies. In some cases, particularly with high Sevin concentrations, there is progressive darkening of the dorsal part of the body; this is most marked at a concentration of 5 ppm active ingredient, and spreads to the ventral side of the body. At higher concentrations (10 ppm), the fishes died within half to 1 h, females being more sensitive than males. The rates at which carbaryl is applied in the field result in concentrations which are toxic to fishes. On the basis of these observations, Aspöck and an der Laan (1963) consider that carbaryl should be placed in both the groups, B and C, proposed by von Bauer (1961): B includes substances which must be handled with extreme care when used in the vicinity of waters inhabited by fishes; C includes substances which may be dangerous to fishes in shallow waters.

Fishes do not seem to be very sensitive to **mexacarbate,** but this substance should nevertheless not be allowed even accidentally to reach waters inhabited by fishes.

HERMANN (1971) described the side effect of Schneckenkorn-Mesurol on insects as of great value, since the active ingredient, **mercaptodimethur,** is an extremely effective insecticide. A higher mortality of *Tipula* and *Nephrotoma* larvae was achieved with this preparation than with Lindan, DDT or Aldrin. It is known also to affect woodlice(*Porcellio scaber* Latr.) and other isopods, millipeds (*Polydesmus* sp.), beetles (Carabidae and Staphilinidae), phalangids and tipulids (MARTIN and FORREST 1969).

Contact of 10 seconds with Schneckenkorn-Mesurol had no effect on earthworms (Lumbricidae), while 30 and 120 s contact rendered the worms incapable of burrowing into the soil, as was the case with carbaryl-poisoned animals. It took 2 days for 50% of animals placed in contact with the preparation to disappear into the soil. Contact for 120 s had, however, no effect on *Eisenia foetida* Sav., which shows that this species is less sensitive than *Lumbricus* sp.; it was, further, more resistant than *Allolobophora* sp., *Dendrobena subrubicunda* Eisen and *Octolasion cyaneum* Sav. According to HERMANN (1971), not more than 5% of the earthworm population is affected. Blackbirds consumed mercaptodimethur-poisoned earthworms only with reluctance; they were not harmed, however, even when as many as 18 worms were eaten by a single bird (HERMANN 1968).

There are, nevertheless, reports of side effects of mercaptodimethur on birds, including fowls and pheasants. Losses were observed amongst birds after they had fed on sown seeds dressed with the substance. Birds may also find slug pellets and eat them; to make the pellets repellent, they are coloured blue, because, as is shown in experiments with pheasants (MARTIN and FORREST 1969), blue and violet are the least attractive colours to birds. Symptoms of poisoning were not observed in pheasant chicks after eating the preparation itself or treated slugs which still contained the toxicant in their bodies. It became apparent, however, that after only 2 days they ate fewer of these slugs. The disinclination could be regarded as a natural safety mechanism in the birds (WALKER 1968). This may be the reason why, in field experiments in which mercaptodimethur formulation was applied at a rate of 8.5 kg/ha, wild birds were not harmed. MARTIN and FORREST found no obvious signs of damage to moles, rabbits, hedgehogs, dogs, cats or toads.

Trifenmorph (Frescon) at 9 kg/ha proved to be completely harmless to soil mites, while mercaptodimethur at 11 kg/ha was heavily destructive to a mite population (EDWARDS and THOMPSON 1973). Frescon concentrations of 0.20 ppm, 0.15 ppm and 0.10 ppm in water resulted in almost 100% mortality of fishes (DEGRÉMONT et al. 1972). A method for analysis of triphenmorph residues has been described by BEYNON and WRIGHT (1975).

Niclosamide (Bayluscid) is used in the rice fields of Indonesia for the control of Lymnaeidae and Planorbidae. At the usual rate of application this compound is harmless to rice plants. But whenever water bodies are treated with chemicals, there is a danger of harmful side effects on the phyto- and zooplankton and also on those higher animals which feed on the plankton. In ponds, fishes reared for human consumption may be affected, while in rice fields the fish *Lebistes reticulatus* Peters, used in the biological control of mosquito larvae, may be harmed. A thorough investigation of this has been carried out by HOLZ and ONG BIE HWA (1963). Niclosamide was found to be toxic to snails [*Planorbis planorbis* (L.)] at concentrations above 0.5 ppm, while fishes could still exist at these concentrations; *L. reticulatus* withstood three times the dose which is lethal for aquatic snails. The zoo- and phytoplankton are reduced almost as rapidly as the molluscs, but unlike the latter, regenerate their populations when the concentration is 2 ppm; resistant stages such as cysts and spores are able to survive concentrations much higher than 5 ppm. The zooplankton which survived 24 h treatment of the water in a rice field at 2 ppm were *Euplodes*, *Paramaecium*, *Loxophylum* and other ciliates, but amongst the phytoplankton only *Oscillatoria* and diatoms. At 5 ppm only *Vorticella*, *Chlorococcus*, *Kirchneriella*, *Ancistrodesmus*, diatoms and bacteria survived. Nevertheless, the stocks recovered after a few weeks.

The hardness of the water influences the toxicity of plant protection products to fishes, which died more rapidly in irrigation canals with soft water (29 to 28.5 $CaCO_3$ mg/l) as in hard (163 $CaCO_3$ mg/l) (HARRISON 1966). A population of *Barbus trimaculatus* Peters in soft water was completely destroyed 5 to 10 min after treatment, while in hard water the species *Barbus paludinosus* Peters survived for nearly 2 h, *B. multimaculatus* Worthington for three and *Clarias gariepinus* (Burchell) for six. Crabs (*Potamon sidneyi* Rathbun) and gyrinids (*Dineutus* sp.) proved to be insensitive. At 60 ml/l nymphs of the insects *Baetis* sp. and *Oecetis* sp. were only poisoned in soft water. Aquatic snails like *Bulinus africanus* (Krauss), *B. forskali* (Ehrenberg), *Biomphalaria pfeifferi* (Krauss), *Gyraulus costulatus* (Krauss) and *Lymnaea natalensis* (Krauss) were greatly reduced in numbers, an effect which lasted for a year or more, whereas their predators were relatively resistant. Cladocerae and copepods proved to be more susceptible than the other Crustacea, but their population size was rapidly restored.

As a result of the death of fishes, however, both *Culex* and *Anopheles* larvae, which are eaten by fishes, survived in greater numbers, so that 2 or 3 months after a control campaign against snails, there was an extraordinarily heavy plague of mosquitoes. This has brought a new problem to light, namely, the necessity to intensify mosquito control when action against snails has been completed. According to HARRISON and MASON (1967), niclosamide destroys snails, but is harmless to other invertebrates in the water.

STRUFE (1968) has discussed the side effects of niclosamide in detail, including its effect on water plants, and also the influence of sunlight, salt content of the water and mud on the bottom.

Isochlorthion, which is used in cress cultivations against the amphibious *Lymnaea* species to good effect, has an immediate toxic effect on fishes and Crustacea (*Gammarus*, amongst others); these animals die as soon as the chemical is introduced into the water. Chlorthion, however, has no side effects provided it is rapidly carried away by flowing water. Isochlorthion, by contrast, has a tendency to sink to the bottom of a canal and can remain effective for a long period (RICOU and MARRE 1960). This substance is also highly lethal to *Tipula* larvae, collembolans, Crustacea (*Gammarus* sp., *Cypris* sp.) and earthworms. Frogs and fishes are also sensitive to it. It has the advantage, however, of being harmless to plants.

Sodium pentachlorophenolate (**NaPCP**) is harmless to terrestrial vertebrates at the application rates normally used in practice, but causes exceptional damage to fishes, tadpoles and Entomostraca. Aquatic arthropods survived a concentration of 10 ppm for at least 3 days. The ciliate faune (*Paramaecium* sp. amongst others) is reduced after 48 h at 10 ppm, but not totally destroyed. The freshwater fauna largely recover within three weeks as a result of the fall in toxicity of NaPCP following its adsorption by mud and plants and also its decomposition by UV radiation.

A disadvantage of NaPCP is its high toxicity to fishes, which reduces its use, since water stocked with fishes may not be treated with it. Red carp survived 0.5 ppm for 13 h without apparent effect, but at 0.8 ppm most of the fishes were floating on their sides on the surface of the water within only 6 h; 10 ppm led to death after only 3 h (ENIGK and DÜWEL 1960). Tadpoles are almost as sensitive, but mature frogs are relatively resistant; so, too, are river crayfish (*Potamobius astacus* L.) which remained unaffected by 50 h at 10 ppm. Water insects, such as Mayfly larvae, back swimmers (*Notonecta* sp.), larvae and adults of the Large Water Beetle (*Dytiscus marginalis* L.), dragonfly nymphs, water boatmen (Corixidae), water bugs (Cryptocerata), water scorpions (Nepidae), whirligig beetles (Gyrinidae) and culicid larvae are all unaffected by three days at a concentration of 10 ppm NaPCP, although specimens of the very sensitive *Daphnia* started to die within 3 h, and after 13 h the entire population had been destroyed.

Terrestrial insects and their larvae are markedly resistant. Dung beetles and house flies withstood direct contact (spraying with a 1% solution) and *Tipula* larvae survived 3 h in a solution of 10 ppm. After a pasture with a large population of *Tipula* larvae had been sprayed with NaPCP at the rate of 6 g/m^2 – three times the normal rate for Liver Fluke control – a few dead larvae were found the next day, but most of them survived.

Fowls showed no ill effects from this substance, but did not ingest it voluntarily, either in the form of pellets or when added to drinking water.

ENIGK and DÜWEL (1960) tested the phytotoxicity of NaPCP to grazing grasses: when applied to dishes 25 × 25 cm in size, at a rate which represented 125 mg/m^2 (12.5 ml solution per dish), all the species of grass tested showed retarded growth compared with the untreated area, and spraying also turned the grass yellow. Timothy was the most strongly affected in comparison with perennial ryegrass, French rye grass, orchard grass and Kentucky blue grass. The reduced growth rate continued for 14 to 17 days. Broad-leaved plants were more strongly affected than the grasses, but the effect was localized. According to HUNTER et al. (1968), dicotyledonous plants are more sensitive than are monocotyledons. Damage to terrestrial plants can be avoided in practice by spraying during rainy weather in early spring. STRUFE (1968) gives further details on the side effects of NaPCP on mammals, fishes and equatic fauna and flora. He also discusses residues, mainly in water bodies and their substrate.

DNOC (dinitro-ortho-cresol) is not toxic to water plants even at a concentration of 10 ppm. Crustaceans and water insects are also largely resistant to the concentrations formerly used in gastropod control (KUNTZ and WELLS 1951). DNOC is, however, harmful to terrestrial organisms such as collembolans and soil mites, but after a certain time their numbers recover – to the point of over-population;

this effect is independant of application rate and resistance of the particular species, and the new population may be 300% of the initial population. KARG (1964) studied the influence of DNOC on the following types of soil down to depths of 0 to 15 cm: sand with little humus, sand with humus and loam with a high chalk content. Tests were conducted in the autumn, the time when gastropod control is normally undertaken, and the effects of a heavy application on the soil fauna in the following year were observed. As well, the immediate effect of a normal dose (6 kg/ha) and its further influence during the growth period of the same year were observed.

The results showed a 60% reduction of the collembolan population about 8 months after treatment. This had been restored to normal 1 month later and the population had doubled 10 months later. Mites recovered more rapidly, so that after only 9 or 10 months a population increase of 100% was observed, but this fell again a year later. The normal application rate did not lead to losses of collembolans or mites, but to a 30% increase in numbers one month after treatment. In loamy soil rich in chalk, in which the soil fauna differed from those of either of the sandy soils, there was a 40% reduction in collembolans and 50% reduction in the mites 10 days after treatment; the population had recovered four months later and in the case of some species over-population followed. Recovery and over-population occurred more rapidly in sandy than in loamy soil. GOULD (1962) found many dead earthworms after application of DNOC and PCP (pentachlorophenol), while Paris Green appeared to have no such undesirable side effects.

Copper sulphate also appears to be toxic to earthworms, as the earthworm population in a field treated with this substance over several years was small in comparison with that in a neighbouring field which had always been kept free of copper sulphate.

Analysis of the soil in the field poor in earthworms showed that the copper content of a surface sample was very high, at 1,500 ppm Cu, while at 10 cm depth it was 120 ppm Cu (HIRST et al. 1961). Copper sulphate makes *Tubifex* sp. incapable of curling up, furthermore, islands of necrotic tissue develop, while there are also changes to the segments similar to those caused by carbaryl (LIEPOLT and WEBER 1968).

Copper sulphate further appears to be toxic to algae: *Enteromorpha intestinalis*, *Mougeotia* sp., *Oedogonium* sp., *Spirogyra* sp. and *Vaucheria* sp. are all very sensitive even to concentrations of 0.25 tp 1 ppm. Other aquatic microflora and microfauna are affected only by concentrations of 20 ppm and above (SHIFF and GARNETT 1961). However, the application of copper sulphate year after year as a prophylactic measure against *Biomphalaria glabrata* (Say) impaired the growth of the nitrogen-binding blue-green algae, the presence of which is a prerequisite for adequate fertility for rice plants. Their absence results in reduced yields and consequent economic losses (LEVINE 1970). Furthermore, insect larvae (apart from those of Diptera, Trichoptera and Coleoptera) as well as populations of Gyrinidae, Gerridae and Dytiscidae are reduced, as are those of the Copepoda and Ostracoda amongst the Crustacea. The Decapoda and, amongst vertebrates, the Amphibia, are not affected.

The toxicity of copper sulphate to fishes depends on the temperature and the hardness of the water. STRUFE (1968) reproduces a table from LLOYD (1961) which

Table 42. Survival times of Rainbow Trout in both hard and soft water containing copper sulphate. (After STRUFE 1968 cited by LLOYD 1961)

	Survival time in h	
$CuSO_4$ (ppm)	Hard water (320 ppm)	Soft water (15 to 20 ppm)
1.2	500	750
1.5	180	500
2.5	80	180
5	–	80
10	–	62
20	–	35

gives the survival times of Rainbow Trout in water containing copper sulphate solutions (Table 42).

In hard water this fish is more affected than in soft. The safe upper limit for the Rainbow Trout (*Salmo iridens* Gibb.) is 7 ppm, and the critical limit is 8 ppm. Copper sulphate, organic compounds containing tin and niclosamide are admittedly toxic to fishes at high concentrations, but can be made harmless by being bound to a matrix from which the molluscicide is only slowly released into the surrounding water (Anon. 1974). Some substances also kill the parasites in the vector snail at concentrations which are not lethal for the host. Another copper salt, copper chloride, is also relatively strongly toxic to fishes (CHU et al. 1968).

Copper sulphate is not toxic to rice plants at an application rate of 20 g/ 1,000 m^2 (IWATA and WATANABE 1960). CHANCELLOR et al. (1958) investigated the phytotoxicity of this substance on various water plants, to establish its suitability for use in controlling water weeds.

In this connection it should be pointed out that **herbicides** and **calcium cyanamide** inhibit the development of fungal diseases in plants, such as that caused by *Erysiphe graminis* DC. (HEITEFUSS 1972; HEITEFUSS and BRANDES 1970). These herbicides are triazine and various derivatives of urea, of which simazin and linuron also have a certain molluscicidal effect. Calcium cyanamide, which was formerly frequently used in gastropod control because of its caustic effect, but is now only rarely employed for this purpose, causes a more long-lasting reduction of fungal infection of plants than the herbicides do. Its action is twofold: as a source of nitrogen, it affects the plant in such a way as to reduce the amount of mildew, while the release of cyanamide during its degradation has a certain fungicidal effect. Calcium cyanamide is lethal to gastropods not only through its caustic effect, but also as a result of the cyanamide phase.

3.2.5.5 Maximum Residue Limits

Small amounts of pesticides used for plant protection are permissible in food stuffs, provided the quantity does not exceed the allowed maximum which is regarded as harmless. Each substance has its own tolerable maximum, which in the Federal Republic of Germany is stipulated in ppm: thus a tolerance of 1 ppm means that maximally 1 mg of the active ingredient is permitted in 1 kg of food stuff.

According to the Höchstmengenverordnung Pflanzenschutz, pflanzliche Lebensmittel of November 30, 1966, updated on February 4, 1976, BGBI. I. page 264, the permissible concentrations for the residues of molluscicides, which are here of interest, were as follows:

Metaldehyde:	1.0 ppm for cabbage, lettuce 0.4 ppm for other vegetable food stuffs
Mercaptodimethur:	No maximum amount stipulated
Carbaryl:	2.5 ppm for apples, apricots, pears, plums, grapes, cabbage, lettuce 1.2 ppm for all other fruit and vegetables 0.8 ppm for rice 0.1 ppm for other vegetable food stuffs
Isolan:	No maximum tolerance level
Mexacarbate:	No maximum tolerance level
Copper sulphate:	20 ppm for vegetables, fruit
(Category: copper salts):	10 ppm for other vegetable food stuffs
DNOC:	0.05 ppm for all vegetable food stuffs
PCP (NaPCP):	Use forbidden in water protection areas under the directives for protected drinking water areas
Methylbromide:	(Used in the USA for destroying snails in cargo) use permitted only for fumigation of warehouses, storerooms, small airtight silos, transport vehicles and containers and, under gas-proof tarpaulins, against storage pests; for soil treatment in ornamental plant cultivations and the production of seed potatoes, only with permission under state law by the official responsible; vegetables may subsequently not be grown earlier than 3 years after soil treatment.
Calcium arsenate:	As an arsenic compound its use forbidden.

The three last-named compounds, which were used in earlier times and are little used as molluscicides today, fall under the „Verordnung über Anwendungsverbote und -beschränkungen für Pflanzenbehandlungsmittel" (Regulations on the Prohibition and Limitation of Use as Plant Protection Substance) of 19th December 1980 (BGBl. I. page 2335). Metaldehyde and mercaptodimethur have today the keyword „gesundheitsschädlich" according to the danger symbol X n.

3.2.5.6 Time Intervals

The Federal Biological Research Centre for Agriculture and Forestry in Brunswick has issued a „Verzeichnis der Wartezeiten für Pflanzenschutzmittelanwendungen" (Register of Time Intervals in the use of Plant Protections substances) (Instructional Pamphlet 40, 2nd ed., 1974). This stipulates, amongst other things, that: "The time intervals apply when the substances are used at the recommended rates and in accordance with such other directives for their use as are laid down in the register of plant protection substances. Every alteration of these conditions leads to a change in the size of the residue, and can result in an agricultural product containing more than the permitted residues ... the length of the time interval is no indication of the toxicity or possible toxicity of the substance listed. The time intervals serve only to ensure that the stipulated maximum amounts are not exceeded ..."

In this register, the following time intervals have been listed for molluscicides:
Metaldehyde: 21 days regardless of whether it is used as spray, dust or bait pellets, for all field and vegetable cultivations. In the case of strawberries, "already determined by the season of growth and the conditions of registration."
Mercaptodimethur: 14 days when used as dusting powder on field crops mentioned under metaldehyde, including the regulation specific to strawberry "cultivation under glass".
Carbaryl: (the only other molluscicidal carbamate listed): 28 days for vineyards, cherries and other fruit, 21 days for potatoes, 14 days for field-grown strawberries "the time interval is already determined by the growth season and the conditions of registration, and does not need to be especially listed." The conditions are thus the same as those for metaldehyde and mercaptodimethur.
Calcium cyanamide*: 4 days for cucumbers, tomatoes and greenhouse lettuce. For field-grown crops, vegetables and vineyards the time interval is determined by the growth seasons and the conditions of registration.
DNOC*: for cereals and post-germination in spring, as well as grass seed production and viticulture the time interval is determined by the season of growth and conditions of registrations of the compound. No registration is granted for "under glass cultivation" of plants grown as food.
Ziram (zinc-dimethyl-dithiocarbamate) is used occasionally for the control of amphibious snails and is therefore listed here: 56 days for viticulture, 28 days for field grown leaf vegetables, 21 days for onions, leeks and root vegetables, 14 days for peas, 10 days for tomatoes, cucumbers and stone fruit. No registration for "under glass cultivation" of plants for food.

3.3 Protection of Stored Products

Terrestrial pulmonates make their presence unpleasantly felt by feeding on and soiling stored potatoes and root vegetables, and by preparing the way for fouling by bacteria.

Gastropods reduce the value of stored crops not only through their feeding activity, but also to a considerable extent by soiling the goods with faeces and the slime trails that are left behind as they creep about. The Cellar Slug, **Limax flavus** L. is of especial importance as a pest to stored products (PLATE 1959). It makes its appearance in cellars, warehouses and storerooms and, south of the Alps, in fountains. This slug is known as a cave-dweller in Hungary and is now found in the wild only in thinly populated areas of eastern Europe. FRÖMMING (1951) pointed out that *Limax flavus* L., which by 1950 was still little recognized as a storage pest, had attached itself to man and could therefore, "like the clothes' moth, bed bug, house

* Use as a molluscicide not permitted in the Federal Republic of Germany

mouse, etc. be regarded as a domestic animal". It is widespread in central, western and southern Europe and is carried everywhere with export goods. It has been known for more than a hundred years from storehouses containing various wares, such as potatoes, carrots, fruit, vegetables, wine, beer and, as it has spread, its numbers, too, have increased. It consumes cucumbers, potatoes, pumpkin, melons, pears, swedes, turnips, radishes, parsley roots, etc. and often in quantities which represent 40% to 60% of its own body weight (FRÖMMING 1951). As a rule it does not eat green plants; from the leaves of vegetables it consumes the central vein, which is poor in chlorophyll and has a high water content; less frequently it may feed on the chlorophyll-containing parts. Bulbs which are infected with bacteria and fungi are eagerly eaten. Since the young slugs bore their way right into the vegetables or fruit, making holes and passages, they remain at first unnoticed, so that the infestation can spread undisturbed. FRÖMMING regards as considerable the indirect damage caused by *Limax flavus* L. as a result of its spreading fouling agents during its nightly wanderings. The slug can cover 20 to 40 m in one night, feeding from several different potatoes. In this way, a single slug can destroy 2 $^1/_2$ kg of potatoes in a month, mainly because those potatoes which have been scraped at become fouled. In the last 15 years *Limax flavus* L. has been found repeatedly in the storerooms of bakeries and cheese dairies. FRÖMMING (1957) showed the slug to be capable of living on flour alone, which, however, causes a change in its body colour. Such "flour animals" were yellowish-orange after 8 to 10 weeks; with increasing age they became yellow to reddish-brown, while the normal colouration is yellow-grey-brown, and in old animals greenish-grey-brown. In old and particularly damp houses these slugs creep along the waterpipes and into the living areas. *Limax maximus* L., too, is found in cellars and storerooms. This slug is spread throughout southern and central Europe, but is most numerous in the west; it is found almost everywhere in Germany. It lives mainly in parks, yards and gardens and is only found occasionally in cellars and storerooms. It feeds mainly on fungi, but also green plants, from which it eats, as in the case of vegetables, the juicy midvein of the leaf. **Small snails** are also found in cellars; these are *Oxychilus cellarius* (Müller), *Oxychilus alliarus* (Miller) and *Oxychilus draparnaudi* (Beck). Even dried plants, fruits and roots in storage are not rejected (FRÖMMING 1950). Many such medicinal herbs are hygroscopic, and if not stored properly, absorb water, becomming sufficiently moist to enable the slugs to live on them.

In the USA special importance is accorded to the snails *Theba pisana* (Müller) and *Cochlicella barbara* (L.) which occur not only in the open but also in storerooms and are carried in ships from one country to another. Both species have been declared to be quarantine pests, together with *Achatina fulica* Bowdich. The latter is the most important of the snails found in the storerooms and holds of ships during official quarantine inspections and regularly appears in the lists of Intercepted Plant Pests.

Table 43 shows the snail finds on ships for the years 1968/69 and 1971/72. These are mainly *Helicella* sp. and to a lesser extent other Helicidae. In 1971/72 a slug, *Deroceras caruanae* (Pollonora) was added to the list.

Spreading of lime in storerooms and cellars is effective against *Limax flavus* L. The rooms must be clean, free of rubbish, airy and above all dry. Metaldehyde pellets can also be used.

Table 43. Gastropods introduced to USA in storerooms and containers of ships

Species (family)	Country of origin
Cochlicella barbara (Linnaeus) (Helicidae)	As in Table 44
Cochlicella conoidea (Draparnaud) (Helicidae)	Morocco
Cochlicella ventricosa (Draparnaud) (Helicidae)	As in Table 44
Deroceras caruanae (Pollonera) (Limacidae)	England Italy
Helicella amanda (Rossmaessler) *Helicella apicina* (Lamarck) *Helicella caperata* (Montagu) *Helicella cespitum* Draparnaud *Helicella cretica* (Férussac) *Helicella maritima* (Draparnaud) *Helicella obvia* (Hartmann) *Helicella virgata* (Da Costa) (Helicidae)	France Greece Italy Spain Turkey
Helix aperta Born (Helicidae)	France Italy
Helix aspersa Müller (Helicidae)	Azores Greece
Helix pomatia Linnaeus (Helicidae)	Italy Mexico Portugal Spain
Monacha cantiana (Montagu) (Helicidae)	Italy Turkey
Monacha cartusiana (Müller) (Helicidae)	Italy
Monacha obstructa (Férussac) (Helicidae)	France Greece Italy Libanon Portugal Turkey
Monacha schotti (Pfeiffer) (Helicidae)	Unknown
Otala vermiculata (Müller) (Helicidae)	Europe
Theba pisana (Müller) (Helicidae)	As in Table 44

It may be that pheromones, which are still to be discussed, could be used to advantage in the protection of stored goods.

Control measures in general will be discussed in connection with quarantine. *Cochlicella barbara* (L.) has proved to be far more resistant than *Theba pisana* (Müller). The relationship between sensitivity to molluscicides and both species and degree of activity has already been discussed in the relevant sections.

3.4 Plant Quarantine Regulations

In Europe – at least in the Federal Republic of Germany – there are no regulations governing plant inspection and quarantine which have a specific bearing upon gastropods and their control. Such regulations have, however, been introduced in the United States of America and Canada, for many important gastropod pests have been spread to these countries, particularly from Europe, and the process still continues today.

3.4.1 Dispersal of Pest Gastropods

The most dangerous and therefore the most important **terrestrial pulmonate pest** is the African Giant Snail *Achatina fulica* Bowdich (Anon. 1969; Dun 1967; Mead 1961, 1977; Müller 1962; van Dinther 1973; Voelker 1959, 1975). It reached the USA as early as 1951 and was declared by law to be an important *quarantine pest;* in 1977 it was proclaimed to be a quarantine pest in Australia too (Quarantine Proclamation No. 74 P, Australian Government Gazette No. S 115, June 24, 1977). From its origin in East Africa this snail has spread by land and sea, east and west, within 150 years. In 1936, for example, it reached the Hawaiian Islands, where it was destroyed by intensive control measures. Later it nevertheless managed to reestablish itself here, and might do so in the USA as well, for it was found in southern Florida in 1966/67 and in northern Miami in 1969. The possibility thus exists, should the campaign of control which was immediately initiated fail, of this species dispersing over the whole of North, Central and South America (Mead 1973), although according to an FAO report published in 1975, no specimen of *Achatina* has been found in the USA since 1973. Complete absence of this snail can, however, only be regarded as a certainty provided not even a single specimen is found over the next few years.

The whole population, but particularly tourists and travellers, are the main agents responsible for the spread of this snail. Thus a boy returning to Miami from Hawaii brought two or three *Achatina* with him, and in only 3 years the animals managed to spread throughout the town of Miami, with the result that in 1969 the gardens were stripped almost bare and the African Giant Snail had become a much feared pest (Communication in News Bull. Florida Dep. Agr. and Consumer Serv. 11, No. 4, 1969). This story illustrates how easily imported pulmonates can become a plague. The ever-increasing scale of global tourist traffic constitutes in this respect a serious threat, which necessitates strict official vigilance. Because of its large size and most interesting appearance, *Achatina fulica* Bowdich seems to many tourists a most attractive souvenir.

Once the African Giant Snail has established itself in a country, it can easily spread further by way of air, rail and road traffic. If both climate and available food are suitable, large pest populations may be built up, mainly because those parasites and predators which held it in check in its area of origin do not occur in the newly settled area.

It is not simply from ignorance of the damage they can cause, or even by accident, that gastropods are distributed with tourists' luggage; they are also introduced as a source of food. A disastrous event of this nature occurred during the second world war, when the Japanese introduced *Achatina* to the Pacific islands (Guam, Phillipines, etc.), with the admittedly praiseworthy intention of providing the population there with an improved source of nourishment. Indeed, in its African home, this snail is valued as a tasty food and has never been regarded as an agricultural or horticultural pest, but on the islands it rapidly became a truly exceptional pest; at the same time its new home served as a springboard for wider dispersal, particularly to North America. The first free-living specimens in Guam were found in 1943 (Peterson 1957).

Since then much has been written about *Achatina fulica* Bowdich; the most important publication is the book by Mead (1961) in which the references extend over 41 pages.

Another important quarantine pest is the Mediterranean White Snail, *Theba pisana* (Müller), which has been introduced by man into many countries. Well-defined sub-species have developed in many areas, not only on islands, as for example *Theba pisana ustulata* (Lowe) on the Salvage Islands and *Theba pisana alboranensis* (Webb and Berthelot) on the neighbouring Canary Islands, but also on the continent (Backhuyis 1972). Although it readily disperses in areas characterized by lime-rich sand, certain food plants and mild winters, this snail can in fact establish itself almost anywhere, since it has a broad ecological tolerance and there are many ways in which it can be passively transported; it has been found not only in luggage, cargo and plant material, but even in postal packages.

For more than a hundred years, the Brown Snail *Helix aspersa* Müller has been dispersed throughout the Mediterranean countries; it was described in Italy as early as 1774 and has been carried everywhere with plants and food stuffs (Herzberg and Herzberg 1962). It is thought to have been introduced to North America with vines from France and has since established itself there; it reached California in 1850 (Basinger 1931). Since it originated in Europe, *Helix aspersa* Müller is often called the European Brown Snail.

According to Curtin (1963), the following snails from Turkey and Greece are important as **quarantine pests** for the United States of America: *Helicella itala* (L.),*Helix aperta* Born, *Helicella striata* (Müller), *Helix aspersa* Müller, *Monacha schotti* (Pfeiffer), *Otala vermiculata* (Müller), *Oxychilus cellarius* (Müller), *Planorbis planorbis* (L.) and *Theba pisana* (Müller).

Slugs have been transported from Europe to the USA mostly in more recent times; an exception is *Limax maximus* L. which was first identified there shortly before 1904. *Arion ater* (L.) made its appearance in 1942 on the islands of the northwest Pacific and about 1954 in the western part of Washington State (USA). Later, the limacids *Deroceras reticulatum* (Müller) and, in the mid 1960's, *Deroceras caruanae* (Pollonera) established themselves in the northeast of North America. About the same time the arionids, *Arion fasciatus* (Nilsson) and *Arion subfuscus* (Drap.), were spread to this area. Neither *Arion intermedius* Normand nor *Arion hortensis* Fér., which occur there now, belongs to the indigenous molluscan fauna of North America. While plant imports from Europe are the chief way in which gastropods are introduced to North America, they spread from the coastal regions where they first establish themselves to the interior (CHICHESTER and GETZ 1969).

In recent times **freshwater** gastropods such as *Radix auricularia* (L.) have been brought from Europe to the USA (Colorado). Several representatives of the Lymnaeidae have also been introduced: *Galba viridis* Quoy and Gaimard, the vector of the Liver Fluke of cattle, was introduced around 1890 from East Asia, and *Pseudosuccinea columella* Say was spread from Hawaii around 1950 (MORRISON 1968).

There are innumerable ways and means by which pest gastropods can be spread; thus during official control at an entry point in the USA, *Monacha cartusiana* (Müller) was found in a roll of cable wire and *Theba pisana* (Müller) in the straw of horse equipment. It has proved also impossible to find *Achatina fulica* Bowdich hidden amongst emptied straw sacks (Australian Dep. Health 1975).

Table 44 lists gastropods which have been inadvertently transported to the **United States of America** and which have been discovered in seed and plant material (leaves, plants, shrubs, stalks, roots, etc.). In the Lists of Intercepted Plant Pests of the Plant Protection and Quarantine Programs, Animal and Plant Health Inspection Service, US Dep. Agr. such finds are regularly reported. The data given here are from the lists of 1963 to 1972. No such lists are published in the Federal Republic of Germany.

There are many ways in which gastropods may be **transported,** and their dispersal may occur by ship, air, motor vehicles or rail. The molluscs occur not only on plants, on soil still attached to them and on harvest crops, but also in packaging, in containers and in straw or hay; ships and aeroplanes are the most important means by which the animals are spread between far distant countries. Most of the gastropods entering the USA with baggage come from Europe, Asia, Africa, Hawaii and the Philippines.

According to the Lists of Intercepted Plant Pests of the years 1967/68/71 and 1972, the following gastropods were found in:

Aeroplanes: *Achatina achatina* (L.), *Bradybaena serotina* A. Adams, *Helix aspersa* Müller, *Helix lucorum* L.

Cars: *Bradybaena similaris* (Fér.), *Cepaea nemoralis* (L.), *Cochlicella barbara* (L.), *Cochlicella ventricosa* (Drap.), *Helix aspersa* Müller, *Helix radiosa* Ziegler, *Opeas pumilum* (Pfeiffer), *Otala vermiculata* (Müller), *Theba pisana* (Müller), *Subulina octona* (Bruguière), *Veronicella leydigi* Simroth.

Baggage: *Achatina achatina* (L.), *Achatina fulica* Bowdich, *Bradybaena similaris* (Fér.), *Cepaea nemoralis* (L.), *Helicella candidula* (Studer), *Helicella itala* (L.), *Helicella maritima* (Drap.), *Helicella variabilis* (Drap.), *Helicigona (Arianta) arbustorum* (L.), *Helix aperta* Born, *Helix aspersa* Müller, *Helix cincta* (Müller), *Helix mazzulii* Jan, *Helix pomatia* L., *Helix radiosa* Ziegler, *Helix secernenda* (Rossmaessler), *Iberus alonensis* (Fér.), *Monacha rizza* (Aradas), *Monacha syriaca*

Table 44. A list of pest gastropods introduced to the USA with seed and plant imports

Species (family)	Plants		Country of origin
Achatina fulica Bowdich (Achatinidae)	Flowers		Asia
	Seeds		Guam
	Peat	*Zoysia* sp.	Hawaii
			Okinawa
			Philippines
			Vietnam
Arion ater (Linnaeus) (Arionidae)	Cauliflower	*Brassica oleracea* var. *botrytis*	Portugal
Arion circumscriptus Johnston (Arionidae)	Apples	*Malus sylvestris*	Germany
	Cabbage	*Brassica oleracea* var. *capitata*	Spain
Arion hortensis Férussac (Arionidae)	Clover	*Trifolium repens*	England
	Cabbage	*Brassica oleracea* var. *capitata*	France
	Leeks	*Allium ampeloprasum*	Germany
	Moss	*Sphagnum*	Holland
	Garden plants	*Begonia rex*	Ireland
	Cellery	*Apium graveolens* var. *dulce*	Spain
	Violets	*Viola tricolor*	
Bradybaena sieboldiana (Pfeiffer) (Bradybaenidae)	Cauliflower	*Brassica oleracea* var. *botrytis*	Japan
Bradybaena similaris (Férussac) (Bradybaenidae)	Pineapples	*Ananas comosus*	Asia
	Leaves	*Cordyline terminalis*	Australia
	Flowers, dried and mixed		Belgium
			Bermuda
	Blossoms, dried	*Hydrangea* sp.	Brasil
	Bromeliads		Formosa
	Cocoanut	*Cocos nucifera*	Hawaii
	Leeks	*Allium ampeloprasum*	Japan
	Orchids	*Cattleya* sp.	Okinawa
	Plants	*Anthurium* sp.	Portugal
	Cut flowers	*Rosa* sp.	Vietnam
	Cut flowers	*Alpinia purpurata*	
	Cut flowers	*Tagetes* sp.	
	Cellery	*Apium graveolens* var. *dulce*	
	Leaf stalks	*Plumeria* sp.	
Cepaea hortensis (Müller) (Helicidae)	Pine needles		Germany
	Lilac	*Syringa* sp.	
Cepaea nemoralis (L.) (Helicidae)	Moss, dried		England
	Plants	*Calendula* sp.	France
			Germany
			Italy
Cochlicella barbara (L.) (Helicidae)	Leaves	*Mentha* sp.	Algeria
	Pines (young plants)		Australia
			Azores
	Millet	*Panicum* sp.	Belgium
	Immortelle	*Xeranthemum* sp.	Crete
	Cabbage	*Brassica oleracea* var. *capitata*	England
	Leeks	*Allium ampeloprasum*	France

Table 44 (continued)

Species (family)	Plants		Country of origin
	Lilies	*Lilium* sp.	Germany
	Lentils	*Lens culinaris*	Holland
	Bayleaves	*Laurus nobilis*	Ireland
	Plants	*Amaryllis*	Italy
	Plants	*Olea* sp.	Israel
	Plants	*Rosmarinus* sp.	Japan
	Lettuce	*Lactuca sativa*	Libanon
	Seeds	*Cichorium* sp.	Libya
	Seeds	*Helianthus* sp.	Morocco
	Celery	*Apium graveolens* var. *dulce*	Portugal
	Cedars	*Cedrus* sp.	Tunisia
			Turkey
			United Arab. Republic
Cochlicella conoidea (Draparnaud) (Helicidae)	Lentils	*Lens culinaris*	Germany
	Plants		Greece
	Seeds		Italy
			Morocco
			Spain
Cochlicella ventricosa (Draparnaud) (Helicidae)	Leaves	*Mentha* sp.	Azores
	Leaves	*Citrus* sp.	Bermuda
	Beans	*Phaseolus vulgaris*	France
	Immortelle	*Xeranthemum* sp.	Germany
	Cabbage	*Brassica oleracea* var. *capitata*	Greece
	Cork	*Quercus ruber*	Holland
	Carnations	*Dianthus* sp.	Israel
	Palm leaves		Italy
	Parsley	*Petroselinum crispum*	Morocco
	Plants	*Crassula* sp.	Portugal
	Plants	*Sarcocaulon patersonii*	South Africa
	Lettuce	*Lactuca sativa*	Spain
	Cut flowers	*Lilium* sp.	Yugoslavia
	Celery	*Apium graveolens* var. *dulce*	
	Tomatoes	*Lycopersicon esculentum*	
	Seeds, mixed		
Deroceras caruanae (Pollonera) (Limacidae)	Cauliflower	*Brassica oleracea* var. *botrytis*	England
	Fennel	*Foeniculum vulgare*	Italy
	Plants	*Rhododendron* sp.	
	Lettuce	*Lactuca sativa*	
	Celery	*Apium graveolens* var. *dulce*	
Deroceras laeve (Müller) (Limacidae)	Bananas	*Musa* sp.	El Salvador
Deroceras reticulatum (Müller) (Limacidae)	Flowers	*Chrysanthemum* sp.	Columbia
	Bromeliads		England
	Chicory	*Cichorium intybus*	France
	Cabbage	*Brassica oleracea* var. *capitata*	Germany
	Orchids		Mexico
	Lettuce	*Lactuca sativa*	Portugal
			Yugoslavia

Table 44 (continued)

Species (family)	Plants		Country of origin
Euhadra peliomphala (L. Pfeiffer) (Bradybaenidae)	Sweet potato	*Dioscorea* sp.	Japan
Helicella amanda	Acacias		Algeria
(Rossmaessler)	Artichokes	*Cynara scolymus*	Austria
Helicella apicina (Lamarck)	Flowers, fresh		Azores
Helicella armillata (Lowe)	and dried		Belgium
Helicella calcarata (Benson)	Beans	*Vicia faba*	Bulgaria
Helicella candidula (Studer)	Beans	*Phaseolus* sp.	Crete
Helicella caperata (Montagu)	Bougainvillia		Cyprus
Helicella cespitum Draparnaud	Bread fruit tree	*Artocarpus altilis*	Denmark
Helicella conspurcata	Pines, young plants		Dominican
(Draparnaud)	Daisies		Republic
Helicella corrugata (Gmelin)	Gardenias	*Bellis* sp.	England
Helicella derbentina (Krynicki)	Honeysuckle		France
Helicella dominicensis (Pfeiffer)	Geraniums	*Lonicera caprifolium*	Germany
Helicella elegans (Gmelin)	Grapefruit		Greece
Helicella itala (L.)	Carrots	*Citrus paradisi*	Holland
Helicella joppensis (Roth)	Cabbage	*Daucus carota* var. *sativa*	India
Helicella krynickii (Andrz)	Herbs	*Brassica oleracea* var. *capitata*	Iran
Helicella langloisiana	Leeks		Ireland
(Bourguignat)	Lentils	*Allium ampeloprasum*	Israel
Helicella larnacensis (Kobelt)	Marjoram	*Lens culinaris*	Italy
Helicella maritima	Mint leaves	*Majorana hortensis*	Jordan
(Draparnaud)	Carnations	*Mentha* sp.	Libanon
Helicella meda (Porro)	Indian figs	*Dianthus* sp.	Lybia
Helicella neglecta	Parsley		Malta
(Draparnaud)	Peaches	*Petroselinum crispum*	Morocco
Helicella obvia (Hartmann)	Plants	*Prunus persicus*	Norway
Helicella protea (Ziegler)	Plants	*Amaryllis* sp.	Poland
Helicella pyramidata	Pistacio nuts	*Hydrangia* sp.	Portugal
(Draparnaud)	Lettuce	*Pistacia* sp.	Sweden
Helicella striata (Müller)	Salvia	*Lactuca sativa*	Spain
Helicella terrestris (Pennant)	Seeds		Tunisia
Helicella trochoides (Poiret)	Seeds	*Cichorium* sp.	Turkey
Helicella variabilis	Seeds	*Convallaria majalis*	United Arab.
(Draparnaud)	Seeds	*Cotinus coggygria*	Republic
(Helicidae)	Seeds	*Cucurbita maxima*	
	Seeds	*Cynara scolymus*	
	Seeds	*Gleditsia triacanthes*	
	Seeds	*Jasum*	
	Seeds	*Laburnum* sp.	
	Seeds	*Mirabilis jalapa*	
	Seeds	*Rosmarinus officinalis*	
	Seeds	*Sanguisorba* sp.	
	Cut flowers	*Tilia* sp.	
	Sun flowers	*Dianthus caryophyllum*	
	Asparagus	*Helianthus* sp.	
	Spinach	*Asparagus officinalis*	
	Shrubs	*Spinacia oleracea*	
	Vetch	*Althea* sp.	
		Vicia sp.	

Table 44 (continued)

Species (family)	Plants		Country of origin
Helicigona (Arianta) arbustorum (L.) (Helicidae)	Nursery plants		France
	Cabbage	*Brassica oleracea* var. *capitata*	Germany
	Plants	Pines	Greece
	Plants	*Rosa* sp.	Holland
	Lettuce	*Lactuca sativa*	Italy
			Switzerland
Helicina chrysocheila Binney (Helicidae)		*Chamaedorea* sp.	Mexico
Helicina zephyrina Duclos (Helicidae)	Bromeliads		Italy
	Palms		Mexico
Helicodonta obvoluta (Müller) (Helicidae)	Plants		Italy
Helix aperta Born (Helicidae)	Artichokes	*Cynara scolymus*	Azores
	Cauliflower	*Brassica oleracea* var. *botrytis*	Germany
	Kohlrabi	*Brassica oleracea* var. *gongylodes*	Greece
	Lettuce	*Lactuca sativa*	Holland
	Turnips	*Brassica campestris*	Italy
Helix aspersa Müller (Helicidae)	Leaves	*Tropaeolum nasturtium*	Argentinia
	Flowers		Australia
	Cauliflower	*Brassica oleracea* var. *botrytis*	Azores
	Beans	*Phaseolus vulgaris*	England
	Bromeliads		France
	Cacti	*Mammilaria* sp.	Germany
	Chestnuts	*Castanea* sp.	Greece
	Garlic	*Allium sativum*	Haiti
	Cabbage	*Brasscia oleracea* var. *capitata*	Holland
	Coriander	*Coriandrum sativum*	Ireland
	Herbs		Italy
	Leeks	*Allium ampeloprasum*	Libanon
	Pepper	*Capsicum* sp.	Mexico
	Plants	*Begonia* sp.	New Zealand
	Plants	*Rosmarinus* sp.	Peru
	Rose leaves, dried		Portugal
	Seeds	*Tilia americana*	Spain
	Cut flowers	*Dianthus* sp.	
	Spinach	*Spinacia oleracea*	
	Sweet oranges	*Citrus sinensis*	
	Tomatoes	*Lycopersicon esculentum*	
	Onions	*Allium cepa*	
Helix lucorum L. (Helicidae)	Plants		Syria
Helix pomatia L. (Helicidae)	Ivy	*Hedera helix*	France
	Cabbage	*Brassica oleracea* var. *capitata*	Germany
	Lettuce	*Lactuca sativa*	Hong-Kong
			Italy
			Yugoslavia
Lamellaxis clavulinus (Potiez & Michaud) (Subulinidae)	Papaya	*Carica papaya*	Hawaii
			Okinawa

Table 44 (continued)

Species (family)	Plants		Country of origin
Lamellaxis micra (Orbigny) (Subulinidae)	Leaves Orchids	*Chamaedorea* sp.	Mexico Nicaragua Samos
Lehmannia marginata (Müller) (Limacidae)	Flowers Bromeliads Chinese cabbage Leeks Orchids Orchids Plants Lettuce Celery Flower bulbs	*Gardenia* sp. *Brassica pekinensis* *Allium ampeloprasum* *Cymbidium* sp. *Cypripedium* sp. *Platycerium grande* *Lactuca sativa* *Apium graveolens* var. *dulce* *Amaryllis* sp.	Australia Japan Mexico Venezuela
Lehmannia poirieri Mabille (Limacidae)	Orchids Plants		Costa Rica Guatemala
Limax flavus Linnaeus (Limacidae)	Lettuce	*Lactuca sativa*	Massachusetts (USA)
Meghimatium bilineatus (Benson) (Philomycidae)	Mushrooms Plants	 *Cycas* sp.	Hawaii Japan
Milax budapestensis (Hazay) (Milacidae)	Plants	*Rosmarinus* sp.	England
Milax gagates (Draparnaud) (Milacidae)	Begonias Leaves Cauliflower Cardoon Strawberry plants Fennel Green beans Cabbage Leeks Orchid plants Lettuce Celery Flower bulbs	 *Helianthus annuus* *Brassica oleracea* var. *botrytis* *Cynara cardunculus* *Fragaria sp.* *Foeniculum vulgare* *Phaseolus* sp. *Brassica oleracea* var. *capitata* *Allium ampeloprasum* *Lactuca sativa* *Apium graveolens* var. *dulce* *Amaryllis* sp.	Azores Chile Columbia France Greece Holland Italy Japan Mexico Norway Portugal Spain
Milax insularis Lessona & Pollonera (Milacidae)	Cabbage Celery	*Brassica oleracea* var. *capitata* *Apium graveolens* var. *dulce*	Italy
Monacha cantiana (Montagu) (Helicidae)	Artichokes Parsley Plants Plants Lettuce	*Cynara scolymus* *Petroselinum crispum* *Clematis viticella* *Gladiolus* sp. *Lactuca sativa*	France Italy Lebanon
Monacha cartusiana (Müller) (Helicidae)	Potatoes Cabbage Sweet peas Lettuce Seeds	*Solanum tuberosum* *Brassica oleracea* var. *capitata* *Lathyrus sativus* *Lactuca sativa* Boxtree	England France Germany Greece Holland

Table 44 (continued)

Species (family)	Plants		Country of origin
	Seeds	*Clematis viticella*	Italy
	Seeds	*Tilia* sp.	Liberia
	Hollyhocks		Yugoslavia
	Shrubs	*Althea* sp.	
	Sweet oranges	*Citrus sinensis*	
	Wheat straw	*Triticum* sp.	
Monacha obstructa (Férussac) (Helicidae)	Apples	*Malus sylvestris*	France
	Leeks	*Allium ampeloprasum*	Greece
	Plants		Italy
			Lebanon
			Portugal
			Turkey
Monacha olivieri (Férussac) (Helicidae)	Cabbage	*Brassica oleracea* var. *capitata*	Germany
	Herbs		Israel
	Lettuce	*Lactuca sativa*	Italy
			Portugal
			Turkey
			Yugoslavia
Omalonyx unguis (Férussac) (Succineidae)	Fruit	*Citrus* sp.	Iran
Opeas pyrgula Schmacker & Boettger (Subulinidae)	Bananas	*Musa* sp.	Guam
	Plants		Japan
	Plants		Lebanon
			Philippines
			Vietnam
Otala lactea (Müller) (Helicidae)	Garlic	*Allium sativum*	Holland
			Italy
Otala vermiculata (Müller) (Helicidae)	Artichokes	*Cynara scolymus*	Europe
	Leaves	*Mentha* sp.	
	Flowers		
	Cauliflower	*Brassica oleracea* var. *botrytis*	
	Fennel	*Foeniculum vulgare*	
	Pines, young plants		
	Leeks	*Allium ampeloprasum*	
	Parsley	*Petroselinum crispum*	
	Lettuce	*Lactuca sativa*	
	Cut flowers	*Lilium* sp.	
Oxychilus alliarius (Miller) (Zonitidae)	Watercress	*Nasturtium officinale*	Belgium
	Orchids	*Dendrobium* sp.	England
	Plants	*Aloe* sp.	France
	Plants	*Mentha* sp.	Germany
	Plants	*Hydrangia* sp.	Japan
	Sphagnum moss	with *Vinca* sp.	Mexico
			Portugal
Oxychilus cellarius (Müller) (Zonitidae)	Ferns	*Fern* sp.	England
	Plants	*Mentha* sp.	Canada
	Primroses	*Primula* sp.	Italy
	Orchids		Portugal

Table 44 (continued)

Species (family)	Plants		Country of origin
Pallifera costaricensis (Morch)	Leaves	*Chamaedorea* sp.	Guatemala
(Philomycidae)	Bromeliads		Honduras
	Orchids	Catasetum sp.	Mexico
			Nicaragua
Pomacea dolioides Reeve	Water plants	*Echinodorus* sp.	Brazil
(Pilidae)			
Rumina decollata (L.)	Leaves	*Mentha* sp.	Mexico
(Subulinidae)	Cacti		Turkey
	Orchids		
	Plum-seedlings	*Prunus* sp.	
Subulina octona (Bruguière)	Leaves	*Chamaedorea* sp.	Bahamas
(Subulinidae)	Leaves	*Tillandsia* sp.	Brazil
	Bromeliads		Columbia
	Cacti		Ecuador
	Orchids		Ethiopia
	Palm seeds		Guatemala
	Roots	*Dion spinulosum*	Honduras
			Mexico
			Nicaragua
			Peru
			Philippines
			Trinidad
Theba pisana (Müller)	Leaves, dried		Africa
(Helicidae)	Flowers, dried		Australia
	Flowers	*Bellis* sp.	Azores
	Flowers	*Chrysanthemum* sp.	Bermuda
	Flowers	*Gladiolus* sp.	Canary
	Flowers	*Lilium* sp.	Islands
	Flowers	*Strelitzia reginae*	Crete
	Cauliflower	*Brassica oleracea* var. *botrytis*	Cyprus
	Watercress	*Nasturtium officinale*	Denmark
	Chicory	*Cichorium intybus*	England
	Carrots	*Daucus carota* var. *sativa*	France
	Cabbage	*Brassica oleracea* var. *capitata*	Germany
	Herbs		Greece
	Leeks	*Allium ampeloprasum*	Holland
	Lentils	*Lens culinaris*	Israel
	Carnations	*Dianthus* sp.	Italy
	Parsley	*Petroselinum crispum*	Lebanon
	Pepper	*Capsicum* sp.	Libya
	Plants	*Amaryllis* sp.	Monaco
	Plants	*Calla* sp.	Morocco
	Plants	*Citrus* sp.	Poland
	Plants	*Origanum vulgare*	Portugal
	Roses		Spain
	Beetroot	*Beta vulgaris* var. *rubra*	South Africa
	Lettuce	*Lactuca sativa*	Turkey
	Seeds	*Althea* sp.	United Arab.
	Seeds	*Euphorbia terracina*	Republic
	Seeds	*Gossypium* sp.	

Table 44 (ontinued)

Species (family)	Plants		Country of origin
	Seeds	*Mirabilis jalapa*	
	Celery	*Apium graveolens* var. *dulce*	
	Asparagus	*Asparagus officinalis*	
	Spinach	*Spinacia oleracea*	
	Leaf stalks	*Rosmarinus* sp.	
	Sweet oranges	*Citrus sinensis*	
	Grapevines	*Vitis* sp.	
	Cedars	*Cedrus libani*	
Trichia hispida (L.) (Helicidae)	Plants		Bermuda England France Greece Holland Ireland Italy Norway Portugal
Trichia striolata (Pfeiffer) (Helicidae)	Plants		France Italy
Vaginulus linguaeformis Semper (Veronicellidae)	Various plants		Peru
Vaginulus occidentalis (Guilding) (Veronicellidae)	Pineapple Bromeliads Croton plants Fig trees Coriander Orchids Lettuce Celery	*Ananas comosus* *Codiaeum* sp. *Ficus* sp. *Coriandrum sativum* *Lactuca sativa* *Apium graveolens* var. *dulce*	Equador Guatemala Honduras Mexico Panama
Vaginulus olivaceus (Stearns) (Veronicellidae)	Pineapples Bromeliads Orchids	*Ananas comosus*	Costa Rica Honduras Nicaragua
Vaginulus pulcher Colosi (Veronicellidae)	Bromeliads		El Salvador
Veronicella kraussi (Férussac) (Veronicellidae)	Breadfruit tree Cabbage	*Artocarpus altilis* *Brassica oleracea* var. *capitata*	Dominican Republic Puerto Rica
Veronicella moreleti Crosse & Fisher (Veronicellidae)	Leaves Bromeliads Orchids	*Chamaedorea* sp.	Equador Honduras Mexico
Veronicella sloanii (Cuvier) (Veronicellidae)	Bromeliads Orchids		Jamaica
Veronicella stohli Martens (Veronicellidae)	Arazee	*Nephytis* sp.	Honduras
Zenobiella incarnata (Müller) (Helicidae)	Beech	*Fagus* sp.	France

(Ehrenberg), *Otala lactea* (Müller), *Otala vermiculata* (Müller), *Rumina decollata* (L.), *Theba pisana* (Müller) (also as food), *Vaginulus alte* Fér.

Postal packages: *Cochlicella barbara* (L.), *Helix aperta* Born, *Helix aspersa* Müller, *Helix pomatia* L., *Otala vermiculata* (Müller), *Theba pisana* (Müller).

Soil, with or without plants (tubers, onions, roots of *Amaryllis* sp., *Bougainvillia* sp., *Cartus* sp., *Chrysanthemum* sp., *Convallaria majalis*, *Echeveria* sp., *Gardenia* sp., *Hyazinthus* sp., young plants of pines and cedars, *Mentha* sp., *Phaseolus* sp., vines.): *Achatina fulica* Bowdich, *Cecilioides aperta* (Swainson), *Cepaea nemoralis* (L.), *Cochlicella barbara* (L.), *Cochlicella conoidea* (Drap.), *Cochlicella ventricosa* (Drap.), *Cernuella virgata* (DaCosta), *Helicella neglecta* (Drap.), *Helix aspersa* Müller, *Lamellaxis gracilis* (Hutton), *Lamellaxis micra* (Orb.), *Monacha cartusiana* (Müller), *Monacha olivieri* (Fér.), *Monacha unguis* (Fér.), *Opeas pumilum* (Pfeiffer), *Opeas pyrgula* Schmacker and Boettger, *Otala vermiculata* (Müller), *Oxychilus alliarius* (Miller), *Rumina decollata* (L.), *Subulina octona* (Bruguière), *Theba pisana* (Müller).

Cargo (for the single year: 1. July 1968–30. June 1969): *Achatina fulica* Bowdich, *Candidula gigaxii* (C. Pfeiffer), *Cochlicella barbara* (L.), *Cochlicella conoidea* (Drap.), *Cochlicella ventricosa* (Drap.), *Helicella amanda* (Rossmaessler), *Helicella apicina* (Lam.), *Helicella caperata* (Montagu), *Helicella cespitum* Drap., *Helicella cretica* (Fér.), *Helicella conspurcata* (Drap.), *Helicella derbentina* (Kryn.), *Helicella joppensis* (Roth), *Helicella krynickii* (Andrz.), *Helicella langloisiana* (Bourguignat), *Helicella littoralis* (Mousson), *Helicella maritima* (Drap.), *Helicella meda* (Porro), *Helicella neglecta* (Drap.), *Helicella protea* (Ziegler), *Helicella striata* (Müller), *Helicella trochoides* (Poiret), *Helicella variabilis* (Drap.), *Helicella virgata* (DaCosta), *Helix aperta* Born, *Helix aspersa* Müller, *Lejeania darnaudi* (L. Pfeiffer), *Monacha cartusiana* (Müller), *Monacha obstructa* (Fér.), *Monacha syriaca* (Ehrenberg), *Otala lactea* (Müller), *Otala vermiculata* (Müller), *Theba pisana* (Müller), *Vaginulus alte* Fér. The same species are not always involved. The following were added to the list only in 1971/72: *Bradybaena despecta* (Gray), *Bradybaena serotina* A. Adams, *Bradybaena similaris* (Fér.), *Bradybaena tourannensis* (Souleyet), *Cepaea nemoralis* (L.), *Helicella barcinensis* (Bourguignat), *Helicella candidula* (Studer)*, *Helicella itala* (L.), *Helicella pyramidata* (Drap.), *Monacha olivieri* (Fér.), *Subulina octona* (Bruguière).

Canada has also been affected by the passive dispersal of gastropods from overseas and from the United States. WILKINSON (1964) lists the following as species which have established themselves there after being introduced from Europe: *Arion ater* (L.), *Arion circumscriptus* Johnston, *Deroceras reticulatum* (Müller), *Limax maximus* L., *Milax gagates* (Drap.) and *Prophysaon andersoni* (Cooper). *Limax maximus* L. is called the Great Slug of Europe. As in the USA, there is a list of Intercepted Plant Pests (Canadian Plant Quarantine, Canada Department of Agriculture), which shows that during 1963 to 1971 pest gastropods were imported in smaller numbers than were insect pests. The extent and importance of the problem of quarantine control of pulmonates is very much greater in the USA.

Table 45 lists the species introduced into Canada with both their place of origin and of discovery (also luggage and packaging).

---

* *Candidula unifasciata* (Poiret)

Table 45. A list of gastropod pests introduced to Canada with imported goods

Species (family)	Location	Country of origin
Achatina fulica Bowdich (Achatinidae)	Luggage	Hawaii
Arion ater (L.) (Arionidae)	*Acer* sp. Nursery material *Berberis* sp. *Laburnum* sp. *Lonicera* sp. Packaging *Picea* sp. *Potentilla* sp. *Rhododendron* sp. *Rosa* sp. *Spiraea* sp.	Belgium Florida (USA) Holland
Arion circumscriptus Johnston (Arionidae)	Garden plants Greenhouse plants	Columbia Europe Ontario (Canada)
Arion fasciatus (Nilsson) (Arionidae)	Herbs *Rosa* sp.	Belgium Holland Oregon (USA)
Arion hortensis Férussac (Arionidae)	*Rosa* sp.	USA
Arion intermedius Normand (Arionidae)	Herbs	Oregon (USA)
Arion subfuscus (Draparnaud) (Arionidae)	*Lonicera* sp.	Belgium
Cepaea hortensis (Müller) (Helicidae)	Nursery material	Holland
Cepaea nemoralis (L.) (Helicidae)	*Ligustrum* sp. Packaging *Sphagnum* sp.	Holland
Deroceras laeve (Müller) (Limacidae)	Herbs	Oregon (USA)
Deroceras reticulatum (Müller) (Limacidae)	Soil Plants	Europe
Helicella sp. (Helicidae)	*Thuja* sp.	Holland
Helicigona (Arianta) arbustorum (L.) (Helicidae)	*Berberis* sp. Herbs Packaging	Belgium Holland Monaco Europe
Helix aperta Born (Helicidae)	Baggage	Italy
Helix aspersa Müller (Helicidae)	*Cactus* sp. *Chrysanthemum* sp. *Citrus* sp. *Erica* sp. *Hibiscus* sp. *Jucca* sp. *Juniperus* sp. *Pelargonium* sp. *Prunus*- plums Cut flowers Ornamental plants	California (USA)
Helix pomatia L. (Helicidae)	Greenhouse plants	France

Table 45 (continued)

Species (family)	Location	Country of origin
Lehmannia marginata (Müller) (Limacidae)	*Acer* sp. *Dianthus* sp.-carnations	Belgium Italy
Limax flavus L. (Limacidae)	*Citrus* sp. *Rhododendron* sp. *Strelitzia* sp.	Ala (USA) Belgium Florida (USA)
Limax maximus L. (Limacidae)	*Malus* sp.-apple *Hydrangea* sp. *Ligustrum* sp. *Rhododendron* sp.	Belgium Holland Minnesota (USA) Washington (USA)
Limax tenellus Müller (Limacidae)	*Rhododendron* sp.	Belgium
Milax gagates (Draparnaud) (Milacidae)	Soil Plants	Europe
Prophysaon andersoni (Cooper) (Arionidae)	*Citrus* sp.	California (USA)
Theba pisana (Müller) (Helicidae)	*Cactus* sp. Herbs Plants Cut flowers	Malta Monaco
Veronicella floridana Leidy (Veronicellidae)	Soil Rubber tree	Costa Rica Florida (USA)

Germany is involved in the spread of gastropods to the United States, as emerges from Table 44; this has occurred mainly in the last 10 years. In the American lists of introductions for 1963/64 none came from Germany, but in 1967/68 four species [*Helicigona arbustorum* (L.), *Deroceras reticulatum* (Müller), *Helicella candidula* (Studer) and *Helix aspersa* Müller] and in 1970/71 as many as eleven species are listed: *Arion circumscriptus* Johnston, *Cepaea hortensis* (Müller), *Cepaea nemoralis* (L.), *Helicella candidula* (Studer), *Helicella caparata* (Montagu), *Helicella striata* (Müller), *Helix aperta* Born, *Helix aspersa* Müller and *Theba pisana* (Müller). In 1971/72 the following were recorded for the first time: *Arion hortensis* Fér., *Helix pomatia* L., *Monacha cartusiana* (Müller) and *Oxychilus alliarius* (Miller).

Movement of gastropods occurs in the other direction as well. Europe and, in particular, the Federal Republic of Germany is "blessed" with pest gastropod species that are not indigenous. As early as 1938 the danger of the introduction of gastropods with banana imports became acute. The small Bark Snail, *Zonitoides arboreus* (Say), was introduced from North America and found in the Botanical Gardens in Berlin (Plate and Frömming 1953). The snail *Vallonia pulchella* (Müller), now widespread in greenhouses, first caused considerable damage to grass seed growing around 1940. Between 1930 and 1934, *Otala vermiculata* (Müller) was repeatedly found in vegetable imports from Italy. *Opeas mauritianum* (Pfeiffer), which comes from the island of Mauritius, has increased considerably in numbers in greenhouses since 1959; it was first found in the Palm Garden in Frankfurt-am-Main and in greenhouses in the Netherlands and has subsequently spread (Plate and FRÖMMING 1953). The Greenhouse Slug *Milax gagates* (Drap.) has been known to occur in the Federal Republic of Germany since 1963 (KUHNA and SCHNELL 1963), having previously been introduced into Belgium (JAECKEL and PLATE 1967) and England (DOBSON 1963).

The name *Milax gagates* (Drap.) is often used for two different species, namely, *Milax gagates* (Drap.) and *Milax insularis* Lessona and Pollonera. *M. insularis* originates from the Mediterranean area, while *M. gagates* comes from northwestern Europe, and now occurs in England, Scotland, Ireland, France, Holland and Belgium – to some extent passively spread – and always more abundant in coastal areas than in the interior (BOYCOTT 1934; HARTOG and SANDEE 1962; QUICK 1947, 1961; SCHOUTEN 1949 and others). *M. gagates* is today also found in Australia, Canada, the USA, the Canary Islands, Hawaii and Tristan da Cunha (FOX 1962; KARLIN and BACON 1961; KARLIN and NAEGELE 1958). *Milax*,either *gagates* or *insularis*, has been introduced into North and South Africa, New Zealand, Tasmania, Polynesia, Bermuda and South America (BENTHEM-JUTTING 1962; HOFFMANN 1935).

The slug *Boettgerilla pallens* Simroth first appeared in Germany in 1959 and in Poland in 1960 (CLAUSS 1964; PLATE 1965; SCHMID 1962; WIKTOR 1960). The Mediterranean White Snail, *Theba pisana* (Müller), came from the Mediterranean region, and was introduced into US bases in Germany after the Second World War with goods shipped from Morocco and Spain. KNORRE (1969) found *Deroceras rodnae* Grossu and Lupu, which comes from Hungary and Rumania, in Thüringen.

Further examples of passive dispersal to European countries include: *Theba pisana* (Müller), introduced to England from Morocco (pers. comm. G. E. SCHMIDT, US Navy in London, 1963); the spread to Sweden around 1960 of *Limax flavus* L. and *Limax maximus* L. from Denmark, and of *Oxychilus cellarius* (Müller) and *O. draparnaudi* (Beck) from other European countries (NILSSON 1962). *Milax gagates* (Drap.) reached Finland from Holland with ornamental plants (tulip and narcissus bulbs), being first found in greenhouses (VALOVIRTA 1967, 1969), as did *Hyalina lucida* Drap. some time before 1948 (SUNDMAN 1947/48). Finally *Deroceras (Hydrolimax) melanocephalus* (Kaleniczenko) has spread to Rumania from the Caucasus (GROSSU and LUPU 1957). *Deroceras caruanae* (Pollonera), formerly restricted to the west coast of Britain, is found now in eastern England as well. *Milax budapestensis* (Hazay) while still uncommon in France, can be expected to spread. The freshwater snail *Lymnaea stagnalis* (L.) spread in about 1970 from France to Spain (ALVAREZ 1972).

Many European species, such as *Arion fasciatus* (Nilsson), *Arion ater* (L.) and *Deroceras reticulatum* (Müller), have established themselves in America, but, according to RUNHAM and HUNTER (1970), none of the American species has been successful in Europe.

With the expansion of world-wide traffic through import and export of food products, plants and other goods, and not least of all as a result of the steady increase in tourism, ever more gastropods are passively dispersed, and the likelihood of infestation is greater for many countries. Thus the USA quarantine controls found *Theba pisana* (Müller) 163 times in 1963, but 201 times in 1966; *Candidula caperata* (Montagu) was found 173 times in 1965 and twice as often (248) just a year later (1966).

The detrimental effects of the introduction of new gastropod species are, first of all, an increase in the number of pest species in the country concerned: this may result in increased damage in agriculture and horticulture and, associated with this, an increase in the risk of economic losses. Secondly, species which normally remain below the threshold number in population density in their country of origin, and are thus harmless there, may become pests in a newly settled area if the plants and the climate are suitable. This is often characterized by an explosive increase in numbers, for the predators and parasites which normally keep the pests in check do not occur in the new area. Thirdly, the new species may, if climate and available food are favourable, cause considerable upset of the ecological balance of the host country. Thus in South Africa, in the area around Cape Town, the indigenous forest molluscs are disappearing, being displaced by other species accidentally introduced from Europe; the latter lack natural predators and are thus able to increase their numbers to an unusual degree.

3.4.2 Regulations and Control

The most important aspects of official inspection can be illustrated by the example of the United States of America. Pest species are listed under Molluscs in the Rules and Regulations of Plant Pest Control Branch and Plant Quarantine Branch in Part 324, Section 7, Chapter III of the Code of Federal Regulations (US Department of Agriculture, Agriculture Research Service, Washington). The following details from the revised version of December 31, 1955 show the great importance accorded to pest gastropods. In the Federal Republic of Germany this is not the case, which is to some extent inexcusable, for with continuous favourable weather conditions for gastropods and in greenhouses, non-indigenous gastropods can easily establish themselves, the more so since the continuing expansion of world traffic and transport (tourism, imports) increases the danger of new species being introduced. The US regulation states under Means of Transport: "every type of ship, motor vehicle or aircraft;" further, under Regulated Article: "every type of product, baggage, salvaged war material or other goods" and in both cases with the clear entry "which arrive in any part of the USA whatsoever." The possibility remains, however, of obtaining legal authorization to import certain molluscs.

The following are listed as **quarantine snails** of especial importance: *Achatina fulica* Bowdich, *Helix aperta* Born, *Helix aspersa* Müller, *Helix pomatia* L., *Cochlicella* spp., *Helicella* spp., *Monacha* spp., *Bradybaena* spp. [in particular *B. similaris* (Fér.)], *Otala lactea* (Müller), *Otala vermiculata* (Müller), *Theba pisana* (Müller). It further stipulates, without naming any species, "every species of slug which is as destructive to plants as *Achatina*". In the revised versions of September 1974 and October 1975, *Veronicella moreletti* Crosse and Fisher is specifically mentioned as a new pest.

The Plant Quarantine Treatment Manual 1971 contains administrative regulations and also dosages for fumigation of containers and ships to control gastropods; the mollucicides recommended are methyl bromide, or a mixture comprising 10% ethylene oxide and 90% carbon dioxide (carboxide). The duration of treatment depends on the fumigant used and on the temperature; it may be brief (several hours) or prolonged (24 to 72 h). If the container is covered with a gastight tarpaulin, 2 h suffice, in particular for destruction of the slug *Veronicella moreleti* Crosse and Fisher.

Fumigation with **methyl bromide** at a concentration of 130 g/m^3 for 72 h was as a rule satisfactory and its use is allowed for household goods, packing cases and similar objects. RICHARDSON and ROTH (1963, 1965) recommend the **carboxide** fumigation mixture for elimination of *Theba pisana* (Müller) and *Cochlicella barbara* (L.). Used instead of methyl bromide at an application rate of 320 g/m^3 and 72-h fumigation, it was highly effective, especially against the more resistant *Cochlicella* species, such as *C. barbara* (L.), *C. conoidea* (Drap.) and *C. ventricosa* (Drap.). It is also effective against passive, aestivating snails, the control of which is more difficult than that of active animals. Fumigation under high pressure is more effective than under normal pressure; carboxide causes 100% mortality at a concentration of 400 g/m^3 and 72 h treatment, even at low temperatures (around 10 °C). EVERSOLE (1971) recommends fumigation of shipments while they are still in Europe, that is, before they are dispatched to the USA; the reason for this is the possibility of gastropods, particularly *Theba pisana* (Müller), being introduced from US naval field stations around the Mediterranean. Washing or steam-cleaning all vehicles in these bases is inadequate and also uneconomic. In experiments conducted by EVERSOLE, cold treatment of the cargo, for instance, one or several hours at 10 °F (−12.2 °C) or 0 °F (−17.8 °C), has proved to be extraordinarily effective, but mortality was higher amongst the young than amongst the older snails.

MEAD (1961) considers that fumigation of cargo with ethylene bromide is more effective against *Achatina fulica* Bowdich than the use of methyl bromide, cyanide or carbon disulphide; 10 years later, ROTH (1971) recommended the mixture of ethylene and carbon dioxide (carboxide), specified in the legal requirements mentioned above. PETERSON (1957) reports very successful control measures on the island of Guam with bait pellets of methaldehyde (1.6%) and calcium arsenate (5%).

Young *Helicella* snails [*H. candidula* (Studer), *H. conspurcata* (Drap.)], introduced with rosemary seeds, can be killed with methyl bromide at a concentration of 48 mg/l and at reduced pressure (100 mm mercury) for 3 h (ROTH and KENNEDY 1973). The germinating capacity of the seeds is not interfered with, even at an application rate of 128 mg/l and reduced pressure for 16 h, or 240 mg/l and normal atmospheric pressure for 24 h. Sulphuryl fluoride at 48 mg/l for 72 h also proved effective.

3.5 Mechanical and Ecological Control

Apart from molluscicides, mechanical procedures can be employed to control pest gastropods, or care can be taken to ensure that the methods of cultivation do not create an optimal habitat for gastropods – this includes careful preparation of seed or plant beds. Such measures, combined with protection of gastropod predators, are of especial importance.

3.5.1 Mechanical Control

Hand collection with subsequent squashing of the slugs and snails is the oldest mechanical method and was the only control measure used up to the middle ages. Common salt was then employed; this causes an over-production of mucus by gastropods. Later, traps and protective barriers were used.

The following objects served as **traps:** old planks, upturned flower pots, tiles, rhubarb or cabbage leaves, flat stones, damp towels or sacks, and also drain pipes; these had to be searched for slugs or snails in the early morning. Such dark shelters are especially effective as traps when they are baited with raw, fresh, damp potato slices, or fresh lettuce leaves. It is necessary, however, to collect the gastropods daily and to kill them with a strong solution of common salt or in boiling water; if dropped into containers with warm or cold water, they simply crawl out again. Baiting in the evening with raw potato slices and collecting in the morning is recommended as an effective measure for use in greenhouses against *Zonitoides arboreus* (Say) and *Deroceras laeve* (Müller). It is not advisable to use the collected slugs as food for poultry because they are often infested with parasites which in turn infect the birds. Collecting slugs by hand, which is only possible in a small area such as a garden or greenhouse, is today unsatisfctory as a practical measure.

An electrical trapping device has been used against *Helix aspersa* Müller in citrus plantations (CARMAN 1965).

Protective barriers of dehydrating substances such as cattle salt, caustic soda, kainite or completely dry quick lime can act as a barrier. Thus small fields or beds can be kept free of Field Slugs by mounds of kainite. For these slugs, a ridge of finely ground lime around seedbeds or valuable plants represents an almost insurmountable barrier. These protective strips must be renewed frequently, since moist air alone is sufficient to slake the lime and make it ineffective. Barriers of unoiled calcium cyanamide half to 1 m wide have also proved to be effective. A mixture of pulverized copper sulphate and kainite dust in a proportion 1:20 was also found to be successful; it remains effective as a barrier longer than lime and thus does not need to be renewed so often.

TOMASZEWSKI (1949) lists the following as substances which can be used to form a barrier: sodium carbonate, sodium and potassium chloride, sodium nitrate, superphosphate, iron (II) and copper (II) sulphate. Soot, ashes, alum, sawdust, plaster of Paris, barley chaff, chopped straw, pine needles and fine sand can all be used to form ridges. The barrier must on all accounts remain dry, since as soon as it becomes wet through rain, watering or heavy dew, the slugs simply crawl over it. Germination boxes and valuable pot plants should be placed on inverted flowerpots which have been covered with cotton wool. Flowers of orchids can be wound around with cottonwool at the base of the peduncle; this forms an effective barrier as long as it is completely dry.

Some of these procedures, which were frequently used in the past, are today either not used at all, or to a limited extent only, because they are labour-intensive. With the pressure for rationalization in agriculture and horticulture, they are no longer suitable, particularly if repeated application be required.

The following are recommended as more long lasting **mechanical barriers:** smooth copper or zinc sheets (0.8 mm thick) and also zinc slabs; the copper sheets

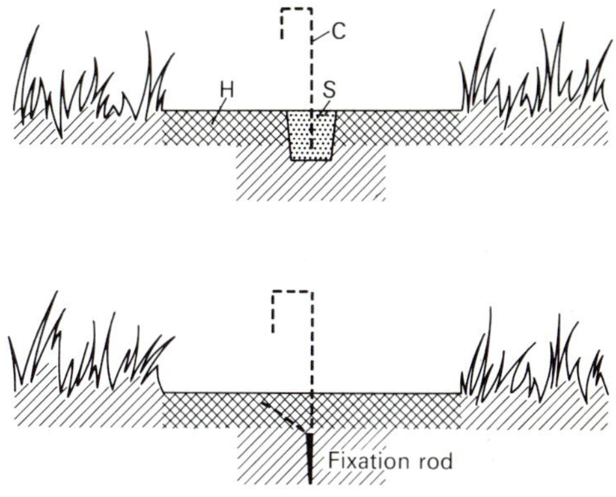

Fig. 103. Mechanical barriers against terrestrial gastropods. *C* red copper screen; *S* sand layer for the copper screen; *H* denuded area (with a total herbicide, fixation rod)

must reach a height of 5 cm above the soil surface, the zinc 20 to 25 cm. The upper edge of the barrier is bent through a right angle twice (Fig. 103). To ensure that slugs do not reach the protected area through the soil beneath the plate, it should be buried to a depth of at least 30 cm below the soil surface. In rainy weather, however, the protective effect diminishes because the metal is covered with a film of moisture which enables slugs to crawl over it. Moens et al. (1967) estimated that 5% managed to get across, and these must be eliminated in some other way. A material with a rough surface is unsuitable for use as a barrier, since the animals can crawl over it right from the start. Metal barriers have proved effective against the slugs *Arion rufus* (L.), *A. hortensis* Fér., *A. circumscriptus* Johnston, *Deroceras reticulatum* (Müller), *Limax maximus* L. and *Milax gagates* (Drap.) and against the snails *Cepaea nemoralis* (L.) and *Helix aspersa* Müller. These climb only a few centimeters up a metal plate, raise the front part of the body and start to crawl down sideways. One condition for the effectiveness of the barrier is the complete absence on either side of vegetation, which would enable the gastropods to surmount the barrier.

Insurmountable barriers cannot be made from plates of glass, polyethylene or sheets of roofing felt, as slugs and snails simply crawl over them. To save labour, protective strips and barriers are today used only to protect valuable plants, although, according to Mead (1961) and van Dinther (1973), they are still frequently used in the tropics to protect trees, shrubs and valuable crops against *Achatina fulica* Bowdich and *Strophocheilus oblongus* (Müller); these barriers consist of fences constructed without gaps from bamboo sticks, wooden poles, wire meshing or from corrugated iron or zinc sheets placed one beside the other.

3.5.2 Ecological Control

Reduction of a gastropod population can be brought about by measures which alter environmental conditions unfavourably for snails and slugs; these include keeping agricultural and horticultural land free of weeds, and in creating an ecological buffer zone by keeping a belt free or all plant refuse and in which all possibility of shelter is removed (van Dinther 1973). A new problem is, however, raised, one which is of particular importance in the tropics, namely the gradual development of erosion and loss of soil fertility; this has happened in inhabited and farmed coral islands which have only a thin covering of humus (MEAD 1961).

Soil cultivation is a very important aspect of ecological control (HUNTER 1967; STEPHENSON 1975). Gastropods are very dependant on availability of adequate shelter, which affords protection from both desiccation and predators. Moderately clumpy fields with large clods afford ideal protection and thus provide conditions leading to especially heavy gastropod damage. A classic example of the way in which inadequate working of the soil can favour gastropod infestation is seen in the reclaimed land of the polders in the Netherlands (WOUTERS 1970). After it has been drained and before being handed over to the tennant, the soil is roughly ploughed, sown with winter wheat and left undisturbed for some time, to avoid the development on the surface of a layer of impervious mud before the soil matures. Most of the wheat seeds remain exposed unless they happen to fall between the earth clods and they are almost all destroyed by Field Slugs.

By contrast, careful tillage of the soil before sowing or planting in the spring resulting in a seedbed with a completely even surface, rids the area of gastropods (GLENDENNING 1952). According to MALLET and BOUGARAN (cited by GODAN 1970), a seedbed of horticultural standard can considerably reduce infestation of rape, because many molluscs are mechanically destroyed by the tools used in the process. The method of rape cultivation recommended at present eliminates ploughing and the seeds are thinly sown; this results in widely spaced plants, and considerably increases the effective damage by slugs. On the other hand, reduction of the size of the clods by intensive working of the soil, dense sowing of the seed and treatment with molluscicidal pellets can keep gastropod infestation in check.

As a phytosanitary precaution, a field should be cleared of all remains of vegetation after harvesting and should also be kept clean during the growing period. Gardens, fields and hedges with long grass, tall weeds, scrub and other plants increase the chances of survival of gastropods and of their eggs; such places are therefore usually more heavily infested than clean areas. On the other hand, it should be appreciated that total removal of plant detritus means that many gastropods which would normally feed on decaying vegetable matter are forced to exploit living plant material instead, damaging the crops as a result (BÖHM 1957).

Burning over is a quick method of clearing land of pests *before* planting, and this is often done in tropical areas. In this way large numbers of *Theba pisana* (Müller) have been destroyed in South Africa (JOUBERT and WALTERS 1951) and California (GAMMON 1943), and of *Achatina fulica* Bowdich in Africa and in Guam (MEAD 1961); this is especially effective during aestivation. Today, however, the use of burning is to be discouraged because of the damage it causes to the environment.

In many parts of the marsh lands of the Federal Republic of Germany there are heavy losses of germinating winter wheat; this is especially the case if, after harvesting winter rape, a luxuriant inter-crop develops from fallen seeds, and after it has been ploughed in, a coarse seedbed is prepared. Slugs destroy cereal seeds and shoots while they are still in the ground and before the seedling has pushed its way above the soil surface. To limit such damage to the germinating winter cereal, the inter-crop should be carefully observed in areas where gastropod infestation is likely to occur, and if damage develops, the inter-crop should immediately be ploughed into the soil. In such areas, broadcasting application of bait pellets once the seeds are sown is not an effective means of gastropod control, since slugs seldom emerge from the soil during or after sowing, particularly during dry weather (Pflanzenschutz 1974/75).

According to RUNHAM and HUNTER (1970), the most effective means of ecological control, particularly of *Deroceras reticulatum* (Müller), *Arion hortensis* Fér. and *Milax budapestensis* (Hazay), is tilling the soil with a rotary plough. In this way populations of Field Slugs and Garden Slugs were reduced to a third or a quarter of their original density. Since *M. budapestensis* (Hazay) lives at depths in the soil not reached by the plough, its numbers were not as markedly reduced; further, its relatively tough skin affords a certain amount of protection against mechanical destruction.

Cracks and spaces in the soil facilitate dispersal of slugs, and enable them to reach drilled seeds more easily. Simply compacting the soil reduces the extent of slug damage, particularly to winter wheat. This is evident in many wheat fields where there is a satisfactory density of plants at the perifery, which often has a more even and compacted surface after ploughing than the centre of the field where the plants are more sparse as a result of the activity of Field Slugs. The effect is also seen in the reduced slug damage in the vicinity of trailer tracks (BREDERLOW 1975).

Furthermore, improvement of soil structure results in exposure of the sensitive eggs and juvenile slugs to direct sunlight and they die of desiccation.

RUNHAM and HUNTER (1970) write that there are two major disadvantages in giving extra cultivations to fields at risk from slug damage. First, cultivations are expensive; it is frequently cheaper to treat with chemicals for slug control than to give an extra pass with a cultivating implement. Secondly on heavy land where slugs are commonest, too many cultivations can lead to over-compaction of the soil and reduce plant root growth. In any case, it is frequently impossible for machinery to travel on heavy land during wet weather.

BREDERLOW (1975) described the significance of cultivation methods in the ecological control of gastropods as follows: "The only method which can be recommended to give even limited protection to cereal seeds in areas threatened by infestation is soil cultivation" ... "Thus inevitably the greatest importance will be attached to the method which most limits the movements of slugs in the soil, yet which also increases the rate of germination and the early stages of development of the cereal. This can be achieved only with careful soil cultivation. One of the most important conditions is that the soil should settle well. Thus special importance should be attached to making a moderately deep seed furrow, at the latest two to three weeks before the seed is sown". The seedbed should be as fine and crumbly as possible. "Spaces in the soil must, as a matter of priority be removed as far as possible by rolling once or several times." Not only is movement of slugs thus limited, but conditions for rapid growth of the young plants are improved in this way. "In favourable weather conditions growth of a cereal crop can be further increased by sowing the seed on well moistened ground and at a depth no greater than two to three cm."

Theba pisana (Müller) can be taken as an example to illustrate the importance of ecological control of snails. It makes its appearance in lucern fields of southeastern France later than slugs and can, according to CAIRASCHI and LECOMTE (1973), be kept below the threshold value by thorough mechanical preparation of the seedbed. Clearing of the edges of fields and irrigation ditches after the harvest can reduce the level of infestation, but does not eliminate it altogether, so that chemical control is nevertheless necessary.

Chemical control of freshwater snails which live in fields which are flooded or which have canals or ditches running through them is inevitably accompanied by such ecological control measures as clearing the canals and ditches of waterplants which serve as food and provide shelter for snails (AL TAHAFI 1973; DESCHIENS 1972; HAIRSTON and SANTOS 1961; LEVINE 1970). Extension of irrigation schemes as a means of increasing the area of arable land, such as those being undertaken, for example, in Africa, encourages the spread of trematode vectors in an alarming manner (MCCULLOUGH 1972); this means that vector control by ecological measures will become more important in future. Bilharzia is the most important endemic disease of man caused by a trematode. Even before completion of the Aswan dam in Egypt it was feared that the disease would spread in the wake of irrigation of newly-won arable land and newly developed rice fields (AYAD 1966); this in fact happened and has required a tremendous effort to control it.

3.6 Biological Control

Terrestrial snails are very difficult to control chemically, using metaldehyde preparations, particularly in a continuously damp environment or during rainy weather; consideration has therefore been given to the possibility of keeping snail populations below the threshold values by biological means (BERG 1973, 1976; FRANZ and KRIEG 1976; HAIRSTON et al. 1975). WRIGHT (1968) has emphasized the difficulties encountered here, since the favourable results obtained by using parasites and predators in the laboratory frequently cannot be repeated in the field.

In the following section the possibilities are discussed of using such environmentally harmless methods of control of molluscs, whether by disease, parasites, predators or by genetical or biotechnical procedures.

3.6.1 Agents of Disease (Viruses, Bacteria, Fungi)

Diseases and their agents are not likely to become important as the basis for intensive biological control measures.

MEAD (1956, 1957, 1979) reported on a disease of unknown genesis in the African Giant Snail (*Achatina fulica* Bowdich). It was highly infectious on contact and was found to be chronic in several snail populations. The symptoms were seen first in the optic tentacles as a proliferation and "cornification" of the skin tissue, thickening and shortening of the tentacles and later the onset of cell disintegration in other organs and parts of the body. 60% of a large sample of snails were found to be infected, and in the part of Ceylon where it occured, there was evidence of a marked decline in the numbers of *Achatina*. No further attention has apparently been paid to this type of disease in molluscs, although His (1970) described a disease of the gills in oysters that seriously limits respiratory exchange.

3.6.1.1 Viruses

Very little is known about the effects of viruses on molluscs. This is a most neglected area of research. Mead (1961) expressed regret that up to 1960 not a single investigation had been undertaken. In the meantime the results of an investigation of viruses in the snails *Zonitoides* sp. and *Murex* sp. has been published by Signoret (1968).

3.6.1.2 Bacteria

Mead (1961) has summarized reports on bacteria in molluscs up to 1960; those which have appeared since then are discussed here. Mead mentions bacteria of the genera *Escherichia*, *Alcaligenes* and *Bacillus;* in France an *Aerobacter* infection occurred on a *Helix* farm. Pan (1956) found an acid-resistant *Bacillus* in various organs of *Biomphalaria glabrata* (Say) and fungal spores or yeast-like microorganisms in nervous tissue of this snail. Spirochaetes live in the hepatopancreas of South American *Galba* species. Dartevelle (1954) found a *Borrelia*-like bacterium with which 10% of the snails of *Achatina* sp. in the Congo were infected. In the laboratory the bacterium *Bacillus pinottii* Cruz and Dias has proved to be extraordinarily pathogenic to *Biomphalaria glabrata* (Say), and has consequently been recommended by Dias and Dawood (1955) for biological control. In field experiments in Egypt this *Bacillus* destroyed all the snails in the ditches to which it was introduced; the addition of pepton as a nutritional supplement is said to increase its rate of multiplication. However, these favourable results were not confirmed by Wright (1968).

Cushing et al. (1971) injected a culture of a Gram-negative, rod-bacterium killed with formalin, into the heart of the marine snail *Haliotis rufescens* Swainson and other *Haliotis* species. These all showed thereafter a high titre of anti-bacterial activity of their haemolymph, indicating that a certain degree of immunity had developed; this would be a serious limitation to the use of bacteria for control purposes. Phagocytosis is an important aspect of the immune system of molluscs (Cheng et al. 1969); after injection of the Butter Clam, *Crassostrea virginica* Gmelin with *Bacillus mycoides* Flügge; Feng et al. (1971) found a more marked granulation of leucocytes than in untreated oysters ($72.19 \pm 4.71\%$ compared with $37.18 \pm 4.48\%$). Payne et al. (1972) discovered a bacterial lysing agent in the alimentary canal of *Scrobicularia plana* (DaCosta). Bayne and Kime (1970) attempted to establish whether in *Helix pomatia* L. there is an immune response to bacterial invasion. There was initially a rapid fall in the haemocyte count, but this rose again, mainly in the region around the point of entry of the injecting needle. From their results the authors concluded, however, that there is in *H. pomatia* no immune mechanism comparable with that of the oligochaetes, crustaceans and insects. Molluscs are, however, capable of eliminating micro-organisms, as Vasconcelos and Lee (1972) established in the Pacific Oyster.

A final conclusion on the presence of an immune mechanism of molluscs has yet to be reached, for the results of other investigations have indicated that it does occur (DAVIS 1972; KUHN and LASNIK 1968; PIETRZAK et al. 1973).

There are bacterial agglutinating mechanisms in molluscs (PROKOP et al. 1968; UHLENBRUCK and PROKOP 1966). UHLENBRUCK et al. (1971) investigated polyvalent proteinase-iso-inhibitors from the albumen gland of *Helix* sp., *Cepaea* sp. and *Achatina* sp. These have the capacity to block several serine-proteinases of animals, plants and bacteria. One of these inhibitors is probably identical with the anti-A-agglutinin extracted from the albumen gland of *Helix aspersa* Müller by ISHIYAMA et al. (1972) and of *Helix pomatia* L. by UHLENBRUCK et al. (1972); anti-A-agglutinin consists of many sub-units (GERBAL et al. 1973). The inhibitory property of the protease inhibitors and the bacterial-agglutinating mechanism are of great importance for the survival of gastropods and their eggs; the authors even considered the possibility of using them therapeutically in man. TSCHESCHE and DIETL (1972) have also studied protease inhibitors from *Helix pomatia* L. and SPRENGER et al. (1972) those of *Helix aspersa* Müller.

For the sake of completeness, mention should be made of *Pseudomonas*, which is regarded as the cause of a yellow fluorescence in the mucus of *Helix aspersa* Müller and also of a harmless species in California (BAUM and RAWLS 1972; RAWLS et al. 1973); it is also found in mussels (PRIEUR 1974). These pigment-forming bacteria live in the mucus on the body surface, in the sub-epithelial mucous glands, the body openings, the fluid of the haemocoel and the reproductive tract (YATES and RAWLS 1972).

3.6.1.3 Fungi

Fungi attack mainly the eggs of gastropods, for example, those of *Deroceras reticulatum* (Müller). MEAD (1961) mentions *Verticillium chlamydosporium* Goddard, but this fungus is not suitable for use in biological control, despite a high infection rate of the eggs. ARIAS and CROWELL (1963) found a fungus of the genus *Arthrobotrys* in *Arion circumscriptus* Johnston.

While fungal infections are important when rearing molluscs in the laboratory, since they may destroy the whole stock, in the field, fungi seldom cause total elimination or even a reduction in gastropod numbers.

3.6.2 Parasites

PELSENEER published in 1928 a comprehensive account of the parasites of molluscs and of molluscan parasites. Formerly, attention was directed to marine and freshwater species (mussels, snails, octopuses). Indeed, as early as 1,200 B.C. the cercariae of *Schistosoma* sp. were discovered in freshwater snails in Egypt.

Ten of the 46 species of invertebrates known to be parasites of molluscs are lethal to 14 host species (STEPHENSON 1968). Some are already important in biological control, others may become so, but have not yet been investigated. The most important of these parasites include protozoa, flat worms, lung worms, carabid beetles, the glowworm larvae of lampyrid beetles as well as the larvae of Sciomyzidae (Diptera).

3.6.2.1 Protozoa

Protozoa are associated with molluscs both parasitically and symbiotically. MEAD (1961) lists the following parasitic species: the ciliate *Balantidium haughwouti* De Leon in freshwater snails in the Philippines; the flagellate *Cryptobia isidorae* (Fantham) and an amoeba in the seminal receptacle of the freshwater snail *Bulinus tropicus* (Krauss); *Trichodina achatinae* in the seminal receptacle of the terrestrial *Achatina zebra* Sganzin. The sporozoan *Klossia* sp. has been found in the kidney in various slugs and snails, and an amoeba is found as an endoparasite in *Arion rufus* (L.). LOWELL (1961) found an *Acanthamoeba* sp. in molluscs. GRASSÉ (1968) lists ciliates, flagellates, coccidians and others. The flagellates *Cryptobia helicis* Leidy and *Trichomonas limacis* Dujardin are found in agriculturally important terrestrial pulmonates. The sporozoan *Klossia helicina* Balbiani *(Coccidiomorpha)* which is found in the kidney of *Helicigona arbustorum* (L.), *Cepaea nemoralis* (L.), *Cepaea hortensis* (Müller), *Bradybaena fruticum* (Müller), *Trichia hispida* (L.), *Succinea putris* (L.) and the species *Klossia loossi* Nabih which occurs in the kidney of *Limax* sp. and *Arion* sp. The coccidian *Isospora rara* Ai. Schneider is found in the hepatopancreas of *Limax* sp. and the microsporidian *Plistiphora husseyi* Mich. is also mentioned. STEPHENSON and KNUTSON (1966) give a further list with the relevant sources. According to LEVINE (1970) the haplosporidians, in particular a species found by BARROW in 1965 *Minchinia pickfordae* (Barrow), are suitable for use in biological control of freshwater gastropods.

Most of the host-specific protozoa are microsporidians; for example, *Plistiphora husseyi* Mich. is host-specific for *Physa* spp. *Physa cubensis* Pfeiffer, the vector of *Fasciola hepatica* L. in Cuba, is highly sensitive to this parasite (MICHELSON 1963).

It is interesting to note that infestation with a protozoan parasite often reduces the susceptibility of the gastropod host to a succeeding trematode infestation. Amoebae can reduce the rate of infestation of *Bulinus* vectors by *Schistosoma*.

The following section, together with Table 46, summarizes information on the protozoan parasites of terrestrial gastropods which are reported in the literature.

Slugs of importance in plant protection are host to a number of ciliates, which, according to BROOKS (1968) could be used for biological control. *Tetrahymena limacis* (Warren) (24.2 µm to 39.6 µm in diameter and 33.0 µm to 62.7 µm in length) and also *Tetrahymena rostrata* (Kahl) (33.0 µm to 57.2 µm in diameter and 50.5 µm to 72.6 µm in length) are parasites of Field Slugs (Fig. 104). These ciliates are distributed throughout the world (Europe, USA, South Africa, New Zealand) and infest the kidney, hepatopancreas and albumen gland of snails and slugs and also mussels. *Tetrahymena limacis* (Warren) has been found in 19 different species and *Tetrahymena rostrata* (Kahl) in 9. BROOKS has published a comprehensive account of the biology of these ciliates as parasites of *Deroceras reticulatum* (Müller), their main host, with a description of the techniques of experimental infection. The parasites influence longevity and fecundity of the slugs. Slugs which are heavily infested with ciliates lay no more eggs. This is thus a case of **parasitic castration.** The way in which this is brought about is not known. Simultaneous infection with both ciliate species seldom occurs, and was observed by Brooks in only a few Field Slugs, and in three Greenhouse Slugs [*Milax gagates* (Drap.)].

Table 46. Parasitic protozoans in terrestrial gastropods

Species	Host	
	Snail	Slug
Amoeba		
Acanthamoeba sp.		*Arion fasciatus* (Nilsson)
Amoeba sp.		*Arion rufus* (Linnaeus)
Flagellata		
Cryptobia helicis Leidy (*Trypanoplasma*)	*Bradybaena fruticum* (Müller) *Cepaea hortensis* (Müller) *Cepaea nemoralis* (L.) *Helix pomatia* L. (in Receptaculum seminis) *Helicigona arbustorum* (L.) *Succinea putris* (L.) *Trichia hispida* (L.)	
Trichomonas limacis Dujardin	Various species, including *Cernuella virgata* (Da Costa) *Helicella variabilis* (Draparnaud) *Otala lactea* (Müller)	*Limax* spp.
Sporozoa		
Isopora incerta Schneider		*Limax cinereoniger* Wolf
Isopora rara Aimé Schneider		*Limax* spp.
Klossia helicina Balbiani	*Bradybaena fruticum* (Müller) *Cepaea hortensis* (Müller) *Cepaea nemoralis* (L.) *Helicigona arbustorum* (L.) *Succinea* spp. *Trichia hispida* (L.)	
Klossia loossi Nabih	*Helix*-, *Cepaea*- und *Succinea*-species	*Arion* spp. *Limax* spp.
Pfeifferinella impudica Leger & Hollande		*Lehmannia marginata* (Müller)
Plistiphora husseyi Michaud	*Achatina zebra* Sganzin	
Trichodina achatinae	*Achatina zebra* Sganzin	
Ciliata		
Colpoda aspersa Kahl		*Deroceras agreste* (L.)
Colpoda steini Maupas		*Deroceras agreste* (L.) *Deroceras reticulatum* (Müller) *Lehmannia marginata* (Müller)
Concophthirus steenstrupi (Stein)	*Helix* spp.	*Arion ater* (L.) *Deroceras agreste* (L.)
Semitricholina sphaeronuclea	*Zonitidae*	

Table 46. (continued)

Species	Host	
	Snail	Slug
Tetrahymena limacis (Warren)	*Trichia lubomirskii* Slósarski	*Arion hortensis* Férussac *Deroceras reticulatum* (Müller) *Deroceras laeve* (Müller) *Lehmannia marginata* (Müller) *Limax flavus* L. *Limax maximus* L. *Milax gagates* (Draparnaud) *Prophysaon andersoni* (Cooper)
Tetrahymena pyriformis (Ehrenberg)		*Deroceras agreste* (L.) *Deroceras reticulatum* (Müller)
Tetrahymena rostrata (Kahl) (= *Paraglaucoma rostrata*)	*Zonitoides nitidus* (Müller)	*Arion intermedius* Normand *Deroceras reticulatum* (Müller) *Deroceras laeve* (Müller) *Lehmannia marginata* (Müller) *Milax gagates* (Draparnaud)

Table 47 lists those pest slugs which have been found in the field to be infected by these two species of ciliate.

Other hosts of *Tetrahymena limacis* (Warren) are the slugs *Prophysaon andersoni* (Cooper) (KOZLOFF 1956), *Milax budapestensis* (Hazay) (CORLISS et al. 1962) and the terrestrial snail *Trichia lubomirskii* Slós. The snail *Zonitoides nitidus* (Müller) is parasitized by *Tetrahymena rostrata* (Kahl) (KAZUBSKI 1958).

The incidence of infection in *Deroceras reticulatum* (Müller) is high, as the following percentages of Field Slugs collected in the vicinity of Berkley show: in the months from August to December, 36.1% to 61.3% were infested with *Tetrahymena limacis* (Warren), in February the figure was 51.6%, while for the other months of the year 1964 averages of 12.5% to 36.1% were measured. The slugs collected in 2 successive years showed low infectivity only during June and July, these being 8.8% in 1964 and 3.3% in 1965. The percentage of slugs found to be infected with *Tetrahymena rostrata* (Kahl) were as follows: in December and March, 1965, 5.16% and 63.3% respectively; from April to November between 18.2% and 38.7%; in January and February, 1966, 47.0% and 41.2% respectively.

Tetrahymena rostrata (Kahl) was isolated from the soil in which the slugs live, showing that contamination occurs through the soil. Slugs are probably infected by contact with cysts in the soil during the winter. It is also possible that the parasites are transfered by way of the eggs, since 10% of the eggs laid by *Deroceras reticulatum* (Müller) were found to be infested with ciliates (Fig. 105). Figure 106 shows schematically the method of transfer of *T. rostrata* to *Deroceras reticulatum* (Müller) by way of the eggs of the slug. BROOKS (1968) found in 114 eggs laid in the field by ciliate-infested Field Slugs the following frequency (%) of egg development:

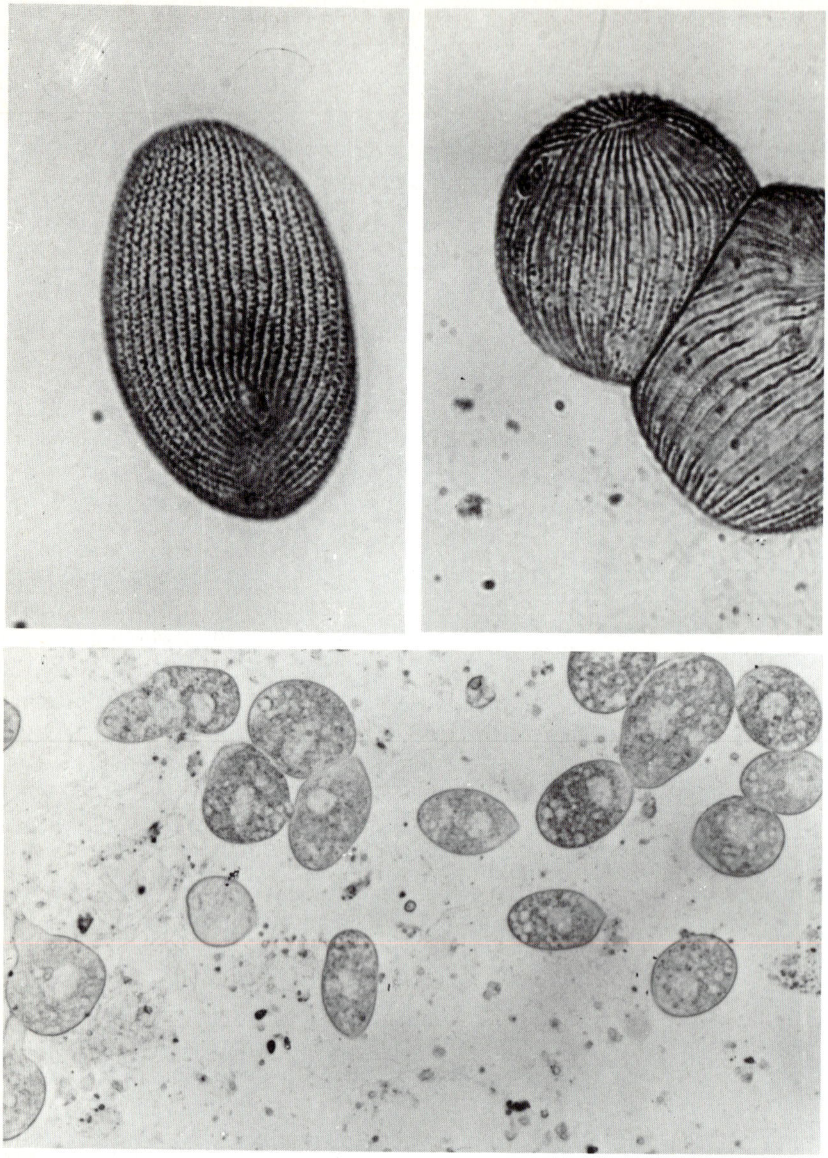

Fig. 104. Ciliates from *Deroceras reticulatum*. *Above left* Klein's dry-fixed silver nitrate preparation of the parasitic phase of *Tetrahymena limacis*. *Above right* Klein's dry-fixed silver nitrate preparation of the parasitic phase of *Tetrahymena rostrata*. *Below* Parasitic phase of *Tetrahymena rostrata* from the kidney of an experimentally infected slug

Table 47. Host record and incidence of *Tetrahymena limacis* and *Tetrahymena rostrata* in slugs other than *Deroceras reticulatum*. (After BROOKS 1968, modified)

Species	Number of Slugs		
	Collected	Infected with	
		T. limacis	*T. rostrata*
Arion circumscriptus Johnston	32	0	0
Arion hortensis Férussac	65	2	0
Arion intermedius Normand	29	0	4
Deroceras laeve (Müller)	57	24	3
Lehmannia valentiana (Férussac)	20	16	0
Limax flavus L.	1	1	0
Limax maximus L.	3	3	0
Milax gagates (Draparnaud)	47	32	3[a]
Prophysaon andersoni (Cooper)	7	1	0

[a] Three of the 12 specimens collected from San Francisco in May 1965 had mixed infections

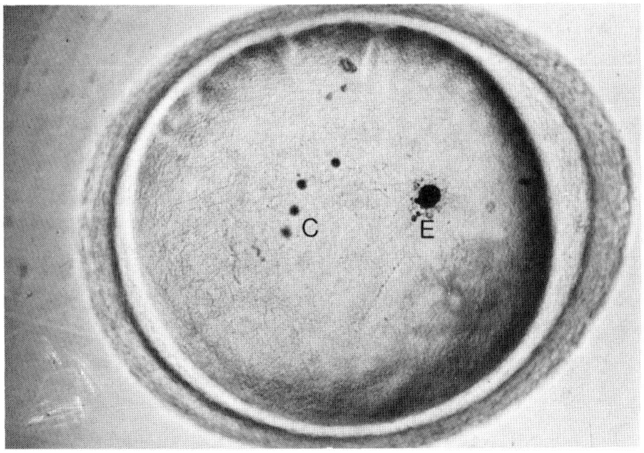

Fig. 105. Ciliate-infected egg with intact remains of a non-viable embryo. *C* ciliates; *E* embryo

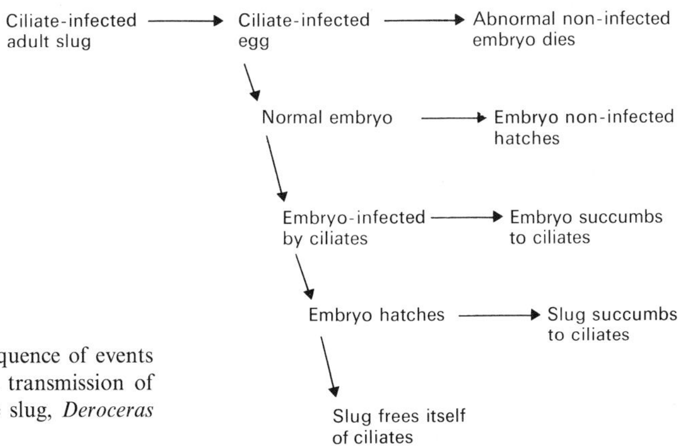

Fig. 106. Diagram of the sequence of events occurring in the transovum transmission of *Tetrahymena rostrata* in the slug, *Deroceras reticulatum*

A. Eggs with
51.1% abnormal embryos, which succumb to the ciliate infection
43.8% normal embryos.
B. Eggs with
3.5% embryos which escape ciliate infestation
40.3% embryos becomming infested by ciliates
C. Eggs with
14.9% infested embryos, which die as a result
25.4% infested embryos, which hatch.
D. Slugs of which
7.0% died as a result of ciliate infestation
18.4% free themselves of ciliates.

Tetrahymena limacis (Warren) is, on the other hand, only seldom found free-living and does not survive long outside the body of its host. Transmission through the soil or by way of the eggs does not occur, for in 987 eggs of *Deroceras reticulatum* (Müller) examined, this ciliate was never found.

Tetrahymena rostrata (Kahl) does not enter the host egg after it has been laid, but penetrates it during its formation in the genital tract of the slug; a large number of the parasites are found in the albumen gland, which of all the organs of the body, is the most heavily infested. The ciliate passes with the secretions of the albumen gland in the egg, which may contain as many as 30, but usually only 5 or 6 of the parasites. The development of the embryo is adversely affected by the presence of the parasite, even though at first it simply coexists with the embryo in the nutritional fluid in the egg. The effect of the parasite is reflected in the fact that 56.1% of the infested eggs developed abnormally, although the parasite was not in the body of the embryo. The latter becomes infested with ciliates at a later stage in its development, apparently through the mouth of the embryo with the uptake of nutritional fluid.

Field Slugs infested with *Tetrahymena rostrata* (Kahl) show the following symptoms: mantle shield swollen, posterior end of the body somewhat elongated and laterally compressed; after death the body shortened, and a tumor-like swelling of the mantle shield (Fig. 107). The histological changes show that *T. rostrata* is highly pathogenic and can easily destroy a slug population. Kozloff (1957) found this protozoan in the kidney of *Deroceras reticulatum* (Müller) in 20% of the specimens examined.

Transmission of parasite by way of the egg occurs only rarely in molluscs. The only other example is that of the nematode *Cosmocercoides dukae* (Holl.) in *Deroceras gracilis* Rafinesque and *Deroceras reticulatum* (Müller), and possibly also the snail *Zonitoides arboreus* (Say) (Anderson 1960). The adult worm lives in the mantle cavity and the genital tract of its host; transmission is effected by the worm penetrating the mantle cavity of the embryo.

From the literature it can be concluded that the following protozoa would be suitable and effective for use in biological control of pest gastropods: the ciliate *Tetrahymena rostrata* (Kahl) – Field Slugs [*Deroceras reticulatum* (Müller)]; the microsporidian *Plistiphora husseyi* Mich. – *Achatina* sp., *Physa* sp.; the haplosporidian *Minchinia pickfordae* (Barrow) – freshwater snails.

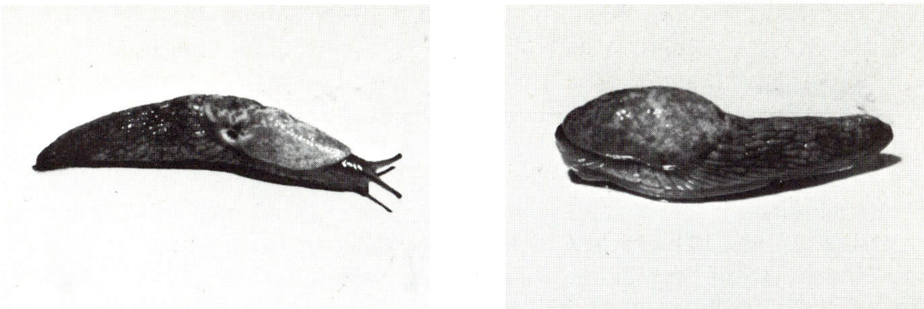

Fig. 107. Healthy and diseased slugs experimentally infected with *Tetrahymena rostrata*. *Left* healthy slug extended in travel; *right* resting slug showing first noticeable signs of infection. Midportion of the slug body covered by mantle is swollen, and tail is slightly elongated and constricted; anterior end (*left*) and posterior end (*right*)

It is pleasing that Field Slugs, the control of which is of especial importance in plant protection, might be kept in check by a highly pathogenic protozoan; more attention should be directed to this parasite, *Tetrahymena rostrata* (Kahl), and its use for the development of a biological control method considered.

3.6.2.2 Helminths

It has long been known that freshwater snails serve as the intermediate hosts for worm parasites of man and other vertebrates (Arvy 1972; Brown and Lemma 1970; Frömming 1932, 1955; Malek and Cheng 1974; Thomas 1973; Vala 1973; Voelker 1975 and others). But it is only in more recent times that terrestrial gastropods have been thoroughly investigated in this respect, and it now appears that there is no species of gastropod which is not infected with some species of worm, either in the larval or adult stage.

In Italy, for example, there are 14 snail species, both terrestrial and freshwater, which are vectors of parasitic worms of man. These include, according to Pampligione and Toffoletto (1971): *Candidula unifasciata* (Poiret), *Helicella itala* (L.), *Euomphalia strigella* (Drap.), *Cochlicella acuta* (Müller), *Cochlicopa lubrica* (Müller), *Abida frumentum* (Drap.), *Zebrina detrita* (Müller), *Galba truncatula* (Müller), *Lymnaea stagnalis* (L.), *Radix peregra* (Müller), *Planorbarius corneus* (L.), *Bulinus truncatus* (Audouin), *Bithynia leachi* (Sheppard). The five trematodes for which these snails act as intermediate hosts are *Schistosoma bovis* (Sonsino), *Fasciola hepatica* L., *Dicrocoelium dendriticum* (Rudolphi) Looss, *Opisthorchis felineus* (Rivolta) Blanchard and *Bilharziella polonica* Kow.. Many of the vectors are also regarded as agricultural and horticultural pests, another reason for paying closer attention to these parasitic helminths, the more so since some of them might be agents for biological control.

Most of the worms which parasitize or prey on molluscs are classified as **Platyhelminthes** or **Nemathelminthes.** Reviews of their importance as parasites or predators to gastropods are given by Chevallier (1973), Grassé (1968), Michelson (1957) and Stephenson and Knutson (1966). The turbellarian platyhelminth *Geoplana septemlineata* (Hyman) is a predator of terrestrial gastropods in Hawaii, mainly of *Achatina fulica* Bowdich and *Deroceras reticulatum* (Müller). This turbellarian grows to a length of 30 to 40 mm and is exceptionally voracious; as many as 51 worms have been found on a single specimen of *Achatina*, their total length being over 2 m. *Geoplana* is a pathogenic agent for meningoencephalitis in man, and consequently may not be used for the destruction of the gastropod pests. Those trematodes which are dreaded as the cause of bilharzia in man and of liver fluke disease in domestic animals and sometimes also of man, are also patyhelminthes, as are the cestodes (tapeworms). The nematodes belong to the Nemathelminthes, while the Annelida (segmented worms) include the Oligochaeta and Hirudinea (leeches). Many of these worms are important to molluscs as parasites or predators. The oligochaete *Chaetogaster limnaei* van Baer lives in the kidney of *Lymnaea stagnalis* (L.), *Radix peregra* var. *ovata* (Drap.), *Physa fontinalis* (L.), and *Physa heterostropha* (Say), up to 100 specimens being found in one snail. MICHELSON (1964) found that this parasite confers on *Biomphalaria glabrata* (Say) a certain immunity from infection with the miracidia of *Schistosoma mansoni* Sambon.

Another class of annelids, the Hirudinea (leeches), includes *Helobdella fusca* (Bosc) which preys preferentially on *Biomphalaria glabrata* (Say) and is quite capable of eliminating an entire population. It has been recommended for biological control of this vector of *Schistosoma mansoni* Sambon (CHERNIN et al. 1956). Other gastropod predators amongst the leeches are *Helobdella punctata-lineata* Moore (MCANNALLY and MOORE 1966), *Limnobdella mexicana* R. Blanchard, which feeds on young molluscs in its early stages, and *Glossosiphonia weberi* Blanchard which is found in Hawaii and attacks species of *Lymnaea* and *Physa* to such an extent that it might well be used in the control of these intermediate hosts (SIMMONDS 1969).

Worm parasites are found in the body cavity the mantle cavity (usually encysted), in the intestine, and in important visceral organs such as kidney, albumen gland and genital tract, more rarely in the heart and pericardium and, according to the most recent research, also in the brain (NASSI 1980).

The following worms occur as *adults* in molluscs: the **trematode** *Proctoeces subtenuis* Linton is found in the kidney of the mussel *Scrobicularia plana* (Da Costa) on the English coast with an infection rate of four to five, and even up to 14, worms per mussel (1969/70), (WHITE 1972). This example shows the infestation of a host with parasitic worms generally can be very considerable. Another example of an adult parasite worm is *Cotylurus flabelliformis* (Faust) found in the hepatopancreas of the freshwater snails *Lymnaea* sp., *Physa* sp. and *Planorbis* sp.. Adult nematodes (sometimes also found as larvae) are: *Angiostoma limacis* Dujardin in the foot muscle of *Deroceras reticulatum* (Müller), *Müllerius capillaris* (Müller)* in various Limacidae and Helicidae; *Ascaroides limacis* Barthélemy in the hepatopancreas of *Deroceras reticulatum* (Müller); *Oxyuris* and *Strongylus* spp. in egg batches of *Lehmannia marginata* (Müller) and *Limax maximus* L.; *Leptodera flexilis* (Dujardin) in the genital tract of *Limax* species and in the body cavity of *Deroceras agreste* (L.); *Rhabditis lambdiensis* (Maupas) in the mantle cavity of *Deroceras reticulatum* (Müller) and *Deroceras laeve* (Müller); and finally *Diplogaster* sp. and *Parogrolaimus* sp. in *Deroceras reticulatum* (Müller). The nematode *Daubaylia potomaca* Chit-

* *Müllerius capillaris* (Müller) today a synonym for *Muellerius minutissimus* (Megnin) (GRASSÉ 1965)

wood and Chitwood is very suitable for biological control of *Biomphalaria glabrata* (Say) and has already been introduced in Puerto Rico for this purpose (LEVINE 1970).

Molluscs harbour relatively few adult worms, whereas their role as the **intermediate hosts** for worm larvae is very great. In this case the molluscan vector is essential to allow the parasitic worm to complete its development to the point at which it is ready to infest either the final host or another intermediate host, which for some species of worms is another gastropod (ZAJÍČEK 1971).

As COMBES and CENS (1973) have stated, the gastropod hosts not only serve as a means whereby parasitic worms increase their numbers, but are also the some extent their means of dispersal and transmission. In rare cases, the whole life cycle of the parasite takes place within a mollusc. In the case of trematodes and cestodes, gastropods play the role of first intermediate host for the development of the larvae, but do not have a significant role in the actual transmission of the parasites to man or his domestic animals.

In Europe, 11 pathogenic worms, 6 trematodes, 2 cestodes and 3 nematodes are known to have freshwater and terrestrial gastropods as intermediate hosts, many of which are at the same time plant pests. The most important are the vectors of the Liver Flukes of cattle and sheep and the Lung Worms of grazing stock, the importance of which is increasing in France, Belgium, England and Germany (JOLIVET 1973; LEIMBACHER 1973; PITOIS 1973 and others). In countries outside Europe, bilharzia, caused by *Schistosoma* spp. remains one of the scourges of mankind, and its incidence is indeed on the increase as a result of recent large-scale irrigation schemes, since an optimal environment for the intermediate hosts is provided by the irrigation canals, their outlet furrows, ponds, flooded fields, lakes with weed and shrub-grown edges and even ponds which are only temporarily filled with water (DESCHIENS 1972).

Since trematodes are unquestionably the main parasites of gastropods (WRIGHT 1968, 1971), and since they are also the most dangerous worm parasites of man and animals, their life cycles will be briefly described.

Worm eggs *(Schistosoma, Fasciola)* reach the water with the faeces or urine of the definite host, or, in the case of lung worms, as a result of coughing or sneezing. Small, ciliated miracidia hatch from the eggs, swim about in search of an intermediate host, penetrate it, and then multiply through various larval stages – sporocysts, rediae, cercariae. As a rule, the miracidia are negatively geotactic and positively phototactic, as in *Schistosoma mansoni* Sambon; they thus reach the upper surface of the water, where the snails are also to be found. Positive geotaxis has been observed in the miracidia of one strain of *Schistosoma haematobium* (Bilharz), and it was then discovered that its intermediate host, *Bulinus truncatus* (Audouin), lives in deep water in the summer. It has already been mentioned that the miracidia are attracted to the snail by substances which it secretes.

After hatching from the sporocysts and rediae, the cercariae leave the host snail and, by means of a swimming tail, move rapidly through the water, when they make contact with a suitable vertebrate (man, mammals, fishes), they rapidly penetrate the skin (*Schistosoma*, *Sanguicola* in fishes). Apart from this direct method of infection, there is also an indirect way in which trematode parasites are ingested by the final host with food. In this case *(Fasciola)* the worm larvae develop in the snail and the cercariae are liberated as metacercariae. In some bird parasites, the larval stage which leaves the snail encysts in arthropods (second intermediate host) which serve as food for the birds. Other worm species require a second intermediate host before they reach the adult stage, may use another species of snail. An example is *Echinostoma murinum* (Tubangui): the first intermediate host is *Gyraulus convexiusculus* (Hutton) and the second, various freshwater snails (Joe 1967). Even a plant may replace the second host if, as in the case of *Fasciola* sp. and *Paramphistomum* sp. the cercariae climb up the grass stalks to their tips, there encyst and are then eaten by grazing stock. With some trematodes, the second intermediate host has to be eaten by a third, before the metacercariae can infect the final host. Once in the vertebrate, the worm larva wanders through the body of its host and then settles in a definite organ or in the vascular system, where it grows and develops into an adult trematode. The eggs pass to the exterior with the faeces or urine of the host.

The trematode *Leucochloridium paradoxum* Carus has an interesting life cycle. It is a parasite of various birds (song birds, crows and others) and has as intermediate host the Amber Snail, *Succinea putris* (L.) (Hecker 1965). Its development in the snail has been redescribed in detail by Rietschel (1970). It is astonishing how well the trematode is adapted to the feeding habits of its host, and by what cunning ways it ensures it is ingested by the bird. Usually the larval stages, cercariae or metacercariae, are taken up by the final host purely by chance. The sporo-

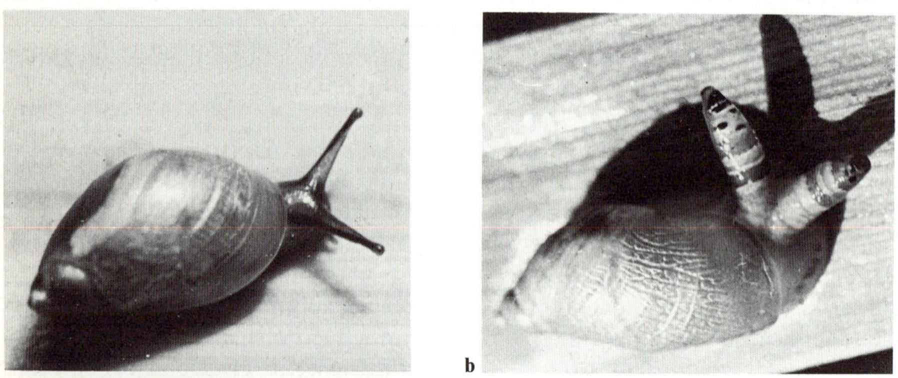
a b

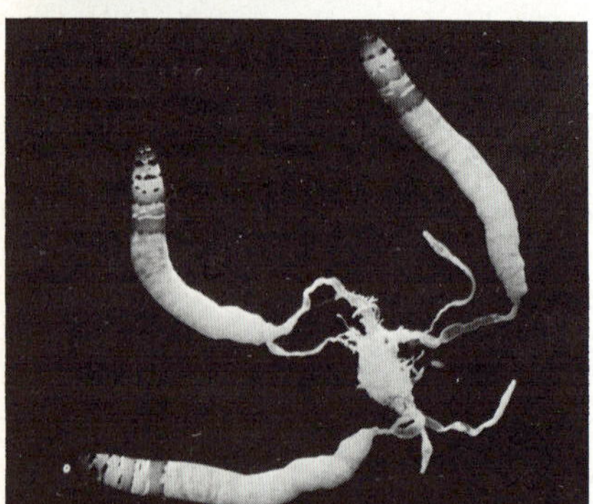
c

Fig. 108. *Leucochloridium paradoxum*: healthy *Succinea putris* (*above left*); snail with "tentacle maggots" (*above right*); sporocyst with tubes including worm larvae (*below left*) cysts of *Leucochloridium*, however, develop brightly coloured tubes which contain the infectious cercariae (Fig. 108). These tubes force their way into the tentacles of the snail and there pulsate regularly and remain obvious as strikingly coloured "maggots". The birds are attracted to these "maggots", tear them out of the tentacles of the snail and eat them. The adult worm lives in the cloaca of many birds; the eggs pass out with the bird's droppings and land on the feeding plants of the Amber Snail. The ciliated larva which hatches from the egg bores through the wall of the intestine of the snail and passes into the hepatopancreas, where it grows into a branched and ramified sporocyst. About 3 weeks after infection of the snail, the "maggots" make their appearance in the tentacles, one tube of which may contain up to 300 cercariae. New tubes are continuously formed in the tentacles. If a snail is not freed of them by birds, it may die of the infestation, or else the tentacle may burst and free itself of the cercariae, which then die in the open.

Ideal intermediate hosts for trematodes of the genera *Echinostoma*, *Schistosoma*, and *Amphistoma* are provided by the freshwater snails *Bulinus truncatus* (Audouin) and *Biomphalaria glabrata* (Say) and to a lesser extent *Physa acuta* Drap. The first named species is so sensitive that Chu et al. (1972) believe that echinostome trematodes could be used for its biological control, particularly since 19.3% of a snail population was found to be infested with these cercariae and there is evidence that infestation with this parasite persists throughout the year, whereas other trematodes *(Amphistoma)* infect snails only during part of the year. In field trials, *Echinostoma recurvatum* Linstow caused a marked reduction in a snail population. The final host of this trematode is as yet unknown, but is probably a rodent. The cercariae are repellent to other trematode larvae in the same snail. Likewise the metacercariae of *Cotylurus flabelliformis* (Faust) are unable to develop in the intermediate host (Lymnaeidae, Planorbidae and Physidae) if it is already infested with sporocysts of the same species. The problem of antagonism between trematodes in the intermediate host and the significance of this phenomenon for the control of schistosomiasis is discussed in Chapter 3.6.2.3.

The investigations of Brunnacker-Daur (1955) on seven species of intermediate host have shown that the degree of infection with trematode parasites varies with the season: in winter and early spring, it is very low, at 0.5% to 3.0%; it increases during spring and summer, with values of 20% and 30%, and is extremely high in autumn with 80%.

The intermediate hosts of the Large Liver Fluke *Fasciola hepatica* L. are freshwater snails, while those of the Small Liver Fluke *Dicrocoelium dendriticum* (Rudolphi) Looss [= *D. lanceolatum* (Rudolphi) Dujardin] are terrestrial snails. There is the risk of infection of people who eat molluscs with the trematode *Echinostoma* sp. while according to Dallibor (1976), eating uncooked shrimps could result in infection whith the Lung Fluke *Paragonimus* sp., for which shrimps act as the second intermediate host.

Olsen (1944) found that between 1940 and 1941, 37.5% of adult cattle and 6% of calves grazing on pastures in the coastal regions of Texas, Louisiana and Florida were infected with the Liver Fluke *Fasciola hepatica* L. Today the infection rate of grazing stock is still extremely high; in south western Colorada, north of New Mexico and in eastern Arizona 49% of sheep and 30% of cattle were found to be infested (Wilson and Samson 1971). In Northern Ireland, fascioliasis is a serious problem in sheep and cattle farming, for it can lead to a 50% mortality of stock if antihelminth drugs are not administered at a sufficiently early stage (MORPHY et al. 1969). Europe is plagued by this parasite from north to south, even in Italy and Portugal, and there is no country with meadows and pastures free of it. Several different freshwater snails can act as intermediate host for this trematode. PAMPIGLIONE and TOFFOLETTO (1971) list 14 species, of which *Galba palustris* (Müller) and *Galba truncatula* (Müller) are the most important. In Poland *Lymnaea stagnalis* (L.) has been recognized as another important intermediate host (FURMAGA and GUNDLACK 1968 and others). In North America (South Colorado) the snail *Stagnicola bulimoides techella* (Haldeman)* has been found to be the intermediate host for the Liver Fluke of grazing stock (WILSON and SAMSON 1971), while in Australia it is *Lymnaea tomentosa* (Pfeiffer) (LYNCH 1965 and others).

Despite the great economic importance of fascioliasis, the ecology and population dynamics of the intermediate host *Galba truncatula* (Müller) have been little studied. This snail is very mobile and, in favourable conditions, can cover a distance of up to 13 m/h. Below 11 °C it becomes immobile and oviposition ceases. HEPPLESTON (1972) studied the influence of weather on *Galba truncatula* (Müller) on the British Orkney Islands. In a constantly damp environment, temperature alone is important, but in changeable weather with occasional sunshine, activity is influenced by humidity and not temperature. According to NICE and WILSON (1974), growth rate of Liver Fluke larvae in the intermediate host is determined by temperature: at low temperatures, which frequently prevail in northern England and in Norway, only one life cycle per year can be completed, but at higher temperatures there may be several. Thus the weather, together with varying degrees of infection of the snails, determine the degree and also the extent of fascioliasis (OLLERENSHAW 1967, 1970).

Galba truncatula (Müller) also occurs in cress cultivations of France and Belgium, where it is not only a phytophageous pest, but is also responsible for transmission of Liver Flukes to man if cress, which may be covered with encysted cercariae, is eaten without being adequately washed. The importance of gastropod control in these cultivations has already been pointed out (RICOU 1966; RICOU and MARRE 1960; RICOU and PICARD 1972 and others).

The cattle Liver Fluke, *Fasciola gigantica* Cobbold, which may, as in Madagascar, also infest man, has as intermediate host *Radix natalensis* (Krauss) *(=Lymnaea n.)*. An optimal environment is afforded for this snail by flooded cultivations, rice fields, banana plantations, streams, and other water courses. Its population density is affected by rainfall, temperature and the collection of water (BOUCHET et al. 1973).

Helicella species, particularly *Helicella itala* (L.) are known to be intermediate hosts to the Small Liver Fluke **Dicrocoelium dendriticum** (Rudolphi) Looss in France, Britain and Germany; in the latter country *Zebrina detrita* (Müller) is another intermediate host of this fluke. The larvae are mainly found in the hepatopancreas of the snail. Ants (*Formica* sp.) play the role of second intermediate host. The cercariae are found in the sub-oesophageal ganglion of the ant, and bring about a change in its behaviour; when the metacercariae are mature, the ant bites

* *Stagnicola bulimoides* (Lea)

into the tip of a grass blade and remains often for days on end in this position, until eaten by the final host, cattle, sheep, as it grazes (HOHORST and GRAEFFE 1961).

Many trematodes of the genus **Schistosoma** live as parasites in the vascular system of man in the tropics and subtropics. *Schistosoma haematobium* (Bilharz) occurs mainly in blood vessels in the wall of the bladder, *Sch. mansoni* Sambon and *Sch. japonicum* Kats. in blood vessels of the mesenteries and intestinal wall. Their main intermediate hosts are *Bulinus* spp. in the case of *Sch. haematobium*, and *Biomphalaria* spp. in the case of *Sch. mansoni* and *Oncomelania* spp. in the case of *Sch. japonicum*. *Bulinus* spp. are limited in their distribution to the Near East, Africa and the southern Mediterranean areas of Europe. *Biomphalaria* spp. occur in Central and South America; of these *Biomphalaria straminea* (Dunker) is the most important. None of the Central and South American species are found in Africa, where south of the Sahara the most widely spread intermediate host of *Schistosoma mansoni* Sambon is *Biomphalaria pfeifferi* (Krauss). In Egypt the vector is *Biomphalaria alexandrina* (Ehrenberg), somewhat further south and extending to east Africa, *Biomphalaria sudanica* (Martens) and several other species of more local significance. *Schistosoma intercalatum* Fisher, is limited to the Congo and the Cameroons and has as its intermediate host *Bulinus forskali* (Ehrenberg). The extensive literature is to be found mainly in the journals of tropical medicine and parasitology, such as the American Journal of Tropical Medicine and Hygiene and the Bulletin of the World Health Organization. WIUM-ANDERSEN (1973) and others deal with identification of the snails.

Apart from the schistosomes which are dangerous to man, there are others which are parasitic in animals; one example is *Schistosoma bovis* (Sonsino), a parasite of grazing stock in Africa and also Southern Europe (Sardinia and Corsica) where schistosomiasis is endemic (COLUZZI et al. 1965). The distribution of the disease corresponds with the distribution of *Bulinus truncatus* (Audouin).

Lymnaea spp. and *Radix* spp. are the intermediate hosts of **Bilharziella** spp. These trematodes are generally harmless to man, with the exception of *Trichobilharzia* = *Bilharziella polonica* Kow. [intermediate host *Lymnaea stagnalis* var. *ocellata* (Brumpf), definitive host, the duck]. The cercariae of this species were the cause of dermatitis in bathers in Lake Como and Lake Maggiore (PAMPLIGIONE and TOFFOLETTO 1971).

Orientobilharzia occurs in the blood of domesticated hoofed animals in India, Asia and the Near East. The intermediate hosts are *Radix auricularia* (L.) and *Radix peregra* (Müller), which were more recently found in Lake Zurich (BURLA and SPEICH 1971); both species act also as intermediate hosts to *Plagiorchis (Multiglandularis) muris* Tanabe, the definitive hosts of which are *Microtus* spp. (ZAJÍČEK 1971). *Radix auricularia* (L.) has been introduced to the USA from Europe, and there has won a bad reputation as an intermediate host of trematode larvae, together with various *Physa* spp., *Galba palustris* (Müller) and *Gyraulus* sp.

The freshwater snail *Pomacea glauca* (L.) is found in the rice fields of Surinam, together with the fish *Lebistes reticulatus* Peters, which is introduced in order to control mosquitoes. The trematode **Guaicaipuria pseudoconcilia** (Nasir, Dias and Lemus de Guevara 1969) encysts in the gills of this fish while the snail acts as its intermediate host (NASIR and SILVA 1972). This example shows how complex the problems associated with biological control can be. The elimination of the snail is desirable both to protect the people working in the rice fields from trematode infestation and to protect the fish from infection with *Guaicaipuria* so that it can continue to perform the function for which it is introduced into the rice fields, namely keeping mosquitoes under control. It is possible to rid the fields of *Pomacea* by introducing the predatory snail *Marisa cornuarietis* (L.), but this snail is not entirely harmless to rice plants (SIMMONDS 1969). This problem is still to be discussed.

The trematode **Diplostomum spathaceum** (Rudolphi) can cause considerable economic losses in trout rearing. Its first intermediate host in the freshwater snail *Radix peregra* (Müller), and its second, the Rainbow Trout *Salmo gairdnerii* Richardson (CROSSLAND et al. 1971). The larvae invade the eyes of the fish and cause blindness. The definitive hosts are various species of gull.

The infestation rate of the intermediate hosts can be very high: 89.1% of a *Radix auricularia* (L.) population in North America, for example, was infected with trematodes. The survival of the snail species, which is constantly threatened by these parasites, is ensured by its high fecundity: OLSEN (1944) recorded *Stagnicola bulimoides techella* (Haldeman) as laying a total of 647 eggs during a period of 129 days and a life span of this species of 5 months.

Oviposition is, however, influenced by *weather conditions*. If these become unfavourable for *Biomphalaria glabrata* (Say) oviposition is suppressed, while the life span of the snail is lengthened. In optimal environmental conditions, the reproductive period lasts, however, only six months and the snail dies soon afterwards. As is to be expected, the development of the parasite larvae in the intermediate host also depends on temperature, since the body of the snail is always at the same temperature as the environment.

Finally, the parasite may be better adapted to the environment than its host, thus trematode miracidia seem to withstand flooding of an area with brackish water better than does *Biomphalaria glabrata* (Say) itself (CHERNIN and BOWER 1971).

ENIGK et al. (1957/58) have made a study of the intermediate hosts of a **cestode,** the short-segmented tapeworm of fowls, *Davainea proglottina* (Davaine). Both slugs and snails represent a serious danger to poultry which in free-range conditions show a preference for these gastropods as food. Field Slugs are the most important intermediate hosts; their infestation and mortality rates can be extremely high, with values for the population studied of 100% infection and 10% to 30% mortality. Infection of the slugs invariably followed feeding them proglottids (segments) of the tapeworm, and on dissection of several specimens of *Deroceras reticulatum* (Müller), up to 1,500 cysticercoids (larvae) were found in a single slug which had fed on only 3 to 5 worm segments. Heavy infestation leads to parasitic castration of the intermediate host. These three features, high infestation rate, relatively high mortality and parasitic castration, might make this cestode a possible agent for biological control of Field Slugs. The risk to fowls could be reduced to a minimum by fencing or caging, which is now largely the practice. Moreover, the agricultural land plagued by Field Slugs is not usually in the vicinity of poultry farms.

The **nematodes** are the third large group of worms which parasitise molluscs; they are mostly of interest to veterinary medicine, but in a few cases also affect man. Comprehensive reviews of snails as intermediate hosts of nematodes, particularly protostrongylids, have been published by KASSAI (1957) and MENGERT (1953);

they include many phytophagous gastropods. The most dangerous nematodes are the lung worms of grazing stock, some wild animals and poultry. A relatively large number of species, such as *Daubaylia potomaca* Chitwood and Chitwood, are already being used for biological control of the trematode vectors, mainly *Biomphalaria glabrata* (Say).

On pastures in Hungary the lung worms of sheep, *Protostrongylus* sp. *Cystocaulus ocreatus* Davtian and *Müllerius capillaris* (Müller), have as intermediate hosts *Lymnaea*, *Helix* and *Helicella* species, *Monacha cartusiana* (Müller) and *Cepaea vindobonensis* (Fér.) (KASSAI 1957). ROSE (1957) established the slugs *Deroceras reticulatum* (Müller), *Arion* spp., *Limax* spp., and the snails *Succinea* sp. and *Monacha* sp. as the chief intermediate hosts of the nematode *Müllerius capillaris* (Müller), the larvae of which live in the foot muscle of these gastropods. This lung worm could undoubtedly be used for biological control because it kills the intermediate host and the larvae are able to survive for a while after the death of the slug or snail. A single Field Slug has been known to harbour as many as 131 larvae in the foot muscle and the rate of infestation in a population may be 20% (STEPHENSON and KNUTSON 1966). In addition to the intermediate hosts of sheep lungworms, mainly *Müllerius capillaris* (Müller), listed in Table 48, *Deroceras agreste* var. *turkestianus* L. and *Deroceras transcaucasicum* Simroth have also been found to act as vectors. Furthermore, *Deroceras agreste* (L.) is an intermediate host of the lung worm *Cystocaulus ocreatus* Davtian (ROSE 1960); this slug and also *Deroceras reticulatum* (Müller) and *Milax sowerbyi* (Fér.) are hosts to the larvae of *Brachylaemidae* (FOSTER 1958). In Britain, some Helicidae have been found to be intermediate hosts of the lung worms of sheep: *Cochlicella acuta* (Müller) is host to both *Müllerius capillaris* (Müller) and *Cystocaulus ocreatus* Davtian, while *Helicella itala* (L.) is the vector of *Protostrongylus rufescens* A. and M. Hobmaier (STEPHENSON 1964).

Physa sp., *Lymnaea stagnalis* (L.) and in particular *Galba palustris* (Müller) have been recognized to be the intermediate hosts of *Angiostrongylus cantonensis* Chen, the cause of an eosinophilic meningo-encephalitis of people living in Hawaii and the Pacific Islands. *Galba palustris* occurs also in North America (KOCAN 1972). WALLACE and ROSEN (1969) have examined several Pacific terrestrial and freshwater gastropods for their receptivity to this lungworm, the definitive hosts of which are rats; of land pulmonates studied, *Laevicaulis alte* (Fér.) was found to be highly sensitive, *Deroceras laeve* (Müller) less so while the snails *Subulina octona* (Bruguière), *Prosopeas javanicum* (Reeve) and *Bradybaena similaris* (Fér.) are relatively resistant. In Thailand *Achatina fulica* Bowdich is regarded as being the most important intermediate host; it will release 33% of its load of worm larvae into water in 50 h (CROOK et al. 1971). People are infected by way of the household water supply from snails which have fallen into the cisterns. Field investigations showed that 18 out of 30 *Achatina* examined were parasitized, and in each of these cases worm larvae were found in the water and survived even if the water were treated with chlorine or iodine. Although *Biomphalaria glabrata* (Say) has been found to be very sensitive to *Angiostrongylus cantonensis* Chen, this parasite cannot be considered for use in biological control because of the risk it constitutes for man.

Agriculturally important pest gastropods can also act as intermediate hosts for the lungworms of wild animals. For example in Canada, the slugs *Deroceras reticu-*

latum (Müller) and *Deroceras laeve* (Müller) and snails like *Zonitoides nitidus* (Müller) and *Zonitoides arboreus* (Say) are the intermediate hosts of *Pneumostrongylus tenuis* Dougherty, a lungworm parasite in the Virginian Deer. The infection rate of *Deroceras laeve* (Müller) is different at different times of the year. From spring to June, in a damp environment, 25% of the slug population was found to be infected, while in July this fell to 1.5%. In the case of *Zonitoides nitidus* (Müller) the infection rate was constant at 4.3% (LANKASTER and ANDERSON 1968). Development of nematode larvae is delayed during aestivation. *Pneumostrongylus* survives the winter mainly in the slug *Deroceras laeve* (Müller) which is very resistant to frost even down to −8 °C, but also in *Zonitoides nitidus* (Müller).

Rhabditis oncomelaniae Jokoo and Okabe might be used effectively for biological control of the intermediate hosts of *Schistosoma japonicum* Katsurada, since in laboratory investigations 80% to 100% of the snails *Oncomelania nosophora* (Robson) were successfully infected(OKABE and SHIRAISHI 1971); in the field mortality is also high.

Those worms **parasitic** in their **larval stages** on pest gastropods are listed in Table 48. No claim is made that this list is exhaustive, as there are certainly other publications in the extensive literature which have not come to the attention of the author.

The list shows, firstly, that not only freshwater, but also a large number of terrestrial gastropods can be intermediate hosts to parasitic worms of man and animals, and secondly, that among the species included are known plant pests, the importance of which as intermediate hosts has previously been ignored or underestimated.

As will be apparent both from what has been said and from the table, Field Slugs are the most frequently parasitised of all pest gastropods. Of the 32 worms listed which are parasitic exclusively on slugs, 28 infest *Deroceras reticulatum* (Müller) and *Deroceras agreste* (L.). Next on the list in order of decreasing importance are: *Arion* spp., *Biomphalaria (=Australorbis) glabrata* (Say), *Bulinus* spp., *Lymnaea stagnalis* (L.) and members of the family Helicidae.

On the basis of present knowledge and experience, the following species could be considered for use in **biological control** of pest gastropods:

Daubaylia potomaca Chitwood and Chitwood	– *Biophalaria glabrata* (Say)
Davainea proglottina (Davaine)	– *Deroceras reticulum* (Müller)
Echinostoma sp.	– *Biomphalaria glabrata* (Say)
	– *Bulinus* spp.
	– *Lymnaea* spp.
	– *Physa* spp.
Glossosiphonia weberi Blanchard	– *Lymnaea* spp.
	– *Physa* spp.
Helobdella fusca (Bosc)	– *Biomphalaria glabrata* (Say)
Müllerius capillaris (Müller)	– *Deroceras reticulatum* (Müller)
Rhabditis oncomelaniae Jokoo and Okabe	– *Oncomelania nosophora* (Robson)
Ribeiroia marini (Faust and Hoffman)	– *Biomphalaria glabrata* (Say)

Other helminths may be found to be suitable for use in biological control of gastropods, but to this end further intensive research is necessary.

Table 48. Gastropod intermediate hosts of worm parasites

Species	Snails	Slugs
Plathelminthes		
Trematoda		
Bilharziella polonica Kowalewsky (= *Trichobilharzia*) (ducks)	*Lymnaea stagnalis* (Linnaeus) *Planorbarius corneus* (L.) *Radix peregra* var. *ovata* (Draparnaud)	
Brachycoelium obesum Nicolli		*Deroceras agreste* (L.)
Brachylaema helicis pomatiae Dies	*Cepaea hortensis* (Müller)	
Brachylaema nicolli (Witenberg)		*Deroceras agreste* (L.)
Brachylaema obesum Nicolli		*Deroceras agreste* (L.)
Brachylaema sp.	*Zonites algirus* (L.)	*Arion ater* (L.) *Arion lusitanicus* Mabille *Deroceras agreste* (L.) *Deroceras reticulatum* (Müller) *Milax sowerbyi* (Férussac)
Brachylaema virginiana (Dickerson)		*Ariolimax columbianus* (Gould)
Cercaria ocellata de la Valette	*Planorbarius corneus* (L.)	
Cotylurus flabelliformis (Faust)	*Lymnaea* spp. *Physa* spp. *Planorbis* spp.	
Cryptocotyle lingua (Creplin) (Marine fishes)	*Littorina littorea* (L.) *Littorina saxatilis* (Olivi)	
Dicrocoelium dendriticum (Rudolphi) Looss (= *D. lanceolatum* (Rudolphi) Dujardin) (Lancet Liver Fluke) (Grazing stock)	*Abida frumentum* (Draparnaud) *Candidula unifasciata* (Poiret) *Helicella obvia* (Hartmann) *Helicella itala* (L.) *Monacha cartusiana* (Müller) *Euomphalia strigella* (Draparnaud) *Cochlicella acuta* (Müller) *Cochlicopa lubrica* (Müller) *Zebrina detrita* (Müller)	*Arion fasciatus* (Nilsson) *Arion subfuscus* (Draparnaud) *Deroceras reticulatum* (Müller) *Limax tenellus* Müller
Diplostomum spathaceum (Rudolphi) (in eye of Rainbow Trout, 2nd interm. host)	*Radix peregra* (Müller)	
Echinoparyphium cinctum (Rudolphi) (pigeon)	*Lymnaea stagnalis* (L.)	
Echinoparyphium elegans (Looss)	*Bulinus truncatus* (Audouin) *Helisoma trivolvis* (Say)	
Echinoparyphium paraulum (Dietz) (definite host, man, 2nd interm. host, fish)	Freshwater snails	
Echinoparyphium sp.	*Helisoma trivolvis* (Say)	
Echinostoma hystricosum Lie & Umáthevy	*Lymnaea rubiginosa* (Michelin)	

Table 48 (continued)

Species	Snails	Slugs
Echinostoma ilocanum (Garrison) Odhner (man, rat, dog)	*Gyraulus convexiusculus* (Hutton) (India, Japan, Indonesia, Philippines)	
Echinostoma lindoense Sandground & Bonne (fowls)	*Gyraulus convexiusculus* (Hutton)	
Echinostoma murinum (Tubangui)	*Biomphalaria glabrata* (Say) *Gyraulus convexiusculus* (Hutton) *Indoplanorbis exustus* (Deshayes) *Lymnaea rubiginosa* (Michelin)	
Echinostoma paraensei sp. n.	*Aplexa rivalis* (Maton & Rackett) *Biomphalaria glabrata* (Say)	
Echinostoma recurvatum Linstow (ducks, pigeons, waterbirds, domestic fowls, chickens)	*Bulinus* spp. *Lymnaea stagnalis* (L.) *Planorbis planorbis* (L.) *Radix ovata* (Draparnaud) *Viviparus viviparus* (L.)	
Echinostoma revolutum (Fröhlich)	*Lymnaea stagnalis* (L.) *Radix peregra* (Müller) *Viviparus viviparus* (L.)	
Fasciola gigantica Cobbold (Giant Liver Fluke)	*Radix natalensis* (Krauss) (Africa) *Lymnaea ollula* Gould *Pseudosuccinea columella* Say	
Fasciola hepatica Linnaeus (sheep, cattle, sometimes man, deer) (Large Liver Fluke)	*Galba palustris* (Müller) *Galba truncatula* (Müller) *Lymnaea stagnalis* (L.) *Lymnaea tomentosa* (Pfeiffer) (in Australia) *Radix peregra* var. *ovata* (Draparnaud) *Stagnicola bulimoides techella* (Haldeman)	
Fascioloides magna Bassi	*Lamnaea stagnalis* (L.)	
Fasciolopsis buski (Lankester) (kidney of pigs and man)	*Hippeutis cantori* (Benson) (East China) *Planorbis* spp. (East China)	
Guaicaipuria parapseudoconsilia sp. n.	*Pomacea glauca* (L.)	
Guaicaipuria pseudoconsilia (Nasir, Dias & L. de Guevara)	*Pomacea glauca* (L.)	
Halipegus ovocaudatus (Vulpian)	*Planorbis planorbis* (L.)	
Hypoderaeum conoideum Bloch (fowls, geese, ducks)	*Lymnaea stagnalis* (L.)	
Leucochloridium paradoxum Carus	*Succinea putris* (L.)	
Microsomacanthus compressa (Linton)	*Lymnaea stagnalis* (L.)	

Table 48 (continued)

Species	Snails	Slugs
Monocerus sp.		*Arion* spp.
Notocotylus attennuatus Rudolphi (geese, fowls)	*Planorbarius corneus* (L.)	
Opisthorchis felineus (Rivolta) Blanchard (dogs, cats, sometimes man)	*Bithynia leachi* (Sheppard)	
Orientobilharzia sp. (hoofed mammals)	*Radix auricularia* (L.) *Radix peregra* (Müller)	
Panopistus pricei Sinitsin	*Zonitoides arboreus* (Say)	*Deroceras agreste* (L.) *Deroceras reticulatum* (Müller)
Parafasciolopsis fasciolaemorpha Ejsmont	*Planorbarius corneus* (L.)	
Paragonimus tricuspidatus	*Radix peregra* var. *ovata* (Draparnaud)	
Paragonimus westermani (Kerbert) Braun (Lungworm of man)	*Melania* spp.	
Paramphistomum cervi (Schrank)	*Biomphalaria pfeifferi* (Krauss) *Bulinus* spp. **Radix natalensis* (Krauss) *Physa fontinalis* (L.) *Planorbis* spp.	
Paramphistomum microbothrium Carmyerius (cattle, sheep in Ethiopia)	*Bulinus truncatus* (Audouin)	
Plagiorchis cirratus (Rudolphi)	*Lymnaea stagnalis* (L.)	
Plagiorchis muris Tanabe (= *Multiglandularis*) (rats, mice)	*Lymnaea stagnalis* (L.) *Radix auricularia* (L.) *Radix peregra* var. *ovata* (Darparnaud)	
Postharmostomum gallinum Witenberg (fowls in Hawaii)	*Bradybaena similaris* (Férussac)	
Pseudoleucochloridium soricis (Soltys)	*Trichia hispida* (L.)	
Schistosoma bovis (Sonsino) (sheep, cattle)	*Bulinus truncatus* (Audouin) *Planorbarius corneus* (L.)	
Schistosoma haematobium (Bilharz) (man)	*Bulinus globosus* (Morelet) *Bulinus truncatus* (Audouin) *Melanopsis dufouri* Férussac	
Schistosoma japonicum Katsurada (man)	*Oncomelania* ssp.	
Schistosoma mansoni Sambon (man)	*Biomphalaria glabrata* (Say) *Biomphalaria pfeifferi* (Krauss) *Biomphalaria sudanica* (Martens)	

Table 48 (continued)

Species	Snails	Slugs
Trichobilharzia ocellata Brumpt (man, dermatitis)	*Lymnaea stagnalis* (L.) *Planorbarius corneus* (L.) *Radix peregra* (Müller)	
Cestoda		
Anomotaenia arionis (Siebold)		*Arion ater* (L.) *Arion rufus* (L.)
Aporina delafondi Railliet		*Deroceras* spp.
Choanotaenia crassiscolex (Linstow)	*Oxychilus cellarius* (Müller) *Vitrina pellucida* (Müller)	*Arion rufus* (L.)
Cysticercoides dukae (Holland)	*Succinea putris* (L.)	
Davainea proglottina (Davaine)	*Candidula caperata* (Montagu) *Cernuella virgata* (Da Costa) *Helicella itala* (L.) *Helicella obvia* (Hartmann) *Helicidae* *Succinea putris* (L.) *Succineidae* *Trichia sericea* (Draparnaud) *Vallonia identata* (Say) *Zonitoides arboreus* (Say)	*Arion ater* (L.) *Arion fasciatus* (Nilsson) *Arion hortensis* Férussac *Arion intermedius* Normand *Deroceras agreste* (L.) *Deroceras caruanae* (Pollonera) *Deroceras laeve* (Müller) *Deroceras reticulatum* (Müller) *Lehmannia marginata* (Müller) *Limax cinereoniger* Wolf *Limax flavus* L. *Limax maximus* L. *Milax budapestensis* (Hazay) *Milax sowerbyi* (Férussac) *Prophysaon andersoni* (Cooper)
Hymenolepis multiformis (Creplin)		*Arion* spp.
Raillietina bonini (Mégnin)	*Cepaea hortensis* (Müller)	*Arion* spp. *Deroceras reticulatum* (Müller) *Lehmannia marginata* (Müller)
Taenia bothrioplitis Filippi	*Monacha cartusiana* (Müller)	
Nemathelminthes **Nematoda**		
Aelurostrongylus abstrusus (Railliet) (Lungworm of cats)	*Helix* spp.	*Ariolimax columbianus* (Gould) *Arion circumscriptus* Johnston *Deroceras agreste* (L.) *Deroceras laeve* (Müller) *Deroceras reticulatum* (Müller) *Limax flavus* L.
Alloionema appendiculatum Schneider		*Arion ater* (L.) *Arion rufus* (L.) *Deroceras agreste* (L.)
Anafilaroides rostratus (Gerichter)		*Deroceras laeve* (Müller)
Angiostoma helicis Conte & Bonnet	*Helix aspersa* Müller	

Table 48 (continued)

Species	Snails	Slugs
Angiostoma limacis Dujardin (= *Leptodera angiostoma* (Dujardin))		*Arion* spp. *Limax* spp.
Angiostrongylus cantonensis Chen (rat Lungworm)	*Achatina fulica* Bowdich *Biomphalaria glabrata* (Say) *Galba palustris* (Müller) *Lymnaea stagnalis* (L.) *Pila ampullacea* L.	*Deroceras laeve* (Müller) *Laevicaulis alte* (Férussac) *Limax* spp.
Angiostrongylus vasorum Baillet (dog, fox)		*Arion lusitanicus* Mabille *Arion rufus* (L.)
Ascaroides limacis Barthélemy		*Deroceras agreste* (L.)
Crenosoma vulpis (Dujardin) (cat, dog, fox)	*Cepaea* spp. *Eulota* spp. *Helicigona arbustorum* (Linnaeus) *Helix* spp. *Succinea* spp. *Zonitoides* spp.	*Arion* spp. *Deroceras reticulatum* (Müller)
Cystocaulus ocreatus Davtian	*Biomphalaria glabrata* (Say) *Cochlicella acuta* (Müller) *Helicidae*	*Deroceras reticulatum* (Müller)
Daubaylia potomaca Chitwood & Chitwood	*Biomphalaria glabrata* (Say)	
Gordius aquaticus Linnaeus (= *G. villoti* Rosa)	*Lymnaea* spp. *Radix peregra* var. *ovata* (Draparnaud)	
Hexamermis albicans (Siebold)	*Succinea putris* (L.)	*Deroceras agreste* (L.) *Deroceras reticulatum* (Müller)
Leptodera flexilis (Dujardin)		*Limax* spp.
Limaconema sp.		*Arion* spp. *Limax* spp.
Mermis nigrescens Dujardin	*Succinea putris* (L.)	*Deroceras agreste* (L.) *Deroceras reticulatum* (Müller)
Müllerius capillaris (Müller) (Lungworm of sheep and cattle)	*Cepaea vindobonensis* (Férussac) *Cochlicella acuta* (Müller) *Helicella obvia* (Hartmann) *Helix pomatia* L. *Monacha cartusiana* (Müller) *Theba pisana* (Müller)	*Arion ater* (L.) *Arion circumscriptus* Johnston *Arion hortensis* Férussac *Arion subfuscus* (Draparnaud) *Deroceras laeve* (Müller) *Deroceras reticulatum* (Müller) *Limax* sp. *Milax sowerbyi* (Férussac)
Pneumostrongylus tenuis Dougherthy (Lungworm of deer)	*Zonitoides arboreus* (Say) *Zonitoides nitidus* (Müller)	*Deroceras laeve* (Müller) *Deroceras reticulatum* (Müller) *Deroceras gracilis* Rafinesque
Protostrongylus capreoli Stroh & Schmid (Lungworm of roe)	*Cepaea* sp.	*Deroceras* sp.

Table 48 (continued)

Species	Snails	Slugs
Protostrongylus rufescens A. & M. Hobmaier (Lungworm of sheep)	*Abida frumentum* (Draparnaud) *Candidula gigaxii* (Pfeiffer) *Cepaea vindobonensis* (Férussac) *Helicella obvia* (Hartmann) *Helix pomatia* L. *Monacha cartusiana* (Müller) *Monacha syriaca* (Ehrenberg) *Zebrina detrita* (Müller)	
Rhabditis lambdiensis (Maupas)		*Deroceras laeve* (Müller) *Deroceras reticulatum* (Müller)
Rhabditis oncomelaniae Jakoo & Okabe	*Oncomelania* spp.	
Syngamus trachea (Montagu) (fowls, trachea)	*Lymnaea stagnalis* (Linnaeus)	

* In the literature usually given as *Lymnaea natalensis* (Krauss)

3.6.2.3 Host–Parasite Interaction

Attention has been focused in recent years on the interaction between parasite and intermediate host; investigations have been made in particular on the aquatic and amphibious hosts of trematode larvae (*Schistosoma* sp., *Fasciola* sp.), because of their medical and veterinary importance. MEULEMANN (1972) gives a comprehensive review with many references to the literature on the reciprocal influence of *Biomphalaria pfeifferi* (Krauss) and *Schistosoma mansoni* Sambon; STURROCK and STURROCK (1970) have investigated the interaction between *Biomphalaria glabrata* (Say) and *Schistosoma mansoni* Sambon, while that between *Bulinus* spp. and *Schistosoma haematobium* (Bilharz) has been studied by WEBBE and JAMES (1972), LO (1972), STIGLINGH and VAN EEDEN (1971).

KURIS (1973) has worked out an analogy between the trophic interaction between pest insect – hymenopteran parasite and snail – trematode larva. The aim of control of the snail-trematode larva system is a reduction of the total mass of worms in man. The controlling factor here is the trematode larva, analogous to the parasite of the insect. Control of bilharziasis could be brought about by competitive displacement of the larvae of *Schistosoma* sp. by those of other trematode species, better both at finding and at establishing themselves in the snail host, even at low population densities of the latter. From this point of view, a major role can be played not only by obligate, but also by facultative snail parasites, such as sciomyzids (Diptera) and the nematode *Daubaylia potomaca* Chitwood and Chitwood. KURIS believes that the trematode larvae themselves, including those of *Schistosoma*, could be an important factor in biological control of snails, because they cause sterility in the host and sometimes even its death. All the *Echinostoma* species studied, with the exception of *Hypoderaeum dingeri* Lie are capable of eliminating *Schistosoma* from the host. The predatory snail *Marisa cornuarietis* (L.) eats the eggs of *Biomphalaria glabrata* (Say) and so contributes to its elimination. This type of control should be successful particularly with small snail populations which constitute a much feared reservoir of bilharzia. The microsporidia of *Perizia helminthorum* Canning and Basch are obligate **hyperparasites** of trematode larvae (LIE and BASCH 1970). Antagonism either between the rediae or between the rediae and sporocysts of two species of trematode in the same snail may also be effective as a control method (LIE 1957; BASCH and UMATHEVY 1965).

Figure 109 shows the model of host-parasite interactions proposed by KURIS as a basis for biological control. The solid arrows indicate trophic interactions in the food web of this aquatic ecosystem; the plus signs refer to the beneficial aspects of the adjacent trophic interaction in terms of proposed biological control programmes. Double headed dashed arrows refer to interactions separated in space and time from the aquatic ecosystem.

The interaction between parasite and snail consists in a physiological dependance of the trematode larva and a certain tolerance on the part of the intermediate host. For example the larvae of *Echinostoma lindoense* Sandground and Bonne, a trematode parasite of fowls, grow rapidly in the intermediate host *Biomphalaria glabrata* (Say) and use a large quantity of organic matter, and if the infection is heavy, cause parasitic castration. HEYNEMAN et al. (1971) believe that the lack of inhibition by the host with the further development of the parasite is the result of some sort of **metabolic mimicry.** DAVIS and FARLEY (1973) investigated the influence of carbon and nitrogen [obtained from the faeces of *Littorina saxatilis* (Olivi)] and found that digestion and assimilation of food is not influenced in the parasitized snail. Here, too, there is a harmonious relationship between parasite and mollusc, the metabolism of which is disturbed as little as possible. This is important for this trematode because its period of development in the snail is relatively long. The sterility caused by the trematode does not affect the snail population, since it occurs towards the end of the reproductive phase. Thus in this case an *ideal* relationship has developed between host and the parasite, with tolerance on both sides. The schistosomes in particular have adapted to their particular intermediate hosts

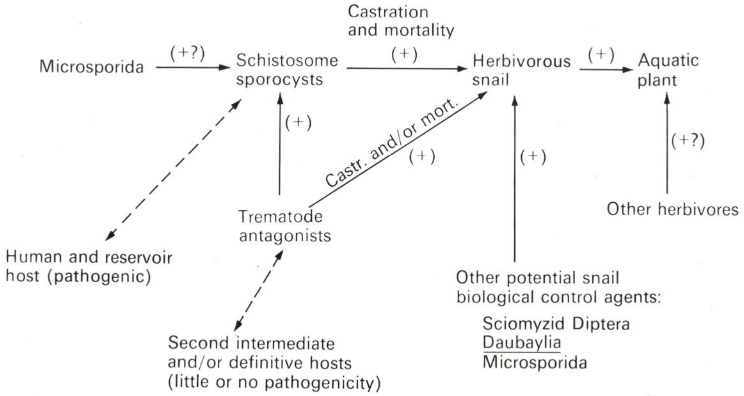

Fig. 109. Generalized theoretical model of the trophic interactions of a schistosome-snail system. *Solid arrows* indicate trophic interactions in the food web of this aquatic ecosystem. Positive (+) and negative (−) signs refer to the beneficial or deleterious aspects of the adjacent trophic interaction in terms of proposed biological control programs. *Double-headed dashed arrows* refer to interactions separated in space and time from the aquatic ecosystem

(KUMAR 1973); either the miracidia are unable to penetrate another species of snail, or they die if they do enter the wrong host.

In general, however, the parasite affects the snail both physically and metabolically, the latter effect usually being to the advantage of the parasite. Thus the lumen of the hepatopancreas becomes blocked mechanically: this is followed by an autolysis of the most distal cells; the tubules and liver tissue between them are mechanically destroyed, with the secondary release of a black pigment.

Endoparasites may cause disturbances in growth, leading to gigantism (ROBSON and WILLIAMS 1971 and others); they may also cause histopathological changes in several organs and also abnormalities in the shell (BRUNNACKER-DAUR 1955; DIX 1973; JOHANNESSEN 1973). According to STURROCK and STURROCK (1970), deformation of the shell begins soon after infection with trematodes, but becomes apparent only around the time when the larvae leave the snail. The total number of cercariae which are released in the life of one snail is very high and is over one million in the case of *Littorina littorea* (L.).

The **haemolymph** of the host may be affected by the parasite (PULLIN 1971). The high consumption of oxygen in parasitized snails, which is reflected in an increased frequency of heart beat (LEE and CHENG 1970, 1971), presumably the result of a rising oxygen requirement of the infected tissues of the host and of the parasite itself as it develops. The temperature tolerance of the snail is also altered by the parasite; thus in addition to an increased rate of heart beat, *Biomphalaria glabrata* (Say) infested with *Schistosoma mansoni* Sambon was found to be less tolerant of higher temperatures than uninfested snails.

Haemoglobin is widely distributed as a respiratory pigment in molluscs (READ 1966). The metabolism of iron, essential for the synthesis of haemoglobin, is also affected by these parasites (HENEINE et al. 1969). LEE and CHENG (1972) have studied this, using radioactive iron (^{59}Fe) and have found that iron is taken up from the surroundings more rapidly and in greater quantities by parasitized than by non-parasitized snails. Radioactive iron was first detected in the liver, later in the gonad of the host snail and finally in the daughter sporocysts of the parasite (*Schistosoma mansoni* Sambon) and the cercariae which develop from them. It is likely that the daughter sporocysts synthesize their haemoglobin from that of the host, for the rate of haemoglobin synthesis was three times greater in infected than in non-infected snails.

Trematode larvae radically affect the **metabolism** of their host in the process of obtaining their own nutritional and metabolic requirements. The developing parasites are a drain on the nutritional reserves of the host (MOORE and HALTON 1973). Although the hepatopancreas is the most important digestive organ in molluscs, little is known about its enzymes or histochemistry. Infection of *Galba truncatula* (Müller) with the rediae of *Fasciola hepatica* L. leads to both histological and physiological changes in the hepatopancreas: there is progressive vacuolization of the cells accompanied by increasing activity of lysosomal enzymes. This probably indicates activation of an intracellular digestive process or autolysis and may be seen as an attempt on the part of the host to meet its own nutritional and metabolic requirements in the face of the stress caused by parasitic infection, since 10 days starvation induces similar changes in the hepatopancreas of healthy, uninfected snails.

Histochemical studies by READER (1971) on *Bithynia tentaculata* (L.) have shown which substances are absorbed from the tissues of the host by developing trematode larvae. The absorptive and secretory cells of the hepatopancreas of uninfected snails contain large masses of glycogen granules, whereas in infected snails these decrease steadily in number as the parasite develops; at the same time, large glycogen deposits are evident in the parenchymatous cells of the metacercariae. This is true also of sporocysts, although rediae contain only small amounts of glycogen. Apart from direct ingestion of host tissue as food, trematode larvae are also able to absorb nutritional substances through their skin. Sporocysts seem to stimulate the activity of the glycolytic enzymes of the host, so that its glycogen is hydrolysed to glucose. Encysted metacercariae probably no longer feed, but they exert pressure on the tubules of the hepatopancreas, causing them to tear apart; these then show in increase in lipid content and a reduction of their phosphatase activity.

In other molluscs infected with parasites, an increase in alkaline phosphatase has been observed (MICHELSON and DUBOIS 1973 and others). This is considered to be the result either of glycolytic activity in the host, or a consequence of autolysis of the cells of the hepatopancreas, so that the enzyme, which is normally intracellularly bound, is released.

Phytophagous and carnivorous molluscs contain **glucose** in their blood (GODDARD and MARTIN 1966). Larval trematodes can obtain the glucose they need either directly from the haemolymph of the snail host or indirectly from the products of hydrolysis of the stored polysaccharides in the hepatopancreas, which are released into the haemolymph; this glucose is then absorbed by the sporocyst stage of the parasite. CHRISTIE et al. (1974) observed that in specimens of *Biomphalaria glabrata* (Say), still alive 25 days after infection, there was an increase in relative weight but a reduction of glycogen content of the hepatopancreas-gonad complex. The relative weight reached a maximum on the 30th day after infection and sank to the normal level on the 40th day. Glycogen content was reduced to zero and no more was synthesized. By contrast, the galactogen content of the albumen gland of the parasitized snails was not much less than in uninfected specimens. Starvation also significantly reduced the glycogen and galactoside content, which was as low as that after 21 days of infection. This experimental observation confirms that the parasite evokes a state resembling starvation in the snail, an aspect of the pathology of infection of the intermediate host which may be of importance. CHENG and LEE (1971) estimated the glucose level in the haemolymph of *Biomphalaria glabrata* (Say) infected with *Schistosoma mansoni* Sambon at seven week intervals; from the third week of infection the values obtained were significantly lower than those for healthy uninfected snails. This reduction seems to be connected with glucose utilisation by the parasite. Snails were found to absorb 95% of ^{14}C-glucose from the surrounding water within 24 h; a larger amount was found in the hepatopancreas than in the foot muscle. The snails released radioactively labelled cercariae from about the fifth day after infection (CHRISTIE et al. 1974). Daughter sporocysts of *Sch. mansoni* also utilize glucose. Experimental observations showed that the development of cercariae could take place only when an adequate amount of glucose was available in the water in which the snails were being kept. It is considered from such experiments that trematode larvae utilize the haemolymph glucose of the intermediate host during their development.

The parasite interferes with the metabolism of the host snail in other ways as well. It leads to an alteration of the **total free amino acid** level of the body fluids and also of the activity of aminotransferase (MANOHAR et al. 1972). WATTS (1951) found that infection of *Littorina littorea* (L.) with rediae of the trematode *Cryptocotyle lingua* (Creplin) caused a 10.9% increase in the total concentration of free amino acids in the head–food muscle; parasitization by rediae of *Himasthla leptosoma* (Creplin), however, resulted in a reduction of 12.7% of the free animo acid pool, and by sporocysts of *Cercaria emasculans* Pelseneer by as much as 57.5%. This effect is dependant on the species, size and mobility of the trematode larvae and also on the degree of damage suffered by the hepatopancreas of the snail. The changes are greatest with the latter two trematodes and are also similar to the effects of starvation. The protein content of the haemolymph is, however, usually lower in infected than in noninfected animals and similar to that found in snails in a state of starvation (LEE and CHENG 1972). Using electrophoresis, 34 protein fractions were isolated from haemolymph serum taken from hearts of *Biomphalaria glabrata* (Say); of these, 17 were found in all the serum samples, 5 in three-quarters of the samples, and the others in various lesser frequencies. Snails each infected with five miracidia of *Schistosoma mansoni* Sambon showed, from the 5th day of infection, a reduction in the total protein content of the serum; from the 44th day this was less than one third of the value recorded from parasite-free snails at the end of the 70-day observation period. GRESS and CHENG (1973) believe this reduction to be connected with the use of protein by the parasites, particularly during the formation of the sporocysts.

SOUTHGATE (1970) has examined the effect of parasitism by *Fasciola hepatica* L. on the **lipid** composition of the hepatopancreas of *Galba truncatula* (Müller), while LUNETTA and VERNBERG (1971) have compared the fatty acid composition of the tissues of parasitized and non-parasitized specimens of *Nassarius obsoletus* (Say), an intermediate host of trematode parasites of fishes and birds. Parasitic infection leads to reduction in the total content of palmitic acid and arachidonic acid in the snail.

The presence of the parasite gradually leads to a decrease in the population density of its host. It can completely inhibit oviposition. In this respect a trematode parasite is a useful means of biological control of the snail vector and thus also of the disease of which it is itself the cause in man or another vertebrate. Oviposition ceases, for example, in *Galba truncatula* (Müller) in the fifth week of infection with Liver Fluke larvae, while mortality of this snail remains normal. **Parasitic castration** in *Helix pomatia* L. in France was reported in 1889. Since that time attention has been directed mostly to host–parasite relationships in the intermediate hosts of the bilharzia parasite and Liver Flukes. The rediae of the parasite cause damage particularly to the hepatopancreas and gonad, as shown in the diagram of HODASI (1972) (Fig. 110). The dark-shaded parts of the liver correspond to the extent of damage, which may lead to complete cessation of liver function. The gonad atrophies and finally no more eggs develop. Although trematode larvae may mechanically affect the acini of the gonad (DUPOUY and MARTINEZ 1973), this is not, as was assumed by NEUHAUS (1941), the cause of sterility. The processes involved in parasitic castration in molluscs are not simple and are still little under stood.

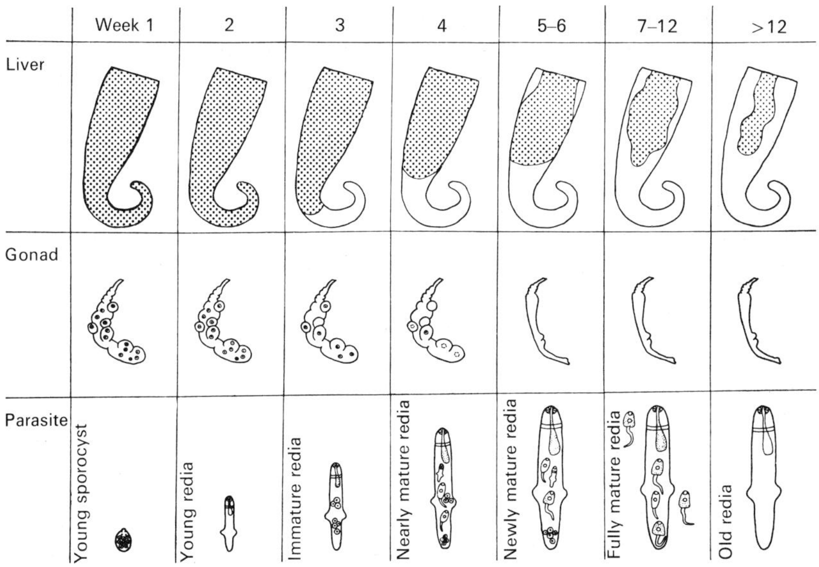

Fig. 110. Diagrams to compare the rate of development of larval stages of *Fasciola hepatica* inside mass-cultured *Galba truncatula* with the effect of the parasite on the liver (hepatopancreas) and the gonad (ovotestis). In the liver *unshaded portions* are damaged. The gonads show gradual degeneration which is associated with loss of reproduction. Drawings not to scale

This sort of castration was first explained as the result of the action of toxins produced by the parasite; later it was thought to be the result of prolonged starvation of the snail caused by the parasite; this suggestion had more justification on the basis of experiments and by analogies with other groups of animals, since it is well-known that the genital apparatus is first affected by longterm undernourishment or starvation (Joosse et al. 1968; Le Breton 1971; Vianey-Liaud 1973) while vital organs such as the nervous system and heart are the least affected. When the period of starvation ends, the changes are reversed if it is not too late. Szidat (1941) speaks of atrophy of the gonad, because it can, as in *Leucochloridium*, become functional again after the trematode infection has ended, whereas the word castration implies a permanent disruption of gonadial function.

The retrogression of the gonad in molluscs during a parasitic infection cannot, however, be explained so simply, and more experimental investigations are necessary. An example of how complicated and many sided the causes of parasitic castration can be is seen in the effect of the trematode *Proctoeces maculatus* (Loos) Odhner, a trematode parasite of fishes (*Labrus* spp.), on its first intermediate host, the mussel *Mytilus galloprovincialis* Lam.; the second is an annelid. Most of the infected mussels retain their normal sexual cycle despite mechanical damage to the acini in the gonad. Only in certain of the infected mussels is development of the gonad inhibited. In fact this only occurs if the time of maximum formation of sporocysts coincides with the onset of gametogenesis in the mussel. If gonadial development has already passed this stage by the time infection takes place, no parasitic castration occurs.

Opinion was at first divided as to whether concurrent infection, that is simultaneous infection with larvae of different species of trematode was possible. The answer seemed at first to be negative, from the investigation of ANTESON (1970), who was unable to infect one snail with several trematode species at the same time. While 15.9% of the specimens of *Lymnaea catascopium pallida* (Adams) collected in the field were found to be infected with five trematode species per snail, all attempts at infecting juvenile gastropods with miracidia of both *Cotylurus flabelliformis* (Faust) and *Diplostomum flexicaudum* (Cort and Brooks) failed. *Cotylurus* was always supressed, although this trematode developed successfully on its own, as did *Diplostomum* alone, the two species showing equally high infection rates of 18.7% and 21.9% respectively. ANTESON believes that the failure of concurrent infection may either be due to a chemical barrier established by the first parasite to infect the snail which repells the miracidia of other species, or that it may be due to the development of real immunity by the snail response to the first parasitic infection.

Later investigations have shown that significant numbers of snails collected in the field could be parasitized by two or several species of trematode (DOLLFUSS et al. 1972; JAIN 1970; STURROCK 1973). Indeed, it seems possible that **superinfection** could be used as a means of gastropod control. Thus, a population of *Lymnaea stagnalis* (L.) collected from one particular area in Canada, was found to be parasitized by cercariae of six different species, and of these, four could infect the snails concurrently. The first infection in fact rendered the snails more sensitive, so that a superinfection by other species followed.

Multiple trematode infection is, however, not the rule, because an **immune reaction** by the intermediate host prevents this (LIE et al. 1965; NOLF and CORT 1933 and others). Antigens produced by the host in response to the parasite have been detected (BASCH 1970; CHENG et al. 1970; HEYNEMAN et al. 1971; RAMADAN 1971 and others). GOLVAN and MOUGEOT (1973) found an immunological effect in *Biomphalaria glabrata* (Say) in which haemolymph cells carry an antigen on their surface which recognizes the species parasitizing the snail. By means of electrophoresis it has been found that the haemocoelic fluid of adult *Littorina saxatilis tenebrosa* (Mont.) contains no less than four proteins, one of which shows similarities to human haemoglobin. The rest are dependant on the physiological state of the snail and others are probably connected with immunity to digenetic trematodes – in this case in the order Digenea (PASCOE 1970).

Survival of the parasite in its intermediate host is secured and interference from other trematode larvae prevented by parasitized snails releasing into the water substances which immobilize free swimming miracidia (MICHELSON 1958, 1963). But the parasite has other means of self-protection. Thus KINOTI (1971) found that snails of the genus *Bulinus*, infected with *Schistosoma haematobium* (Bilharz), have a cellular mechanism of resistance to infection by other *Schistosoma* species, in this particular investigation *Schistosoma mansoni* Sambon. There is a relationship between the survival of the parasite and the density of the host tissue in which it occurs. Miracidia only develop normally in loose tissue, such as a network of collagen and muscle fibres with large blood-filled interstices; in compact body tissues they die within 24 h. This resistance mechanism is different in the two species *Bulinus africanus* (Krauss) and *Bulinus truncatus* (Audouin): in the latter, newly penetrating miracidia are destroyed by amoebocytes, while in the former there is no reaction either to miracidia or to mother or daughter sporocysts, but only to cercariae.

On the other hand it has been suggested that differences in the resistance to trematode infection shown by different snail species is related to the **chromosome** complement: thus, for example, the Mediterranean *Bulinus truncatus* (Audouin), which is very sensitive to *Schistosoma haematobium* (Bilharz) is tetraploid, whereas the resistant *Bulinus tropicus* (Krauss), which lives further south, is diploid (Lo et al. 1970).

Trematode larvae may also prevent infection of their host by annelid larvae. Lie et al. (1973) have found that there may be **synergism** as well as **antagonism** between different trematode species. Thus the presence in *Lymnaea rubiginosa* (Michelin) of sporocysts of *Trichobilharzia brevis* Basch made the snail particularly susceptible to infection by sporocysts of *Echinostoma hystricosum* Lie and Umáthevy – a synergistic effect. Later, once the double infection had been established, the *Trichobilharzia* mother and daughter sporocysts were eaten by the rediae of *Echinostoma* – an antagonistic effect. The latter trematode is thus dominant; in this case the sporocysts of *Tr. brevis* are synergistic to the sporocysts of *E. hystricosum*, but are subordinate to its rediae. The phenomenon of **antagonism** between trematode larvae has been investigated several times with a view to its possible exploitation as an agent for biological control (Basch et al. 1969; Lie 1967; Lie et al. 1973; Lim and Heyneman 1972 and others). The results have shown that this method of control can easily be effected if large numbers of worm eggs of a dominant trematode species are dispersed in small water bodies. Multiple effects come into play: first, intertrematode antagonism, secondly, a steady reduction of the snail population as a result of a higher mortality rate and/or parasitic castration, and thirdly, an epidemic of microsporidia as a result of hyperparasitism by rediae and sporocysts, so that the formation of cercariae no longer occurred. The usefulness of this method depends on whether a dominant parasite can be found and whether other factors in the environment will ensure that the snail is in fact infected by the dominant species. Lie (1973) describes many examples of suppression of a snail population by hyperparasites. In particular, the trematode *Ribeiroia marini* (Faust and Hoffman), which is a natural parasite of *Biomphalaria glabrata* (Say), could be suitable for biological control (Huizkinga 1973).

The host–parasite interaction in the case of **bacterial** or **protozoal** infections mainly results in a reduction of egg-laying and/or parasitic castration of the host snail. Only brief descriptions of this are to be found in the literature, whithout any explanation of the processes involved in, for example, parasitic castration.

Brooks (1968) examined the histological changes in individual organs of Field Slugs brought about by infection with *Tetrahymena limacis* (Warren) and *Tetrahymena rostrata* (Kahl). The pericardial cavity of parasitized snails displayed many tumour-like masses of hypertrophied amoebocytes, most of which were surrounded by a single layer of epicardial cells. Many of the growths also contained one or several ciliates living between the hypertrophied amoebocytes; these ciliates reach the heart of the snail by way of the kidney and pulmonary vein. The parasites subsequently migrate from the heart into the arteries, finally reaching the hepatopancreas and later the gonad, the albumen gland and the genital ducts. Histological studies have shown that *Tetrahymena rostrata* (Kahl) is pathogenic for *Deroceras reticulatum* (Müller), while *Tetrahymena limacis* (Warren) can be regarded as a relatively harmless facultative parasite. The ciliate *Colpoda steini* Maupas gives rise to changes in the organs of the Field Slug similar to those caused by *T. rostrata;* it must therefore be regarded as pathogenic.

Foster (1958) found necrotic tissue in the kidney of *Deroceras reticulatum* (Müller) and *Milax sowerbyi* (Fér.); the cause of this was the presence of encysted Brachylaemidae; *Milax sowerbyi* is the natural host of these trematodes.

A further aspect of host–parasite interaction is the influence of the intermediate host on its parasites. Thus the development of trematode larvae is dependent on the age and size of the host and its nutritional state. Investigations with many different species have shown that snails are not susceptible to trematode infection throughout the whole of their life cycle. Thus *Lymnaea stagnalis* (L.), *Galba palustris* (Müller),*Galba glabra* (Müller), *Planorbis planorbis* (L.) and even the very sensitive *Biomphalaria glabrata* (Say) can only be infected with the miracidia of *Fasciola* and *Schistosoma* in the first few days after hatching from the egg, or at least during the juvenile stage (Kumar 1973; Solitys and Ziomko 1972; Sturrock and Sturrock 1970; Sturrock and Webbe 1971; Ubelaker and Olsen 1970).

If the food supply of the snail is inadequate, many trematode larvae die, only a few reaching the cercaria stage necessary for infection of the final host (Eveland and Ritchie 1972).

Trematode infection is species specific; it is supposed that resistance is genetically determined. This is discussed in Chapter 3.6.4.1

3.6.3 Predators

Attempts have understandably been made to use predators for control of pest gastropods, especially for the biological control of freshwater and amphibious pulmonates (Hairston et al. 1975; Knutson 1976). It is not enough, however, that suitable prey be available in the area involved; other environmental factors such as climate, soil conditions, available shelter, must also be suitable to ensure that the predator introduced does not simply survive, but also reproduces. The African Giant Snail was able to establish itself in new environments because of the absence of its natural predators. In its area of origin in East Africa, this snail is kept on check by many enemies and so does not become a pest. Mead (1961) has given a survey of the predators used in its biological control. By 1960, 19 invertebrates had been introduced into various areas where *Achatina* had established itself.

For molluscs, as for insects, close ecological and biological relationships exist between predator and prey. This was shown by the studies of Paine (1963) on the marine molluscan fauna of a sandbank in west Florida; his findings are summarized in Figure 111. In this habitat, a number of carnivorous gastropods, *Pleuroploca* sp., *Fasciolaria tulipa* L. and *F. hunteria* Perry, *Busycon contrarium* Conrad and *B. spiratum* Lam., *Murex* sp. and *Nassarius* sp. all prey on bivalves. These mussels are shown in the bottom row. The predators are in their turn, together with polychaete worms, preyed on by other carnivorous molluscs. At the apex of the pyramid is *Pleuroploca gigantea* Kiener which has a diameter up to 35 cm. The

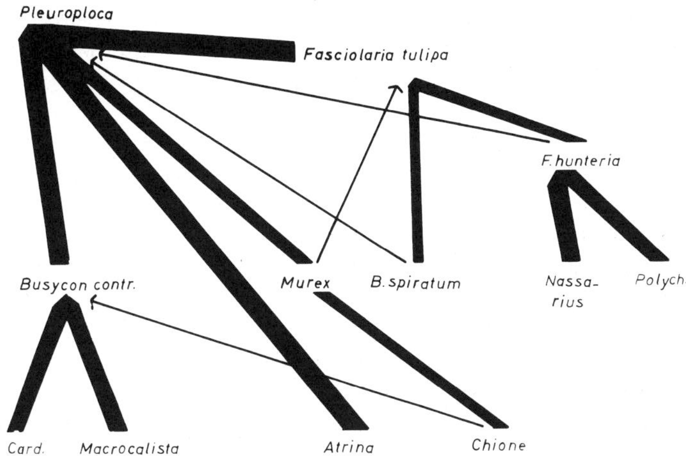

Fig. 111. Predator–prey interaction of some marine carnivorous snails

thickness of the lines in the diagram correspond to the percentage which each prey item represents of the total food of the predator.

Similar food webs are found for terrestrial snails as well. The food web of which *Achatina fulica* Bowdich is a member is described by MEAD (1961) and is even more complex, including the Coconut Crab *Birgus latro* L., hermit crabs, cockroaches, beetles, flies and predatory snails. *Achatina* is also attacked by giant toads, the Indian Monitor, rats and mongooses. The giant toads feed on other phytophagous and carnivorous snails, predatory beetles and flies as well as cockroaches; the rats which feed on predatory snails are attacked by the Indian Monitor, mongooses, dogs and cats; all of which eat other carnivorous and phytophagous gastropods. The last are at the bottom of the food chain because they are preyed on by many *Achatina* enemies including hermit crabs, flies and predatory beetles, giant toads, the Indian Monitor, rats and fowls. The Monitor eats rats, hens and their eggs, coconut beetles and Coconut Crabs. The mongooses, at a high trophic level in the food web of this community, feed not only on *Achatina* but also on rats, hen's eggs, predatory snails and beetles. But there are many more interrelations than those mentioned here.

The predators, together with abiotic factors and also the availability of food and the presence of disease and parasites, control the density of a pest population. OWEN (1974) and YOM-TOV (1970) have emphasized the influence of predators on molluscs. Facultative predators, like birds which feed on gastropods, are more useful than are obligatory predators, because once a gastropod population has been reduced in size they find sufficient other food; thus they do not die of starvation following destruction of their prey, while specialized obligatory predators are thus at a disadvantage.

When, however, a predator feeds on a wide selection of food, damage may recur; the predatory snail *Marisa cornuarietis* (L.), for example, also damages plants under cultivation and the predatory *Gonaxis* attacks other members of the community in the ecosystem of its particular habitat. Thus four years after the introduction of this predator to an island to control *Achatina fulica* Bowdich, a fall in the numbers of hermit crabs, coconut crabs, large monitor lizards and rats was observed and at the same time an increase in the population of wild goats as a consequence of the decline of Indian Monitors and rats (MEAD 1961). Later, changes were also observed in the composition of the flora of the island in that there was an increase in the prevalence of poisonous herbs because the edible plants were devoured by the goats. Such were the consequences of introducing only a single predator into an environment where it had not previously occurred.

It can also happen that a species which is an effective predator in the laboratory or in a confined space such as a greenhouse is ineffective in the field. Thus snails of the genus *Physa* are effective at eradicating *Bulinus* sp. in the laboratory, but in the field, in ditches and irrigation canals, the two snails coexist without troubling each other.

It is thus clear that before the introduction of a predator for the biological control of pest pulmonates, many difficulties have to be overcome and problems solved, especially if the prey–predator relationship forms part of an ecosystem in ways not clearly understood. Furthermore, the predators themselves may prey on each other. Of the many predators of *Achatina fulica* Bowdich, the glowworm *Lamprophorus* attacks the predatory snails *Gonaxis* and *Euglandina*, the beetle *Tefflus* preys on the glowworm; *Euglandina* feeds on *Gonaxis* and the latter cannibalizes members of its own species.

The following sections deal with the most important predators considered as possible biological control agents, namely molluscs, arthropods and vertebrates; the associated problems are discussed.

3.6.3.1 Molluscs

Molluscan predators have two ways of seizing their prey. If the prey is a snail, it is reached through the mouth of the shell, the predator penetrating ever deeper and devouring its prey as it does so: *Gonaxis* feeds on terrestrial snails in this way. *Murex* species, predators on bivalves, have a special boring organ with which the thickest oyster shell can be penetrated (CARRICKER et al. 1972; CARRICKER and VAN ZANDT, 1973; CHETAIL et al. 1971; CHETAIL and FOURNIE 1969).

It has long been known that predatory snails can keep down the numbers of pest pulmonates. The first to which attention was drawn was *Marisa cornuarietis* (L.), about which there is now an extensive literature (CHERNIN et al. 1956; MICHELSON and AUGUSTINE 1957; OLIVIER-GONZALES et al. 1956; RADKE et al. 1961; MEIER-BROOK et al. 1977 and others). This freshwater snail, which is 50 mm long (Fig. 112), has been used against other aquatic snails and is still valued as a means of destroying the intermediate hosts of *Schistosoma*. It has more recently been introduced to tropical and sub-tropical islands and to other areas where it did not previously occur (DEMIAN 1964; DEMIAN and KAMEL 1973; JOBIN et al. 1973; RUIZ-TIBEN et al. 1969). *Marisa* eats eggs and young snails and is capable of totally destroying a population of *Biomphalaria glabrata* (Say) or *Bulinus* sp. in 5 to 8 months. It also feeds voraciously on water plants, thereby depriving lymnaeids and planorbids of their source of food.

Fig. 112. *Marisa cornuarietis*

The effect of *Marisa cornuarietis* (L.) is, however, not always satisfactory, for in some areas it can coexist peacefully with its prey, *Biomphalaria glabrata* (Say). Once the numbers of *Biomphalaria* have been drastically reduced, it is no longer sufficiently abundant to serve *Marisa* as a main source of food, and the vector remains as a reservoir of infection. The predatory snail then itself becomes a pest, attacking plants such as rice or vegetables in flooded fields (Ortiz-Torres 1962; Seaman and Porterfield 1964 and others). This predator, expressly introduced to deal with pest snails, thus became a source of damage.

This risk is always there if, for control of pest snails, **Marisa cornuarietis** (L.) is introduced to flooded or very moist fields under cultivation. Nor can it be considered for control of lymnaean snails in cress cultivations because it feeds on watercress itself. The investigations of Demian and Ibrahim (1972) have shown that while this snail cannot tolerate frost, it can live within a range of 5 °C to 40 °C, and would thus certainly be able to survive in southern Europe, where it could possibly be used against the "Liver Fluke snails" in ditches and irrigation canals of meadows and pastures without damage to the plants under cultivation. Before this can be done extensive investigations are necessary, not only of the extent to which it feeds on water plants considered to be weeds, the incidental destruction of which would serve to keep ditches clear without the use of herbicides, but also on the extent to which it might inflict damage on grazing.

The importance of *Marisa cornuarietis* (L.) for biological control both of the intermediate hosts of *Schistosoma mansoni* Sambon and *Schistosoma haematobium* (Bilharz) and also of water weed is reflected in the extensiveness of recent research into the biology of this snail and the fact that mass rearing methods have been developed, a prerequisite for its use as a predator (Demian 1964; Demian and Ibrahim 1972; Demian and Kamel 1973; Rich and Rouse 1970).

Marisa cornuarietis (L.) is certainly very effective in the destruction of water plants (Andres and Benett 1975; Holm et al. 1969, 1970; Seaman and Porterfield 1964; Simmonds 1969; Yeo and Fisher 1970; Blackburn et al. 1971 and

Fig. 113. *Gonaxis quadrilateralis* amongst eggs of *Achatina fulica* (*below*); three adult *Gonaxis* with a 120-mm *Achatina* (*above*)

others). The South American *Pomacea australis* (Orbigny) is also suitable for this purpose. At this point an interesting case should be mentioned; it concerns the red alga *Cephaleuros virescens*, which causes great damage to citrus plantations in the coastal belt of Florida during very damp weather. FISHER (1974) found that the terrestrial snail *Bradybaena similaris* (Fér.) can be used for biological control of this alga; 22 specimens of this snail per citrus tree so reduced the algal infection within three months that no further sign of it was to be seen. Large snail populations which might develop as a result of a rich supply of algal food can, according to FISHER, be dealt with in the autumn by application of molluscicides.

An important predator for biological control of phytophagous terrestrial pulmonates, particularly of *Achatina fulica* Bowdich, is the "Hunter Snail", **Gonaxis kibweziensis** Smith, a small (20 mm) African snail. It eats its prey from within the shell and has proved useful in the protection of citrus plantations (ANON. 1956; MEAD 1961; PETERSON 1957). Its suitability for control purposes was demonstrated in 1950/51 on the island of Agiguan, one of the Pacific islands north of Guam; in the first year after its introduction it reduced the *Achatina* population by 20% and later in 1954, when it had established itself, brought about a reduction of 60%. This predatory snail was thereupon introduced to most of the islands and other areas where *Achatina* was found (Hawaii, Seychelles, Mauritius, amongst others). In 1955 it was introduced into California for the control of *Helix* spp. in citrus plantations. *Gonaxis* has rapidly multiplied wherever it has been introduced; in one area where 400 specimens were liberated, 21,750 were recovered a year later. The east African Species *Gonaxis quadrilateralis* (Preston) showed itself to be twice as successful as *G. kibweziensis* Smith; this newly discovered predator was introduced to Mauritius together with another predatory snail *Euglandina rosea* (Fér.) to control *Achatina fulica* Bowdich (Anon. 1965) (Fig. 113).

This third promising predator, **Euglandina rosea** (Fér.) has subsequently attracted attention and since 1962 has been used for biological control (Mead 1961; Ranaivosova 1965, 1971). It has been introduced from Hawaii to the Bermudas for use against the snails *Rumina decollata* (L.) and *Otala lactea* (Müller) which were introduced in 1876 and 1928, and which are pests to vegetables and flowers. This was done because the predatory snails previously introduced for this purpose, including *Gonaxis kibweziensis* Smith, were not a success, since their populations either died out or remained small. *Euglandina rosea* (Fér.) was released in this area in 1958 and 1960 and caused heavy reduction in the numbers of *Otala lactea* (Müller), so that biological control could here be considered a great success. This predator was, however, ineffective against *Rumina decollata* (L.) and also against slugs (Simmonds and Hughes 1963).

The search for predatory snails continues. In 1970 *Edentulina ovoidea* (Bruguière), even more aggressive than *Euglandina rosea* (Fér.), was introduced into Madagascar and the Comores to control the African Giant Snail. *Euglandina*, however, prefers *Bradybaena similaris* (Fér.) and eats only the smaller snail if this occurs together with *Achatina*.

Other predators on terrestrial gastropods are *Edentulina affinis* C.R. Boettger and *Haplotrema minimum* Ancey. The latter species reduced the population of *Helix aspersa* Müller which had been passively spread to California; this is a rare example of a newly introduced species encountering an indigenous predator which then succeeds in keeping its numbers down. This was not the case with *Achatina fulica* Bowdich.

Recent research has shown that in Europe, too, biological control with malacophagous snails is a possibility; in particular, destruction of *Galba truncatula* (Müller), the intermediate host of the Liver Fluke *Fasciola hepatica* Linnaeus by *Zonitoides nitidus* (Müller) (Moens 1980, Rondelaud 1975).

Biological control with the predatory **Oxychilus draparnaudi** (Beck) is effective only in greenhouses. In such a confined space, the pest *Oxychilus cellarius* (Müller) is eaten by its related predator, so that after a while only *O. draparnaudi* (Beck) is to be found in the greenhouse; the entire population of *O. cellarius* (Müller) is wiped out (Schmidt 1960). In the field, however, the two species coexist without troubling each other, for in the larger area enough other snails are available as prey, so that the predatory species is not dependant entirely on *O. cellarius* (Müller) as it is in the greenhouse. In the field *O. draparnaudi* also eats the eggs of *Cepaea nemoralis* (L.).

Members of the families **Testacellidae** and **Oleacinidae** are voracious predators (BARNES and STOKES 1951; GRASSÉ 1969; TOMASZEWSKI 1949). These subterranean snails have a very much reduced shell and very small mantle (QUICK 1961), so that they have a worm-like appearance (Fig. 47); they are 8 to 12 cm long and feed on earthworms and slugs. During the day they remain underground, going deeper into the soil during dry weather, and hunt their prey mainly by night. *Testacella scutulum* Sowerby is found in England and in Europe from the Atlantic to the Balkans, while *Polygyra multilineata* (Say) occurs in North America, Bermuda and the Bahamas, where it attacks members of the Succineidae. *Oleacina* spp. prey on *Helicina*, *Helix*, *Rumina* and *Pomatias* spp. Several Helicinidae also fall prey to the giant *Rhytia*.

Testacella can evert its toothed radula, impaling the prey through its end or middle and ingesting it within a few minutes.

MEAD (1961) mentions also as predators *Oleacina oleacina straminea* (Deshayes) from Cuba, *Gulella wahlbergi* (Krauss) from South Africa, *Gonaxis vulcani* (Thiele) and *Gella bicolor* Hutton, both from the Congo region. The last named has proved effective, not against *Achatina fulica* Bowdich, but against *Subulina octona* (Bruguière), an intermediate host of an intestinal fluke of poultry.

Pomacea species can contribute to the reduction in numbers of the freshwater snail *Biomphalaria glabrata* (Say), because it eats the egg masses of the latter, causing a high rate of mortality amongst the newly hatched young snails (PAULINYI and PAULINI 1972). *Pomatias elegans* (Müller) appears to be the preferred prey of *Poiretia algira* Bruguière. *Daudebardia rufa* (Draparnaudi) is another predatory snail.

Helisoma duryi (Wetherby) could be suitable for biological control because it affects the production of cercariae of *Schistosoma mansoni* Sambon in *Biomphalaria glabrata* (Say) (FRANDSEN and CHRISTENSEN 1977), and also reduces the numbers of *Bulinus truncatus* (Audouin).

Many species which are known to be plant pests are also cannibals and may sometimes eat other members of their own species: examples are *Succinea putris* (L.) and *Physa* spp., while Lymnaeidae, Succineidae and *Oxychilus draparnaudi* (Beck) eat the eggs of their own species (GRASSÉ 1968; SCHMIDT 1955/56). Although they are also cannibalistic, the larger specimens eating the smaller ones, slugs are not suitable for use in biological control – at least, nothing is as yet known about their application.

3.6.3.2 Arthropods

Ecological relationships exist between molluscs and arthropods and have been described by PLATE (1951). Commonly arthropods are predatory on molluscs. STEPHENSON (1964) gives a brief review of the parasites and predators of *Deroceras reticulatum* (Müller).

Marine molluscs and freshwater mussels are omitted from this discussion because they are of no interest in plant protection, which is more concerned with terrestrial pest gastropods, and with amphibious snails on agricultural land.

Although the decapod **Crustacea** normally breathe by means of gills, the hermit crab, *Cenobita perlatus* Edwards is able to live for long periods on land and attacks

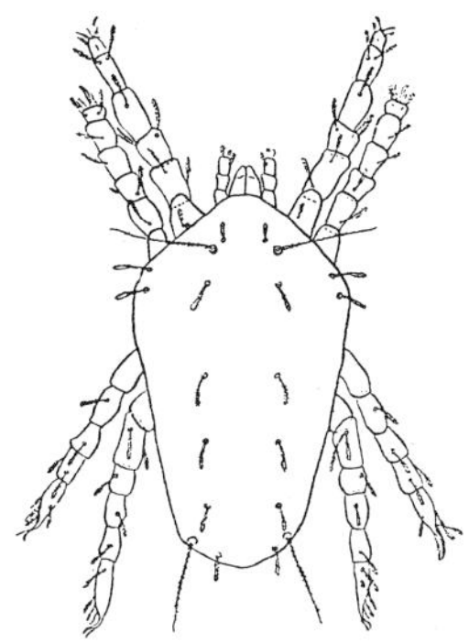

Fig. 114. Adult female of *Riccardoella limacum*, dorsal

Achatina fulica Bowdich, first eating it alive and then occupying its shell; it closes the opening of the shell with its strong claw. This crab keeps whole areas of land completely free of *Achatina* (MEAD 1961). The Coconut Crab, *Birgus latro* L. lives in holes in the ground and also preys on *Achatina*. This crab has, like the pulmonates, true aerial respiration, the wall of the gill chamber being covered with a respiratory network of capillary blood vessels.

Most terrestrial arthropods breathe through trachea, air filled tubes which are kept open by firm chitinous rings in their walls. Among these tracheates, as such arthropods are called, are the millipedes. Some may be up to 25 cm long and have been identified as predators of *Achatina fulica* Bowdich (SRIVASTAVA and SRIVASTAVA 1967).

The **mites** (Acarina) belong with the scorpions and spiders to the Arachnida. Water mites of the genus *Hydrachnella* prey on molluscs (mussels, aquatic snails) in their natural environment (LANCIANI and HARMAN 1968; VIETS and PLATE 1954 and others). There is, however, also a terrestrial species, *Riccardoella limacum* (Schrank) (Fig. 114), which lives in the mantle cavity of terrestrial pulmonates such as *Limax maximus* L., *Limax flavus* L., *Deroceras reticulatum* (Müller), *Deroceras laeve* (Müller), *Lehmannia marginata* (Müller), *Arion rufus* (L.), *Helicigona arbustorum* (L.), *Helix aspersa* Müller, *Helix pomatia* L., *Pomatias elegans* (Müller), *Oxychilus cellarius* (Müller) and *Testacella scutulum* Sowerby. This mite belongs to the family Trombidiidae, and can move extremely swiftly over the moist, slimy body of the slug; when disturbed it disappears immediately into the respiratory cavity. TURK and PHILIPS (1945) have reported on the taxonomy, anatomy, morphology, development and biology of this mite. It has been found in Australia on *Limax maximus* L. and *L. flavus* L.; in North America the species *Riccardoella concolor* Haldeman occurs also in Limacidae (PLATE 1951); MEAD (1961) mentions mites associated with *Achatina fulica* Bowdich. *Riccardoella limacum* (Schrank) is undoubtedly capable of considerably reducing a gastropod population, or at any rate can cause a loss of gastropods which are being artificially reared by weakening the individual animals. According to ANDRÉ and LAMY (1930, 1931) there is more justification for regarding this mite as a parasite than as a predator.

Investigation of the relationship between mites and gastropods has thus far received very little attention.

Harvestmen (Opiliones) are known to be predators of both snails and slugs, such as *Deroceras* spp., *Arion hortensis* Fér. and others. PLATE (1951) has described prey-catching in *Ischyropsalis helwigii* Panzer; when preying on *Vitrina pellucida* (O.F. Müller) or on gastropods with a transparent hyaline shell such as *Retinella* and *Oxychilus* it grasps the edge of the shell and drags it with living contents to its retreat. There, holding the shell edge with one chelicera, it places the shell with its apex on the ground. Fragments of the prey, which has withdrawn into the shell, are torn off and eaten. If the harvestmen cannot reach the snail in the upper coils of the shell, the mouth of the shell is broken open and the prey completely devoured.

The most important predators are, however, the **insects,** of which several species of beetle, bug and fly are of particular significance.

It should also be mentioned that **ants**, which belong to the **Hymenoptera,** also eat gastropods. Thus *Pheidologeton diversus* Jerd. is a predator on snails in Sumatra and drags them to the nest, often over considerable distances – up to 20 m from where they are found; empty shell are removed from the nest (MEAD 1961; VAN DER MEER-MOHR 1931).

Beetles (Coleoptera) are considered to be especially important as predators of molluscs. The characteristic features of certain species which prey on snails are the small, mobile, snout-shaped head with strong mandibles, while the posterior end of the thorax is narrow and flexibly joined to the abdomen. This **"cychrus facies"** was first noticed in snail predators of the genus *Cychrus* and later in carrion beetles and the carabid *Carabus morbillosus* Fab. in northwest Africa, and in a less extreme form even in glowworm larvae. Bristles in front of the spiracles of many beetle larvae are thought to prevent these openings from becoming soiled and blocked. When overpowering and then devouring its prey, *Cychrus* moves the tip of the abdomen so as to produce air movements under the elytra, thus allowing it to continue to respire while eating its slimy prey. This is necessary because the snail's defensive reactions are to secrete large quantities of foamy mucus, to attempt to shake off the predator by powerful movements and by withdrawal into the shell. A slit-like opening of the shell as well as the presence of teeth and folds at its mouth offer some protection against these predators (SOLEM 1972 and others). PLATE (1951) has described several such predator–prey relationships.

Species of the families Silphidae (carrion beetles), Cantharidae (soldier beetles), Carabidae (ground beetles) and Staphylinidae (rove beetles) are snail predators either as larvae or as adults. The most important predators are to be found amongst the Silphidae and Cantharidae.

Among the **Silphidae,** two species, *Ablattaria laevigata* Fab. and *Phosphuga atrata* L., are bound in their development and way of life to snails. In both species the head is elongated in the form of a snout as an adaptation to feeding. According to PLATE (1951), *Phosphuga* finds its prey by smell, the olfactory sensillae being located on the antennae. The beetle climbs onto the shell and bites several times into the snail. Resistance on the part of the snail is generally useless: it withdraws into its shell and secretes vast amounts of mucus, but this is simply eaten by the beetle, which pursues its prey into the coils of the shell and seizes it first by the foot, and then it devours the still living snail. By contrast, *Ablattaria laevigata* Fab. always bites at the head; it positions itself over the mouth of the shell and squirts onto the body of the snail first anal fluid from the tip of the abdomen and then, from its mouth, a secretion from the midgut which soon leads to disintegration of the soft parts of the snail's body. This beetle and its larva thus digest their prey extraintestinally.

The **Cantharidae** include the important snail predators, the Lampyrinae (glow-worms and fireflies) and the Drilinae.

Of the **Lampyrinae** four large genera, *Lampyris*, *Luciola*, *Photinus* and *Photuris* inhabit the palaearctic region and are therefore distributed in America, Europe and Asia, to the north in Finland and the Amur region, to the south as far as Borneo, Sumatra and Java; they live also in Africa, Algeria and Morocco (SCHWALB 1961). The species *Lampyris noctiluca* L. and the American *Photinus frigidus* are the most northerly of the Lampyrinae in distribution. The former is found in Europe, in the east as far as China, in America and around the Mediterranean. It has been introduced from England to New Zealand for biological control of *Helix aspersa* Müller and to various islands for the control of *Achatina fulica* Bowdich (MEAD 1961). O'DONALD (1968) found this species to be an excellent predator of the Banded Snail *Cepaea nemoralis* (L.).

Lampyris noctiluca L. is one of the most fearsome enemies of snails and can overpower prey more than 20 times its own body weight; this is then devoured within 24 to 36 h. The voracity of lampyrid larvae makes them very suitable for use in biological control, as is shown, for example, by the fact that a single larva ate, in the course of its development to a female imago, 40 to 60 live *Achatina fulica* Bowdich. The adult beetle does not feed.

Lampyris noctiluca L. and, in California, *Phausis splendidula* L. prey preferentially on Helicidae, but they also attack slugs such as *Limax* spp., *Arion ater* (L.), *Arion fasciatus* (Nilsson), *Arion hortensis* Fér. and *Arion subfuscus* (Drap.), several larvae often feeding on the same slug (FINCHER 1947; INGRAM 1946; STEPHENSON and KNUTSON 1966 and others). Though species such as *Limax maximus* L., *Limax flavus* L. and *Arion rufus* (L.) are generally avoided and only eaten after long starvation.

Lamprophorus tenebrosus (Walker), *Photuris pennsilvanicus* DeG., *Luciola cruciata* Mot. and *Luciola vittata* Cast. are important predators (BESS 1956; BESS and ALICATA 1953; PETERSON 1957; SIMMONDS 1969). The first, in particular, has been used successfully against *Achatina fulica* Bowdich.

In contrast to the very active larva of *Drilus*, the larva of *Lampyris* is cautious. It crawls onto the shell of a snail and attaches itself with the clasping organ at its hind end (Fig. 115). SCHWALB (1961) has described prey-catching and sexual behaviour in Lampyrinae found in Germany. Glowworm larvae are nocturnal predators and find their prey by following the slime trails left by snails as they creep about. The predator always injects its poison at the anterior end of the prey. The beetle orients itself to the head of the prey not by touch or by sight, but by smell, since the mucus the snail secretes from the anterior parts of its body differs in its chemical composition from that secreted from the posterior end; the larva is able to distinguish between the two. The cerebral and pedal ganglia of the snail are situated anteriorly, so injection of poison at this end leads more rapidly to paralysis than injection at the posterior end of the snail.

With head thrust forward and mandibles open, a *Lampyris* larva bites several times into the tentacles and later other soft parts of the prey, which becomes

Fig. 115. *Lampyris* larvae attacking a *Cepaea* (*below left*), biting into anterior end of retreating snail (*above left*), and attacking a slug (*above right*)

paralyzed and finally dies. The poison gland of the larva is probably situated in the head, since injection of extracts from the tiny head have a lethal effect.

The mucus secreted by the snail is eaten by *Lampyris* and *Phausis* larvae before they attack the soft parts of the body. The flesh of the snail is torn off by the mandibles, passed to the mouth and swallowed. It has frequently been assumed that extraintestinal digestion occurs in glowworms. This idea must be rejected, as SCHWALB has observed that pieces of food remain in the mouth before being swallowed for far too short a time for any digestion to occur.

The larvae of the **Drilinae,** which are closely related to the Lampyrinae, not only eat the snail, but also take possession of the shell. The adults do not feed. According to the description by PLATE (1951), a larva climbs onto the shell (mainly of Helicidae), attaches itself with its clasping organ, and bites the snail several times as it emerges from the shell. The larva then drags the snail under cover by digging its jaws into the ground and pulling up its body with the attached snail shell and its still living contents. If it encounters an obstacle, it lifts its load and throws it forwards. Once the prey has been brought to the sought after shelter, the larva devours it alive, and later even the rotting remains of the viscera. The larva then cleans out the empty shell and creeps backwards into it, either to moult or to pupate if it is in the last larval instar. The shell is closed with the shed larval skin to protect the inhabitant. If there are not enough snails in a locality with many newly hatched *Drilus* larvae, then two of them may occupy the same shell, but cannibalism often follows. SIMMONDS (1969) also describes the drilinid *Malacogaster* spp. as snail predators.

The formation of two larval types, one inactive and the other active, is characteristic of *Drilus flavescens* Fourer and *Drilus concolor* Ahr. The inactive form has poorly developed, small jaws and legs; it appears after the third moult in autumn and pupates in spring of the following year; the active larval phase occurs during the summer and has strong mandibles and legs for prey catching.

Plate (1951) has described the peculiarities of the development of adult Lampyrinae and Drilinae.

Among the **Carabidae,** large ground beetles, such as *Carabus violaceus* L., *Scaphinotus interruptus* (Menetries), *Scaphinotus striatopunctatus* (Chaudoir) and *Thermophilum hexasticum* Gerstaecker are predators of *Achatina fulica* Bowdich (Mead 1961), and of slugs such as *Arion hortensis* Fér., *Deroceras reticulatum*(Müller), *Milax gagates* (Drap.). *Abax striola* (Fab.), *Feronia melanaria* (Illiger) and *Carabus violaceus* L. attack *Deroceras reticulatum* (Müller) (Stephenson 1965, 1968). In feeding experiments, *Carabus violaceus* L. was observed to overpower five adult Field Slugs in 4 days, but *Feronia nigrita* (Fab.) and *Feronia madida* (Fab.) each ate only one slug in 2 weeks, while starved *Harpallus rufipes* (DeG.) and *Nebria brevicollis* (Fab.) ate neither slugs nor their eggs. The large carabid, *Calosoma frigidum* Kirby is, according to Poulin and O'Neil (1969) a voracious predator of *Arion ater* (L.) in Canada, attacking and eating prey several times its size. The beetles *Pterostichus Bonelli* sp., *Diplocheila striatopunctata* Lec., *Necrophorus pustulatus* (Hers.) and *Carabus morbillosus* Fab. are mentioned as predators; Wolda (1972) regards *Pterostichus vulgaris* L. as a significant predator of *Cepaea nemoralis* (L.). Mienis (1980) found that in Israel *Procrustes impressus* Klug preys on nine different species of snail, including *Theba pisana* (Müller), which is becoming a pest in this country, and also *Cochlicella barbara* (L.).

The large black species of **Tefflus,** such as *Tefflus zanzibaricus alluaudi* Sternberg (45 mm) and *Tefflus purpureipennis wituensis* Kolbe (25 mm) attack *Achatina fulica* Bowdich both as larvae and as adults; they are so effective in its eradication that an attempt was made to introduce the first into Hawaii to control *Achatina;* but difficulties arose. According to Mead (1961), the species *Damaster blaptoides* Kollar from Japan is much more suitable, and it was brought into Hawaii in 1958. Although *Tefflus* is an excellent snail predator, several factors reduced its efficacy. *Tefflus* is nocturnal as is its prey, *Achatina*, but the beetle itself was preyed upon by the nocturnally active toad *Bufo marinus* L.; furthermore, the beetle and its larva did not invariably prefer *Achatina*, but attacked the smaller predatory snails *Gonaxis* and *Edentulina*, which had also been introduced to control *Achatina.* The result was that *Achatina* remained unaffected and was even able to recover its numbers to the point of being once again a pest.

Of the **Staphilinidae,** the beetle *Ocypus olens* Müller has proved to be an effective predator on *Helix aspersa* Müller in California, and is recommended for biological control of this extraordinarily harmful pest of citrus plantations (Fisher et al. 1976; Orth et al. 1975) (Fig. 116).

Amongst the **Hemiptera,** water bugs like the 40 to 50 mm long *Limnogeton fieberi* Mayr (family Belostomatidae) from north Africa is specialised as a predator of freshwater and amphibious snails. In all the other belostomatids, the first legs are adapted for prey capture, with broad thick femora. In *L. fieberi*, however, the first legs are even more weakly developed than those of the two posterior pairs of legs. This is because a snail is slow-moving prey, not capable of much resistance and therefore easy to overpower; the use of force by the predator is not necessary as it is in the case of other carnivorous bugs which prey on fast-moving

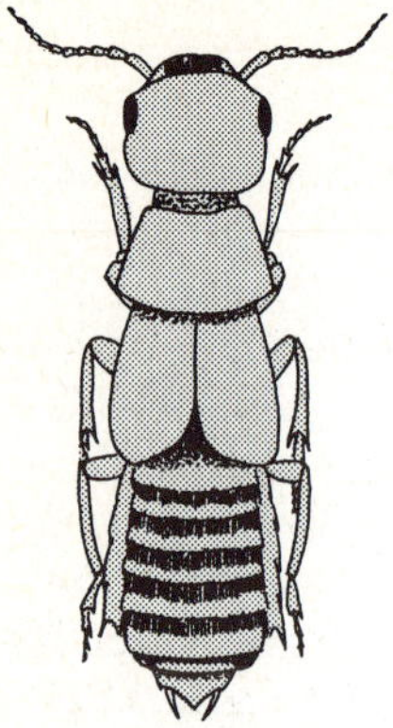

Fig. 116. Adult *Ocypus olens*, the Large Black Rove Beetle, which feeds on *Helix aspersa*; in three week a single animal consumes more than 20 Brown Garden Snails

animals. VOELKER (1966, 1968) believes that *L. fieberi* Mayr can certainly be used successfully against the intermediate hosts of *Schistosoma* sp. If a snail is found by a *Limnogeton*, it is stung by the bug, thus causing it to release its hold of the substratum and withdraw into the shell. The predatory bug now turns the snail over until the mouth of the shell is facing it and it then pierces the soft body of the snail with bristly stylets. The tissues of the snail are digested extraintestinally by a fluid from the midgut, and the products of digestion are sucked up through the proboscis. This is done with the shell mouth facing upwards so that the liquid food does not pour out. The voracity of *Limnogeton* larvae is remarkable: during their developmental period of 43 to 51 days, 28 specimens consumed 3,547 snails – mainly *Biomphalaria glabrata* (Say) and *Physa acuta* Drap., that is, an average of 2.7 snails per larva per day. An adult bug, at water temperatures of 27 °C to 29 °C eats about four *Biomphalaria alexandrina* (Ehrenberg) (15 to 18 mm) to two *Biomphalaria glabrata* (Say) (20 to 22 mm) or one large *Lanistes bolteniana* (Chemnitz) (30 to 40 mm) per day. The daily consumption of snails by female bugs is 2 to 3.5 times greater than that by males; the females ate at least 2 to 4, maximally 11 to 20 specimens of medium size, while the males only 0 to 8 snails. VOELKER estimated that during the course of its life span, a female *Limnogeton* destroyed 1,800 snails and a male 670.

The male *Limnogeton fieberi* Mayr broods the eggs; these are stuck by the female to the upper surface of his wings, which the male keeps in a horizontal position at the surface of the water (Fig. 117), where the developing embryos can obtain oxygen of the air, for they die if immersed in water, even if the male submerges for too long a period. *Diplonychus urinator* Dufour, a belostomatid only 14 mm in length, is a facultative snail predator, not an obligatory one like the species mentioned earlier.

According to SOMASUNDARAO (1963), the water bug *Sphaerodema rusticum* Fab. is suitable for use in India for biological control of the intermediate host of the Liver Fluke *Fasciola gigantica* Cobbold.

The order **Diptera,** which can be subdivided into midges and flies, includes a number of gastropod predators and parasites, many of which can be used for biological control, especially of the Helicidae. Two families of flies, the Phoridae and Sciomyzidae, are of particular interest in this respect; their larvae are at first parasitic, then predatory and finally, in the third stage, they are saprophagous.

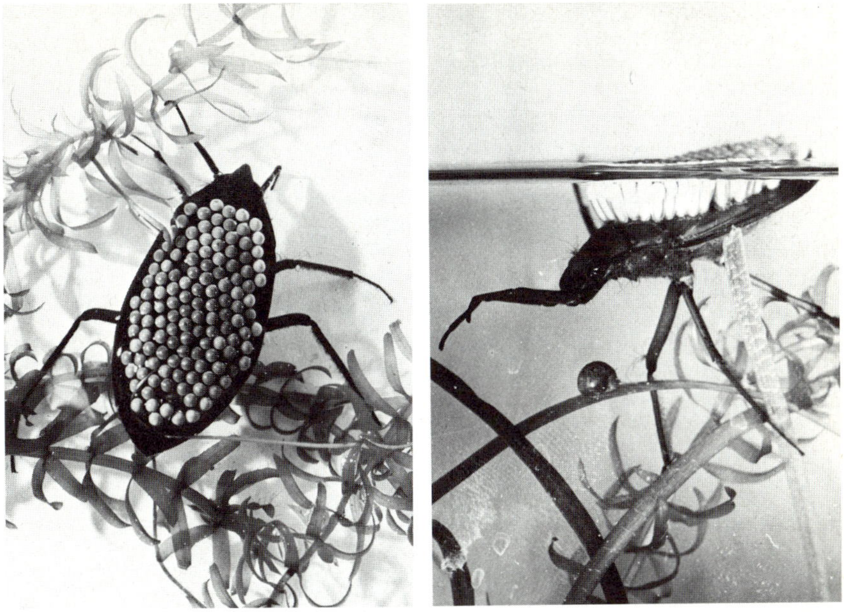

Fig. 117. Breeding male Waterbug, *Limnogeton fieberi*: eggs deposites on back (*left*), animal "suspended" from water surface to aerate eggs (*right*)

Three hundred and sixty eight species of 46 genera of the family **Phoridae** are known to occur in the nearctic region. Their mode of life and, in particular, the feeding habits of their larvae are not well known; the larvae usually develop in the eggs of insects and spiders and only a few species (Table 49) need gastropod eggs. There are some Phoridae which prey on living molluscs in the larval stage. Those of *Paraspiniphora* sp., however, develop in dead snails; the larvae have on the upper surface of the maxillary palp an olfactory organ which is used to detect carion. This structure differs from species to species and may consist of either a simple or a much subdivided indentation, containing many sensillae. A relatively broad nerve with a thickened end passes into the palp (PLATE 1951). *Hypocera* larvae, which also feed on dead gastropods, have a similar olfactory organ.

Megaselia (Aphiochaeta) aequalis (Wood), widely distributed in the holarctic region, can greatly reduce the numbers of a population of slugs. According to ROBINSON and FOOTE (1968), the female lays her eggs either directly in a snail's egg or close to it. Immediately after hatching, the larva penetrates through the outer covering and gelatine matrix, and enters the host egg where it feeds on the fluid surrounding the embryo, without, at this stage, attacking the embryo itself. Only in the second and third instars does the larva become predatory, eating the embryo, and usually completely destroying the whole of a host egg mass before it pupates. It spends the winter as a pupa. Adult *Megaselia aequalis* (Wood) are found in Ohio from the end of March until the beginning of September.

Table 49 lists the Phoridae which live on gastropods or on their eggs.

There is an extensive literature dealing with the **Sciomyzidae** or "marsh flies", from the time (BERG 1953, 1959, 1960, 1964, 1973, 1976; BERG et al. 1959) drew at-

Table 49. Phoridae found in living pest gastropods and their eggs. (Modified after ROBINSON and FOOTE 1968)

Species	Larvae feed on
Phorinae	
Chaetopleurophora bohemanni Becker	*Helix pomatia* L.
Pericyclocera javicola Beyer	*Achatina fulica* Bowdich
Plethysmochaeta sp.	Snails
Spiniphora sp.	*Achatina fulica* Bowdich
Spiniphora bergenstammi (Mik)	*Cepaea hortensis* (Müller)
	Cepaea nemoralis (L.)
	Helicigona arbustorum (L.)
	Helix pomatia L.
Spiniphora excisa Becker	*Cepaea hortensis* (Müller)
	Cepaea nemoralis (L.)
	Helix lapicida L.
	Planorbarius corneus (L.)
Spiniphora helicivora Dufour	*Helix* spp.
	Planorbis spp.
Spiniphora maculata Meigen	*Helix aspersa* Müller
Metopininae	
Megaselia sp.	Eggs of *Deroceras laeve* (Müller)
Megaselia achatinae	*Achatina fulica* Bowdich
Megaselia aequalis (Wood)	Eggs of *Deroceras laeve* (Müller)
Megaselia biformis Brues	*Achatina fulica* Bowdich
Megaselia ferruginea Brunetti	Molluscs
Megaselia ruficornis Meigen	Snails
Megaselia scalaris (Loew)	Terrestrial snails
Megaselia spiracularis Schmitz	*Bradybaena sieboldiana* L. Pfeiffer
Megaselia stuntzi Bohart	Molluscs
Megaselia tertia (Brues)	Eggs of *Deroceras laeve* (Müller)
Puliciphora borinquenensis Wheeler	Snails
Puliciphora wymani Bohart	Molluscs
Wandolleckia cooki Brues	*Achatina* sp.

tention to this group of flies as predators of the intermediate hosts of Liver Flukes, *Fasciola hepatica* L. and *Fasciola gigantica* Cobbold, and recognized their potential in biological control of these snails (Fig. 118). Like *Coremacera marginata* F. (KNUTSON 1973), most could be used in biological control of freshwater and terrestrial pulmonates. The reader is referred in particular to the reviews of BERG and KNUTSON or KNUTSON and BERG (1971 and others), and also to the studies of European Sciomyzids by DAVIS et al. (1961), ELBERG (1968), FERGUSON and RUIZ-TIBEN (1971), FOOTE (1959, 1961), GECKLER (1971), KNUTSON (1963, 1966, 1970, KNUTSON (1976), KNUTSON et al. (1970, 1975), NEFF (1960, 1964), ROZKOŠNY (1965), STEPHENSON and KNUTSON (1966), VERBEKE (1964) and others. Up to 1960, BERG and his co-workers had found, from rearing the flies, 84 species of 12 American and 15 European genera which prey on gastropods. Of these, 49 species have tetanocerine larvae and belong to the genera *Sepedon*, *Dictya*, *Hedroneura*, *Elgiva*, *Hydromya*, *Pherbina*, *Protodictya*, *Renocera* and *Tetanocera*. The main characteristic of tetanocerine larvae is their cephalo-pharyngeal skeleton; a similar arrangement is

Fig. 118. Marsh fly (*Sepedon macropus*) in the typical head-downward position, just above water line on emergent vegetation (*above left*); fly larva attacks snail (*right*), pursues when it retreats into shell, and begins big meal (*below*)

also found in larvae of Salticellinae, but is not developed to the degree found in the Sciomyzinae.

There are over 600 species of Sciomyzidae, but detailed knowledge of the biology of less than half of these is available (Knutson 1976; Simmonds 1969). While no species is cosmopolitan, the family itself is represented in all the continents, and shows a wide range of larval behaviours. Some species have aquatic larvae, which eat freshwater snails; others are semi-terrestrial and attack amphibions as well as

Fig. 119. *Left* two eggs of *Antichaeta analis* on an egg mass of *Galba truncatula*; *centre* an egg of *Salticella fasciata* in the umbilicus of *Cernuella virgata*, a terrestrial snail; *right* eggs of *Colobaea bifasciella* on the shell of the semi-aquatic *Lymnaea* snail

stranded pulmonates, while others are entirely terrestrial and prey on land gastropods. We still lack information, however, on the influence of sciomyzid larvae on a snail population as a whole.

The use of sciomyzids against *Oncomelania* spp., the intermediate hosts of *Schistosoma japonicum* Katsurada in Japan, and against other operculate snails was doomed to failure from the outset. These snails are, admittedly, attacked by the fly larvae, but they remain unharmed because the larva is unable to penetrate between the shell and the operculum; the defensive reaction of the snail is to withdraw into the shell, and the larva may be trapped between the shell and operculum in the process.

In Europe, the Sciomyzid **Coremacera marginata** Fab. is widely distributed; its biology has been studied by KNUTSON (1973). It belongs to the western palaearctic region; its larvae live in terrestrial pulmonates, mainly in *Cochlicopa* species, but also in *Bradybaena fruticum* (Müller), *Helicella itala* (L.), *Cernuella virgata* (DaCosta), *Oxychilus* spp., *Discus rotundatus* (Müller) and others. Species like *Monacha cantiana* (Montagu), which have a tough, relatively firm layer of mucus on their bodies, are, however, protected from the attack of fly larvae, because the latter die in the mucus.

The systematic subdivision of the Sciomyzidae has not yet been finally settled, but KNUTSON et al. (1970) divide them into Sciomycinae, Tetanocerinae and Salticellinae. Females of the first of these subfamilies usually lay their eggs on the shell of living snails (Fig. 119). The larva migrates from the shell to the prey and finally pupates inside the shell of the snail it has devoured. The Tetanocerinae commonly lay their eggs in moss, grass and weeds in a very damp environment or on floating plants close to the water surface; the newly hatched larva immediately attacks a snail, killing several during the course of its development; it pupates in a nest in the soil.

Salticella fasciata (Meigen) the first sciomyzid recognized as a molluscan predator when, in 1850, its pupae were found in *Theba pisana* (Müller). A thorough

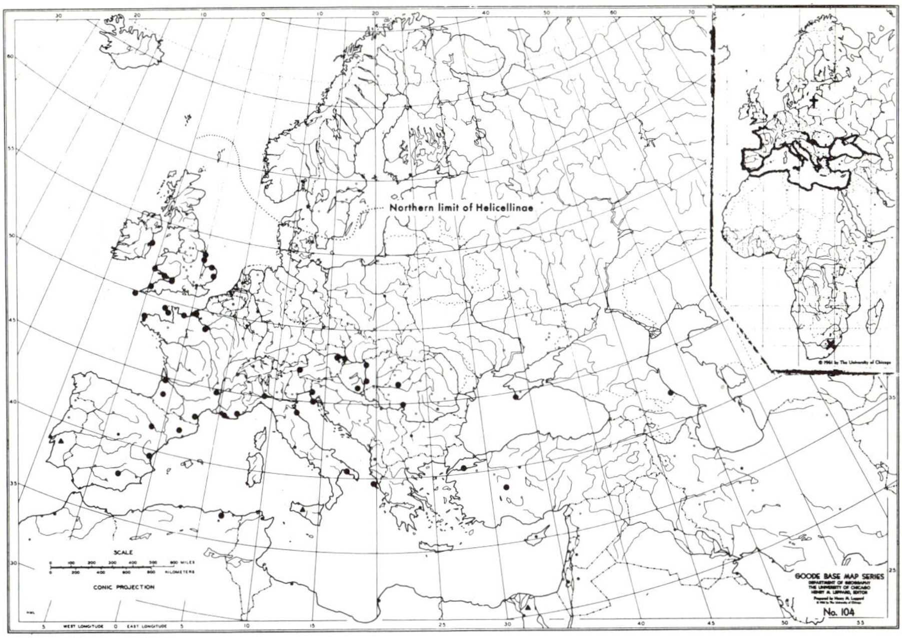

Fig. 120. Distribution of *Salticella fasciata*; ● = precise locality; ▲ = general locality: *right map* distribution of Salticellinae: + = *Prosalticella succini*; *blackened area* = *Salticella fasciata*; × = *Salticella stuckenbergi*

study has been made by KNUTSON et al. (1970). The fly develops in living land pulmonates, mainly *Theba pisana* (Müller) and is quite capable of keeping this dangerous pest of dry areas in check. The distribution map shows that it does not yet occur in the Federal Republic of Germany (Fig. 120); its most northerly limit does not correspond with that of the distribution of its prey, the Helicidae, but is probably determined by temperature. The female lays her eggs on the umbilicus of the shell of a living snail (Fig. 121), but dead snails do not stimulate egg laying. The newly hatched larva penetrates between the shell and mantle of the snail, which remains alive during the first and second larval instars, but then dies and is completely devoured by the third instar larva.

Both development and mode of life in relation to snails have been studied in the following genera of Tetanocerinae; *Antichaeta* (KNUTSON 1966), *Elgiva* (KNUTSON and BERG 1964), *Hoplodictya* (NEFF and BERG 1962), *Hydromya* (KNUTSON and BERG 1963), *Pherbina*(KNUTSON 1963), *Protodictya* (NEFF and BERG 1961), *Sepedon* (NEFF 1960; KNUTSON et al. 1967) and *Tetanocera* (FOOTE 1961; KNUTSON 1963; KNUTSON et al. 1965).

The genus **Elgiva** is of particular importance. There are three species in Europe, *E. rufa* (Panzer), *E. cucularia* (L.) and *E. divisa* (Loew); these all occur in Germany, as the distribution map shows (Fig. 122). *Elgiva rufa* (Panzer) survives the winter in the adult stage. The larvae kill *Biomphalaria glabrata* (Say), *Helisoma trivolvis* (Say), *Gyraulus* sp., *Lymnaea humilis* Say, *Galba palustris* (Müller), *Lymnaea stagnalis* (L.), *Physa* sp. and *Planorbis planorbis* (L.). The larvae have relatively large

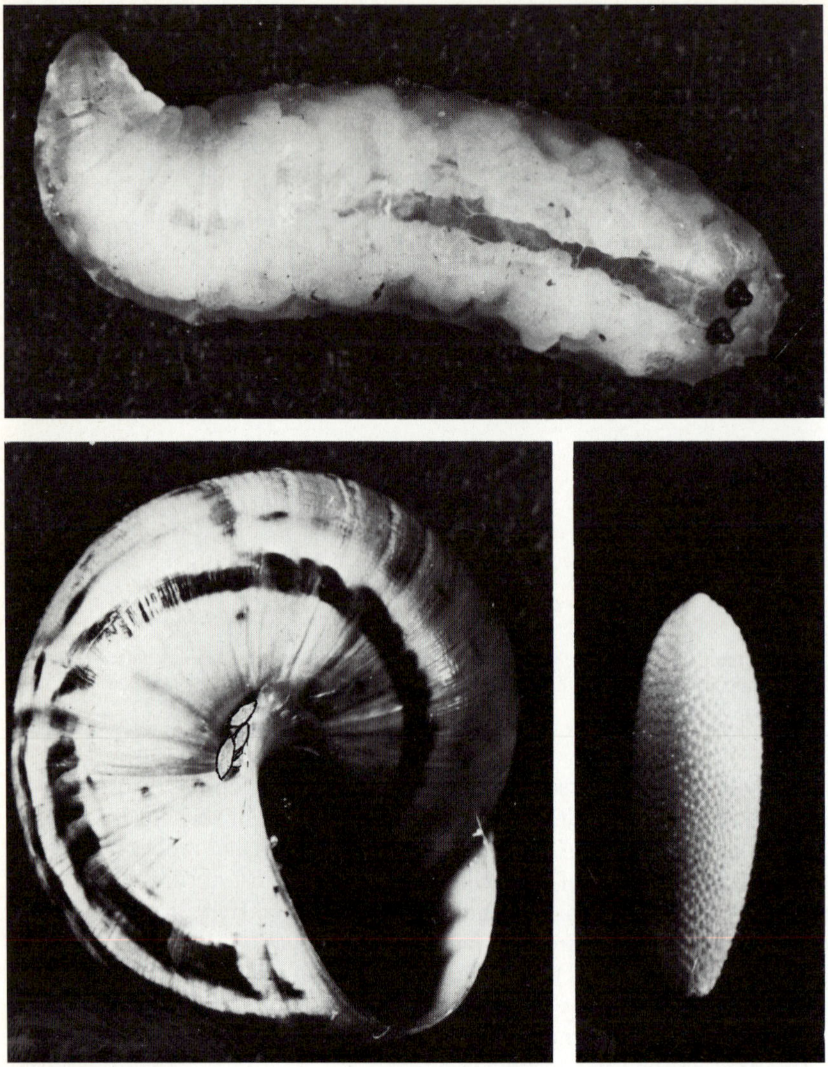

Fig. 121. *Salticella fasciata. Above* mature third-instar larva, dorsal view; *below left* eggs of *Salticella fasciata* in umbilicus of living *Cernuella virgata*; *below right* egg of *Salticella fasciata*, dorsal view

hook-like mandibles with accessory teeth with which they tear off soft parts of their prey (Fig. 123). They feed for only a few hours on one snail and then kill another, even if the first has not yet been consumed completely. *Elgiva connexa* (Steyskal) was first described in 1954.

The central American **Sepedon macropus** Walker [=*Sepedomerus macropus* (Walker)] lays its elliptical eggs one behind the other, up to about 15 at a time, on water plants in the vicinity of the prey, which are freshwater snails. This fly has been introduced into several Hawaiian islands and used with success in biological control of *Lymnaea ollula* Gould, an intermediate host of the Liver Fluke *Fasciola*

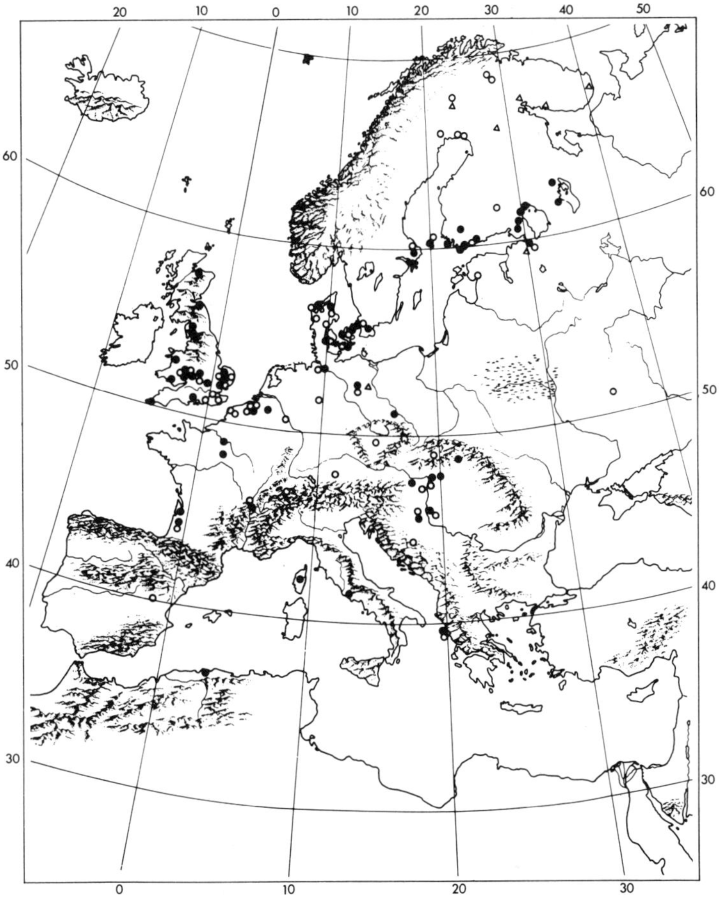

Fig. 122. Distribution in Europe: *Elgiva cucularia* = ●, *E. rufa* = ○ and *E. divisa* = △

gigantica Cobbold (Chock et al. 1961; Davis et al. 1961). Large numbers of flies can be reared on a mixture of powdered snails – an essential ingredient – water and honey. An attempt to establish *Sepedon sauteri* Hendel from Japan in Hawaii failed, however; after several years, no surviving individuals of this fly were to be found. The species *Atrichomeline pubera* (Loew), *Pherbellia grisescens* Meigen and *Sepedon praemiosa* Giglio-Tos, all from the USA, were not held to be very successful. *Sepedon tenuicornis* Cresson in Canada and *Sepedon fuscipennis* Loew in the USA have, however, been recommended for use for biological control (Geckler 1971; McCoy and Joy 1977).

Four species of **Tetanocera** which are obligatory gastropod predators have been described by Foote (1963), Knutson et al. (1965), Knutson (1966), Trelka and Foote (1970). Flies of the genus *Tetanocera*, indigenous to North America and Central Europe, certainly could be used for biological control of gastropods. The larvae of *Tetanocera plebeia* Loew live under the mantle shield of *Deroceras laeve* (Müller). First instar larvae of *Tetanocera valida* Loew have been observed to eat

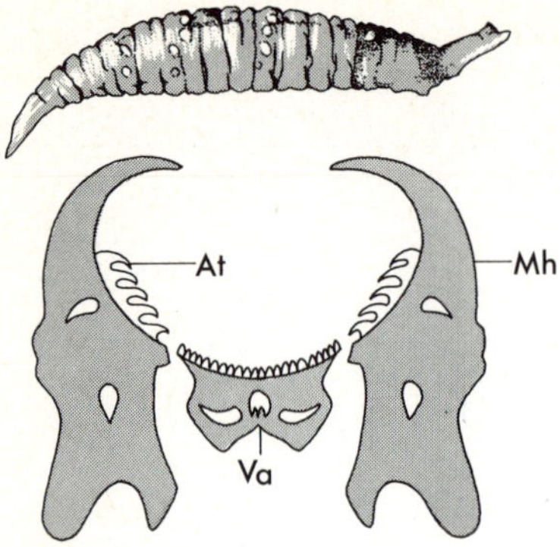

Fig. 123. *Elgiva rufa*, third-instar larva (*above*); mouth-hooks and ventral arch of third-instar larva (*below*). *Mh* mouth-hooks; *Va* ventral arch; *At* accessory teeth

the optic tentacles of *Deroceras laeve* (Müller), while older larvae pushed their way into the mouth of another slug. The larvae of *Tetanocera clara* (Loew) do this in the first instar. *Tetanocera elata* Meigen is distributed throughout North America and Central Europe and its larvae attack the slugs *Arion subfuscus* (Nilsson), *Arion intermedius* Normand, *Limax flavus* L., *Limax tenellus* Müller also *Milax budapestensis* (Hazay) and *Milax sowerbyi* (Fér.) (Stephenson 1968). The voracity of the larvae during development is enormous, a single specimen killing as many as 4 to 6 Field Slugs of 8 to 25 mm length. Older larvae are the more rapacious predators and paralyse their prey before they eat it (Trelka 1972).

Hoplodictya setosa (Coquillet) has a typical tetanocerine larva. On hatching it attacks a snail, such as *Littorina littorea* (L.) and starts to feed on its prey, with many larvae possibly feeding off a single snail. It has been established experimentally that one larva can devour 19 to 23 snails – in this particular case *Physa* sp. – during its development to an adult.

Other species of Tetanocerinae found to be parasites or predators of snails are *Protodictya hondurana* Steyskal, *Melanophora helicivora* Goureau and *Melanophora roralis* L..

Table 50 lists those Sciomyzidae mentioned in the literature which could be considered as agents for biological control of pest gastropods. It is noteworthy that the larvae which attack slugs are all of the genus *Tetanocera*.

Of other families of Diptera, several species of **Sarcophaga** develop in Helicidae (Berner 1973) and in slugs. An example is *Sarcophaga melanura* (Meigen) which deposits its larvae on the back of *Arion hortensis* Fér.; these penetrate the skin and tissues of the slug which finally dies as a result (Stephenson and Knutson 1966). The larvae of *Sarcophaga helicis* Townsend and *Sarcophaga haemorrhoa* Meigen prey on *Cepaea hortensis* (Müller), although these fly larvae are usually found on moribund and dead snails (Mead 1961). Ramachandran (1968) found two species of *Sarcophaga* on *Achatina* sp. *Theria muscaria* (Meigen) which preys on *Helicigona arbustorum* (L.) should also be mentioned.

Chironomids develop on or in snails or their eggs (Grassé 1968; Guibe 1942). Wirth (1971) found the species *Platygymnopa helicis* Wirth in the physid snail *Aplexa hypnorum* (L.) in North America. Among the **Calliphoridae,** *Xerophilophaga* spp. are parasites of *Helicella itala* (L.) and *Candidula unifasciata* (Poiret); the snails die as a result within ten to twelve days. The larvae of *Onesia cognata*

Table 50. Larval development of scyomicids in pest gastropods

Species	Snail	Slug
Antichaeta analis (Meigen)	*Galba truncatula* (Müller)	
Coremacera marginata (Fabricius)	Terrestrial snails, especially *Cochlicopa* sp.	
	Discus rotundatus (Müller)	
	Monacha cantiana (Montagu)	
Coremacera tristis (Harris)	*Monacha cantiana* (Montagu) (Palaearctic region)	
Dictya brimleyi Steyskal	*Biomphalaria glabrata* (Say)	
Dictya expansa Steyskal	*Biomphalaria glabrata* (Say)	
Dictya pistipes (Loew)	*Bulinus africanus* (Krauss)	
Dictya stricta Steyskal	*Biomphalaria pfeifferi* (Krauss)	
Elgiva connexa (Steyskal)	*Biomphalaria glabrata* (Say)	
Elgiva cucularia (L.)	*Gyraulus* sp.	
Elgiva rufa (Panzer)	*Helisoma trivolvis* (Say)	
Elgiva divisa (Loew)	*Lymnaeidae*	
	Lymnaea humilis Say	
	Galba palustris (Müller)	
	Lymnaea stagnalis (L.)	
	Physidae	
	Planorbidae	
	Planorbis planorbis (L.)	
Hoplodictya setosa (Coquillet)	*Littorina littorea* (L.)	
Melanophora helicivora Goureau	*Helicidae*	
Melanophora roralis Linnaeus	*Helicidae*	
Pherbellia grisescens Meigen	*Lymnaeidae*	
	Planorbidae	
Pherbina coryleti (Scopoli)	Freshwater snails	
Psacadina verbekei Rozkošný (semi-aquatic)	Freshwater snails	
Protodictya hondurana Steyskal	*Helisoma trivolvis* (Say)	
	Biomphalaria glabrata (Say)	
Salticella fasciata (Meigen)	*Candidula caperata* (Montagu)	
	Cepaea nemoralis (L.)	
	Cernuella virgata (DaCosta)	
	Cochlicella acuta (Müller)	
	Cochlicopa lubricella (Porro)	
	Helicella itala (L.)	
	Lymnaeidae	
	Monacha granulata (Alder)	
	Oxychilus aliarius (Miller)	
	Pupilla muscorum (L.)	
	Theba pisana (Müller)	
	Trichia hispida (L.)	
	Vallonia pulchella (Müller)	
	Vitrina pellucida (Müller)	
Sciomyza Aristalis (Coquillet)	*Succinea* spp.	
Sciomyza varia Coquillet	*Galba palustris* (Müller)	
Sciomyza dorsata Zetterstedt	*Lymnaea* spp.	
	Planorbidae	
Sepedon macropus Walker	*Biomphalaria glabrata* (Say)	
	Lymnaea ollula Gould	
	Pseudosuccinea columnella Say	

Table 50. (Continued)

Species	Snail	Slug
Sepedon praemiosa Giglio-Tos	Intermediate hosts of *Fasciola hepatica* L.	
Sepedon tenuicornis Cresson	*Biomphalaria glabrata* (Say)	
Tetanocera sp.		*Arion* spp. *Deroceras laeve* (Müller) *Deroceras reticulatum* (Müller) *Limax flavus* Linnaeus *Limax tenellus* Müller
Tetanocera clara (Loew)		*Arion* spp.
Tetanocera elata Meigen		*Limax* spp. *Milax* spp.
Tetanocera ferruginea Fallén	*Biomphalaria glabrata* (Say) *Lymnaea* spp.	
Tetanocera plebeia Loew		*Deroceras laeve* (Müller)
Tetanocera valida Loew		*Deroceras laeve* (Müller)
Tetanocera vicina Macquart	*Bulinus africanus* (Krauss) *Bulinus globosus* (Morelet) *Bulinus truncatus* (Audouin)	

Meigen are parasites of terrestrial snails such as *Cernuella virgata* (DaCosta), *Discus rotundatus* (Müller), *Trichia hispida* (L.) and *Retinella* sp.

3.6.3.3 Vertebrates

Gastropod predators are to be found in each of the vertebrate classes – fishes, amphibians, reptiles, birds and mammals.

Fishes, of course, prey on aquatic snails. *Gambusia* sp., as well as *Astatoreochromis alluaudi*, can be used for biological control of the intermediate host of the bilharzia trematode, *Schistosoma* sp. (LEVINE 1970; MVOGO and BARD 1964); the teleost *Heteropneustes fossilis* (Bloch) is also recommended for this purpose (ZAKARIA 1963). Of three species tested in the Sudan, *Lates niloticus* L., *Synodontis* sp. and *Protopterus aethiopicus* Haeckel, the lungfish *Protopterus* proved to be a voracious predator: a single specimen devoured 450 *Bulinus* and *Planorbis* in four days, but only 11 out of 139 of its normal food, the fish *Tilapia*. Worms were admittedly preferred to snails, but *Protopterus* nevertheless ate sufficient snails for it to be considered for use in biological control. MAHDI and AMIN (1966) in fact emphasize that the absence of bilharzia in the human population of the Upper Nile is attributable to the presence of these lung fishes in the waters of this region. Guppies (*Lebistes reticulatus* Peters) eat large quantities of cercariae (KNIGHT et al. 1968), so they can be used for the control of the stage of *Schistosoma* infectious for man, although this is not achieved through destruction of the intermediate host. Trout eat *Radix peregra* (Müller) and Goldfishes (*Cyprinus auratus* L.) eat *Physa* spp. (GRASSÉ 1968).

Amongst the **Amphibia,** frogs and toads prey on snails, and toads were formerly kept in greenhouses, which is still sometimes done today, to eliminate slugs.

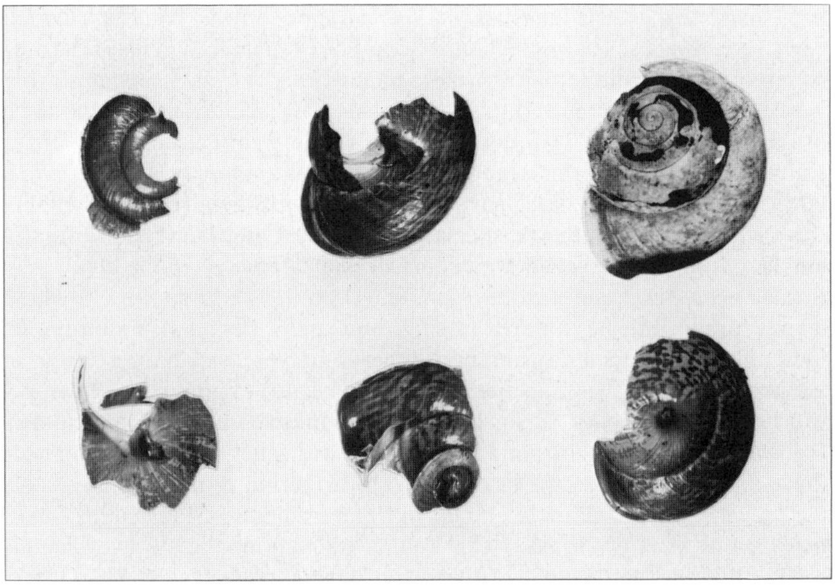

Fig. 124. Shell remains of *Helicigona arbustorum*, which have been eaten by *Sitta europaea caesia*

Crested Newts and Grass Frogs also devour molluscs (FALKNER 1970; MATYSIAK 1970). In America, *Bufo* sp. and *Ceratophrys* sp. attack *Veronicella tuberculata* (Martens). About 1937, the Central American Giant Toad *Bufo marinus* L. was introduced from Hawaii into Guam for biological control of the damaging black *Veronicella leydigi* Simroth; it destroyed these snails in such vast numbers that the Japanese followed suite and introduced the toad into their area. Since then it has been introduced into many of the islands of Micronesia. According to MEAD (1961), *Bufo marinus* L. is not very effective against the dangerous *Achatina fulica* Bowdich: opinion is, however, divided, and there are others who are convinced of the value of the Giant Toad for control of *Achatina* (ICCP 1953). The frog *Rana tigrina* Daud. and other amphibians eat the eggs of this snail.

The significance of **reptiles** for biological control of gastropods, and especially of *Achatina fulica* Bowdich by the Indian Monitor, is discussed in Chapter 3.6.3. *Varanus niloticus* (L.) is a noted predator of the African Giant Snail and its teeth are adapted for cracking the shell of the snail (MEAD 1961). MIENIS (1979, 1980) has established that the Golden Skink, *Eumeces schneideri pavimentatus* (Geoffroy-St. Hilaire) is a predator on *Theba pisana* (Müller) and *Monacha haifaensis* (Pallary), lizards like the blindworm and also snakes attack gastropods.

Pulmonates are preyed on also by **birds,** such as blackbirds, *Sitta europaea* L., thrushes, starlings, poultry, turkeys, ducks, geese and the passeriform birds and, in particular, most of the soft-bodied animal feeders (Fig. 124). Thrushes and poultry are particular predators of slugs and snails, especially of *Cepaea nemoralis* (Linnaeus) and *Helicigona arbustorum* (Linnaeus) (REICHHOLF 1979). Free range poultry farming has a devastating effect on gastropod populations. In such circumstances, only those snail species which live beneath the shelter of stones and cracks, like *Pupilla muscorum* (Linnaeus) and *Vallonia* spp., show no reduction in their numbers. According to Wolda (1970), the Song Thrush, *Turdus ericetorum* Turton, plays an important role in keeping down the populations of *Cepaea nemoralis* (L.) in the Netherlands. Blackbirds and thrushes smash snails against a stone and the piles of broken shells occasionally found near stones are indications of such feeding places. Vultures and owls eat the large Arionidae in great numbers (Grassé 1968). Song birds may accidentally carry gastropods away with them (Brandes 1951) and in this way contribute to their dispersal. The Oyster Catcher, which is found, for example, on the British coast, does considerable damage to edible bivalves, such as *Cardium edule* (L.) and *Mytilus edulis* L. (Dare 1970). Graf (1959), who recommends turkeys to clear an area of slugs and snails, says that the birds should be used only two hours a day and not longer; otherwise they have a surfeit of slugs and snails and lose their appetite for them. The value of poultry for gastropod control is, however, debatable, since the birds do more damage to a vegetable garden by trampling on the plants and soiling them with their droppings than is done by the snails and slugs (van Dinther 1973); furthermore, poultry may become infected with worm parasites. In its indigenous home, Madagascar, *Anastomus lamelligerus madagascariensis* Ch. Temminck, one of the storks (Ciconiidae) is called "snail-breaker" because it feeds on *Achatina lamelligera* (Mead 1961). The beak of this East African bird is adapted to feeding on snails of the genus *Lanistes* Monfort, which are its main food. Mienis (1978, 1980) mentions the following predators: *Turdus philomelos*, which feeds on snails, including *Cepaea nemoralis* (Linnaeus), *Cepaea hortensis* (Müller), *Helicigona arbustorum* (Linnaeus), *Helix aspersa* Müller and *Theba pisana* (Müller); *Burhinus oedicnemus* (Linnaeus), which preys on *Theba pisana* (Müller) and, finally, the Carrion Crow, *Corvus corone corone* L. and the Hooded Crow, *Corvus corone cornix* Brehm, which feed on various species of snails, including *Melanopsis tuberculata* (Müller).

The insectivors are **mammals** which may eat molluscs. *Suncus murinus* (L.) eats young *Achatina;* it was introduced into Guam in 1953 for control purposes and rapidly increased its numbers there (Peterson 1957); the Beaver Rat, *Ondatra zibethicus* L. eats mussels (Akkermann 1972). Moles (*Talpa europaea* L.), hedgehogs (*Erinaceus europaeus* L.) and badgers (*Meles meles* L.) are natural enemies of gastropods; these animals should be protected in gardens and parks and not driven away or killed, as moles are because their mole hills disfigure lawns. These mounds are, however, an indication that the mole has found an adequate supply of worms and slugs as food in the area, and by eating them, prevents the damage they can do. Rats will feed on snails, as has already been mentioned in discussing *Achatina;* another example is the use of *Rattus rattus rattus* (L.) in forestry in New Zealand (Best 1969).

Shrews, moles and hedgehogs are all insectivors, the badger is a mustelid and rats are rhodents.

3.6.4 Genetic Methods

These methods involve genetic manipulation of the pest, which is then used to destroy its own species. Such methods may possibly be of value in the future for gastropod control, in the same way as they are now used against insect pests (WHITTEN and FOSTER 1975). JONES (1974) discusses natural selection in molluscs in the light of their ecological genetics.

3.6.4.1 Hybrid Gastropods

Biological control using **hybrids** involves crossing a species of snail which is very susceptible to infection with worm parasites with a resistant and dominant species, with the result that the sensitive species is eliminated.

Parasitic worms are generally adapted to their host; in extreme cases, such as the trematode *Leucochloridium paradoxum* Carus, the parasite can live in only one intermediate host species; others can parasitize a number of different host species. BOŽKO (1970) has analyzed the significance of the range of host specificity for parasites, also with respect to molluscs.

On the basis of experiments and field observations, a number of resistant species has been found (CRIDLAND 1970; WILSON and SAMSON 1971 and others). KENDALL (1950) observed in Britain that while the miracidia of *Fasciola hepatica* L. could penetrate *Radix peregra* (Müller), they reached only the sporocyst stage, or at the most the redia stage in this "unsuitable" host; cercariae never developed in this snail, even *Radix auricularia* (L.) was not infected at all. By contrast, *Galba truncatula* (Müller) proved to be extremely susceptible to infection with *Fasciola hepatica* L. and is in fact its most important intermediate host. *Bulinus* provides another example: although the whole genus is susceptible to infection by *Schistosoma haematobium* (Bilharz), there are differences from species to species, the trematode larvae developing more successfully in *Bulinus globosus* (Morelet) than in *Bulinus truncatus* (Audouin), so that the former releases more cercariae than the latter. The parasite is adapted to the species of snail most common in its habitat. Hybridization is not possible in this case because there is a genetic barrier between the sensitive and the resistant species (PRINGLE et al. 1971).

The investigations of CHI et al. (1971) on the infection of *Oncomelania* sp. by *Schistosoma japonicum* Katsur. open up interesting possibilities for control with hybrid snails. It was found that the hybrids which arose from crossing two of four species of *Oncomelania* were susceptible to infection if one of the parent snails was sensitive, but the infection rate was lower than that of the parent, and the release of cercariae occurred much later. This was also observed when snails were infected with parasites indigenous to a different geographical area from that of the snail, a very significant finding from the viewpoint of biological control. BORAY (1973) has emphasized the significance of geographically and even microgeographically different races of Planorbidae and Lymnaeidae as a limiting factor for *Schistosoma* sp. and *Fasciola* sp.

The field observations of BARBOSA (1973) have resulted in closer attention being paid to the possibilities of biological control with hybrids. A population of *Biomphalaria glabrata* (Say) had lived undisturbed for many generations in a certain pond. For some reason, another species, *Australorbis stramineus* Dunker arrived in this pond, multiplied, and within three years had eliminated the *Biomphalaria.* What is interesting is that a year after the arrival of the new species, hybrids of the two species appeared. These proved to be unsuitable as the intermediate hosts for the worm larvae which parasitize *Biomphalaria glabrata* (Say) because they were resistant to infection by the miracidia of *Schistosoma mansoni* Sambon, as is the dominant species *Australorbis stramineus* Dunker. The hybrid snail did not reproduce and therefore disappeared in succeeding years. In this way the vector of *Schistosoma*, the sensitive species *Biomphalaria glabrata* (Say), was eliminated from this pond.

Amongst terrestrial pulmonates, too, hybrid forms may arise naturally. Thus hybrids of *Cepaea nemoralis* (L.) and *C. hortensis* (Müller) have been found; it is not, however, known whether they are capable of reproduction.

It may be possible gradually to eliminate a gastropod population by breeding hybrids, which are then sterile. Many investigations into such a method are, however, necessary, and particularly into genetic barriers which exist in some cases and prevent hybridization.

3.6.4.2 Self-Destruction

The so-called **autocidal** method of biological control has not yet been used for molluscs. Investigations should be made of this method using pest gastropods. Sterile animals might be used where the infected area is isolated. But the "production" of sterile terrestrial gastropods and their successful use would become possible only once information was available on mode of life and behaviour (in, on or above the ground) of the various species, while sterilization procedures would also have to take into account species-specific differences in sexual development.

Gastropods are hermaphrodite; either the spermatozoa mature before the egg cells, as in the pulmonates, or the animal is first male and then female, as in the prosobranchs. The animal is hermaphrodite in the transformation period, which differs in length from species to species and during which changes occur in the gonad and genitalia as well as behaviour of the animal (Chap. 1.2.6). The process of sexual transformation, particularly in terrestrial pest pulmonates, is not yet completely understood. Recent experiments on *Indoplanorbis exustus* (Deshayes) are worth mentioning here: egg-laying was inhibited by injection of adult snails with the blood of juveniles, but was elicited two hours after injection with the blood of sexually mature animals (NAGABHUSHANAM, India, pers. comm. 1977).

It is not yet known whether the chemosterilants, such as Tepa, Metepa and others, which are used for insects, can be used for molluscs as well. These substances would be effective only if applied during the infantile or juvenile phases. In the first phase, the gonad is not yet developed and during the second or juvenile phase it is differentiated as a male gonad. The snail has then reached sexual maturity and during this adult phase transformation of the genital organs from male to female occurs. Sterilization during this adult phase has no point because earlier copulation with a fertile partner could mean that there were healthy spermatozoa in the seminal receptacle of the now artificially sterilized gastropod, and these could still fertilize eggs.

Sexual rhythms must also be taken into account. These are not uniform, but influenced by many factors, both extrinsic, such as climate, state of the water (for aquatic snails and mussels) and nutrition, and intrinsic, such as genetic, nervous, endocrine and metabolic factors (LUBET and CHOQUET 1971). It does not by any means follow that different species occupying the same area will all have identical sexual rhythms, since there are genetically determined differences from species to species.

There is also the possibility of **self-fertilization,** whether this occurs within the genital tract as the eggs and sperm pass outwards, or as a result of insertion of the copulatory organ into the animal's own vagina. Self-fertilization has been observed amongst the Basommatophora, in the aquatic Lymnaeidae, Physidae and Planorbidae, in *Ancylus fluviatilis* Müller and two *Bulinus* species, and amongst the Stylommatophora, in several terrestrial pulmonates of the families Arionidae, Limacidae and Helicidae and in one species each of the Zonitidae and Vitrinidae. Self-copulation occurs in *Planorbis planorbis* (L.) (NOLAND and CARRIKER 1946 and others).

Many pest gastropods are, however, **autosterile;** these include *Deroceras reticulatum* (Müller), *Helix pomatia* L., *Helix aspersa* Müller, *Cepaea vindobonensis* (Fér.), *Cepaea nemoralis* (L.), *C. hortensis* (Müller) and *Helicigona arbustorum* (L.) (FRETTER and GRAHAM 1964). The way in which self-fertilization is prevented is interesting. In the hermaphrodite gland of *Helicigona arbustorum* (L.), for example, eggs and sperm are formed at the same time, but the eggs are covered with a layer of cells during the development of the spermatozoa. At a later stage, the spermatozoa and spermatocytes degenerate and only then are the eggs released from their protective cell layer. In *Deroceras reticulatum* (Müller), *Arion rufus* (L.) and *Arion circumscriptus* Johnston, the male and female gametes develop in physiologically isolated compartments (LUCHTEL 1972).

The hormones and steroids involved in gonadal development and sexual transformation have been investigated in recent years and experiments conducted on the effects of such substances on the gonad after its extirpation. The effects of implantation of the tissues of the genital tract have also been investigated. Hormones and steroids in molluscs were dealt with in Chapter 1.2.6.

The following section describes the results of castration and implantation experiments because they allow conclusions to be drawn about the possibilities of sterilization in gastropods and the degree to which it can be considered permanent. To facilitate understanding of the surgical steps involved, the reader is referred to the diagram of the anatomy of the genital tract (RUNHAM and HUNTER 1970) shown in Figure 10. Methods of tissue culture in invertebrates with special reference to molluscs are described by BAILEY (1970, 1973), GOMOT and GUYARD (1964, 1968), KURSTAK and MARAMOROSCH (1976), LUTZ (1970), QUIOT et al. (1973), STREIFF (1970), STREIFF and PEYRE (1963), and also WOLFF and HAFFEN (1962).

LAVIOLETTE (1950, 1951, 1954) found that implantation of a functional gonad into juvenile terrestrial slugs [*Arion rufus* (L.), *Arion subfuscus* (Drap.)], gave rise to precocious maturation of the hermaphrodite gland and the genital tract. Complete extirpation of the gonad of *Limax maximus* L. caused a degeneration of the albumen gland within three months.

In the freshwater snails *Bulinus* sp. and *Planorbis* sp. removal of the gonad caused first disturbance of and then complete cessation of oviposition; but there was, however, no associated atrophy of the accessory glands of the female genital tract which continued to secrete fluid and even hypertrophied. The same effect was observed when the gonad was isolated by cutting of the hermaphrodite duct. Implantation of gonad and hermaphrodite duct together restored normal oviposition.

Regeneration of the gonad after castration can occur if a portion of the hermaphrodite duct is implanted. Thus 90% of castrated *Aeolidiella alderi* (Cooks) developed a functional gonad (TARDY 1967, 1969, 1971). The epithelial cells of the hermaphrodite duct probably differentiate as gametes (BRISSON 1970). TARDY in fact found that a new hermaphrodite gland develops as a result of intensive proliferation of the epithelium. Extirpation of the hermaphrodite duct, however, prevents regeneration of the gonad. Furthermore, the capacity of the epithelial cells of the hermaphrodite duct to form a new gonad is inhibited as long as the old one is still present and it develops fully only after removal (by castration) or destruction (by irradiation) of the original gonad. The hermaphrodite duct is, however, not able to form a new gonad autonomously, but can do so only under the influence of the cerebral ganglion and its hormones (Chap. 1.2.6).

When mice infected with *Schistosoma mansoni* Sambon are treated with a chemosterilant, gametogenesis of the trematode is inhibited (DAVIES and JACKSON 1970). It might be possible to apply the same method of sterilization to pest snails, for YARINSKY and FREELE (1970) report that **Hycanthon** and **Lucanthon** inhibit gametogenesis not only of the intermediate host, but also of its parasite.

Ionizing radiation kills the eggs, but even sublethal doses can reduce growth and fertility, provided the gastropods are irradiated during the very sensitive juvenile phase. When 9- to 10-day eggs of *Biomphalaria glabrata* (Say) were irradiated with a high dose (over 40,000 r) the young snails died soon after hatching. When 3-day-old eggs were irradiated with 2,000 r, only 40% hatched and of these young, only 10% reached an age of 6 weeks, while 75% of the untreated snails survived (PERLOWAGORA-SZUMLEWICZ 1964; PERLOWAGORA-SZUMLEWICZ and BERRY 1964). Exposure of young *Biomphalaria* sp. and *Lymnaea stagnalis* (L.) to sublethal doses of X-rays resulted in a reduction of growth rate and the onset of sexual maturity was delayed (JOOSSE et al. 1968).

While these observations may favour the idea of the sterile-partner method, the radiation should not be so strong that the gonad is destroyed, because the epithelium of the hermaphrodite duct is then able to form a completely new normal gonad.

3.6.5 Biotechnical Control Methods

According to the definition of FRANZ and KRIEG (1976), biotechnical methods make use of the natural reactions of pest animals to physical or chemical stimuli, which are deliberately introduced for a purpose other than that which they normally serve; this involves the use of biological substances which are attractants, growth stimulants, inhibitors or repellents. Their effect is not to kill the pest pulmonates, but to keep the population density below the threshold limit. Such methods have not yet been employed in controlling pest gastropods. They could become important, however, because recently substances have been found which are released by gastropods into their environment and which influence other members of the same species or members of other species. These substances enable "male" and "female" molluscs to find each other, are involved in the flight or other defensive reactions of phytophagous snails to attack by predators, and include a "density factor" which regulates the size of a population. It may be possible in the future to isolate these substances and then produce them in such quantities that they can be used to disrupt the normal chemical interrelationships of molluscs with their biotic environment.

DINTER and MANOS (1972) have discovered a **sex pheromone** in the marine periwinkle *Littorina littorea* (L.). They found that an adult specimen – but not a juvenile – is attractive to other mature snails and its presence releases seeking behaviour. The experimental conditions were such that it was not possible for the presence of the "attractive" snail to be detected by other adult snails either visually or by its slime trails. Young experimental snails did not release such a secretion which attracts the male; snails in both the male and the female phase were attracted to each other.

Steroid hormones have also been detected in molluscs in connection with development of the gonad and sexual transformation (BARDON et al. 1971). Sex hormones, like oestrogen, belong to the steroid hormones and they are mainly secreted by the gonad. It is very likely that such substances occur in gastropods, too, since ROHLACK (1959) found an oestrogenic hormone in the ovary of *Littorina littorea* (L.); this is, however, not identical with the oestrogen of vertebrates. GOTTFRIED et al. (1967) found an oestrogen to be produced by the slug *Arion rufus* (L.).

It is possible that this offers a line of attack for biotechnical control measures. **Steroids** and **sterols** (NEUMÜLLER-RÖMPP 1975) could possibly disturb or inhibit the rhythm of gonadial development into "male" and "female" phases, and in this way limit the development of a pest snail population.

Finally, it would be interesting to investigate **sterol metabolism inhibitors** in species which do or do not require sterols (LEVINSON, Seewiesen, pers. comm. 1976), and thus to create a basis for the use of these substances for the protection of valuable plant cultivations. A sterol (zoosterine) and a sterol ester have been found in primitive molluscs, the chitons (TESHIMA and KANAZAWA 1973), as well as in oysters, prosobranchs and in freshwater, amphibious and terrestrial pest gastropods such as *Helix* sp., *Helicigona arbustorum* (L.), *Succinea putris* (L.), *Lymnaea stagnalis* (L.), *Arion rufus* (L.) and *A. ater* (L.) (DI PRISCO et al. 1973; GOTTFRIED and LUSIS 1966; IDLER and WISEMAN 1972; VAN DER HORST and VOOGT 1972; VOOGT 1967, 1971, 1972; VOOGT and VAN DER HORST 1972). Several sterol antagonists known to be effective in inhibiting insect growth (insectistatica), for example, 25-azacholesterol, 3β-methoxy-25-azacholesterol, N,N-dimethyl-11-methoxy-3,7,11-trimethyldodecanamin, N-ethylpentadecanamin (LEVINSON 1976), could be used in appropriate experiments on pest gastropods.

On the basis of their investigations, LEVY et al. (1973) regard the use of **natural inhibitors** as a possible means of keeping an existing population below the threshold limit. Thus, according to AYAD et al. (1970), the population density of *Bulinus* decreases in the presence of *Helisoma duryi* (Wetherby); this snail has the effect of inhibiting maturation of the eggs and the development of juvenile *Bulinus*. *Helisoma* always dominates; this inhibition is possibly effected by way of an unknown secretion released into the water. It has been called a **"density factor"**. It determines growth, fertility and mortality; however, only intermediate hosts of the helminth parasites of man and his domestic animals have been investigated in this respect (CHERNIN and MICHELSON 1957 and others). NAJARIAN (1961) found that isolated specimens of *Bulinus truncatus* (Audouin) lay eggs more actively than when kept together with other members of the species; he concludes that single animals left surviving a control campaign are reproductively more active, and that it is in this effect, rather than migration of snails from elsewhere, that gives rise to the recovery of the snail population – apart, of course, from individuals hatched from surviving eggs if a non-ovicidal plant protection product was used.

The distribution of freshwater snails is determined by a complex of factors. Experimental and field observations have both shown that the "density factor" plays an important role. Thus among populations of 5, 15, 30 and 60 pairs of the amphibious *Oncomelania nosophora* (Robson), the intermediate host of *Schistosoma japonicum* Katsur., fertility over a six month period was greatest in the population with 30 pairs (MATSUDA 1972). What is remarkable is that the "density factor" also influences the following (F_1) generation. Fecundity and longevity of the offspring of the smallest population (5 parent pairs) were in no way reduced, but the F_1 individuals from the group with 60 parent pairs were very strongly affected. An inhibitory factor has in fact been found to regulate the population density of *Biomphalaria sudanica* (Martens) living in a small pond. BERRIE and VISSER (1963) isolated it in the form of a yellow oil, which at high concentrations had a lethal effect after three days on the experimental snails, while at lower concentrations, like those encountered in the natural habitat, growth and probably also fertility were inhibited. What is important is that this inhibitor was released by the snails themselves into the pond water. It was first regarded as a **pheromone,** but later, on account of its lethal effect at high concentrations, described as a toxin.

GAZZINELLI et al. (1970) describe a method which could be used for assessing the density factor by labelling individual animals with ^{59}Fe. Here, too, it was shown that over-crowding leads to inhibition of growth and fertility in *Biomphalaria glabrata* (Say). Thus growth of the shell and associated uptake of ^{59}Fe were higher in the smallest population (10 animals) in comparison with the values in a dense population (20 to 40 individuals). The presence of a chemical substance was demonstrated both in the faeces and in the water in which the snails lived; this substance, because of its effect at high population density, was called an **inhibitory factor.** LEVY et al. (1973) found a naturally occuring inhibitor which regulates population density in *Fossaria cubensis* Pfr., the intermediate host of *Fasciola hepatica* L. in Texas. This substance was isolated from the water in which the snails had been living for a month, and was found to inhibit growth in young animals and reduce oviposition in adults. Egglaying over a period of 2 weeks was reduced to 56% of the normal rate. Furthermore, although the rate of early embryonic development was normal, the embryo died if the egg was in water containing the inhibitor released by snails of the same species.

The effect of the isolated inhibitor on growth of *Fossaria cubensis* Pfr. became apparent 24 h after 2 days exposure to the substance. Its effect is specific, a ten times greater dose being without effect on the planarian *Dugesia tigrina* (Girard) or the Sword-tailed *Xiphophorus*. Further, the substance is stable at 22 °C for at least a month and for seven months at a concentration of 1 part per 1,000 kept at 5 °C.

Inhibitors could be combined with carrier substances and, as in chemical control of pest gastropods, made attractive as a granular bait formulation, for it is quite possible that they will become important in biotechnical control.

SHEANON and TRAMA (1972) have found that **phenol** in concentrations of less than 2 mg/l inhibited oxidative metabolism in the freshwater snail *Helisoma trivolvis* (Say); this inhibition is also temperature dependant.

It is possible that **alarm substances** could be used to protect wet cultivations such as rice, cress or valuable aquatic plants in greenhouses; such substances released into the water by wounded snails induce flight reactions in other members of the species which happen to be in the vicinity. CLARK (1958) observed a reaction of this sort in phytophagous marine snails, on the approach of predators like starfishes or carnivorous snails. The South American *Helisoma nigricans* Spix moves off and buries itself in the substratum when another member of the species in the vicinity is attacked and eaten by, for example, the Painted Turtle, which shows a preference for this snail (KEMPENDORFF 1942). But the alarm reaction is no longer observed when satiate turtles cease to attack the snails, nor while the turtles hunt mosquito larvae or small crustacea. Experiments have shown that neither auditory nor mechanical stimuli, such as shaking, release a mass flight reaction; only a chemical secreted by a member of the same species can elicit the response which takes the form of either digging into the sand, leaving the water, or hiding in the tangle of water plants. Fluid expressed from other species of gastropods does not induce the flight response if taken from harmless species, but does elicit mass flight if taken from carnivorous snails which are predators on the phytophagous species under observation.

The *behaviour* of *predatory gastropods* towards their prey is also affected by *chemical substances*. Thus proteins which occur in human plasma and also an extract from oysters release strong food-seeking behaviour in the marine carnivorous snail *Nassarius obsoletus* (Say). A glycoprotein was isolated from the oyster extract and found to be effective at contrations of $1–2 \times 10^{-10}$ M (GURIN and CARR 1971).

Special substances regulate the **behaviour** of intermediate hosts *towards their parasites* and vice versa, as when the parasites have attached themselves to the host and are about to penetrate it. Infection of the intermediate host with *Schistosoma* miracidia is influenced not only by complex interactions with physico-chemical environmental factors, such as water turbidity, salinity, temperature, pH and the size of the water body (STURROCK and UPATHAM 1973), but the snails themselves facilitate host-finding by the miracidia (CHERNIN 1972; KINOTI 1971). In combating bilharzia, it is important to know how the miracidia find a vector snail. Freshwater and amphibious snails like *Biomphalaria glabrata* (Say), *Galba palustris* (Müller) and *Helisoma caribaeum* (d'Orbigny) secrete a chemical into the water which influences the swimming behaviour of the miracidia of *Schistosoma mansoni* Sambon, enhancing their aggressiveness and capacity to penetrate the snail. Investigation has shown that this substance is not specific; its effectiveness is, however, dependant on whether the snail is already infested with a parasite or not and whether the snail is sensitive or resistant to infections. Figure 125 shows impressively the stimulating effect of this still unknown substance on trematode larvae. SHIFF (1968) concluded, from experiments with *Bulinus (Physopsis) globosus* (Morelet), the intermediate host of *Schistosoma haematobium* (Bilharz) that the miracidia were

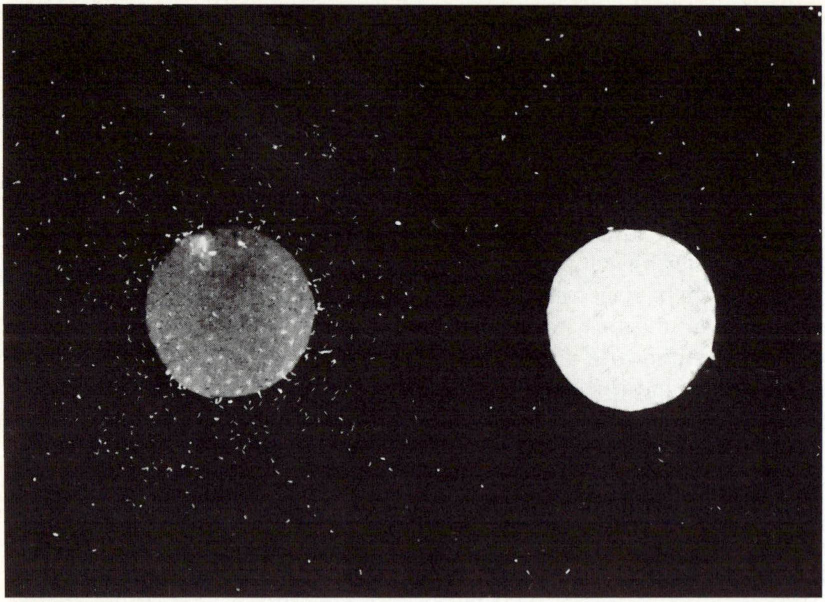

Fig. 125. Aggregation of *Schistosoma mansoni* miracidia around an agar disc (left) impregnated with "snail-conditioned water" from *Biomphalaria glabrata*; the control disc (*right*) had been soaked in well water. Results similar to those with *B. glabrata* were also obtained using discs impregnated with "snail-conditioned water" prepared from *Galba palustris* and *Helisoma caribaeum*. (The photograph was taken, using a 0.001-s flash of an electronic ring-light against a dark background, 5 min after the discs had been placed in a 5 cm petri dish containing numerous miracidia. Original diameter of the discs = 5 mm. The apparent difference in brightness of the discs is an artefact created by the flash. Note that the photograph illustrated the general phenomenon of miracidial response to a "snail-conditioned water" – impregnated disc)

guided to the snail by chemotaxis in response to its mucus secretion, and that infection in moderately flowing water was possible only when there was a turbulence in the immediate vicinity of the snail. However, the freshwater snail releases into the water amino acids which are attractive to the rhodent blood fluke *Schistosoma douthitti* (Cort); investigations of MacInnis et al. (1974) have shown that *Biomphalaria glabrata* (Say) releases **amino acids** which probably elicit **chemotactic** and **chemokinetic** responses from the miracidia of *Schistosoma mansoni* Sambon. The following amino acids were found in water in which the snail had lived for 18 h after defaecation: (concentrations in μM) aspartic acid 0.075; threonine 0.075; serine 0.382; asparagine/glutamine traces; proline 0.052; glutamic acid 0.046; citrulline 0.028; glycine 0.124; alanine 0.153; valine 0.055; methionine 0.008; isoleucine 0.033; leucine 0.046; tyrosine 0.025; phenylalanine 0.027; ornithine 0.015; lysine 0.019 and histidine 0.014. *Snail-conditioned water* acted as a stimulator attractive to 1- to 4-h-old miracidia of *Schistosoma* 1 to 4 h after hatching; the response was strongly marked after an exposure of 20 min. The same effect was obtained with an artificially prepared mixture of amino acids. In addition to this group of substances, a peptide fraction was isolated, but found to have no particular attractive action. The attractiveness of amino acids could be utilized in the control of bilharzia by adding them to molluscicides and larvicides, so that both snails and miracidia would be attracted to and confined to a limited area and there could be more easily destroyed.

The **effect** which some freshwater snails have **of protecting** other, more susceptible species from trematode infection may also depend on the secretion of substances which are not attractive but repellent. An investigation was undertaken by UPATHAM (1972) to establish whether each of the species normally found in the same habitat associated with the sensitive *Schistosoma* intermediate host *Biomphalaria glabrata* (Say) was alone able to afford the snail protection, or whether this effect depend on the whole group of species which included snails like *Physa marmorata* Guilding and *Drepanotrema surinamensis* (Cressin), the guppy *Poecilia reticulata* Garmon and tadpoles of *Bufo marinus* L. Field observations showed that both of the snail species, even if present alone, afforded protection, while the guppies and tadpoles, even in large numbers, had a negligible effect. The protective effect was greater in stagnant than in flowing water – further evidence for the presence of a chemical substance secreted by *Physa* and *Drepanotrema* into the water. The following species were fond to be effective as "bait snails" protecting *Biomphalaria glabrata* (Say) against infection by the miracidia of *Schistosoma mansoni* Sambon: *Bulinus truncatus* (Audouin), *Helisoma caribaeum* (d'Orbigny) and *Helisoma anceps* (Menke), which reduced the rate of infection to between 0% and 7%. Five "bait snails" reduced the rate of infection of the sensitive *Biomphalaria glabrata* (Say) to 38%, 10 or more, however, only to 12%, so that 88% of the snails were infested. When used as "bait snail", species like *Galba palustris* (Müller), *Marisa cornuarietis* (L.) and *Polypylis hemisphaerula* (Benson), which are less susceptible than *Biomphalaria* but not wholly resistant to infection, reduced the infection rate of *Biomphalaria glabrata* (Say) to between 20% and 39%. In experiments with the very small species *P. hemisphaerula* (Benson) and small specimens of *B. glabrata* (Say) and *G. palustris* (Müller), CHERNIN (1968) demonstrated that the size of the "bait snail" is important. He concluded from his studies that given a suitable population density and the right species in a particular environment, some species were able to reduce the rate of infection of sensitive snail species, resulting in a biological balance being established in the habitat between trematode and its vector snail. The use of molluscicides for control of schistosomiasis can, however, injure this useful gastropod fauna, thus destroying the natural limitation to infection of the sensitive species. Most such investigations have, however, been conducted on freshwater pulmonates, because of their medical significance as intermediate hosts.

In this connection it is worth noting that snails can also be protected from miracidial infection by toxic substances secreted by other groups of animals. Thus CHERNIN and PERLSTEIN (1971) have established that **planarians** and **mosquito larvae** protect *Biomphalaria glabrata* (Say) from infection by miracidia, even though these invertebrates are not themselves subject to infection by trematode larvae. Interestingly, the planarian *Dugesia tigrina* (Girard) releases a slimy secretion into the water; this immobilizes free-swimming miracidia and kills them within a few minutes. In these experiments, as few as five planarians placed in water together with miracidia and snails protected most of the snails from infection, while the presence of 15 planarians no snails were infected. In the absence of planarians,

93% of the snails were infected. This protection is, however, provided only by fresh secretions (up to 6 h) and not by planarians which had been in the water 23 to 30 h previously. Mosquito larvae, too, but not pupae, reduce the infection rate; *Culex pipiens* L. showed itself to be three times as effective as *Aedes aegypti* L. In field conditions, a large number of the miracidia present in the water probably perish in the presence of planarians and mosquito larvae, with the result that the chances of infection of the vector snails are reduced. This illustrates the problem involved when a natural environment is tampered with, in this case by continued chemical control of mosquitos.

According to GLAUDEL and ETGES (1973), planarians can be used for biological control of *Schistosoma* sp. because they are toxic to trematode larvae. Fifty percent of the larvae died after 30 s in water containing secretions of *Dugesia dorotocephala* Marg. MURRAY, while in the absence of the planarians more than 85% were still alive after half an hour. The nematode *Daubaylia potomaca* Chitwood and Chitwood also releases a substance which immobilizes the miracidia of *Schistosoma mansoni* Sambon (MICHELSON 1963) and thus protects sensitive intermediate hosts to some degree.

Since many of the substances mentioned are capable of suppressing growth or reproduction in pest gastropods, they could, extending the use of the term "insectistatica" coined by LEVINSON (1975, 1976), be regarded as **"molluscistatica".** The use of these substances, possibly to be expected in the future, could serve to prevent population increase of pest gastropods, keeping their numbers below the threshold limit; in such a way it would be possible to avoid the disadvantageous selective pressures exerted by pesticides, chosen for their ability to kill rapidly and in large numbers.

Sources of the Illustrations

ANT, H., Hamm, originals: Fig. 34, 35, 36, 37, 38, plate V/2.
BARRY, D., Columbia, original: Fig. 72.
BERG, CL., Ithaca, 1959, original: Fig. 118.
BROOKS, W.M., Raleigh, 1968, originals: Fig. 104, 105, 107.
BROOKS, W.M., drawing Alex, Berlin, from copy: Fig. 106.
BUHL and SCHÜTTE, 1971, original drawing by F. Schütte, Heikendorf (Federal Biological Research Centre): Fig. 79.
CHERNIN, E., Boston, 1957, drawing Alex, Berlin, from copy: Fig. 100.
CHERNIN, E., 1972, drawing Alex, Berlin, from copy: Fig. 125.
CHEVALLIER, D., Paris, photograph Dinkloh, Berlin-Dahlem (Federal Biological Research Centre), from original: plate IV/3, IV/4.
COE, W.R. (from Durchon 1967), drawing Alex, Berlin, from copy: Fig. 9.
CROWELL, H.H., Corvallis, originals: plates VII/3, VII/4, VII/5.
DAXL, R. Managua, drawing Poeschel, Stuttgart, from original: Fig. 92.
DAXL, R., 1969, original drawing: Fig. 54, 55.
DAXL, R., 1970, original drawing: Fig. 26.
DAXL, R., 1970, drawing Alex, Berlin, from copy: Fig. 97, 98.
DIRLBECK, K., Prague, originals: Fig. 67, 70, 124.
Federal Biological Research Centre for Agriculture and Forestry, Berlin-Dahlem, originals: Fig. 13, 17, 18 (centre, below), 28, 31, 32, plates I/2, I/3, III/1, III/2, III/3, III/4, III/5, III/6, IV/1, IV/2, VI/3, VII/2.
FISHER et al., original by Fisher, T.W., Riverside: Fig. 68.
FISHER et al., 1976, drawing Poeschel, Stuttgart, from copy: Fig. 116.
FRÖMMING, E., drawing Poeschel, Stuttgart, from copy: Fig. 41.
FRÖMMING and PLATE, 1951 [Anz. Schädlingskunde 24 (6), originals from publisher Parey, Berlin]: Fig. 62, 63, 64, 65, 66.
GERHARDT, U., 1933, drawing Alex, Berlin, from copy: Fig. 16.
GODAN, D., Berlin-Dahlem, drawing Alex, Berlin, from original: Fig. 2, 49, 51.
GODAN, D., drawing Poeschel, Stuttgart, from original: Fig. 1, 50.
GODAN, D., 1973, drawing G. Koch, Berlin, from original: Fig. 59, 60, 61, 101, 102.
GODAN, D., photograph Alex Berlin, from original: plate II/6.
GODAN, D., photograph Dinkloh, Berlin-Dahlem (Federal Biological Research Centre), from original: Fig. 3, 7 (centre), plates VII/6, VII/7.
GODAN, D., photograph Schälow, Berlin-Dahlem (Federal Biological Research Centre), from original: Fig. 18 (above), 27, plates I/1, I/4, I/5, I/6.
GODAN, D., 1966, original drawing: Fig. 82, 83.
GODAN, D., 1961, original drawing: Fig. 84, 85.
GODAN, D., 1965, original drawing: Fig. 93, 94.
GODAN, D., 1965, drawing Alex, Berlin, from original: Fig. 89.
GODAN, D., 1967, original drawing: Fig. 90, 91.
GODAN, D., 1965, photograph Schälow, Berlin Dahlem (Federal Biological Research Centre), from original: Fig. 86, 95, 96.
GÖTTLE, L. and DOSTLER, M., Backnang, original: plate II/1.
GUYARD, A., Pointe-à-Pitre, 1971, drawing Alex, Berlin, from original: Fig. 21, 22.
HERING, M., Bernkastel-Kues (Federal Biological Research Centre), 1966, originals: Fig. 73, 74, 75, 76, 77, 78.
HODASI, J.K., Legon, 1972, (Parasitology 65), drawing Alex, Berlin, from copy: Fig. 110.

HOLM et al., 1970, drawing Poeschel, Stuttgart, from original photograph by Holm, L., Madison: Fig. 112.
HUNTER and SYMONDS, 1971, drawing Alex, Berlin, from copy: Fig. 57, 58.
JUNGBLUTH, H., Heidelberg, original: Fig. 7 (below).
KAISER, W., Traisa, original: Fig. 14.
KAUFMANN, I., Brunswick, photograph Dinkloh, Berlin-Dahlem (Federal Biological Research Centre) from original: plates VI/1, VI/2.
KILIAS, R., Berlin, 1960, drawing Alex, Berlin, from copy: Fig. 19, 24 (B), 25.
KNUTSON, L.V., Beltsville, drawing Poeschel, Stuttgart, from original photograph: Fig. 119.
KNUTSON et al., 1970, original drawing by Knutson, L.V., Beltsville: Fig. 120.
KNUTSON et al., 1970, original photograph by Knutson, L.V., Beltsville: Fig. 121.
KNUTSON and BERG, 1964, drawing Alex, Berlin, from original by Knutson, L.V. Beltsville: Fig. 122.
KNUTSON and BERG, 1964, drawing Poeschel, Stuttgart, from original photograph by Knutson, L.V., Beltsville: Fig. 123.
KOCH, W., Brunswick (Federal Biological Research Centre), original: plate V/1, VI/4.
KURIS, A., Santa Barbara, 1973, original drawing: Fig. 109.
MARTIN and FORREST, 1969, drawing G. Koch, Berlin, from copy: Fig. 88.
MAZEK-FIALLA, K., 1933, drawing Alex, Berlin, from copy: Fig. 24 (A).
MEAD, A.R., Tucson, original: Fig. 113.
MEIER-BROOK, CL., Tübingen, photograph Alex, Berlin, from original: plate VI/5, VI/6, VI/7, VI/8.
MOENS, R., Gembloux, 1972, original drawing: Fig. 87.
MOENS et al., 1967, original drawing by Moens, R., Gembloux: Fig. 103.
PAINE, R.T., 1963 (Ecology 44), drawing from copy: Fig. 111.
PLATE, H.P., Plant Protection Office, Berlin, photograph Alex, Berlin, from original: Fig. 29, 30, plates: V/3, V/4, V/5, V/6.
POLLARD, E., Huntington, 1973, drawing Alex, Berlin, from copy: Fig. 81.
PROTA, R., Sassari, 1961, original: Fig. 69.
RICOU, G., Rouen, original: Fig. 71, plate VII/1.
RIEDEL and WIKTOR, 1974, drawing Alex, Berlin, from copy: Fig. 4, 5, 42.
RIETSCHEL, G., Giessen, 1970, original: Fig. 108.
RUNHAM, N., Bangor, original: Fig. 7 (above), 15, 47.
RUNHAM and HUNTER, 1970, (Terrestrial Slugs, Hutchinson and Co) (Publishers) drawing Alex, Berlin, from copy: Fig. 6, 8, 10, 12.
RUNHAM, BAILEY and LARYEA, 1973, drawing Alex, Berlin, from copy: Fig. 20.
SCHWALB, H., Grünstadt, 1961, original: Fig. 115.
THOMPSON, T.E., Bristol, 1973, drawing Alex, Berlin, from copy: Fig. 11.
TUKE, E., London [F. and M. Jones: Pests of field crops. Slugs and Snails, Edward Arnold (Publishers) London], original: plates II/2, II/3, II/4, II/5.
TURK and PHILLIPS, 1945, drawing Alex, Berlin, from copy: Fig. 114.
VAN DEN BRUEL, E., Gembloux, 1968, original: Fig. 48.
VAN DINTHER, J., Wageningen, 1957, original: Fig. 33.
VAN DINTHER, J., 1956, drawing Poeschel, Stuttgart, from original: Fig. 39, 40.
VOELKER, I., Hamburg, original drawing: Fig. 23.
VOELKER, I., 1968, original photograph: Fig. 117.
VOELKER, I., 1975 (Nauck: Lehrbuch der Tropenkrankheiten, Georg Thieme, Stuttgart), original: plate IV/5, IV/6.
WALDÉN, H.W., Göteborg, 1965, drawing Alex, Berlin, from copy: Fig. 46.
WEBLEY, D., Llanished, 1965, drawing G. Koch, Berlin, from copy: Fig. 99.
WHITE, A.R., 1959 (with permission of the Controller of Her Majesty's Stationary Office), drawing G. Koch, Berlin, from copy: Fig. 52, 53.
WIKTOR, A., Wroclaw, original drawing: Fig. 44, 45.
WIKTOR, A., 1973, drawing Alex, Berlin, from original: Fig. 43.
WOLDA, H., Balbao, 1970, drawing Alex, Berlin, from copy: Fig. 56.
WOLDA and KREULEN, 1973, drawing G. Koch, Berlin, from copy: Fig. 80.

References

ABOLINS-KROGIS, H.: Shell regeneration in *Helix pomatia* with special reference to the elementary calcifying particles. Symp. Zool. Soc. London 22, 75–92, 1968.

ADAM, W.: Mollusques 1. Mollusques terrestres et dulcicoles. Faune de Belgique, Brussels, 1–402, 1960.

ADEGOKE, D.: Mineralogy and biochemistry of calcareous operculi and shells of some gastropods. Malacologia 14, 39–46, 1973.

ADLUNG, K.-G., und KAUTH, H.: Prüfung der molluskiziden Eigenschaften einiger insektizider Wirkstoffe. Z. ang. Zool. 43, 301–305, 1956.

AGUILAR, J., und PACQUÉTIAN, B.: Dégâts de limaces sur tournesol. Phytoma 15, 27–28, 1963.

AKKERMANN, R.: Süßwassermuscheln als tierische Zukost des Bisam, *Ondatra zibethica*. Bonner Zool. Beiträge 23, 61–65, 1972.

AKRAMOVSKII, N. N. und BABAYAN, G. S.: Controlling slugs in hotbeds. Zashch. Rast. Nr. 8, 32, 1977.

ALBONICA, R.: Die Farbvarietäten der großen Wegschnecken. Rev. Suisse Zool. Genève 55, (25), 347–425, 1948.

ALICATA, J. E.: Observations on the biology and control of a garden slug injurious to orchids in Hawaii. Pacific Orchid Soc. Hawaii 8, 279–285, 1950.

ALLOWAY, T., KRASNE, F. R., LAHUE, R. P., WELLS, P., und WILLOWS, A. O. C.: Arthropods and Gastropod/Mollusks, Bd. 2. In: Invertebrate Learning (Ed. CORNING, W. C., und DYAL, J. A.). Cambridge 284 p., 1973.

ALONSO, M. R.: Fauna malacologica terrestre de la depresion de Granada (España). I. Pulmonados desnudos. Cuad. C. Biol. 4, 71–88, 1975.

– Fauna malacologica terrestre de la depresion de Granada (España). II. El Genero *Helicella* Férussac, 1821. Cuad. C. Biol. 4, 11–28, 1975.

– und IBAÑEZ, M.: Nota sobre la presencia de *Oxychilus (Oxychilus) draparnaudi* (Beck) *(Mollusca, Zonitidae)* en la depresion de Granada (España). Cuad. C. Biol. 4, 29–31, 1975.

AL-TAHAFI, N.: An Approach to the Control of *Schistosoma haematobium* with a Special Reference to Iraq. Bull. endem. Disease 14, 113–132, 1973.

ALVAREZ, J.: *Lymnaea stagnalis* (L.) en Espagne. Haliotis 2, 41–42, 1972.

ALZONA, C.: Malacofauna italica, Catalogo e bibliografia dei Molluschi viventi, terrestri e d'acqua dolce. Atti Soc. It. Sc. Nat. e Mus. Civ. St. Nat. Milano 111, 1–433, 1971.

Amtl. Pflanzenschutzmittelverzeichnis u. Pflanzenschutzgeräteverzeichnis 1970 der Bundesanstalt für Pflanzenschutz, Wien. Der Pflanzenarzt (Wien) 23, 1. Sondernr. März 1970.

AN DER LAAN, H., und ASPÖCK, H.: Zur Wirkung von Sevin auf Regenwürmer. Anz. Schädlingskunde 35, 180–182, 1962.

ANDERSON, K. C.: On the development and transmission of *Cosmocercoides dukae* of terrestrial mollusces in Ontario. Can. J. Zool. 38, 801–825, 1960.

ANDERSON, W., und PERSONNE, P.: The fine structure of the neck region of spermatozoa of *Helix aspersa*. J. Microsc. 6, 1033–1042, 1967.

ANDERSSON, K., und NILSSON, I.: [Ackerschnecken, ein Pflanzenschutzproblem auf schweren Böden]. Vaxtskyddsnotiser 31, 67–70, 1967.

ANDRÉ, M., und LAMY, E.: Les acariens parasites des mollusques. J. de Conchyl. 74, 199–221, 1930.

– und LAMY, E.: Les acariens parasites des mollusques (notes complémentaires). J. de Conchyl. 75, 322–327, 1931.

ANDRES, L. A., und BENETT, F. D.: Biological control of Aquatic Weeds. Ann. Rev. Entomol. 20, 31–46, 1975.

ANDREWS, E., und LITTLE, C.: Ultrafiltration in the gastropod heart. Nature (London) 234, 411–412, 1971.

– und LITTLE, C.: Structure and function in the excretory systems of some terrestrial prosobranch snails *(Cyclophoridae)*. J. Zool. Proc. Zool. Soc. London 168, 395–422, 1972.

ANONYM: Großschäden durch Ackerschnecken. Bauernblatt für Schleswig-Holstein 7/103, 1687–1688, 1953.

– The hunter snail. Time 67, 28–29, 9. Januar 1956.

– Les Molluscicides. II. Rapport du Commité d'experts de la Bilharziose. Bull. Org. mond. Santé, Série rapports techniques 214, 1–57, 1961.

– Ziram: Molluskizid, Fungizid, Larvazid und Herbizid [Le Zirame: molluscicide, fongicide, larvicide et herbicide]. Chronique OMS 16 (10), 409–410, 1962.

– Other people's pest problems. I. Mauritius. PANS 11, 285–294, 1965.

– Snails and Slugs. Planters Bull. Rubber Res. Inst. Malaya 93, 284–287, 1967.

– Giant African Snail invades Florida. News Bull. 11, 1–2, 1969.

– Snail Pest Fouls Wheat Harvest. World Crops 21, 362, 1969.

– Pflanzenschutz 1974/75. Herausgeber: Pflanzenschutzamt d. Landwirtschaftsk. Weser-Ems, Oldenburg. 8. Aufl., 1–125, 1975.

– Giant African Snail eradicated from continental United States. FAO Plant Protect. Bull. 23, 67–68, 1975.

ANSARI, N.: Epidemiology and Control of Schistosomiasis (Bilharziasis). Verlag Karger, Basel, 1–752, 1973.

ANSUL SCHUTH, C. K., ISENSEE, A. R., WOOLSON, E. A., und KEARNEY, P. C.: Distribution of ^{14}C and Arsenic Derived from (^{14}C) Cacodylic Acid in an Aquatic Ecosystem. J. Agr. Food Chem. 22, 999–1003, 1974.

ANT, H.: Faunistische, ökologische und tiergeographische Untersuchungen zur Verbreitung der Landschnecken in Nordwestdeutschland. Abh. Landesmus. Naturk. Münster 25, 1–125, 1963.

– Eine neue Nacktschnecke, *Boettgerilla (pallens?)* vermiformis, in Westfalen. Natur und Heimat (Münster) 26, 71–74, 1966.

– Volkstüml. Namen für einheimische Weichtiere. Natur und Heimat 28 (4), 166–168, 1968.

– Bestimmungstabellen für einheimische Mollusken. Hamm 1973, 1–32, 2. Aufl.

– Erfassung der europäischen Wirbellosen. Kartierungsanweisungen. European Invertebrate Survey. Biological Records Centre Huntingdon und Hamm, 1973, 23 p.

– Schutz bedrohter Weichtiere. Naturwissensch. Rundschau 26, 157–158, 1973.

ANTESON, R. K.: On the resistance of the snail, *Lymnaea catascopium pallida* (Adams) to concurrent infection with sporocysts of the strigeid trematodes, *Cotylurus flabelliformis* (Faust) and *Diplostomum flexicaudum* (Cort and Brooks). Am. trop. Med. Parasit. 64, 101–107, 1970.

ARCH, S.: Biosynthesis of the egg-laying hormon (ELH) in the bag cell neurons of *Aplysia californica.* J. Gen. Physiol. 60, 102–119, 1972.

ARIAS, R. O., und CROWELL, H. H.: A contribution to the biology of the grey garden slug. Bull. Calif. Acad. Sci. 62, 83–97, 1963.

ARNOLD, R.: Studies on *Cepaea:* VII Climatic selection in *Cepaea nemoralis* (L.) in the Pyrenées. Phil. Trans Roy. Soc. London, Ser. B. Biol. Sci 253, 549–593, 1968.

– The effects of selection by climate on the landsnail *Cepaea nemoralis* (L.). Evolution 23, 370–378, 1969.

– A comparison of populations of the polymorphic land snail *Cepaea nemoralis* (L.) living in a lowland district in France with those in a similar district in England. Genetics, 64, 589–604, 1970.

ARNOLD, W. J.: Snail and lawn pest control with Zectran pesticide. Down to Earth 18, 20–24, 1962.

Arnoux, J., und Ritter, M.: Etude expérimentale des modalités d'application du metaldehyde contre les limaces. Meded. Landbouwhog. Opz. Gent 19, 571–580, 1954.

Arun Shanker, N.: The amoebocytes of lamellibranch molluscs, with special reference to the circulating amoebocytes. Malacol. Rev. 6, 1–12, 1973.

Arvy, L.: Données sur les parasitoses cercariennes des mollusques et leurs corrélations avec les fluctuations des populations. Bull. Lab. mar. Dinard 1, 207–216, 1972.

Asher, P.: Inhibitory and excitatory effects of dopamine on *Aplysia* neurones. J. Physiol. (London) 225, 173–209, 1972.

Aspöck, H., und An der Laan, H.: Ökologische Auswirkungen und physiologische Besonderheiten des Pflanzenschutzmittels Sevin (1-Naphthyl-N-methylcarbamat). Z. ang. Zool. 50, 343–380, 1963.

Aubry, R.: Les élements nourriciers dans la glande hermaphrodite de *Limnaea stagnalis* adulte. C. R. Soc. Biol. Fr. 148, 1626–1629, 1954.

– La structure de l'acinus et la lignée femelle dans la glande hermaphrodite de *Limnaea stagnalis* adulte. C. R. Soc. Biol. Fr. 148, 1498–1500, 1954.

– La lignée mâle dans la glande hermaphrodite de *Limnaea stagnalis.* C. R. Soc. Biol. Fr. 148, 1856–1859, 1954.

– L'action de l'oestradiol et de la progestérone sur la gonade des Pulmonés. C. R. Acad. Sc. Paris, 248, 1225–1227, 1959.

– Etude de l'hermaphrodisme et de l'action pharmacodynamique des hormones de Vertébrés chez les Gastéropodes pulmonés. Arch. Anat. micr. Morphol. exp. suppl., 50, 521–602, 1962.

Australian Dep. of Health Cargo containers and Unit loads, Quarantine aspects & procedures. Canberra 1975.

Ayad, M.: Die Bekämpfung der Bilharziose in der Vereinigten Arabischen Republik. Pflanzenschutz-Nachr. „Bayer" 19, 1–7, 1966.

–, Mousa, A. H., Ishak, M. M., Yousif, F., und Zaghloul, S.: A preliminary study on biological control of the snail intermediate hosts of Schistosomiasis in U.A.R. by *Helisoma duryi* snails. Hydrobiologia (Den Haag) 35, 196–202, 1970.

Baalawy, S. S.: Laboratory evaluation of the molluscidal potency of a butanol extract of *Phytolacca dodecandra* (endod) berries. Bull. Wld. Hlth. 47, 422–425, 1972.

Backhuys, W.: Notes on *Theba pisana ustulata* (Lowe, 1852), the landsnail of the Salvages Islands. Basteria 26, Nr. 2–5, 1972.

– und Ripken, T. E. J.: *Cernuella (Xerocincta) neglecta* (Draparnaud, 1805), een nieuwe soort voor Nederland. Basteria 30, 42–47, 1966.

Badino, G.: I fattori della gametogenesis di *Arion rufus* studiati con il metodo della cultura in vitro. Arch. Zool. Ital. 52, 271–275, 1967.

Bailey, T. G.: Studies on organ cultures of slug reproductive tracts. Ph. D. Thesis Univ. of Wales, 1970.

– Haemolymph proteins of the slug *Agriolimax reticulatus.* Comp. Biochem. Physiol. B. Comp. Biochem. 40, 699–707, 1971.

– The in vitro culture of reproductive organs of the slug *Agriolimax reticulatus* (Muell.). Neth. J. Zool. 23, 72–85, 1973.

Baker, R. E.: The ecology of the wrinkled snail, *Helicella caperata* Mont., on the Braunton Burrows sand dune system. Proc. Malacol. Soc. (London) 38, 41–54, 1968.

Barbosa, F. S.: Possible competitive displacement and evidence of Hybridization between brazilian species of Planorbid snails. Malacologia 14, 401–408, 1973.

Bardon, I., Drosdowsky, M., De Longcamp, D., und Lubet, P.: Recherches recentes sur les stéroïdes des Mollusques. Activité stéroïdogene dans les gonades de *Mytilus edulis* L. et *Crepidula fornicata* Phil. Haliotis 1, 15–16, 1971.

–, Lubet, P., und Drosdowsky, M. A.: Biosynthese des stéroïdes chez un mollusque gastéropode marin, *Crepidula fornicata* Phil. Steroidologia 2, 366–377, 1971.

BARETT, R. C., GIRARD, C., und RIOU, J.: Sur certaines propriétés des arginases du tissue hepatopancreatique d'*Helix pomatia* Lin. et d'*Helix aspersa* Muell. Biochim. (Paris) 54(4), 421–430, 1972.

BARNES, H. F.: Discussion on slugs. I. Introduction. Seasonal activity of slugs. Ann. appl. Biol. 31, 160–163, 1944.

– und WEIL, J. W.: Baiting slugs using metaldehyde mixed with various substances. Ann. appl. Biol. 29, 56–58, 1942.

– und WEIL, J. W.: Slugs in gardens: Their numbers, activities and distribution. Part 1. J. anim. Ecol. 13, 140–175, 1944.

– und WEIL, J. W.: Slugs in gardens: Their numbers, activities and distribution, Part 2. J. anim. Ecol. 14, 71–105, 1945.

– und STOKES, B. M.: Marking and breeding *Testacella* slugs. Ann. appl. Biol. 38, 540–545, 1951.

BARNETT, J. E. G.: An acid ß-galactosidase from the albumen gland of *Helix aspersa*. Comp. Biochem. Physiol. B. Comp. Biochem. 40, 585–592, 1971.

BARRETT, B. W.: *Theba pisana* in Guernsey, 1971. J. Conchol. 27, 391–396, 1972.

BARROW, J. H., Jr.: Observations on *Minchinia pickfordae* (Barrow 1961) found in snails of the Great Lake region. Trans. Amer. Microsc. Soc. 84, 587–593, 1965.

BARRY, B. D.: Evaluation of chemicals for control of slugs on field corn in Ohio. J. econ. Ent. 62, 1277–1279, 1969.

BARRY, M. M., und YEVICH, P. P.: Incidence of gonadal cancer in the quahaug *Mercenaria mercenaria*. Oncology (Basel) 26, 87–96, 1972.

BASCH, P. F.: Relationships of some larval strigeids and echinostomes (Trematoda): Hyperparasitism, antagonism, and „Immunity" in the snail host". Exp. Parasit. 27, 193–216, 1970.

– LIE, K. J., und HEYNEMAN, D.: Antagonistic interaction between strigeid and schistosome sporocysts within a snail host. J. Parasit. 55, 752–758, 1969.

BASINGER, A. J.: The European Brown Snail in California. Bull. 515. Univ. California, Berkeley, Calif, 1–22, 1931.

– A valuable snail poison. J. econ. Ent. 16, 456–458, 1953.

BATTE, E. G., und SWANSON, L. E.: Laboratory evaluation of organic compounds as molluscides and ovocides. J. Parasit. 38, 65–68, 1952.

BAUER, K.: Studien über Nebenwirkungen von Pflanzenschutzmitteln auf Fische und Fischnährtiere. Mitt. Biol. Bundesanst. Land- u. Forstwirtschaft, Berlin-Dahlem, 105, 1–72, 1961.

BAUM, J. M., und RAWLS, H. C.: Fluorescence caused by *Pseudomonas* in the mucus of *Anguispira kochi* (Pfeiffer). Sterkiana 45, 1–13, 1972.

BAYNE, B. L., und THOMPSON, R. J.: Some physiological consequences of keeping *Mytilus edulis* in the laboratory. Helgoländer wiss. Meeresunters. 20, 526–552, 1970.

BAYNE, C. J.: A study of the dessiccation of egg capsules of eight gastropod species. J. Zool. (London) 155, 401–411, 1968.

– On the reported occurence of an ecdysonelike steroid in the freshwater snail, *Biomphalaria glabrata (Pulmonata; Basommatophora)*, intermediate host of *Schistosoma mansoni*. Parasitology 64, 501–509, 1972.

– Physiology of the pulmonate reproductive tract: Location of spermatozoa in isolated, self-fertilizing succinid snails, with a discussion of pulmonate tract terminology. Veliger 16, 169–175, 1973.

– und KIME, J. B.: In vivo removal of *bacteria* from the haemolymph of the land snail *Helix pomatia (Pulmonata: Stylommatophora)*. Malacol. Rev. 3, 103–113, 1970.

BAYNE, Ch.: Organization of the spermatozoon of *Agriolimax reticulatus*, the Grey Field Slug *(Pulmonata, Stylommatophora)*. Z. Zellforsch. 103, 75–89, 1970.

BECKER, A.: Ackerschnecken bedrohen die Wintersaaten. Landw. Zt. Nord.-Rheinprov. 118, 996, 1951.

BECKER, K. F.: Metaldehyd; Kohlrabi, Kopfsalat, Möhren, Rotkohl, Wirsing. In: Methodensammlung zur „Rückstandsanalytik von Pflanzenschutzmitteln". Herausg.: DFG 151(1–4). Verlag Chemie, Weinheim, 1969.

BECKER, W.: The glucose content in haemolymph of *Australorbis glabratus*. Comp. Biochem. Physiol. A. Comp. Physiol. 43, 809–814, 1972.

BEEDHAM, G. E.: Identification of the British Mollusca., Hulton Group Keys, Amersham, 1–236, 1972.

BEEMAN, R. D.: An autoradiographic and phase contrast study of spermatogenesis in the anaspiedean opistobranch *Phyllaplysia taylori* Dall, 1900. Arch. Zool. Exp. Gen. 111, 5–22, 1970.

BELL, E.-J., ETGES, J., und JENELLE, L. J.: The identity and source of a molluscicide-degrading bacterium. Amer. J. Trop. Med. 15, 539–543, 1966.

BENJAMIN, P. und WALKER, T.: Two pigments in the brain of a freshwater pulmonate snail. Comp. Biochem. Physiol. B. Comp. Biochem. 41, 813–821, 1972.

BENTHEM-JUTTING, W. S. S.: Mollusca 1. A. *Gastropoda Prosobranchia* et *Pulmonata*. Fauna van Nederland, 1–387; Leiden 1933.

– *Theba pisana* (Müller) levend in Zeeland. Basteria 26, 57–58, 1962.

– und VAN REGTEREN-ALTENA, C. O.: Naamlijst van de Nederlandse land- en zoetwatermollusken. Corr. Blad Ned. Malac. Ver. 99, 1019–1023, 1962.

BERG, C. O.: Biology of snail-killing Sciomyzidae *(Diptera)* of North America and Europe. XI. Int. Kongress Entomol. Wien, 17.–25. Aug. 1960, I. 197–202, 1960.

– Can malacophagous diptera control trematode disease. Wld. Hlth. Org. 24, 1–9, 1964.

– Sciomyzid larvae *(Diptera)* that feed on snails. J. Parasitol. 39, 629–636, 1953.

– New approach – to control of flukes. Farm Res. 25, 8–9, 1959.

– Snail control in trematode diseases: The possible value of Sciomyzid larvae, snail-killing *Diptera*. Advanc. in Parasitol. 2, 259–309, 1964.

– Biological Control of Snail-Borne Diseases: A Review. Exp. Parasitology 33, 318–330, 1973.

– Medically Important Snails. In Huffacker, C. B.: Theory and Practice of Biological Control. Univ. of California, Berkeley 1050, Albany, Calif. 1976.

– FOOTE, B. A., und NEFF, S. E.: Evolution of predatoryprey relationships in snail-killing sciomyzid larvae *(Diptera)*. Amer. Mal. Union Bull. 25, 10–11, 1959.

BERGERARD, J.: Facteurs écologiques et cycle sexuel de *Littorina saxatilis* (Olivi)*(Mollusca, Gastropoda)*. Cah. Biol. Mar. 12. 187–193, 1971.

BERKIUS, R.: The circulatory system of *Lymnaea stagnalis* (L.). Neth. J. Zool. 22, 1–58, 1972.

BERNER, L.: Sur le parasitisme des Helicidés par des mouches du genre *Sarcophaga*. Bull. Mus. Hist. Nat. Marseille 33, 87–93, 1973.

BERRIE, A. D.: On the life cycle of *Lymnaea stagnalis* (L.) in the west of Scotland. Proc. Malacol. Soc. (London) 36, 283–295, 1965.

– und VISSER, S. A.: Investigations of a growth-inhibiting substance affecting a natural population of freshwater snails. Physiol. Zool. 36, 167–171, 1963.

BERRIOS-DURAN, L. A., und RITCHIE, L. S.: Molluscicidal activity of bis (tri-n-butyltin)oxide formulated in rubber. Bull. Wld. Hlth. Org. 39, 310–312, 1968.

BERRY, A. J., und CHAN, L. C.: Reproductive condition and tentacle extirpation in Malayan *Achatina fulica (Pulmonata)*. Austr. J. Zool. 16, 849–855, 1968.

BESS, H. A.: Ecological notes on *Lamprophorus tenebrosus* (Walker) *(Coleoptera: Lampyridae)*, an enemy of the giant African snail. Proc. Hawaiian Ent. Soc. 16, 24–29, 1956.

– und ALICATA, J. E.: Japanese lampyrids introduced. Proc. Hawaiian Ent. Soc. 15, 12, 1953.

BEST, L. W.: Food of the roof-rat, *Rattus rattus rattus* (L.), in two forrest areas of New Zealand. New Zealand J. Sci. 12, 258–267, 1969.

BETT, J. A.: The breeding seasons of slugs in gardens. Proc. Zool. Soc. London 135, 559–568, 1960.

BEYNON, K. I., und WRIGHT, A. N.: Breakdown of the Molluscicide N-tritylmorpholine in Soil and Rice. Bull. Org. mond. Santé 37. 65, 1967.

– WRIGHT, A. N., BOSIO, P., MCGILL, A. E. J., und ROBINSON, J.: The fate of the molluscicide N-Tritylmorpholine following its use on pasture for the control of liver fluke. Pesticide Sci. 3, 689–703, 1972.
– und WRIGHT, A. N.: An improved method for the analysis of Trifenmorph in water residues, molluscicide, bilharziosis vectors, control. Pestic. Sci. 6, 515–523, 1975.
BIERBAUER, J., und TÖRÖK, L. J.: Cytological and neurosecretory investigations on the optic tentacle of *Pulmonata.* Acta biol. Acad. Sci. hung. 15, suppl. 6, 39–40, 1964.
– TÖRÖK, L. J., und TEICHMANN, J.: Cytologische und histochemische Untersuchungen am neurosekretorischen System der Augententakel von Pulmonaten. Zool. Jb. Physiol. 71, 545–551, 1965.
– und MOLNAR, J.: [Experimentally influenced regulation of gametogenesis in *Pulmonata* during the hibernation period]. Allatani Kozl. 59, 36–38, 1972.
BINGHAM, Fr.: Several aspects of the reproductive biology of *Littorina irrorata (Gastropoda).* Nautilus 86, 8–10, 1972.
– The influence of environmental stimuli on the direction of movement of the supra littoral gastropod *Littorina irrorata.* Bull. mar. Sci. 22, 309–335, 1972.
BLACKBURN, R. D., TAYLOR, T. M., und SUTTON, D. L.: Temperature tolerance and necessary stocking rates of the *Marisa cornuarietis* for aquatic weed control. Proc. Eur. Weed. Res. Counc., 3rd Int. Symp. Aquatic Weeds, Oxford. 79–86, 1971.
BLAIR, D. M.: Dangers in Using and Handling Sodium Pentachlorophenate as a Molluscicide. Bull. Wld. Hlth. Org. 25, 597–601, 1961.
BLAŽKA, P.: Temperaturadaption des Gesamtmetabolismus bei der Weinbergschnecke *Helix pomatia* L. Zool. Jahrb. Abt. Allg. Zool. Physiol. Tiere 65, 430–438, 1955.
BLOCK, M. R.: Epiphragmas: some observations. J. of Conch. 26, 388–409, 1971.
– und HEW, H. Y. C.: Schedule of spermatogenesis in the Pulmonate Snail *Helix aspersa* with special reference to histone transition. J. biophys. biochem. Cytol. 7, 515–535, 1960.
BLOWERS, J. W.: Slug damage to orchids. Gardeners Chronicle Illustr. 153, 31, 1963.
BOBACK, A. W.: Zur Frage eines forstlichen Schadens durch die Waldschnecke *Arion subfuscus* Drap. Anz. Schädlingskde. 25, 189, 1952.
BOER, H. H.: A cytological and cytochemical study of neurosecretory cells in *Basommatophora* with particular reference to *Lymnaea stagnalis* L. Arch. Néerl. Zool. 16, 313–386, 1965.
– und JOOSSE, J.: Endocrinology. In: FRETTER, V., und PEAKE, J.: Pulmonates, Vol. I. Chapt. 6, 245–307. Academic Press 1975.
– SLOT, J. W., und VAN ANDEL, J.: Electron microscopical and histochemical observations between medio-dorsal bodies and neurosecretory cells in the basommatophorean snails *Lymnaea stagnalis, Ancylus fluviatilis, Australorbis glabratus* et *Planorbarius corneus.* Z. Zellforsch. mikr. Anat. 87, 435–450, 1968.
BOERNER, F.: Taschenwörterbuch der botanischen Pflanzennamen. 1–435. Parey, Berlin-Hamburg 1966, 2. Aufl.
BOETTGER, C. R.: Schnecken in Sendungen von Kamerun-Bananen. Sitz. ber. Ges. naturf. Freunde zu Berlin, 83–84, 1938.
– Die Nordgrenze der Nacktschnecke *Milax rusticus* Millet in Westdeutschland. Arch. Moll. 78, 53–56, 1949.
– Größenwachstum und Geschlechtsreife bei Schnecken und pathologischer Riesenwuchs als Folge einer gestörten Wechselwirkung beider Faktoren. Zool. Anz. Suppl. Nr. 17, 468–487, 1953.
BOGUSCH, G.: Zur Histochemie der Cholinesterase im Penisretraktor von *Helix pomatia.* Z. Mikrosk. Anat. Forschg. (Leipzig) 85, 35–49, 1972.
BÖHM, O.: Über Schnecken und Schneckenbekämpfung. Pflanzenschutz Berichte Wien 19, 111–129, 1957.
– Schädliche Schnecken im Pflanzenschutz. Pflanzenarzt Wien, 10, 40–43, 1957.
– Erfolgreiche Schneckenbekämpfung mit einem Spritzmittel. Pflanzenarzt Wien. 14(9), 83–84, 1961.

Booth, Ching-Chieh Yu, G. M., Hansen, D. J., und Larsen, J. R.: Fate of Carbofuran in a Model Ecosystem. J. Agr. Food Chem. 22, 431–434, 1974.

Boray, J. C.: Studies on the ecology of *Lymnaea tomentosa,* the intermediate host of *Fasciola hepatica* L., History, geographical distribution, and environment. Austral. J. Zool. 12, 217–230, 1964.

– The role of the relative susceptibility of snails to infection with helminths and of the adaptation of the parasites in the epidemiology of some helminthic diseases. Malacologica 14, 125–127, 1973.

Borg, A.: Bekämpningsförsök mot åkersniglar på fältgrödor. Växtskyddsnotiser 17, 74–80, 1953.

– Ackerschnecken auf Herbstölpflanzen. Växtskyddsnotiser 26, 75–77, 1962.

Borthwick, P. W., Cook, G. H., und Patrick, Jr. J. M.: Residues in fish, wildlife and estuaries. Mirex Residues in Selected Estuaries of South Carolina – June 1972. Pestic. Monit. J. 7, 144–145, 1974.

Boss, K. J.: Critical estimate of the number of recent *Mollusca.* Occas. Pap. Mollusks Mus. Comp. Zool. Harv. Univ. 3, 81–135, 1971.

Bottke, W.: Zur Morphologie des Ovars von *Viviparus contectus* (Millet 1813) *(Gastropoda, Prosobranchia):* 1. Die Follikelzellen. Z. Zellforsch. Mikrosk. Anat. 133, 103–118, 1972.

Bouchet, A., Daynes, P., und Ramalanjaona, Ch.: Etude des variations saisonnières des populations de *Lymnaea natalensis* dans le Moyen Ouest de Madagascar. Haliotis 3, 81–88, 1973.

Bouillon, J.: Influence of temperature on the histological evolution of the ovotestis of *Cepaea nemoralis* L. Nature 177, 142–143, 1956.

Boulange, J. M., De Leersnyder, M., und Hoerstland, H.: Installation du Gastéropode méditerranéen *Cochlicella ventricosa* Draparnaud sur les côtes de la Mer du Nord. Bull. Soc. zool. Fr. 83, 44–47, 1958.

Bowen, J. D., und Davies, P.: The fine structural distribution of acid phosphatase in the digestive gland of *Arion hortensis* (Fér.). Protoplasma 73, 73–81, 1971.

Boyce, C. B. C., Crossland, N. O., und Shiff, C. J.: A New Molluscicide, N-Tritylmorpholine. Nature (London) 210, 1140–1141, 1966.

Boycott, A. E.: The habitats of land mollusca in Britain. J. Ecol. 22, 1–38, 1934.

Božkov, D. K.: Die polyvalenten Wirte bei den Helminthen und ihre biologische Bedeutung. Angew. Parasitol. 11, 152–256, 1970.

Brahmachary, R. L., Mallik, B., und Tapaswi, P. K.: Transcription and degradation of heavy (pre 28 S) RNA and 4 S RNA during embryogenesis in *Limnaea.* Z. Naturforsch. Teil B. 27, 1264–1266, 1972.

Brahmanadam, V., und Krishnamoorthy, R. V.: Hemochemistry and hematology of the aestivating pond snail *Pila globosa (Gastropoda, Ampullariidae).* Veliger 16, 176–180, 1973.

Brandes, J.: Verschleppung von Landschnecken durch einen Singvogel. Arch. Molluskenkde. 80, 85, 1951.

Brandt, A.: Fischerei auf Schnecken und Muscheln. Protok. Fischereitechn. 11, 128–148, 1968.

Branson, B. A.: The slugs (Gastropode: *Pulmonata*) of Oklahoma and Kansas with new records. Trans. Kans. Acad. Sci. 65, 110–119, 1962.

– und Batch, D. L.: Notes on exotic mollusks in Kentucky. Nautilus 82, 102–106, 1969.

Brederlow, H.: Schneckenbekämpfung in der ostfriesischen Marsch – ein ungelöstes Problem. Gesunde Pflanzen 27, 218–221, 1975.

Breucker, H.: Cytologische Untersuchungen des Zwittergangs und des Spermioviduktes von *Helix pomatia* L. Protoplasma 58, 1–41, 1964.

Briel, G., Neuhoff, V., und Osborne, N.: Determination of amino acids in single identificable new cells of *Helix pomatia.* Int. J. Neurosic. 2, 129–136, 1971.

Brien, P.: Gamétogenèse et sexualisation. Ann biol. 9, 373–397, 1970.

BRINK, M., und BOER, H. H.: An electron microscopical investigation follicle gland (cerebral gland) and some neurosecretory cells in the lateral lobe of the cerebral ganglion of the pulmonate Gastropod *Lymnaea stagnalis* L. Z. Zellforsch. mikr. Anat. 79, 230–243, 1967.

BRISSON, P.: La castration chirurgicale chez *Bulinus* (*contortus* Michaud) *truncatus* (Audouin) (Mollusque Gastéropode Pulmoné). C. R. Acad. Sc. Paris, 264, 131–133, Sér. D. 1967.

– Développements intragonadiques observés à la suite de diverses interventions chez *Bulinus* (*contortus* Mich.) *truncatus* (Aud.) (Mollusque Gastéropode Pulmoné). C. R. Acad. Sc. Paris, 266, 403–405, 1968.

– Développement de l'embryo et de ses annexes et étude en culture in vitro chez les achatines (gastéropodes pulmonés). Arch. Anat. Microsc. Morph. exp. 57, 345–368, 1968.

– Contribution à l'étude des corrélations entre les différentes régions de l'appareil génital, par castration, ablation, implantation, chez quelques Mollusques Gastéropodes des Pulmonés Basommatophores et principalement chez *Bulinus truncatus* (Audouin). Thèse Sci., Poitiers, Arch. orig. Centre documentation C.N.R.S. 4320, 1–154, Juin 1970.

– Castration chirurgicale et régénération gonadique chez quelques Planorbidés (Gastéropodes Pulmonés). Ann. Embryol. morphogen. 4, 189–210, 1971.

– Développements embryonnaires intragonadiques observés à la suite de diverses interventions sur le tractus génital chez quelques gastéropodes pulmonés et principalement chez *Bulinus truncatus* (Audouin). Arch. Anat. Microsc. Morphol. Exp. 60, 365–388, 1971.

BRODTMANN, Jr. N. V.: Studies on the Assimilation of 1,1,1-Trichloro-2,2-bis(p-chlorophenol)ethane (DDT) by *Crassostrea virginica* Gmelin. Bull. Environem. Contam. Toxicol. 5, 455–462, 1970.

BROOKS, W. M.: Tetrahymenid Ciliates as Parasites of the Gray Garden Slug. Hilgardia 39, 205–276, 1968.

BROWN, A. C.: Elimination of Foreign Particles by the Snail, *Helix aspersa*. Nature 213, 1154–1155, 1967.

BROWN, D. S., und LEMMA, A.: The molluscan fauna of the Awash river, Ethiopia, in relation to the transmission of schistosomiasis. Ann. trop. Med. Parasit. 64, 533–538, 1970.

BROWN, E. B.: Some current British soil pest problems. In: KEVAN, D. K.; MC E. (Ed.) Soil Zoology London: Butterworths 256–268, 1955.

BRULINS, M. F. M., VAN REGTEREN-ALTENA, C. O. und BUTOT, L.: The Netherlands as an environment for land molluscs. Basteria 23 (suppl.). 132–162, 1959.

BRUNNACKER-DAUR, M.: Die Einwirkungen von Trematoden auf Wasserschnecken. Ein Beitrag zum Wirt-Parasit-Verhältnis. Z. Parasitenkde. 17, 193–216, 1955.

BRUSSARD, P. F.: Population size and natural selection in the land snail *Cepaea nemoralis*. Nature (London) 251, 713–715, 1974.

BRYAN, G. W.: The occurence and seasonal variation of trace metals in the scallops *Pecten maximum* (L.) and *Chlamys opercularis* (L.). J. Mar. Biol. Assoc. U K 53, 145–166, 1973.

BUHL, CL. und SCHÜTTE, F.: Prognose wichtiger Pflanzenschädlinge in der Landwirtschaft. 1–264, Parey, Berlin-Heidelberg 1971.

BUNDY, J. W.: The significance of pest and disease damage in the production and marketing of processed vegetables. Proc. Brit. Insect. and Fungic. Conf. Brighton 1967, 2, 534–544, 1967.

BURCH, J. B.: Some ecological factors of the soil affecting the distribution and abundance of land snails in Eastern Virginia. Nautilus 69, 62–69, 1955.

BURLA, H. und SPEICH, C.: *Lymnaea auricularia* and *Lymnaea ovata* im Zürichsee. Rev. Suisse Zool. 78, 549–556, 1971.

BURNETT, A. L.: Control of polarity and cell differentation through autoinhibition A model. Exp. Biol. Med. 1, 125–140. Karger ed. Basel, New York 1967.

BURTON, J. D., und LEATHERLAND, T. M.: Mercury in a Costal Marine Environment. Nature (London) 231. 440–441, 1971.

BURTON, R. F.: Natural variations in cation levels in the blood of three species of land snail *(Pulmonata; Helicidae)*. Comp. Biochem. Physiol. 39 (2 A), 267–275, 1971.

– The storage of calcium and magnesium phosphates and of calcite in the digestive glands of the *Pulmonata (Gastropoda)*. Comp. Biochem. Physiol. A. Comp. Physiol. 43(3), 655–663, 1972.

Butler, P. A.: Residues in Fish, Wildlife and Estuaries. Part I. General Summary and Conclusions. Pestic. Monitor J. 6, 238–262, 1973.

Butot, L. J. M.: De geschiedenis en de verspreiding van de Wijngaardslak in 's Gravenhage. De levende natuur 73, 275–282, 1970.

– De geschiedenis en de verspreiding van de Wijngaardslak langs de duinzoom. De levende natuur 75, 29–40, 1972.

– und Kiauta, B.: Notes on the cytology of *Rissoacea:* I. Cytotaxonomical conditions in some *Hydrobiidae* and *Assimineidae (Gastropoda Streptoneura)*. Basteria 30(2), 21–24, 1966.

Cadart, I.: Les Escargots (*Helix pomatia* L. et *Helix aspersa* M.). Savoir en histoire naturelle XXIV, Lechevalier Paris-VI, 1–420, 1955.

Cain, A. J.: Studies on *Cepaea.* V. Sand-dune populations of *Cepaea nemoralis* (L.). Phil. Trans. Roy. Soc. London Ser. B. Biol. Sci. 252, 499–513, 1968.

– und Currey, J. D.: Area effects in *Cepaea* on the Larkhill artillery ranges Salisbury Plain. J. Linnean Soc. Lond., Zool. 45, 1–15, 1963.

– und Currey, J. D.: Studies on *Cepaea.* III. Ecogenetics of a population of *Cepaea nemoralis* subject to strong area effects. Phil. Trans. Roy. Soc. London, B. 253, 447–482, 1968.

Cairaschi, E. A.: Problème de la protection sanitaire des luzernes-graines. Intern. Pflanzenschutzkongr. (Paris) 7, 588–589, 1970.

– und Lecomte, V.: Observations sur *Euparypha (Theba) pisana* Müller en luzernières à graines. Haliotis 3, 33–37, 1973.

– und Lecomte, V.: Dégâts d'un escargot *Euparypha (Theba) pisana* Müller en luzernières à graines. Phytoma 25, Nr. 250, 13–16, 1973.

Cairncross, G.: Combating slugs and snails. Gardeners chronicle 141, 307, 1957.

Cameron, R. A. D.: The survival, weight-loss and behaviour of three species of land snail in conditions of low humidity. J. Zool. London 160, 143–157, 1969.

– The effect of temperature on the activity of three species of Helicid snail (*Mollusca: Gastropoda*). J. Zool. Lond. 162, 303–315, 1970.

– The distribution of *Helicodonta obvoluta* (Müller) in Britain. J. of Conchology 27, 363–369, 1972.

– und Palles-Clark, M. A.: *Arianta arbustorum* (L.) on chalk downs in Southern England. Proc. malacol. Soc. London 39, 311–318, 1971.

Cardot, J.: Variations saisonnières de la 5-hydroxytryptamine dans les tissues nerveux et cardiaque chez le mollusque *Helix pomatia.* C. R. Séances Soc. Biol. Fil. 165, 338–341, 1971.

Carlson, G. P.: Detoxification on foreign organic compounds by the quahaug, *Mercenaria mercenaria.* Comp. Biochem. Physiol. B. Comp. Biochem. 43, 295–302, 1972.

Carman, G. E.: Electrical trapping device for land snails. J. Econ. Entom. 58, 786–787, 1965.

Carpenter, D. G., Breese, G., Schanberg, S., und Kopin, I.: Serotonin and dopamine: Distribution and accumulation in *Aplysia* nervous and non-nervous tissues. Int. J. Neurosci. 2, 49–59, 1971.

Carrick, R.: The grey field slug *Agriolimax agrestis* L. and its environment. Ann. appl. Biol. 29, 43–55, 1942.

Carriker, M. R., Person, P., Libbin, R. und Van Zandt, D.: Regeneration of the proboscis of muricid gastropods after amputation, with emphasis on the radula and cartilages. Biol. Bull. (Woods Hole) 143, 317–331, 1972.

– und Van Zandt, D.: Regeneration of the accessory boring organ of muricid gastropods after excision. Trans Am. Microsc. Soc. 91, 455–466, 1972.

Casida, J. E.: Mode of action of Carbamates. Ann. Rev. Ent. 8, 39–58, 1963.

Cavallero, R. und Ravera, O.: Biological indicator of Manganese-54 Contamination in terrestrial environments. Nature 209 (5029), 1259, 1966.

Cedar, H. und Schwartz, J. H.: Cyclic adenosine monophosphate in the nervous system of

Aplysia californica: II. Effect of serotonin and dopamine. J. gen. Physiol. 60, 570–587, 1972.

CHANCELLOR, R. J., COOMBS, A. V. und FOSTER, H. S.: Control of aquatic weeds by copper sulphate. Proc. 4th Brit. World Control Conf., 1–80, 1958.

CHARNIAUX-COTTON, H.: Endocrinologie et génétique de la différenciation sexuelle chez les Invertébrés. C. R. Soc. Biol. Fr. 161, 6–9, 1967.

CHATFIELD, J.: The natural history of Slapton Ley Nature Reserve: V. Non-marine-mollusca. Field Stud. 3(4), 579–588, 1972.

– Aspects of Feeding and Growth in Land Snails. Malacologia 14, 391–392, 1973.

CHENG, T. C.: The Effects of *Echinoparyphium* larvae on the structure of an glycogen deposition in the hepatopancreas of *Helisoma trivolvis* and glycogenesis in the parasite larvae. Malacologia 1–2, 291–301, 1963.

– THAKUR, A. S., und RIFKIN, E.: Phagocytosis as an internal defense mechanismus in the Mollusca: with an experimental study of the role of leucocytes in the removal of ink particles in *Littorina scabra* Linn. Proceed. Symp. Mollusca Part II, 546–563, 1969.

– SMITHERS, S. R., und TERRY, R. J.: „Host“ antigens associated with schistosomes: observations on their attachement and their nature. Parasitology 61, 87–94, 1970.

– und LEE, F. O.: Glucose levels in the mollusc *Biomphalaria glabrata* infected with *Schistosoma mansoni.* J. Invert. Pathol. 18, 395–399, 1971.

– und SULLIVAN, J. T.: A Comparative Study of the Effects of Two Copper Compounds on the Respiration and Survival of *Biomphalaria glabrata (Mollusca: Pulmonata).* Comp. Gen. Pharm. 4, 315–320, 1973.

– SULLIVAN, J. T., und HARRIS, K. R.: Parasitic castration of the marine prosobranch gastropod *Nassarius obsoletus* by sporocysts of *Zoogonus rubellus (Trematoda):* Histopathology. J. Invertebr. Pathol. 21, 183–190, 1973.

– und SULLIVAN, J. T.: Alterations in the osmoregulation of the pulmonate gastropod *Biomphalaria glabrata* due to copper. J. Invertebr. Pathol. 29, 101–104, 1977.

CHERNIN, E.: Studies on the control of Schistosome-bearing snail. VII. Streptomycin-induced inhibition of growth and reproduction in the schistosome-bearing snail, *Australorbis glabratus.* Amer. J. Hyg. 66, 321–330, 1957.

– Interference with the capacity of *Schistosoma mansoni* miracidia to infect the molluscan host. J. Parasit. 54, 509–516, 1968.

– Penetrative activity of *Schistosoma mansoni* miracidia stimulated by exposure to snail-conditioned water. J. Parasit, 58. 209–212, 1972.

– und BOWER, C.: Experimental transmission of *Schistosoma mansoni* in brackish waters. Parasitology 63, 31–36, 1971.

– und PERLSTEIN, J. M.: Protection of snails against miracidia of *Schistosoma mansoni* by various aquatic invertebrates. J. Parasitol. 57, 217–219, 1971.

– MICHELSON, E. H., und AUGUSTINE, D. L.: Studies on the biological control of schistosome-bearing snails. I. The control of *Australorbis glabratus* by the snail *Marisa cornuarietis* under laboratory conditions. Amer. J. Trop. Med. Hyg. 5, 297–307, 1956.

– MICHELSON, E. H., und AUGUSTINE, D. L.: Studies on the biological control of schistosome-bearing snails, Part II. The control of *Australorbis glabratus* populations by the leech, *Helobdella fusca* under laboratory conditions. Amer. J. Trop. Med. 5, 308–314, 1956.

– und MICHELSON, E. H.: Studies on the biological control of schistosome-bearing snails. III. The effect of population density on growth and fecundity in *Australorbis glabratus.* Amer. J. Hyg. 65, 57–70, 1957.

– und MICHELSON, E. H.: Studies on the biological control of schistosome-bearing snails. IV Further observations on the effects of crowding on growth and fecundity in *Australorbis glabratus.* Amer. J. Hyg. 65, 71–80, 1957.

CHETAIL, M.: Etude de la régénération du tentacule oculaire chez un *Arionidae (Arion rufus* L.). et un *Limacidae* (*Agriolimax agrestis* L.). Arch. Anat. micr. morphol. exp. 52, 1 (Suppl.). 130–203, 1963.

– und FOURNIE, J.: Shell-boring mechanism of the gastropod, *Purpura (Thais) lapillus:* a physiological demonstration of the role of carbonic anhydrase in the dissolution of Ca CO_3. Amer. Zoologist 9, 983–990, 1969.

– DERER, M., und FOURNIE, J.: Mécanisme de perforation chez les gastéropodes Prosobranches muricides: Transferts ioniques au niveau de l'organe de perforation en activité chez *Thais lapillus* L. Haliotis 1, 13–14, 1971.

CHEVALLIER, H.: Taxonomie et biologie des grands *Arion* de France *(Pulmonata: Arionidae).* Malacologia 9, 73–78, 1969.

– Les Limaces de Bretagne. Penn ar Bed (Brest) 7, 371–389, 1970.

– Cycle biologique des grands *Arion* en France. Atti Soc. It. Sc. Nat. e Museo Civ. St. Nat. Milano 112, 316–320, 1971.

– Les Mollusques continentaux de France d'intérêt économique. Haliotis 3, 9–18, 1973.

– Répartition en France et importance économique de l'Escargot de Bourgogne, *Helix pomatia* Linné. Haliotis 3. 177–183, 1973.

– Répartition en France de *Deroceras caruanae* (Pollonera, 1891). Haliotis 3, 205–207, 1973.

– Les Grands *Arion* de France *(Mollusca, Pulmonata).* Taxonomie, Biogéographie, Ecologie, Polymorphisme, Croissance et Cycle biologique. Thèse, Doctorat Univ. Paris VI, 1–234, 1974.

– La marché des escargots en France et les recherches sur leur élevage. L'Escargot écologiste 1, 1976.

– Présence de *Helix lucorum* dans la banlieue lyonnaise. Elona 2, 1–30, 1975.

CHI, L. W., und WINKLER, L. R.: The effects of copper sulfate, sodium pentachlorophenate and sulfonated hydrocarbon on the eggs of *Oncomelania.* Amer. J. Trop. Med. Hyg. 11, 851, 1962.

– WAGNER, E. D., und WOLD, N.: Susceptibility of *Oncomelania* hybrid snails to various geographic strains of *Schistosoma japonicum.* Amer. J. trop. Med. 20, 89–94, 1971.

CHIANG, P. K., BOURGEOIS, J. G., und BUEDING, E.: Histochemical distribution of acetylcholinesterase in the nervous system of the snail *Biomphalaria glabrata.* Int. J. Neurosci 3, 47–60, 1972.

CHICHERY, R.: Action de l'acetylcholine et de diverses substances curarisantes sur le systeme nerveux d'un Cephalopode *(Eledone).* C. R. Séances Soc. Biol. Fil. 165, 1225–1229, 1971.

CHICHESTER, L. F., und GETZ, L. L.: The zoogeography and ecology of Arionid and Limacid slugs introduced into northeastern North America. Malacologia 7, 313–346, 1969.

CHING-CHIEH YU, BOOTH, G. M., HANSEN, D. J., und LARSEN, J. R.: Fate of Carbofuran ni a Model Ecosystem. J. Agr. Food Chem. 22, 431–434, 1974.

CHOCK, O., C., DAVIS, C. J., und CHONG, M.: *Sepedon macropus (Diptera: Sciomyzidae)* introduced into Hawaii as a control for the liverfluke snail, *Lymnaea ollula.* J. econ. Entom 54, 1–4, 1961.

CHOQUET, M.: Contribution à l'étude du cycle biologique et de l'inversion du sexe chez *Patella vulgata* L. (Mollusque Gastéropode Prosobranche). Thèse Doct. Sc. Nat., Univ. Lille, 185, 1–234, 1969.

CHRISTIE, J. D., PRENTICE, M. A., UPATHAM, E. S. und BARNISH, G.: Laboratory and field trials of a slow copper molluscicide in St. Lucia. Amer. J. Trop. Med. Hyg. 27, 616–622, 1978.

– FOSTER, W. B., und STAUBER, L. A.: The Effect of Parasitism and Starvation on Carbohydrate Reserves of *Biomphalaria glabrata.* J. Invert. Pathol. 23, 55–62, 1974.

– FOSTER, W. B., und STAUBER, L. A.: ^{14}C Uptake by *Schistosoma mansoni* from *Biomphalaria glabrata* Exposed to ^{14}C-Glucose. J. Invert. Pathol. 23, 297–302, 1974.

CHRISTOFFERSEN, G. R.: Study state contribution of the Na, K-pump in the membrane potential in identified neurons of a terrestrial snail, *Helix pomatia.* Acta Physiol. Scand. 86, 498–514, 1973.

CHU, K. Y.: Trials of ecological and chemical measures for the control of *Schistosoma haematobium* in a Volta Lake village. Bull. WHO. 56, 313–322, 1978.

– MASSOUD, J., und ARFAA, F.: Comparative Studies of the Molluscicidal Effect of Cuprous

Chloride and Coppersulfat in Iran. Bull. Wld. Hlth. Org. 39, 320–326, 1968.

– und Dawood, I. K.: Cercarial production from *Biomphalaria alexandrina* infected with *Schistosoma mansoni*. Bull. Wld. Hlth. Org. 42. 569–574, 1970.

– Dawood, I. K., und Nabi, H. A.: Seasonal abundance of trematode cercariae in *Bulinus truncatus* in a small focus of schistosomiasis in the Nile Delta. Bull. Wld. Hlth. Org. 47. 420–422, 1972.

Clampitt, P. T.: Comparative ecology of the snails *Physa gyrina*. Ecology 36, 40–44, 1970.

Clark, W. C.: Escape Response of Herbivorous Gastropodes when stimulated by Carnivorous Gastropods. Nature 181, 137–138, 1958.

Clauss, E.: *Boettgerilla vermiformis* Wiktor im östlichen Erzgebirge. Zool. Abh. Ber. Staatl. Mus. Tierkde. Dresden 26(18), 285–287, 1964.

Coats, J. R., Metcalf, R. L., und Kapoor, P.: Metabolism of the Metoxychlor Isomere, Dianisylneopentane, in Mouse, Insects, and a Model Ecosystem. Pestic. Biochem. Physiol. 4, 201–211, 1974.

Coe, W. R.: Nutrition and sexuality in protrandric gastropods of the genus *Crepidula*. Biol. Bull. 94, 158–200, 1948.

– Influences of association, isolation and nutrition on the sexuality of snails of the genus *Crepidula*. J. Expt. Zool. 122. 5–19, 1953.

Coles, G. C.: Snail „metabolic hormone" and snail parasite metabolism. Comp. Biochem. Physiol. 34, 213–220, 1970.

Collatz, K. G., Mommsen, G., und Mommsen, T.: Emulsifiers of the alkylsulfuric acid and partial glyceride types in the digestive juice of the snail *Helix pomatia* L. J. Comp. Physiol. 2, 123–129, 1975.

Coluzzi, A., Nuvole, A., Orecchia, P., und Paggi, L.: Indagini su *Bulinus truncatus* e *Schistosoma* bovis della Sardegna. Parasitologia (Roma) 7, 173–228, 1965.

Combes, C., und Cens, B.: Rôle des mollusques dans la transmission des helminthiases en domaine continental. Haliotis 3, 59–62, 1973.

Comperse, E. Jr., und Bates, J.: Determination of calcite: Aragonite ratios in mollusc shells by infrared spectra. Limnol. Oceanogr. 18, 326–331, 1973.

Cook, L. M.: Genetical studies on *Helix aspersa* and other helicid snails: A review of work by A. W. Stelfox. Irish Naturalist's J. 16, 1970.

Cooke, A. S., und Polland, E.: Shell and Operculum Formation by Immature Roman Snails *Helix pomatia* L. when treated with pp'-DDT. Pesticide Biochemistry and Physiol. 3, 230–236, 1973.

Corliss, J. O., Smith, A. C., und Foulkes, J.: A species of *Tetrahymena* from the British Garden Slug *Milax budapestensis*. Nature 196, 1008–1009, 1962.

Corning, W. C., Dyal, J. A., Alloway, T. und Krasne, F. O.: Invertebrate Learning. Bd. 2: Arthropods and Gastropod Mollusks. Cambridge, 1–284, 1973.

Cragg, J. B., und Vincent, M. H.: The action of metaldehyde on the slug *Agriolimax reticulatus* (Müller). Ann. appl. Biol. 39, 392–406, 1952.

Crawford-Sidebotham, T. J.: Differential Susceptibility of Species of Slugs to Metaldehyde-Bran and to Methiocarb Baits. Oecologia (Berl.) 5, 303–324, 1970.

– The influence of weather upon the activity of slugs. Oecologia (Berl.) 9, 141–154, 1972.

Cridland, C. E.: Ecological factors affecting the numbers of snails in temporary bodies of water. J. Trop. Med. Hyg. 60, 287, 1957.

– Susceptibility of the snail *Biomphalaria alexandrina* from the UAR and the Sudan to infection with a strain of *Schistosoma mansoni* from Tanzania. Bull. Wld. Hlth. Org. 43, 809–815, 1970.

Crook, J. R., Fulton, S. E., und Supanwong, K.: The infectivity of third stage *Angiostrongylus cantonensis* larvae shed from drowned *Achatina fulica* snails and the effect of chemical agents on infectivity. Transact. Royal Soc. Trop. Med. Hyg. 65, 602–605, 1971.

Crossland, N. O.: A large experiment in the control of aquatic snail by the use of molluscicides on a sugar estate in the northern region of Tanganyika. Bull. Wld. Hlth. Org. 29,

515–524, 1963.
– Field trials to evaluate the effectiveness of the molluscicide, N-tritylmorpholine. Nature (London) 210, 1140–1141, 1967.
– The Effect of the Molluscicide N-Tritylmorpholine on Transmission of *Fasciola hepatica,* Liver Fluke Infection in Sheep. Vet. Rec. 98, 45–48, 1976.
– PEARSON, A. J. A., und BENNETT, M. S.: A field trial with the molluscicide Frescon for control of *Lymnaea peregra* (Müller), snail host of *Diplostomum spathaceum* (Rudolphi). J. Fish Biol. 3, 297–302, 1971.
CROWELL, H. H.: Slug and snail control with Experimental poison baits. J. econ. Entom. 60, 1048–1050, 1967.
– Laboratory study of calcium requirements of the Brown Garden Snail, *Helix aspersa* Müller. Proc. malac. Soc. London 40, 491–503, 1973.
– Preserving Terrestrial Slugs by Freeze-Drying. Veliger 15, 254–256, 1973.
– Chemical control of terrestrial slugs and snails. Agric. Experim. Stat. Oregon State Univ., Corvallis, Bull. 628, 69 p., July 1977.
CUMIN, R.: Normentafel zur Organogenese von *Limnaea stagnalis (Gastropoda, Pulmonata)* mit besonderer Berücksichtigung der Mitteldarmdrüse. Rev. Suisse Zool. 79, 709–774, 1972.
CURTIN, TH. J.: Snails from Turkey, Crete and Greece. Nautilus 76, 132–135, 1963.
CUSHING J. E., EVANS, E. E., und EVANS, M. L.: Induced bactericidal responses of abalones. J. Invertebr. Pathol. 17, 446–448, 1971.
DAINTON, B. H.: The activity of slugs: I. The induction of activity by changing temperatures. J. exp. Biol. 31, 165–187, 1954.
– The activity of slugs: II. The effect of light and air currents. J. Exp. Biol. 31, 188–197, 1954.
DALLIBOR, K.: Die Entschlüsselung des Lungenegels. Umschau 1, 3, 1976.
DARE, P. J.: The movements of oystercatchers *Haematopus ostralegus* L., visiting or breeding in the British Isles. Fish Invest. Min. Agr. Fish Food Gt. Brit. Ser. II. Salmon Freshwater Fish 25(9), 1–137, 1970.
DARTEVELLE, E.: Report on giant African snail prepared at request of Hawaiian Board of Agriculture and Forestry. Trans. by Philip Weber, 1954, 14 pp. zit. bei Mead (1961).
DAVID, W. A., METCALF, R. L., und WINTON, M.: The systemic insecticidal properties of certain carbamates. J. econ. Ent. 53. 1021–1025, 1960.
DAVIDSON, R. E.: Experiments with Zectran to control snails attacking orchids. Ohio Florist's Ass. Bull. Nr. 397, 6, 1962.
– Experiments with Zectran to control snails attacking orchids. Down to Earth 18, 2 und 23, 1962.
DAVIES, P., und JACKSON, H.: Experimental studies on the chemosterilization of *Schistosoma mansoni.* Parasitology 61, 167–176, 1970.
DAVIS, B. N. K., und FRENCH, M. C.: The accumulation and loss of organochlorine insecticide residues by beetles, worms and slugs in sprayed fields. Soil Biol. Biochem. 1, 45–55, 1969.
DAVIS, C. J., CHOCK, O. C., und CHONG, M.: Introduction of liver-fluke snail predator, *Sciomyza dorsata (Sciomyzidae, Diptera)* in Hawaii. Proc. Hawaii Ent. Soc. 17, 395–397, 1961.
DAVIS, D. S., und FARLEY, J.: The effect of parasitism by the trematode *Cryptocotyle lingua* (Creplin) on digestive efficiency in the snail host, *Littorina saxatilis* (Olivi). Parasitology 66, 191–197, 1973.
DAVIS, G. M.: The effect of a graded dose of antigen on the production of hyperimmune antimesogastropod ascitic fluid. Comp. Biochem. Physiol. A. Comp. Physiol. 41, 339–347, 1972.
DAWOOD, J. K., DAZO, B. C., und FAROOQ, M.: Large-scale application of bayluscide and sodium pentachlorophenate in the Egypt – 49 Project area. Bull. Wld. Hlth. Org. 35, 357–367, 1966.
DAXL, R.: Ein Beitrag zur Biologie von *Boettgerilla vermiformis* Wiktor 1959. Z. ang. Zool. 54, 227–231, 1967.

– Die Abhängigkeit der Wirkung molluskizider Substanzen (Metaldehyd, Isolan, Ioxynil) von endogenen und exogenen Faktoren auf Nacktschnecken. Diss. Berlin D 83, 252, 1–162, Fakultät Landbau der TU 1968.
– Beobachtungen zur diurnalen und saisonellen Aktivität einiger Nacktschneckenarten. Z. ang. Zool. 56, 357–370, 1969.
– Der Einfluß von Temperatur und relativer Luftfeuchte auf die molluskizide Wirkung des Metaldehyd, Isolan und Ioxynil auf *Limax flavus* und dessen Eier. Z. ang. Ent. 67, 57–87, 1970.
– Das Verhalten der molluskiziden Verbindungen Metaldehyd, Isolan und Ioxynil gegen Nacktschnecken unter Freilandbedingungen. Z. ang. Zool. 58, 203–241, 1971.

Day, J. C. L., und Dowdeswell, W. H.: Natural selection in *Cepaea* on Portland Bill. Herdity 23, 169–188, 1968.

De Andrade, R. M. und Carvalho, O. S.: [Alimentation and fecundity of planorbids grown in the laboratory: 1. *Biomphalaria tenagophila* (d'Orbigny, 1835) (*Pulmonata, Planorbidae*)]. Rev. Bras. Biol. 32, 225–233, 1972.

Degrémont, A. A.: Aspects practiques de la lutte contre les schistosomiases dans les périmètres irrigués: Méthodes, organisation, rentabilité. Acta Trop. 24, 362–400, 1972.
– Geigy, R., und Perret, P.: Résultats préliminaires du Projet de lutte et de prévention contre les schistosomiases dans le Bas-Mangoky (République Malagasy). Epidémiologie et Médicine tropicale. Acta Trop. 29, 101–137, 1972.

De Jong-Brink, M.: The effects of desiccation and starvation upon the weight, histology and ultrastructure of the reproductive tract of *Biomphalaria glabrata,* intermediate host of *Schistosoma mansoni.* Z. Zellforsch. Mikrosk. Anat. 136, 229–262, 1973.

Delhaye, W., und Bouillon, J.: L'évolution et l'adaptation de l'organe excrétant chez les Mollusques Gastéropodes Pulmonés: II. Histophysiologie comparée du Rein chez les Stylommatophores. Bull. Biol. T, C VI 2, 123–142, 1972.

De Longcamp, D., Drosdowsky, M., und Lubet, P.: Steroid biosynthesis in mollusca, Demonstration of 17 ß-hydroxy-steroid dehydrogenase in the gonads of *Mytilus edulis* (bivalve mollusk.). C. R. Acad. Sci. Sér. D. 271, 1564–1566, 1970.
– Drosdowsky, M., und Lubet, P.: Biosynthèse des stéroides chez les Mollusques. Mise en évidence d'une 17 β-hydroxy-stéroide déshydrogénase dans les gonades de *Mytilus edulis* L. (Mollusque bivalve). C. R. Acad. Sc. Paris, Sér. D., 271, 1561–1564, 1970.

Del Rivero, J. M. und Roca, M.: Preliminary Screening Trials of Pesticides for the Control of Snails. J. AVC 10 (6), 108–110, 1969.

Demian, E. S.: Prospects of the use of *Marisa cornuarietis* in the biological control of *Lymnaea cailliaudi* in the U.A.R. Proc. Egypt. Acad. Sci. 18, 48–52, 1964.
– und Ibrahim, A. M.: Tolerance of the snail *Marisa cornuarietis* (L.) to extreme temperatures under laboratory conditions. Zool. Soc. Egypt. Bull. 24, 20–30, 1972.
– und Ibrahim, A. M.: Sexual dimorphism and sex ratio in the snail *Marisa cornuarietis* (L.). Zool. Soc. Egypt. Bull. 24, 52–63, 1972.
– und Kamel, E. G.: Effect of *Marisa cornuarietis* on *Bulinus truncatus* populations under semi-field conditions in Egypt (abstract). Malacologia 14, 439, 1973.

De Martinez, N., Garcia, M., Salas, M., und Candela, J.: Proteins with insulinlike activity isolated from oyster (*Ostrea edulis* L.) hepatopancreas. Gen. Comp. Endochrinol. 20, 305–311, 1973.

Den Boer, W.: – ohne Titelangabe –. Zitiert bei Hermann 1971: Groenten en fruit, 14. V. 1969, p. 1939.

Deschiens, M. R.: Comportement écologique et épidémiologique de *Bulinus forskalii* vecteur de la bilharziose à *Schistosoma intercalatum.* C. R. Acad. Sc. Paris Sér. D. 1972, 1394–1395.
– und Lang, L.: Action prédatrice des écrevisses des genres *Astacus* et *Cambarus* sur les mollusques vecteurs des bilharzioses. Bull. Soc. Pathol. Exot. 47, 809, 1954.
– und Adetonah, A.: Conditions d'élevage de *Bulinus (Pyrgophysa) forskalii* en laboratoire. Bull. Soc. Path. Exot. 62, 1096–1101, 1969.

De Souza, C. P., und Paulini, E.: [Influence of organic matter in suspension on the toxicity of some molluscicides]. Rev. Brasil. Malariol. Doencas. Trop. 20, 155–160, 1968.
Dexheimer, L.: Beiträge zum Kalkstoffwechsel der Weinbergschnecke *(Helix pomatia).* Zool. Jahrb. Abt. Allg. Zool., Physiol. 63, 129–152, 1951.
Dexter, R.: A colony of the introduced European snail, *Cepaea nemoralis* at Lynn, Massachusetts. Sterkiana 21, 41–43, 1966.
Dias, E., und Dawood, M. M.: Preliminary trials on the biological snail control with *Bacillus pinotti* in Egypt. Mem. Inst. Cruz. 53, 13–29, 1955.
Dinamani, P., und Wolf, P. H.: Multiple tumors in the pericardial cavity of an Australian rock oyster, *Crassostrea commercialis* (Iredale & Roughley). Int. J. Cancer 11, 293–299, 1973.
Dindal, D. L., und Wurzinger, K.-H.: Accumulation and Excretion of DDT by the Terrestrial Snail, *Cepaea hortensis.* Bull. Environm. Contam. Toxicol. 6, 362–371, 1971.
Dinter, I., und Manos, P. J.: Evidence for a pheromone in the marine periwinkle, *Littorina littorea* Linnaeus. Veliger 15, 45–47, 1972.
Di Prisco, Lupo, C., Dessi'Fulgheri, F., und Tomasucci, M.: Identification and biosynthesis of steroids in the marine mollusc *Aplysia depilans.* Comp. Biochem. Physiol. B. Comp. Biochem. 45, 303–310, 1973.
Dirlbeck, K.: [Some carbamates for the protection of garden cultures against snails]. Ochr. rostlin. (Praha) 5, 271–276, 1969.
– Molluskocidni účinnost některých připrávků ve zkouškách UKZUZ v Praze. Agrochémia Časop. 6, 187–188, 1970.
Dix, T. G.: Mesenchymal tumors in a pearl oyster, *Pinctada margaritifera.* J. Invertebr. Pathol. 20, 317–320, 1972.
– Mantle changes in the pearl oyster *Pinctada maxima* induced by the pea crab *Pinnotheres villosus.* Veliger 15, 330–331, 1973.
Dobson, R. M.: Observation on the occurence of keeled slugs (*Milax* Gray) in the Glasgow area. Glasgow Naturalist 18, 246–248, 1963.
Dolezalova, H., Giacobini, E., und Stepita-Klauco, M.: An attempt to identify putative neurotransmitter molecules in the central nervous system of the snail. Int. J. Neurosci. 5, 53–59, 1973.
Dollfus, R. P., Camerlynck, P., Benmansour, N., und Soussi, C.: Infection mixte de *Bulinus truncatus* (Audouin, 1827) par des formes larvaires de *Schistosoma haematobium* (Bilharz, 1852) et d'un *Echinostomatidae.* Ann. Parasit. (Paris) 47, 499–504, 1972.
Dorough, H. W.: Metabolism of insecticidal methylcarbamates in animals. J. Agr. Food Chem. 18, 1015–1022, 1970.
Dossaji, S. F., Kairu M. G., Gondwe A. T., und Ouma J. H.: On the Evaluation of the Molluscicidal Properties of *Polygonum senegalense* forma *senegalense.* J. Natur. Prod. 40 (3), 290–293, 1977.
Doucette, C. F.: An Introduced Slug, *Arion ater* (L.) and its Control with Metaldehyde. J. econ. Entom. 47, 370, 1954.
Drees, H.: Der Pflanzenschutzmittelmarkt im Wandel der Zeit. Gesunde Pflanzen 26, 39, 1974.
Ducros, C.: Histochemical detection of cholinesterases in the posterior salivary glands of cephalopods. C. R. Hebd. Seances Acad. Sci. Ser. D. Sci. Natur (Paris) 272, 3318–3321, 1971.
Duhm, B., Maul, W., Medenwald, H., Patzschke, K., und Wegner, L. A.: Radioaktive Untersuchungen mit einem neuen Molluskizid. Z. Naturforschung 16b, 509–515, 1961.
Dun, G. S.: The giant snail. Papua New Guinea Agr. J. 18, 123–125, 1967.
Duncan, C. J.: The life cycle and ecology of the freshwater snail *Physa fontinalis* (L.). J. animal Ecol. 28, 97–117, 1959.
– The genital system of the freshwater *Basommotaphora.* Proc. Zool. Soc. London 135, 339–356, 1960.
– Brown, N. und Dunlop, R. W.: The uptake of the molluscicide, 4'-Chloronicotineanilide into *Biomphalaria glabrata* (Say) in a flowing water system. Pestic. Sci. 8, 345–353, 1977.

DUNNING, B. B., und MACHINE, X.: pH optimum for the rate of acetylcholine action on neurons. Agents actions 2, 131–137, 1972.

DUPOUY, J., und MARTINEZ, J.-Cl.: Action de *Proctoeces maculatus* (LOOSS 1901) *(Trematoda, Fellodistomatidae)* sur le dévelopement des gonades chez *Mytilus galloprovincialis* Lmk. C. R. Acad. Sc. Paris, 277. 1889–1892, 1973.

DURCHON, M.: L'endocrinologie des vers et des mollusques. In: GRASSÉ: Les grands problèmes de la biologie. 1–242. Masson, Paris 1967.

DUTHOIT, C. M.: Assessing the activity of the field slug in cereals. Plant. Pathol. 10, 165, 1961.

– Slugs and food preferences. Plant. Pathol. 13, 73–78, 1964.

DUVAL, D. M.: Some aspects of the behaviour of slugs. J. Conch. (London) 27, 163–170, 1970.

EAKIN, R. M., und MILLER FERLATTE, M.: Studies on eye regeneration in a snail, *Helix aspersa*. J. exp. Zool. 184, 81–96, 1973.

EDWARDS, C. A., und THOMPSON, A. R.: Pesticides and soil fauna. Residue Reviews 45, 1–79, 1973.

EFFERSON, J. N.: A Rice Seed Farm in Central America. Rice J. 71. 6–10, 1968.

EHRMANN, P.: Mollusca. In: „Die Tierwelt Mitteleuropas“, Herausgeber BROHMER, P., EHRMANN, P., ULMER, G. II (1), 1–264; Quelle & Meyer, Leipzig 1956/1933.

EISENBERG, R. M.: The regulation of density in a natural population of the pond snail, *Lymnaea elodes*. Ecology 47, 889–906, 1966.

ELBERG, K.: A preliminary list of snail-killing flies *(Sciomyzidae, Diptera)* of Lithuania. Izvest. Akad. nauk. Est. SSR. Biol. 17, 56–60, 1968.

ELLIS, A. E.: British snails. Oxford, 1–275, 1926.

– Census of the distribution of British non-marine *Mollusca*. (7th edit.). J. Conchol. 23, 171–244, 1951.

– *Arion lusitanicus* Mabille in Cornwall. J. Conchyl. (London) 25, 285–287, 1964.

– The non-marine *Gastropoda* of Great Britain and Ireland. Pleistocene to Recent. 1–298. Oxford Univ. Press 1969.

EMPSON, D. W.: Cereal pests. Bull. Min. Agr. Fish. Food 186, 1–97, 1965.

ENIGK, K., STICINSKY, E. und ERGÜN, H.: Die Zwischenwirte von *Davainea proglottina (Cestoidea)*. Z. Parasitenkde. 18, 230–236, 1957/1958.

– und DÜWEL, D.: Zur Bekämpfung von *Galba truncatula (Limnaeidae)* durch Natrium-pentachlorphenolat. Z. Trop. Med. Parasit. 11, 134–151, 1960.

ERNST, W., SCHAEFER, R. G., GOERKE, H., und EDER, G.: Aufarbeitung von Meerestieren für die Bestimmung von PCB, DDT, DDE, DDD, γ-HCH und HCB. Z. Anal. Chem. 272, 358–363, 1974.

– GOERKE, H., EDER, G. und SCHAEFER, R. G.: Residues of chlorinated hydrocarbons in marine organisms in relation to size and ecological parameters. I. PCB, DDT, DDE, and DDD in fishes and molluscs from the English channel. Bull. Environ. Contam. Toxicol. 15, 55–65, 1976.

ETGES, F. J., BELL, E. J., und GILBERTSON, D. E.: Bacterial degradation of some molluscidal chemicals. Amer. J. Trop. Med. Hyg. 14, 846–851, 1965.

– und GILBERTSON, D. E.: Repellent action of some chemical molluscicides on Schistosome vector snails. Amer. J. Trop. Med. 15, 618–624, 1966.

EVANS, J. G.: Changes in the composition of land molluscan populations in North Wiltshire during the last 5000 years. Symp. Zool. Soc. London 22, 293–317, 1968.

EVANS, H. C.: Invertebrate vectors of *Phytophthora palmivora*, causing black pod disease of cocoa in Ghana. Ann. appl. Biol. 75, 331–345, 1973.

EVELAND, L. K., und RITCHIE, L. S.: Infectivity of cercariae of *Schistosoma mansoni* from snails on inadaequate diets. Parasitol. 64, 441–444, 1972.

EVERSOLE, J. E.: Mortalities of *Cochlicella barbara* and *Theba pisana* land snails exposed to low temperatures in Marocco. J. econ. Entom. 64, 667–670, 1971.

EXPONDA, P., und STOCKERT, J. C.: Evolution of the synaptonemal complex in *Helix aspersa* spermatocytes. Chromosoma 36, 150–157, 1972.

EYSTER, W. H.: Snails and slugs in the garden. Oregon Garden 18, 38–39, 1951.

FALKNER, G.: Beobachtungen über die Schneckennahrung des Alpenkammolchs *(Triturus cristatus carnifex* Laur.) während des Landaufenthaltes. Faun.-ökol. Mitt. 3, 1, 1970.

FARRIS, V. K.: Molluscan cells: Dissociation and reaggregation. Science 160, 1245–1246, 1968.

FENG, S. Y., FENG, J. S., BURKE, C. N., und KHAIRALLAH, L. H.: Light and electron microscopy of the leucocytes of *Crassostrea virginica (Mollusca: Pelecypoda).* Z. Zellforsch. Mikrosk. Anat. 120, 222–245, 1971.

FENWICK, A.: Laboratory Evaluation of Biocidal Rubber Pellets 633 B as a Residual Molluscicide Against *Biomphalaria pfeifferi.* Bull. Wld. Hlth. Org. 41, 326–329, 1969.

– The Development of Snail Control Methods on an Irrigated Sugar-Cane Estate in Northern Tanzania. Bull. Wld. Hlth. Org. 42, 589–596, 1970.

– und LIDGATE, H. J.: Attempts to eradicate snails from impounded water by the use of N-Tritylmorpholin. Bull. Wld. Hlth. Org. 42, 581–588, 1970.

FERGUSON, M., RICHARDS, C., und PALMER, J.: Control of *Australorbis glabratus* by acrolein in Puerto Rico. Public. Health Rept. 76, 461, 1971.

FERGUSON, F. F. und BUTLER, J. M., Jr.: Ecology of *Marisa* and its potential as an agent for the elimination of aquatic weeds in Puerto Rico. Proc. South. Weed. Conf. 19, 468–476, 1966.

– und RUIZ-TIBÉN, E.: Review of biological control methods for schistosome-bearing snails. Ethiop. Med. J. 9, 95–104, 1971.

FIGUEIREDO, E. R., und LEPAGE, H. S.: Snails injurious to orchids. Biologico 19, 31–33, 1953.

FINCHER, F.: *Lampyris noctiluca* (L.) *(Col. Lampyridae)* attacking a slug. Entom. Monthly Mag. 83, 149, 1947.

FISCHER, P.-H.: Vie et moeurs des Mollusques. Paris, 1–312, 1950.

FISH, J. D.: The breeding cycle and growth of open coast and estuarine populations of *Littorina littorea.* J. Mar. Biol. Assoc. UK 52, 1011–1019, 1972.

FISHER, F. E.: Biocontrol of *Cephaleuros virescens* on Tahiti lime in Florida. Phytopathology 64, 768, 1974.

FISHER, T. W., MOORE, I., LEGNER, E. F., und ORTH, R. E.: *Ocypus olens,* a Predator of Brown Garden Snail. California Agriculture, March, 20–21, 1976.

– und ORTH, R. E.: Differential susceptibility of brown garden snail to Metaldehyde. Calif. Agric., Berkeley, 29, 7–8, 1975.

FLASÀR, I.: *Boettgerilla vermiformis* Wiktor 1959 in Böhmen. Zool. listy (Brno) 11, 93–94, 1962.

– und KROUPOVÁ, V.: Die Malakofauna der Gewächshäuser in Bratislava (Tschechoslowakei)*(Gastropoda).* Malakol. Abh. Staatl. Mus. Tierkunde Dresden 5, 139–154, 1976.

FLASAROVÁ, M., und FLASAR, I.: [Isopoden und Gastropoden der Treibhäuser in Teplice Lázne, Böhmen]. Zool. listy (Brno 11), 71–76, 1962.

– und FLASAR, I.: [Isopoden und Gastropoden der Treibhäuser im Nordböhmischen Kreis, Böhmen]. Zool. listy (Brno) 14, 251–260, 1965.

FLETCHER, F.: Vergleichsversuche mit den drei Molluskiziden Na PCP, Cu SO_4, Bayluscid. Pflanzenschutz-Nachr. „Bayer" 18, 153–156, 1965.

FOCARDI, S., und QUATTRINI, D.: Structure of the reproductive apparatus and life cycle of *Milax gagates* (Draparnaud): *Mollusca Gastropoda Pulmonata.* Boll. Zool. 39, 9–27, 1972.

FOLEY, D. A., und CHENG, TH, C.: Interaction of molluscs and foreign substances: The morphology and behavior of hemolymph cells of the American oyster, *Crassostrea virginica,* in vitro. J. Invertebr. Pathol. 19, 383–394, 1972.

FOOTE, B. A.: Biology and life history of the snail killing flies belonging to the genus *Sciomyza* Fallén *(Diptera, Sciomyzidae).* Ann. Ent. Soc. Amer. 52, 31–43, 1959.

– Biology and immature stages of the snail-killing flies belonging to the genus *Tetanocera (Diptera: Sciomyzidae).* Ph. D. Thesis, Cornell Univ. Ithaca, N.Y. (L.C. Card. No. Mic. 62–105) 190 pp. Univ. Microfilms, Ann. Arbor, Mich./Diss. Abstr. 22, 3302, 1961.

– Biology of slug-killing *Tetanocera (Diptera: Sciomyzidae).* Proc. North-Central Branch,

Entomol. Soc. Amer. 18, 97, 1963.

FORCART, L.: Die Verbreitung der Limaciden und Milaciden in der Schweiz. Arch. Moll. 74, 114–119, 1942.

FOSTER, G. N.: Problems in cucumber crops caused by Slugs, Cuckoospit insect, Mushroom Cecid, Hairy Fungus Beetle and the House Mouse. Plant. Pathol. 26, 100–101, 1977.

FOSTER, R.: The effects of trematode metacercaria *(Brachylaemidae)* on the slugs *Milax sowerbii* Férussac, and *Agriolimax reticulatus* Müller. Parasitology 48, 261–268, 1958.

– Infestation of the slugs *Milax sowerbii* Férussac and *Agriolimax reticulatus* Müller by trematode metacercariae *(Brachylaemidae)*. Parasitology 48, 304–311, 1958.

FOX, C. J.: First record of the keeled slug, *Milax gagates* (Drap.) in Nova Scotia. Canad. Field Naturalist 76, 122–123, 1962.

– The effects of five herbicides on the numbers of certain invertebrate animals in grassland soil. Can. J. Plant. Sci. 44, 405–409, 1964.

– RITCHIE, L. S., und FRICK, L. P.: Einfluß des pH-Wertes auf die molluskizide Wirksamkeit von Bayer 73 gegen *Australorbis glabratus* von Puerto Rico unter Laborbedingungen. Pflanzenchutz-Nachr. „Bayer“, 16, 269–277, 1963.

FRANÇOIS, E., RIGA, A., und MOENS, R.: Labelling the Grey Field Slug *Agriolimax reticulatus* by means of radionuclides. Parasitica 21, 139–151, 1965.

– MOENS, R., und RIGA, A.: Utilisation du tantale radioactif dans l'étude de l'écologie d'un mollusque ravageur: *Agriolimax reticulatus* Müller. Parasitica 24, 10–15, 1968.

– RIGA, A., und MOENS, R.: Estimation des populations de *Agriolimax reticulatus* Müller aux moyens de la technique de marquage au radiophosphore ^{32}P et recapture. Parasitica 24, 63–78, 1968.

FRANDSEN, F. und CHRISTENSEN, N. O.: Effect of *Helisoma duryi* on the survival, growth and cercarial production of *Schistosoma mansoni* – infected *Biomphalaria glabrata*. Bull. WHO 55, 577–580, 1977.

FRANZ, J.: Schadensschwellen bei forstschädlichen Insekten. Z. Pflanzenkrankh. 77, 642–647, 1970.

– und KRIEG, A.: Biologische Schädlingsbekämpfung. Reihe Pareys Studientexte 12. 1–222. Parey, Berlin-Hamburg 1976, 2. Aufl.

FREDEEN, F. J. H., und DUFFY, J. R.: Insecticide Residues in some components of the St. Lawrence River Ecosystem following Applications of DDD. Pesticid. Monit. J. 3, 219–226, 1970.

FRETTER, V.: Experiments with P^{32} und I^{131} on Species of *Helix, Arion* and *Agriolimax*. Quart. J. Microsc. Sci. 93. (2), 123–146, 1952.

– und GRAHAM, A.: Reproduction. In: Physiology of *Mollusca*, ed. WILBUR, K. M. und YONGE, C. M., I (4), 127–164. Acad. Press, New York-London 1964.

– und PEAKE, J.: Pulmonates, I und II. Acad. Press, London 1975, 1979.

FRICK, L. P., RITCHIE, L. S., FOX, I., und JIMENEZ, W.: Molluscidal Qualities of Copper Protoxide (Cu_2O) as Revealed by Tests on Stages of *Australorbis glabratus*. Bull. Wld. Hlth. Org. 30, 295–298, 1964.

FRIED, G., und LEVIN, N.: Enzymatic activity in hepatopancreas of *Nassarius obsoletus*. Comp. Biochem. Physiol. B. Comp. Biochem. 45, 153–157, 1973.

FRÖMMING, E.: Die Zooparasiten unserer Süßwasserschnecken. Arch. Molluskenkde. 64, 154–159, 1932.

– Quantitative Untersuchungen über die durch Landlungenschnecken als Nahrung aufgenommenen Heil- und Gewürzpflanzen. Die Pharmazie 4, 339–341, 1949.

– Untersuchungen über aufgenommene Nahrungsmengen und DDT-Wirkung bei zwei Nacktschneckenarten. Z. Hyg. Zool. 37, 320–325, 1949.

– Nacktschnecken als Drogenschädlinge. Die Pharmazie 5, 402–404, 1950.

– Die postembryonale Entwicklung einer Nacktschnecke *(Limax flavus* L.). Z. Naturforsch. 5 (b), 398–399, 1950.

– Untersuchungen über die Farbvarietäten und die Ernährung der Nacktschnecke *Arion em-*

piricorum. Arch. Moll. 79, 117–126, 1950.
– Ein bisher wenig beachteter Vorratsschädling, die Nacktschnecke *Limax flavus* L. Z. Pflanzenkr. 58, 322–326, 1951.
– Quantitative und allgemein-physiologische Untersuchungen über die Nahrungsaufnahme der Kellerschnecke *Limax flavus* L. Anz. Schädlingskde. 24, 13–14, 1951.
– Über die Nahrung von *Limax (Heynemannia) maximus* L. Anz. Schädlingskde. 25, 41–43, 1952.
– Über das Verhalten einheimischer Landlungenschnecken zu den Wildfrüchten. Die Pharmazie 7, 448–449, 1952.
– Unsere Wegschnecke *(Arion)* als Überträger von Krankheiten für Mensch und Pflanze. Z. Hyg. Zool. 40, 376–379, 1952.
– Neuere Erkenntnisse über unsere Nacktschnecken als Schädlinge. Ang. Bot. 26, 94–99, 1952.
– Quantitative Untersuchungen über die Nahrungsaufnahme der Süßwasserlungenschnecke *Lymnaea stagnalis* L. Z. Fischerei u. Hilfsw. II, 451–456, N. F. 1953.
– Unsere Ackerschnecken *(Deroceras)* als Überträger von Krankheiten für Mensch, Tier und Pflanze. Die Pharmazie 8, 962–964, 1953.
– Biologie der mitteleuropäischen Landgastropoden. 1–440. Duncker & Humblot, Berlin 1954.
– Über die Wirkung von Insektiziden auf Gastropoden. Mitt. III. Anz. Schädlingskde. 27, 156–158, 1954.
– Die ökologischen Beziehungen zwischen unseren Landlungenschnecken und den Würmern. Zool. Anz. 154, 253–259, 1955.
– Ein neuer Kulturpflanzenschädling in Mitteleuropa: die Nacktschnecke *Milax (Tandonia) budapestensis* Hazay. Anz. Schädlingskde. 28, 42–46, 1955.
– Die Rolle unserer Schnecken bei der Ausbreitung von infektiösen Pflanzenkrankheiten. I. Fäulekrankheiten bei Erntegut. Z. ang. Zool. 42, 433–439, 1955.
– Die Rolle unserer Schnecken bei der Ausbreitung von infektiösen Pflanzenkrankheiten. II. Rost- und Brandpilze. Gesunde Pflanzen 7, 70–72, 1955.
– Der heutige Stand der Landschnecken-Bekämpfung mit chemischen Mitteln. Gesunde Pflanzen 7, 261–266, 1955.
– Biologie der mitteleuropäischen Süßwasserschnecken. 1–313. Duncker & Humblot, Berlin 1956.
– Unsere Weinbergschnecke als Schädling in den Tabakpflanzungen. Anz. Schädlingskde. 29, 73–74, 1956.
– Quantitative Untersuchungen über die Bedeutung bodenbewohnender Landschnecken für den Abbau des Fallaubes. Biol. Zbl. 75, 705–711, 1956.
– Das Verhalten unserer Landlungenschnecken gegenüber einigen angebauten Nachtschattengewächsen. Pharm. Ztg. 101, 912–913, 1956.
– Bericht über neue Versuche mit Metaldehydköder. Anz. Schädlingskde. 30, 43–44, 1957.
– Über den jetzigen Stand unserer Kenntnis von der Lebensweise der einheimischen Nacktschnecken. Angew. Bot. 23, 23–33, 1957.
– Quantitative Untersuchungen über die Nahrungsaufnahme der Schnecke *Helix aperta* Born. Anz. Schädlingskde. 30, 123–124, 1957.
– Nacktschnecken als Schädlinge in Mehl-Vorratskellern und über den Einfluß dieser Ernährung auf die Körperfarbe. Z. ang. Zool. 44, 349–357, 1957.
– Ein bisher wenig beachteter Kulturpflanzenschädling, die Gehäuseschnecke *Arianta arbustorum* L. Gesunde Pflanzen 10, 123–126, 1958.
– Die Rolle unserer Landschnecken bei der Stoffumwandlung und Humusbildung. Z. ang. Zool. 45, 341–350, 1958.
– Das Verhalten unserer Schnecken zu den Pflanzen ihrer Umgebung. 1–348. Duncker & Humblot, Berlin 1962.
– und Plate, H.-P.: Fraßschäden an Gewächshauspflanzen durch Schnecken. Anz. Schäd-

lingskde. 24, 87–91, 1951.
– und Plate, H.-P.: Schneckenfraßschäden an *Datura.* Die Pharmazie 6, 123–124, 1951.
– und Plate, H.-P.: Die wichtigsten einheimischen Nacktschnecken in ihrer Bedeutung für den Schädlingsbekämpfer. Schädlingsbekämpfung 43, 104–107, 1951.
– und Plate, H.-P.: Sind die Metaldehydköder zuverlässige Schneckenbekämpfungsmittel? Schädlingsbekämpfung 44, 130–131, 1952.
– und Plate, H.-P.: Über das Verhalten pflanzenfressender Schnecken zu unseren wirtschaftlich wichtigen Gräsern. Z. ang. Zool. 41, 71–75, 1954.
– und Plate, H.-P.: Schnecken als Schädlinge an unseren kultivierten Ölfruchtgewächsen zugleich ein Beitrag zur Sortenwahl wie auch zum Einfluß des Alters der Pflanzen. Pflanzenschutz (München) 7, 158–161, 1955.
– und Plate, H.-P.: Weitere Untersuchungen zur Frage der Sortenwahl unserer blattfressenden Landlungenschnecken. Pflanzenschutz (München) 8, 61–64, 1956.
– Peter, H., und Reichmuth, E.: Beitrag zur Frage der pathologischen Gestaltsveränderung und der Geschwülste bei unseren Nacktschnecken. Zool. Anz. 166, 139–147, 1961.
Furmaga, St., und Gundlach, J. L.: *Lymnaea stagnalis* L. as one more intermediate host of *Fasciola hepatica.* Acta Parasitol. Pol. 15, 231–235, 1968.
– und Gundlach, J. L.: *Lymnaea stagnalis* L. as a source of infection with *Fasciola hepatica.* Acta Parasitol. Pol. 15, 237–239, 1968.
Furness, G. O.: Survey of snails on citrus in the riverland of South-Australia. Australian J. Exp. Agr. Animal Husbandry 17, 1036–1039, 1977.
Gabbott, P. A., und Holland, D. L.: Growth and Metabolism of *Ostrea edulis* Larvae. Nature 241, 475–476, 1973.
Galangau, V.: Le cycle sexuel annuel de *Milax gagates* Drap. (Gastéropodes pulmonés) et ses deux pontes. Bull. Soc. zool. Fr. 89, 510–513, 1964.
– Etude en microscopie électronique de la gamétogenèse de *Milax gagates* Draparnaud 1801 (Gastéropodes-Pulmonés-*Limacidae*); Evolution des ultrastructures au cours de la spermatogenèse chez differents types de Mollusques. Thèse Doct. Sc. Nat., 1–151. Univ. Montpellier 1969.
Galtshoff, P. S.: Physiologie of reproduction of *Ostrea virginica.* II. Stimulation of spawning in the female oyster. Biol. Bull. 75, 286–307, 1938.
– Physiologie of reproduction of *Ostrea virginica.* III. Stimulation of spawning in the male oyster. Biol. Bull. 78, 117–135, 1940.
Gammon, E. T.: Helicid snails in California. Calif. Dept. Agric. Bull. 32, 173–187, 1943.
Gancarz, Z.: Studies on schistosomiasis. 3. Resistence of *Oncomelania* snails to sodium pentachlorophenate. Clin. Med. J. 77, 242, 1958.
Garcia, M. C.: Recherches sur l'échauffement de *Cepaea nemoralis* (L.) par l'énergie rayonne. Malacologia 14, 385–389, 1973.
– Composition de la lumière et activité reproductrice chez l'Escargot *Otala lactea* Müller (Moll. Gast. Pum.). Bull. Soc. Zool. Fr. 100, 147–152, 1975.
Gardner, D. R. und Moreton, R. B.: Disinhibition caused by the molluscide N-Trityl Morpholine (Frescon) in the nervous system of a freshwater snail. Pestic. Biochem. Physiol. 8, 225–240, 1978.
Gazzinelli, G., Ramalho-Pinto, F. J., Pellegrino, J., und Gilbert, B.: Uptake of ^{59}Fe as a tool for study of the crowding effect in *Biomphalaria glabrata.* Amer. J. Trop. Med. 19, 1034–1037, 1970.
Gebhardt-Dunkel, E.: Die Trockenresistenz bei Gehäuseschnecken. Zool. Jahrb. Abt. Allg. Zool. Physiol. 64, 235–266, 1953.
Geckler, R. P.: Laboratory studies of predation of snails by larvae of the marsh fly, *Sepedon tenuicornis (Diptera: Sciomyzidae).* Canad. Entomologist 103, 638–649, 1971.
Gerbal, A., Liberge, G., Maslet, C., und Salmon, C.: Réactivité des Anti-A d'*Helix pomatia* et d'*Helix aspersa* vis-à-vis de phénotypes à erythrocytaires. Haliotis 3, 155–166, 1973.
Gerhardt, U.: Zur Kopulation der Limaciden. Z. Morph. Ökol. Tiere 27, 401–450, 1933.

– Weitere Untersuchungen zur Sexualbiologie der Limaciden. Z. Morph. Ökol. Tiere 32, 518–541, 1937.

– Neue biologische Untersuchungen an Limaciden. Z. Morph. Ökol. Tiere 35, 183–202, 1939.

– Neue biologische Nacktschneckenstudien. Z. Morph. Ökol. Tiere 36, 557–580, 1940.

GERMAIN, L.: Mollusques terrestres et fluviatiles. Faune de France 21, 1–477 und 22, 478–897. P. Lechevalier, Paris 1930/1931.

GERSCH, M.: Vergleichende Endokrinologie der wirbellosen Tiere. 1–535. Geest & Portig, Leipzig 1964.

GETZ, L. L.: Notes on the Ecology of Slugs: *Arion circumscriptus, Deroceras reticulatum,* and *D. laeve.* Amer. Midland Naturalist 61, 485–498, 1959.

GETZIN, L. W.: Control of the Gray Garden Slug with Bait Formulations of a Carbamate Molluscicide. J. Econ. Entomol. 58, 158–159, 1965.

– und COLE, S. G.: Evaluation of Potential Molluscicides for Slug Control. Washington Agric. Exp. Stat. Wash. St. Univ. Bull. 658, 1–9, 1964.

GEVREY, J., MICHEL, S., und EUZEBY, J.: Mise en evidence de la toxicité d'un complexe algal sur la faune aquatique: 2. Toxicité à l'encontre de diverses espèces de limnées. Bull. Soc. Sci. Vet. Med. Comp. Lyon 74 (2), 191–194, 1972.

GHIRETTI, F.: Molluscan Hemocyanins. In: WILBUR, K., und YONGE, C.: Physiology of Mollusca. II, Chapt. 7, 233–248. Academ. Press, New York 1966.

GHOSE, KR. CH.: XVII. Embryogenesis and larval organs of the giant land snail *Achatina fulica* Bowdich. Proc. roy. soc. Edinb. sect. B Biol. 68 (17), 237–260, 1965.

GIANG, P. A., und SMITH, F. F.: Colorimetric determination of Metaldehyd Residues in plants. J. Agric. Food Chem. 4, 623–624, 1956.

GILBERT, B., PAES LEME, L. A., FERREIRA, A. M., BULHÕES, M. S., und CASTLETON, C.: Field tests of hexabutyldistannoxane (TBTO) in slow-release formulations against *Biomphalaria* spp. Bull. Wld. Hlth. Org. 49, 633–636, 1973.

GIROD, A.: La malacofauna della breccia ossifera pre-würmiana di Zandobbio (Bergamo). Natura-Soc. It. Sc. Museo Civ. St. Nat. Aquario Civ. Milano 64, 57–93, 1973.

GISH, C. D.: Organochlorine insecticide residues in soils and soil invertebrates from agricultural lands. Pest. Monit. J. 3, 241–253, 1970.

GITTENBERGER, E., BACKHUYS, W., und RIPKEN, TH. E. J.: De Landslaklen van Nederland. Kon. Ned. Nat. Veren. édit. Amsterdam I. 1970. 177 p.

GLAUDEL, R. J., und ETGES, F. J.: Toxic Effects of Freshwater Turbellarians on Schistosome Miracidia. J. Parasit. 59, 74–76, 1973.

GLENDENNING, R.: Slug control in Canada. Proc. Publ. Series No. 85, 1–5. Entomol. Sci. Serv. Dep. Agric. Canada 1952.

GODAN, D.: Über den Repellent- und Attraktiveffekt insektizider Pflanzenschutzmittel. Nachrichtenbl. Deutsch. Pflanzenschutzd. (Braunschweig) 10, 105–111, 1958.

– Über die Massenzucht von Nacktschnecken nebst einigen Bemerkungen zur Biologie. Z. ang. Zool. 45, 497–504, 1958.

– Untersuchungen über den Einfluß organischer Phosphorpräparate auf das Verhalten von Insekten. Z. Pflanzenkrankh. 66, 338–353, 1959.

– Bestimmungstabelle der schädlichen Schneckenarten. Gesunde Pflanzen 12, 26–33, 1960.

– Untersuchungen über die Wirksamkeit von Metaldehydködern auf Nacktschnecken unter Berücksichtigung ihrer Verhaltensreaktionen. Nachrichtenbl. Deutsch. Pflanzenschutzd. (Braunschweig) 13, 113–120, 1961.

– Untersuchungen über den Einfluß tierischer Nahrung auf die Vermehrung der Maulwurfsgrille (*Gryllotalpa gryllotalpa* L.). Z. angew. Zool. 51, 207–223, 1964.

– Untersuchungen über die molluskizide Wirkung der Carbamate. I. Teil: Ihre Toxizität auf Nacktschnecken. Z. Pflanzenkrankh. 72, 398–410, 1965.

– Untersuchungen über die molluskizide Wirkung der Carbamate. II. Teil: Abhängigkeit von Art, Größe und Ernährung der Schnecke. Z. ang. Zool. 53, 417–430, 1966.

– Untersuchungen über die Wirkung der Carbamate auf Gehäuseschnecken. Z. ang. Ent. 59, 385–396, 1967.
– Untersuchungen über den Einfluß der Herbizide auf *Drosophila melanogaster* Meig. Teil I: Populationsdichte, Entwicklungsdauer und Körpergröße im Verlauf von zehn Generationen. Z. ang. Zool. 56, 89–117, 1969.
– Untersuchungen über den Einfluß der Herbizide auf *Drosophila melanogaster* Meig. Teil II: Das Verhalten legeaktiver Weibchen. Z. angew. Ent. 66, 225–235, 1970.
– Bericht über die Teilnahme an dem „Internationalen Kongreß über Raps" vom 25.–30. Mai 1970 in Paris. Nachrichtenbl. Deutsch. Pflanzenschutzd. (Braunschweig) 22, 156–157, 1970.
– Die ökologischen Grundlagen der Prüfungsmethoden von Molluskiziden. Nachrichtenbl. Deutsch. Pflanzenschutzd. (Braunschweig) 24, 35–37, 1972.
– IV. Kongreß Europäischer Malakologen. Nachrichtenbl. Deutsch. Pflanzenschutzd. (Braunschweig) 24, 44–45, 1972.
– Schadwirkung und wirtschaftliche Bedeutung der Schnecken in der Bundesrepublik Deutschland. Nachrichtenbl. Deutsch. Pflanzenschutzd. (Braunschweig) 25, 97–101, 1973.
– L'influence des herbicides sur les limaces. Haliotis 3, 125–130, 1973.
– Common names von Schadgastropoden in 12 Sprachen. Mitt. Biol. Bundesanstalt, Berlin-Dahlem, 159, 1–91, 1974.
– Mosebach, E., und Plate, H.-P.: Prüfungen von Molluskiziden gegen Nacktschnecken. In „Richtlinien für die amtliche Prüfung von Pflanzenschutzmitteln" Reihe 8, 1–10, herausgegeben von der Biologischen Bundesanstalt für Land- und Forstwirtschaft, Berlin-Dahlem, 1965, erschienen 1967.
Goddard, C.: A study of the reproductive tract of *Helix aspersa* Müller after partial gonadectomy. Austral. J. Biol. Sc. 13, 378–386, 1960.
– und Martin, A.: Carbohydrate Metabolism. In: Wilbur, K., und Yonge, C.: Physiologie of *Mollusca* II (9). 275–308. Academ. Press, New York 1966.
Gönnert, R.: Bayluscid, ein neues Mittel zur Bekämpfung medizinisch wichtiger Süßwasserschnecken. Pflanzenschutz-Nachr. „Bayer" 15, 4–25, 1962.
– und Webbe, G.: Molluskizide zur Bekämpfung der Bilharziose und neue Erkenntnisse über den Wert dieser Maßnahme. Pflanzenschutz-Nachr. „Bayer" 19, 36–42, 1966.
Götting, K.-J.: Malakozoologie (Grundriß der Weichtierkunde). 1–320. Gustav Fischer (Stuttgart) 1974.
Golvan, J. J., und Mougeot, G.: Mise en évidence des propriétés cytoadhérentes des cellules de l'hémolymphe de *Biomphalaria glabrata* parasité par *Schistosoma mansoni*. C. R. Acad. Sc. Paris, 276. Sér. D. 1229–1231, 1973.
Gomot, L.: La culture organotypique chez les Invertébrés autres que les Insectes. In Vago. C.: „Invertebrate tissue culture". Acad. Press, Paris-New York-London 1971.
– Etude de l'activité géneitale de Mollusques Gastéropodes (Paludines et Escargots) par la mé-des cultures d'organes. Arch. Anat. Hist. Embr. norm. et exp. 56, 131–160, 1973.
– Culture in vitro de la glande hermaphrodite de l'Escargot *Helix aspersa* en présence de stéroïdes. C. R. Sc. Soc. Biol. 168, 837–843, 1974.
– Etude de l'activité génitale de Mollusques Gastéropodes (Paludines et Escargots) par la méthode des cultures d'organes. Bull. Soc. Zool. France Suppl. 101, Fasc. 2. 23–34, 1975.
– und Guyard, A.: Evolution en culture in vitro de la glande hermaphrodite de jeunes escargots de l'espèce *Helix aspersa* Müller. C. R. Acad. Sc. Paris 258, 2902–2905, 1964.
– und Guyard, A.: La culture organotypique appliquée à l'endocrinologie d'un Mollusque Gastéropode Pulmoné: *Helix aspersa* Müll. 22–31. 2. Int. Coll. Tissue Culture, Inst. Lombardo, Fondazione Baselli (1967). 1968.
Gottfried, H., und Lusis, O.: Steroids of Invertebrates: The in vitro production of 11-Kerostone and other steroids by the eggs of the slug *Arion ater rufus* (Linn.). Nature (London) 212, 1488–1489, 1966.
– Dorfman, R. I., und Wall, P. E.: Steroids of invertebrates: Production of oestrogens by an

accessory reproductive tissue of the slug *Arion ater rufus*. Nature (London) 215, 409–410, 1967.

– DORFMAN, R. I., FORCHIELLE, E., und WALL, P. E.: Aspects of the reproductive endocrinology of the giant land slug *Ariolimax californicus (Stylommatophora: Gastropoda)*. Gen. comp. Endocr. 9, 454–470, 1967.

– und DORFMAN, R. I.: Invertebrate hormone II. Excerpt. med. intern. Congr. Ser. Pays-Bas 184, 268–376, 1969.

– und DORFMAN, R. I.: The steroid biochemistry of the Molluscan ovotestis: A general concept of reproductive control mechanisms. Excerpt. med. intern. Congr. Ser. Pays-Bas 184, 368–376, 1969.

und DORFMAN, R. I.: Steroids of invertebrates. V. In vitro biosynthesis of steroids by the male-phase ovotestis of the slug *(Ariolimax californicus)*. Gen. Comp. Endocr. 15, 120–138, 1970.

GOULD, H. J.: Observations on slug damage to winter wheat in East Anglia 1957–1959. Plant Pathol. 10, 142–146, 1961.

– Trials on the control of slugs on arable fields in autumn. Plant Pathol. 11, 125–130, 1962.

– Tests with seed dressings to control grain hollowing of winter wheat by slugs. Plant Pathol. 11, 147–152, 1962.

– Observations on the susceptibility of maincrop potato varieties to slug damage. Plant Pathol. 108–111, 1965.

– und WEBLEY, D.: Field Trials for the Control of Slugs in Winter Wheat. Plant Pathol. 21, 77–82, 1972.

GRABER, M. M.: Mollusques du sous-genre *Bulinus* s. s. *(Basommatophora: Planorbidae)*, hôtes intermédiaires de divers parasites des ruminants domestiques d'Ethiope. C. R. Acad. Sc. Paris, Sér. D 278, 2037–2039, 1974.

GRAF, A.: Die Schneckenbekämpfung durch Truthühner. Mitt. dt. Landw. Ges. 74, 83–84, 1959.

GRAHAM, A.: British Prosobranchs and other operculate Gastropod molluscs. In: Synopses of the British Fauna 2. 1–112; Acad. Press, London-New York 1971.

GRASSÉ, P.-P.: Traité de Zoologie. Tome IV: Nemathelminthes, Rotifères, Gastrotriches, Kinorhynques. Masson, Paris 1965.

– Traité de Zoologie, Tome V. Mollusques, Gastéropodes et Scaphopodes, (F. III, 1083 pp.). III. Parasites; Prédateurs. 510–517. Masson, Paris 1968.

GRESS, F. M., und CHENG, TH, C.: Alterations in Total Serum Proteins and Protein Fractions in *Biomphalaria glabrata* Parasitized by *Schistosoma mansoni*. J. Invert. Pathol. 22, 382–390, 1973.

GRÉTILLAT, S.: Un nouveau molluscicide, le diméthyldithio-carbamate de zinc (zirame). Bull. Wld. Hlth. Org. 25, 581–588, 1961.

GRIFFOND, B.: Survie et évolution, en culture in vitro, des testicules de *Viviparus viviparus* L., Gastéropode Prosobranche à sexes séparés. C. R. Acad. Sc. Paris, 268, 963–965, Sér. D. 1969.

– und GOMOT, L.: Cultivation in vitro of gonads of the Fresh Water Gastropod Prosobranch Mollusc *Viviparus viviparus* L. In Vitro, 10, (3 & 4), 206–215. 1974.

GRIGOR'EVA, G. M.: Propionylcholinesterase in the nervous ganglia of the fresh-water pulmonate mollusc *Lymnaea stagnalis*. Comp. Biochem. Physiol. A. Comp. Physiol. 44, 1341–1352, 1973.

GRIME, J. P., MCPHERSON-STEWART, S. F., und DEARMAN, R. S.: An investigation of leaf palatability using the snail *Cepaea nemoralis*. J. Ecol. 56, 405–420, 1968.

– und BLYTHE, G. M.: An investigation of the relationships between snails and vegetation at the Winnats Pass. J. Ecol. 57, 45–66, 1969.

– BLYTHE, G. M., und THORNTON, J. D.: Food selection in the snail *Cepaea nemoralis* L. In: WATSON, A.: Animal populations in relation to their food resources. 73–99. Blackwell; Oxford-Edinburg 1970.

GROSSU, A. V.: *Mollusca (Gastropoda, Pulmonata).* In: Fauna Republici Populae Romîne 3, 1–518. Ed.: Acad. Republ. Popul. Romîne, Bucuresti 1955.

– *Boettgerilla vermiformis* Wiktor *(Gastropoda-Pulmonata, Limacomorpha)* în România. Si. Cerc. Biol. Ser. Zool. 22 (4), 289–291, 1970.

– [Révision des espèces du genre *Arion* Férussac en Roumanie (Gastéropodes, *Arionidae*). – Soc. Stjinte Bio. Rep. Soc. România, Com. Zool. 61–71, 1970.

– Le polymorphisme des gastéropodes et la possibilité d'identifier et de limiter l'espèce avec des méthodes biochimiques. Malacologia 16, 15–19, 1977.

– und LUPU, D.: *Deroceras (Hydrolimax) melanocephalus,* eine kaukasische Limacide in Rumänien. Arch. Molluskenkde. 86, 171–172, 1957.

– und LUPU, D.: Un nou gasteropod în fauna R.P.R.: Limax flavus L. Com. Acad. R.P. Romîne (Bukarest) 9, 561–566, 1959.

– und LUPU, D.: Revision der Gattung Limax aus Rumänien, nebst Beschreibung einiger neuer Arten. Arch. Moll. 89, 157–165, 1960.

– und LUPU, D.: Zur Kenntnis der Gattung *Lehmannia* Heynemann *(Limacidae)* und deren Variabilität. Arch. Moll. 91, 191–201, 1962.

– und LUPU, D.: Variabilité du genre *Deroceras* (Fam. *Limacidae,* Gastéropodes terrestres). Trav. Mus. Hist. Nat. „Gr. Antipa" (Bucuresti) 4, 193–200, 1963.

– und TESIO, C.: Suggestions for species grouping within the family *Limacidae (Gastropoda, Pulmonata)* by biochemical methods. Proc. Malac. Soc. Lond. 41, 321–329, 1975.

– und TESIO, C.: Systematic studies on some species of the genus *Helicigona (Pulmonata: Helicidae).* Malacol. Rev. 8, 111–113, 1975.

– und TESIO, C.: Contributions biochimiques (électrophorèse) à la détermination de l'unité du genre *Deroceras (Gastropoda, Limacomorpha).* Malak. Abh. Staatl. Mus. Tierkde. Dresden 4 (19), 178–183, 1975.

GUERRUCCI, M. A.: Intérêt des espèces *Cepaea nemoralis* (L.) et *Cepaea hortensis* Müller pour l'étude des facteurs du polymorphisme. Haliotis 3, 143–148, 1973.

GUIBÉ, J.: Chironomes parasites de mollusques gasteropodes. Bull. biol. Fr. et Belg. 76, 283–297, 1942.

GUISTI, F.: *Oxychilus (Ortizius) alliarius* (Miller) nuovo reperto per l'Italia appeninica (Notulae malacologicae VII). Mem. Mus. Civ. Stor. Nat. Vérone XVI. 375–378, 1968.

GURIN, S., und CARR, W. E.: Chemoreception in Nassarius obsoletus: The role of specific stimulatory proteins. Science (Washington) 174, 293–295, 1971.

GUYARD, A.: Féminisation de la glande hermaphrodite juvenile d'*Helix aspersa* Müll. associée in vitro au ganglion cérébroide d'Escargot adulte ou de *Paludina* femelle. C. R. Acad. Sc. Paris 265, 145–149, 1967.

– Autodifférenciation femelle de l'ébauche gonadique de l'Escargot *Helix aspersa* Müll. cultivée sur milieu anhormonal. C. R. Acad. Sci., Paris 268, 966–969, 1969.

– Elaboration d'un milieu synthétique enrichi destiné à la culture d'organes de Mollusques. C. R. Acad. Sc., Paris, 268, 162–164, 1969.

– Aspects ultrastructuraux de la différenciation gonocytaire au début de l'organogenèse de la glande hermaphrodite d'*Helix aspersa* Müll. (Gastéropode pulmoné). Bull. Soc. Zool. France 95. 471–474. 1970.

– La différenciation gonocytaire des Mollusques Gastéropodes en culture in vitro. Ann. Biol. 9, 401–408, 1970.

– Etude de la différenciation de l'ovotestis et des facteurs controlant l'orientation sexuelle des gonocytes de l'escargot *Helix aspersa* Müller. Thèse Univ. Besançon C.N.R.S. A.O. Nr. 5939 (56), 1–187, 1971.

– und GOMOT, L.: Survie et différenciation de la gonade juvénile d'*Helix aspersa* en culture organotypique. Bull. Soc. Zool. Fr. 89, 48–56, 1969.

HABERKERN, CH.: Chlorprothixen-Behandlung von Metaldehydvergiftungen beim Hund. Tierärztl. Umschau 32, 427–431, 1977.

HAIRSTON, N. G.: An evaluation of techniques used in estimating snail population. Bull. Wld.

Hlth. Org. 19. 1958, 661.
– und SANTOS, B. C.: Ecological control of the snail host of *Schistosoma japonicum* in the Philippines. Bull. Wld. Hlth. Org. 25, 603–610, 1961.
– WURZINGER, K.-H., und BURCH, J. B.: Non-chemical Methods of Snail Control. WHO IV BC 75. 1975. 573 WHO/Schisto/ 75, 40, 1975.
HALAWANI, A.: Effect on the delta isomer of benzene hexachloride on the snails *Bulinus* and *Planorbis*. Roy. Egypt. med. Assoc. J. 30, 35–41, 1947; C. A. 1947, 4606.
HALEY, G. F., und GIBSON, M. A.: Calcium storage in the soft tissues of freshwater gastropods: The influence of environmental calcium concentrations. Can. J. Zool. 49, 1001–1004, 1971.
HAMEURY, M. P.: Sur la présence en France de *Deroceras caruanae* (Pollonera, 1891). Vie et Moeur 9, 81–87, 1958.
HAMMER, O. H.: Zectran pesticide – Some chemical and physical properties, and some resultats from use on pests on ornamental plants. Down to Earth 17, 9–13, 1962.
HARANGHY, Z.: Wirkung des 3,4-Benzpyrens auf Süßwassermuscheln. Acta biol. Budapest. 7, 101–108, 1956.
HARLEY, P., und HARLEY, E.: A technique for cutting or accessing nerves and connectives in *Aplysia californica*. Comp. Biochem. Physiol. A. Comp. Physiol. 45, 389–392, 1973.
HARPAZ, J., und OSERI, J.: Control of the White Garden Snail in Israel. Pest technology 4, 41–43, 1961.
HARRISON, A. D.: The effects of Bayluscid on gastropod snails and other aquatic fauna in Rhodesia. Hydrobiologia (Den Haag) 28, 371–384, 1966.
– und MASON, M. H.: The effects on the fauna of natural waters of surveillance treatment with bayluscide in Rhodesia. Hydrobiologia 29, 149–155, 1967.
HARTOG, C., und SANDEE, A. J. J.: *Milax gagates* en *Limax valentianus*, twee zeldzame naakte landslakken in Yerseke. Basteria 26, 83–85, 1962.
HARVEY, P.: Populations of *Cepaea nemoralis* from southwestern France and northern Spain. Heredity 27, 353–363, 1971.
HASAN, S., und VAGO, C.: Transmission of *Alternaria brassicola* by slugs. Pl. Dis. Reptr. 50. 764–767, 1966.
HASSLEIN, L.: Neue Funde von *Boettgerilla vermiformis* Wiktor 1959 im Siebengebirge und am Rodderberg. Decheniana 125, 111–112, 1972.
HAYWARD, J. F.: A further note on *Agriolimax caruanae* Pollonera. J. Conch. 23 (12), 403–405, 1954.
HECKER, U.: Zur Kenntnis der mitteleuropäischen Bernsteinschnecken *(Succineidae)*. Arch. Moll. 94, 1–45, 1965.
HEILBRONN, E., HAUSE, S., und LUNDGREN, G.: Chemical identification of acetylcholine in squid-head ganglion. Brain Res. 33, 431–437, 1971.
HEIN, G.: Die Weinbergschnecke *(Helix pomatia* L.), Lebensweise, Verbrauch, Handel und Zucht. Z. Hyg. Zool. 40, 225–248, 1952.
HEINZE, K.: Können Schnecken pflanzliche Virosen übertragen? Z. Pflanzenkrankh. 65, 193–198, 1958.
HEITEFUSS, R.: Ursachen der Nebenwirkungen von Herbiziden auf Pflanzenkrankheiten. Z. Pflanzenkrankh. Sonderh. 6, 79–87, 1972.
HEITEFUSS, R. und BRANDES, W.: Nebenwirkungen von Herbiziden und Kalkstickstoff auf den Befall von Weizen mit *Erysiphe graminis*. Nachrichtenbl. Dtsch. Pflanzenschutzd. (Braunschweig) 22, 40–43, 1970.
HEKSTRA, G. P., und LEVER, J.: Some effects of ganglion extirpation in *Limnaea stagnalis*. Koninkl. Ned. Akad. Wetenschap. C, 63, 271–282, 1960.
HELMY, M.: The effect of environment factors on the distribution and control of *Planorbis boissyi* the intermediate host of *Schistosoma mansoni* in the valley of the Nile. J. Trop. Med. Hyg. 56, 25, 1953.
HENDERSON, J. F.: Molluscicides. In: Rep. Rothamst. exp. Stat. for, 173–174, 1965.
– Laboratory methods for assessing the toxicity of contact poisons to slugs. Ann. appl. Biol. 62,

363–369, 1968.
– A laboratory method for assessing the toxicity of stomach poisons to slugs. Ann. appl. Biol. 63, 167–171, 1969.
– The fumigant effect of metaldehyde on slugs. Ann. appl. Biol. 65, 507–510, 1970.
HENDERSON, N. L., und PELLUET, D.: The effect of visible light on the ovotestis of the slug *Deroceras reticulatum* Müll. Canad. J. Zool 38, 173–178, 1960.
HENEINE, I. F., GAZZINELLI, G., und TAFURI, W. L.: Iron metabolism in the snail *Biomphalaria glabrata:* uptake, storage, and transfer. Comp. Biochem. Physiol. 28, 391–399, 1969.
HENRARD, J. B.: On the occurence of *Helisoma anceps* (Menke) in Italy. Basteria 32, 2–3, 1968.
HEPPLESTON, P. B.: Life history and population fluctuations of *Lymnaea truncatula* (Müll.), the snail vector of fascioliasis. J. appl. Ecol. 9, 235–248, 1972.
HERFS, W.: Über einige internationale Aktivitäten im Bereich der zoologischen Mittelprüfung. Gesunde Pflanzen 26, 100–106, 1974.
HERING, M.: Über Schneckenfraß an Reben des Weinbaugebietes Mosel-Saar-Ruwer. Weinberg Keller 13, 491–500, 1966.
HERLIN-HOUTTEVILLE, P., und LUBET, P. E.: The sexuality of pelecypod molluscs. In: REINBOTH, R.: Intersexuality in the Animal Kingdom. 179–187. Springer 1975.
HERMANN, G.: Sekundärgefährdung von Amseln durch Regenwürmer nach Anwendung von Schneckenkorn Mesurol. Biol. Institut Farbenfabriken Bayer AG. Leverkusen; Unveröffentlichter Bericht vom 29. 11. 1968; Zit.: Martin & Forrest 1970.
– Bekämpfung von Landschnecken mit Methiocarb. Meded. Facult. Landb. Wetensch. Rijksuniv. Gent 36, 197–206, 1971.
– und KOLBE, W.: Wirkung der Saatgutinkrustierung mit Mesurol zur Vogelabwehr im Maisanbau unter Berücksichtigung der Sortenverträglichkeit und der Nebenwirkungen. Pflanzenschutz-Nachr. Bayer 24, 283–323, 1971.
HERZBERG, F., und HERZBERG, A.: The effect of cold on the growth of *Helix aspersa*. J. Expt. Zool. 145, 191–196, 1960.
– und HERZBERG, A.: Observations on reproduction in *Helix aspersa.* Amer. Midland Naturalist 68, 297–306, 1962.
HEYNEMAN, D., FAULK, W. P., und FUDENBERG, H. H.: *Echinostoma lindoense:* larval antigen from the snail intermediate host, *Biomphalaria glabrata.* Exp. Parasit. 29, 480–492, 1971.
HILL, R., und WELSH, J. H.: Heart, Circulation, and Blood Cells. In: WILBUR, K. M., und YONGE, C. M.: Physiology of *Mollusca,* II, (Chapt. 4) 125–174; Acad. Press New York 1966.
HIRA, P. R., und WEBBE, G.: The Effect of Sublethal Concentrations of the Molluscicide Triphenyl Lead Acetate on *Biomphalaria glabrata* (Say) and on the Development of *Schistosoma mansoni* in the Snail. J. Helminthology 46, 11–26, 1972.
HIRST, J. M., LE RICHE, H. H., und BASCOMB, C. L.: Copper Accumulation in the soils of Apple orchards near Wisbeck. Plant Pathol. 10, 105–108, 1961.
HIRWE, A. S., METCALF, R. L., LU, P. Y., und CHIO, L. C.: Comparative Metabolismus of 1,1-Bis(p-Chlorophenyl)-2-Nitropropane(Prolan) in Mouse, Insects, and in a Model Ecosystem. Pestic. Biochem. Physiol. 5, 65–72, 1975.
HIS, E.: Recherche d'un test permettant de comparer l'activité respiratoire des huîtres au cours de l'évolution de la maladie des branchies. Rev. Trav. Int. Peches Mar. 33, 171–175, 1970.
HODASI, J. K. M.: The effects of *Fasciola hepatica* on *Lymnaea truncatula.* Parasitol. 65, 359–369, 1972.
Höchstmengen für Pflanzenschutzmittel-Rückstände. Gesunde Pflanzen 26, 21–36, 41–46, 1974.
HOFFMANN, H.: Südamerikanische Nacktschnecken. Zool. Jahrb. (Syst.) 67, 213–224, 1935.
HOGBEN, L., und KIRK, R. L.: Studies on temperature regulation: The Pulmonata and Oligochaeta. Proc. Roy. Soc. Ser. B. Biol. Sci. 132 (868). 239–252, 1944.
HOHORST, W., und GRAEFE, G.: Ameisen – obligatorische Zwischenwirte des Lanzettegels *(Dicrocoelium dendriticum).* Die Naturwissenschaften 48, 229–230, 1961.

HOLM, L. G., WELDON, L. W., und BLACKBURN, R. D.: Aquatic Weeds. PANS 16, 576–589, 1970.

– WELDON, L. W., und BLACKBURN, R. D.: Aquatic weeds. The rampant quality of aquatic weeds has become one of the symptoms of our failure to manage our resources. Science 166, 699–709, 1969.

HOLZ, J.: Über die Wirkung von Kupfersulfat, Pentochlorphenolnatrium und „Bayer 73" auf parasitenübertragende Schnecken Indonesiens. Anz. Schädlingskde. 35, 36–38, 1962.

– und ONG BIE HWA: Über Einflüsse des Molluskizids Bayluscid auf Plankton indonesischer Fischteiche und Reisfelder. Pflanzenschutz-Nachr. „Bayer" 16, 253–257, 1963.

HOPF, H. S., und MÜLLER, R. L.: Laboratory breeding and testing of *Australorbis glabratus* for molluscidal screening. Bull. Wld. Hlth. Org. 27, 783–789, 1962.

HORNE, F.: The utilization of foodstuffs and urea production by a land snail during estivation. Biol. Bull. (Woods Hole) 144, 321–330, 1973.

HORTON, W. J.: Preliminary report on investigations of molluscicide components of *Phytolacca dodecandra.* WHO Moll. Information Ser. 24, 1968.

HOSKIN, F. C., und LONG, R. J.: Purification of a DFP-hydrolyzing enzyme from squid heated ganglion. Arch. Biochem. Biophys. 150, 548–555, 1972.

HOWESS, N. H., und WELLS, G. P.: The water relations of snails and slugs: II Weight rhythmus in *Arion ater* L. and *Limax flavus* L. J. Exp. Biol. 11, 344–351, 1934.

HOWITT, A. J.: The slug problem on strawberries. Proc. Vth Ann. Meet. Western Washington Hort. Assoc. 1960.

– Chemical control of slugs in Orchard Grass-ladino White Clover Pastures in the Pacific Northwest. J. econ. Ent. 54, 778–781, 1961.

– und COLE, G.: Chemical control of slugs affecting vegetables and strawberries in the Pacific Northwest. J. econ. Ent. 55, 320–325, 1962.

HOYLE, G.: Muscle and neuromuscular physiology. In: WILBUR, K. M., und YONGE, C. M.: Physiology of *Mollusca* I (10). 313–351. Acad. Press New York-London 1964.

HUBENDICK, B.: Recent lymnaeidae. Their variation, morphology, taxonomy, nomenclature, and distribution. Kungl. Svenska Vetenskap. Handl. 3, 1–223, 1951.

HÚDEC, V., und VASÁTKO, J.: Beitrag zur Molluskenfauna Bulgariens. Acta Sc. Nat. Brno 5 (2), 1–38, 1971.

HUIZINGA, H. W.: *Ribeiroia marini:* Pathogenesis and Larval Trematode Antagonism in the Snail *Biomphalaria glabrata.* Exp. Parsit. 33, 350–364, 1973.

HUNTER, P. J.: The distribution and abundance of slugs on an arable plot in Northumberland. J. Anim. Ecol. 35, 543–557, 1966.

– The effect of cultivations on slugs of arable ground. Plant Pathol. 16, 153–156, 1967.

– Studies on slugs of arable ground: I. Sampling methods. Malacologia 6, 369–377, 1968.

– Studies on slugs of arable ground: II Life cycles. Malacologia 6, 379–390, 1968.

– Studies on slugs of arable ground: III. Feeding habits. Malacologia 6, 391–399, 1968.

– An estimate of the extent of slug damage to wheat and potatoes in England and Wales. N.A.A.S. q. Rev. 85, 31–36, 1969.

– Slugs and their control. Proc. 5th Brit. Insect. Fung. Conf. 3, 715–719, 1969.

– und JOHNSTON, D. L.: Screening carbamates for toxicity against slugs. J. econ. Ent. 63, 305–306, 1970.

– SYMONDS, B. V., und NEWELL, P. F.: Potato Leaf and Stem Damage by Slugs. Plant Path. 17, 161–164, 1968.

– und SYMONDS, B. V.: The distribution of bait pellets for slug control. Ann. appl. Biol. 65, 1–7, 1970.

– und SYMONDS, B. V.: The Leap-frogging Slug. Nature 229, 349, 1971.

HUNTER, W. R.: Annual variations in growth and density in natural populations of freshwater snails in the west of Scotland. Proc. zool. Soc. Lond. 136, 219–253, 1961.

– Physiological Aspects of Ecology in Nonmarine Molluscs. In: WILBUR, K. M., und YONGE, C. M.: Physiology of *Mollusca* I, 83–126. Acad. Press New York-London 1964.

HURLE, K., und RADEMACHER, B.: Untersuchungen über den Einfluß langjähriger wiederholter Anwendung von DNOC und 2,4-D auf ihren Abbau im Boden. Weed Res. 10, 159–164, 1970.

HYMAN, L. H.: The invertebrates *Mollusca* I, 6, 1–729. McGraw-Hill Book Company, New York-London 1967.

ICCP, Invertebrate Consultants Committee for the Pacific Pac. Sci. Board; Minutes of seventh annual meeting Febr. 18. 1953. Hawaiian Sugar Planters Association Exp. Stat. Honolulu 1953, 1–21.

IDLER, D. R., und WISEMAN, P.: Molluscan sterols: A review. J. Fish Res. Board Can. 29, 385–398, 1972.

IMHOF, G.: Der Einfluß von Temperatur und Photoperiode auf den Lebenszyklus einiger Süßwasserpulmonaten. Malacologia 14, 393–395, 1973.

INABA, A.: Cytotaxonomic studies of lymnaeid snails. Malacologia 7, 143–168, 1969.

INAGAKI, H.: Sénescence dans la croissance chez le male de *Littorina saxatilis* var. *rufus* (Olivi), Mollusque Gastéropode, mise en evidence par la regression sexuelle et la diminution de l'intensité respiratoire. C. R. Hebd. Séance Acad. Sci. Sér. D, Sci. Nat. (Paris) 274, 1828–1831, 1972.

INGRAM, M. W.: Mollusk food of the beetle *Scaphinotus interruptus* (Men.). Bull. Southern Calif. Acad. Sci. 45, 34–36, 1946.

IRISAWA, H., WILKENS, L. A., und GREENBERG, M. J.: Increase in membrane conductance by 5-hydroxytryptamine and acetylcholine on the hearts of *Modiolus demissus* and *Mytilus edulis (Mytilidae Bivalvia)*. Comp. Biochem. Physiol. A. Comp. Physiol. 45, 653–666, 1973.

ISHAK, M. M., SHARAF, A. A., und MOHAMED, A. H.: Studies on the mode of action of some molluscicides on the snail, *Biomphalaria alexandrina:* II. Inhibition of succinate oxydation by Bayluscide, sodium pentachlorphenate, and copper sulphate. Comp. Gen. Pharmacol. 3, 385–390, 1972.

ISHIYAMA, J., UHLENBRUCK, G., und HERMANN, G.: Isolation of an anti-A-agglutinin from *Helix aspersa.* Blut. Z. ges. Blutforschung 24. 1972, 178–179.

IWATA, S., und WATANABE, S.: Studies on eradication of the intermediate host of liverfluke. Japan. Vet. Med. Assoc. 13, 110, 1960.

IZZATULLAEV, Z.: Slugs in Tadzhikistan Zashch. Rast. 12, 44–45, 1977.

JAECKEL, S. H.: Praktikum der Weichtierkunde. 1–87, VEB Gustav Fischer, Jena 1953.

– Die Land- und Süßwasser-Mollusken der nördlichen Balkanhalbinsel. Mitt. Zool. Mus. Berlin 30, 54–85, 1954.

– Mollusca – Weichtiere. In: STRESEMANN, E.: Exkursionsfauna von Deutschland, Wirbellose I. Volk und Wissen (Berlin), 1967, 3. Aufl.

– und MEISE, W.: Über Land- und Süßwasserschnecken Jugoslawiens und Albaniens. Mitt. Hambg. Zool. Mus. Inst 54, 21–32, 1956.

– und PLATE, H.-P.: Beitrag zur Molluskenfauna Griechenlands. Zool. Abh., Abh. u. Ber. Staatl. Mus. Tierkde. Dresden 26 (1), 3–19, 1961.

– und PLATE, H.-P.: Land- u. Süßwasserschnecken aus den Gewächshäusern des Botanischen Gartens Berlin-Dahlem. 1. Gehäuseschnecken. Z. ang. Zool. 54, 361–371, 1967.

JAIN, S. P.: Double infection of larval trematodes in Molluscs. Agra Univ. J. Res. Sci. 19, 47–48, 1970.

JANUS, H.: Unsere Schnecken und Muscheln. 1–124, Kosmos Naturführer, Franckh'sche Verlagsbuchh. Stuttgart 1958.

– Molluscs. The Young Specialist looks at ... 1–180 Burke, London 1965.

JEFFERSON, R. N.: Metaldehyde slug control. Florist's Rev. 112 (2887), 86–87, 1953.

JESSOP, N. H., und READING, M. K.: The effects of simulated slug damage on the Yield of Winter Wheat. Plant Pathol. 18, 1969. 172–175.

JEWSBURY, J. M., und HOMEWOOD, C. A.: Radio-labelling of schistosome cercariae. Ann. Tropic. Medic. Parasit. 65, 273–274, 1971.

JEZEWSKA, M.: Purinotelism in slugs *Limacidae* and *Arionidae.* Bull. Acad. Pol. Sci. Ser. Sci

Biol. 20, 365–368, 1972.

JOBIN, W. R., FERGUSON, F. F., BERRIOS DURAN, L. A., und SAN JUAN, P. R.: Effect of *Marisa cornuarietis* on populations of *Biomphalaria glabrata* in Fann ponds of Puerto Rico. Amer. J. Trop. Med. Hyg. 22, 278–284, 1973.

JOE, L. K.: Studies on *echinostomatidae (trematoda)* in Malaya. XV. The life history of *Echinostoma murinum* (Tubangui, 1931). Proc. helminth. Soc. Wash. 34, 139–143, 1967.

JOHANNESSEN, O. H.: Deformations of the inner shell surface of *Venerupis pullastra* (Montagu) (*Lamellibranchia)* as a result of infection by a trematode metacercaria: With a note of parasitism leading to parasitic castration. Sarsia 52, 117–122, 1973.

JOHNSON, A. L., und KINGSTON, R. J.: Structure Elucidation and Molluscicide Evaluation of the Major Saponin from the Berries of *Phytolacca americana.* Diss. Abst. Intern. 335 Nr. 2731. Kingston, R. I., USA 1974.

JOLIVET, G.: Importance de la Fasciolose du Betail en France. Haliotis 3, 39–42, 1973.

JONES, F. G. W., und JONES, M.: Pests of field crops. Kap. 10: Slugs and Snails – *Mollusca.* Edward Arnold Ltd., London 1964.

JONES, J. S.: Ecological genetics and natural selection in molluscs. Science 145, 376–377, 1974.

JONES, T. P.: The effect of aldrin on yield and slug damage in carrots in South Wales. Plant Pathol. 14, 39–40, 1965.

JOOSSE, J.: Dorsal bodies and dorsal neurosecretory cells of the cerebral ganglia of *Lymnaea stagnalis.* Arch. Neerl. Zool. 16, 1–103, 1964.

– Structural and Endocrinological aspects of Hermaphroditism in Pulmonate snails, with particular reference to *Lymnaea stagnalis* (L.). In: REINBOTH, R.: Intersexuality in the Animal Kingdom. 158–169. Springer 1975.

– BOER, H. H., und CORNELISSE, C. J.: Gametogenesis and oviposition in *Lymnaea stagnalis* as influced by γ-irridiation and hunger. Symp. Zool. Soc. London 22, 213–235, 1968.

– und REITZ, D.: Functional anatomical aspects of the ovotestis of *Lymnaea stagnalis.* Malacologia 9, 101–109, 1969.

JORDAN, P., BARNISH, G., BARTHOLOMEW, R. K., CRIST, E. und CHRISTIE, J. D.: Evaluation of an experimental mollusciciding programme to control *Schistosoma mansoni* transmission in St. Lucia. – Bull. WHO. 56, 139–146, 1978.

JOUBERT, C. J., und WALTERS, S. S.: The Control of Snails and Slugs. Farming in South Africa 26, 379–380, 1951.

JUDGE, F. D.: Preliminary Screening of candidate Molluscicides. J. econ. Ent. 62, 1393–1397, 1969.

– und KUHR, R. J.: Laboratory and Field Screening of Granular Formulations of Candidate Molluscicides. J. econ. Ent. 65., 242–245, 1972.

JUILLET, J.: Notes sur la dispersion de la limace *Arion ater* (L.). Phytoprotection 49, 38–40, 1968.

JUNGSTAND, W.: Untersuchungen über die Neurosekretion und deren Abhängigkeit von verschiedenen Außenfaktoren bei der Lungenschnecke *Helix pomatia* L. Zool. Jb. Physiol. 70, 1–23, 1962.

JUORIO, A. V., und KILLICK, S. W.: The effect of drugs on the synthesis and storage of monoamines in nervous tissues of molluscs. Int. J. Neurosci 4(5), 195–202, 1972.

JURCHENKO, O. P., VULFIUS, C. A., und ZEIMAL, E. V.: Cholinesterase activity in ganglia of *Gastropoda, Lymnaea stagnalis* and *Planorbarius corneus:* I. Effect of anticholinesterase agents on giant neurone depolarization by acetylcholine and its analogues. Comp. Biochem. Physiol. A. Comp. Physiol 45, 45–60, 1973.

– KULTAS, K. N., und VULFIUS, C. A.: Cholinesterase activity in ganglia of *Gastropoda, Lymnaea stagnalis* and *Planorbarius corneus:* II. Histochemical investigation. Comp. Biochem. Physiol. A. Comp. Physiol. 45, 61–68, 1973.

KAISER, P.: Bau, Entwicklung und Regeneration des Haarkleides von *Trichia hispida* (Linnaeus). Arch. Moll. 95, 111–122, 1966.

KANAZAWA, J., ISENSEE, A. R., und KEARNEY, P. C.: Distribution of Two Methylcarbamate Insecticides in a Aquatic Microecosystem. Abstr. Papers, Amer. Chem. Soc. 167 Meet. Pest 66, 1974.

KAPUR, S. P.: Storage of calcium in the tissues of the aquatic pulmonate, *Helisoma duryi eudiscus* (Pilsbry). Microscope 18, 139–141, 1970.

KARG, W.: Untersuchungen über die Wirkung von Dinitroorthokresol (DNOC) auf die Mikroarthropoden des Bodens unter Berücksichtigung der Beziehungen zwischen Mikroflora und Mesofauna. Pedobiologia 4, 138–157, 1964.

KARLIN, E. J., und NAEGELE, J. A.: Slugs and snails in New York greenhouses. N. Y. State Coll. Agric. Ext. Bull. 1004, 1–156, 1958.

– und BACON, CH.: Courtship, mating and egg-laying behaviour in the *Limacidae (Mollusca)*. Trans. Amer. Micr. Soc. 80, 399–406, 1961.

KASSAI, T.: Schnecken als Zwischenwirte der Protostrongyliden. Z. Parasitenkde. 18, 5–19, 1957.

KATO, G.: Acetylcholinesterase: I. A study by nuclear magnetic resonance of the binding of inhibitors to the enzyme. Mol. Pharmacol. 8, 575–581, 1972.

– Acetylcholinesterase: II. A study by nuclear magnetic resonance of the acceleration of acetylcholinesterase by atropine and inhibition by eserine. Mol. Pharmacol. 8, 582–588, 1972.

– TAN, E., und YUNG, J.: Acetylcholinesterase: Kinetic studies on the mechanism of atropine inhibition. J. Biol. Chem. 247, 3186–3189, 1972.

KAWAZOE, U., ALCANTARA, F. G., HADLER, W. A., und MAGALHAES, L. A.: Granuloma do canal deferante de *Biomphalaria tenagophila (Mollusca, Planorbidae)*. Ciec. Cult. 23, 167–169, 1971.

KAZUBSKI, S. L.: *Trichia lubomirskii* Slós *(Helicidae)*, a new host of *Tetrahymena limacis* (Warren, 1932) Kozloff, 1946 *(Ciliata)* and *Zonitoides nitidus* Müll. *(Zonitidae)*, a new host of *T. rostrata* (Kahl, 1926) Corliss, 1952, in Poland. Bull. Acad. pol. Sci. Cl. II. Sér. Sci. biol. 6, 247–252, 1958.

KEHDE, P. M., und WILHM, J. L.: The Effects of Grazing by Snails on Community Structure of Periphyton in Laboratory Streams. Amer. midland Naturalist 87, 3–24, 1972.

KEMPENDORFF, W.: Über das Fluchtphänomen und die Chemorezeptoren von *Helisoma (Taphius) nigricans* Spix. Arch. Molluskenkde. 74, 1–27, 1942.

KENAGA, E., DOTY, A., und HARDY, J.: Laboratory insectical tests with 4-Dimethyl-amino-3,5-Xylyl-Methylcarbamate. J. econ. Ent. 55, 466–469, 1962.

KENDALL, S. B.: Snail hosts of *Fasciola hepatica* in Britain. J. Helminth. 24, 63–74, 1950.

– The life history of *Limnaea truncatula* under laboratory conditions. Helminth. 27, 17–28, 1953.

KERKUT, G. A., NIALL HORN, und WALKER, R. J.: Long-lasting synaptic inhibition and its transmitter in the snail *Helix aspersa*. Comp. Biochem. Physiol. 30, 1061–1074, 1969.

KERNEY, M. P.: Atlas of the non-marine mollusca of the British Isles. European Invertebrate Survey. Herausg.: Conch. Soc. Great Britain and Ireland, Biolog. Records Centre, 1–208, 1976.

KESSEL, E.: Über die Schale von *Viviparus viviparus* L. und *Viviparus fasciatus* Müll. Ein Beitrag zum Strukturproblem der Gastropodenschale. Z. Morph. Ökol. Tiere 27, 129–198, 1933.

KEYL, M. J., MICHAELSON, J. A. und WHITTACKER, V. P.: Physiologically active choline esters in certain marine gastropods and other invertebrates. J. Physiol. 139, 434–454, 1957.

KHAN, M. A., KAMAL, A., WOLIN, R. J., und RUNNELS, I.: In Vivo and In Vitro Epoxidation of Aldrin by Aquatic Food Chain Organisms. Bull. Envir. Contam. Toxicol. 8, 219–228, 1972.

KILIAS, R.: Weinbergschnecken. 1–94. VEB Deutsch. Verlag Wissensch., Berlin 1960.

– SCHNITZLER, S., KOTHBAUER, H., STOBER, D., und PROKOP, O.: Weitere Haemagglutinin-Untersuchungen bei Landlungenschnecken (Haemagglutinintests an vier weiteren Arten, Biotopvergleiche mit schon untersuchten Arten und Tests an Organen von *Helix pomatia* Linnaeus). Z. Immunitätsforschg. Exp. Klin. Immunol. 144, 157–166, 1972.

KINOTI, G. K.: The Attachment and Penetration Apparatus of the Miracidium of *Schistoso-*

ma. J. Helminth. 57, 229–235, 1971.
– Observations on the infection of bulinid snails with *Schistosoma mattheei.* II. The mechanism of resistance to infection. Parasit. 62, 161–170, 1971.
KISS, T., und ROZSA, K. S.: Effect of ions and resting and action potentials of the myocardial fibers of *Helix pomatia* L. Acta Physiol. Acad. Sci. Hung. 40, 27–35, 1971.
– und ROZSA, K. S.: Effect of biologically active substances on the spontaneous electrical activity of the heart muscle cells of *Helix pomatia* L. Ann. Inst. Biol. (Tihany) Hung. Acad. Sci. 39, 29–38, 1972.
– und ROZSA, K. S.: The role of mono- and divalent cations in the spike generation of myocardial cells in the snail, *Helix pomatia* L. Comp. Biochem. Physiol. A. Comp. Physiol. 44, 173–181, 1973.
KITTEL, R.: Untersuchungen über den Geruchs- und Geschmackssinn bei den Gattungen *Arion* und *Limax (Mollusca: Pulmonata).* Zool. Anz. 157, 185–195, 1956.
KLAPPENBACH, M. A., und OLAZARRI, J.: [Notes on *Strophocheilidae (Moll. Gastr.).* I. *Strophocheilus parodizi* n. sp. of the province of Corrientes, Argentinia]. Communicationes Mus. Argent. Cienc. Natur. Bernardino Inst. Nac. Invest. Cienc. Natur. Cienc. Zool. 4 (1), 1–5, 1965.
KLEMM, W.: Mollusca. In: Catalogus faunae Austriae Wien 7, 1–59, 1960.
– Die Gehäuseschnecken. In: BEIER, M.: Zoologische Studien in West-Griechenland. 10. Teil. Sitzber. Oest. Akad. Wiss. Math.-Naturw. Kl. (1). 171, 73–84, 1962.
– Die Verbreitung der rezenten Land-Gehäuseschnecken in Österreich. Österr. Akad. Wiss. Math.-naturw. Kl., Denkschr. 117, 1–503, 1974.
KLESSEN, CH.: Die proteolytische Aktivität gebräuchlicher Diastasepräparate. Histochemie 33, 291–300, 1973.
KLIMMER, O. R.: Pflanzenschutz- und Schädlingsbekämpfungsmittel. Abriß einer Toxikologie und Therapie von Vergiftungen. 1–216. Hundt-Verlag Hattingen 1971, 2. Aufl.
KLIMOWICZ, H.: The molluscs of impermanent water bodies in the environs of Warsaw. Pol. Arch. Hydrobiol. 10, 271–285, 1962.
KNIGHT, W., LIARD, F., RITCHIE, L. S., PELLEGRINO, J., und CHIRIBOGA, J.: Labeling of *Biomphalaria glabrata* and Cercariae of *Schistosoma mansoni* with radioselenium. Exp. Parasit. 22, 309–315, 1968.
KNIPRATH, E.: Formation and structure of the periostracum in *Lymnaea stagnalis.* Calcif. Tissue Res, 9, 260–271, 1972.
KNORRE, D. von,: *Deroceras rodnae* Grossu und Lupu, ein Neufund für die deutsche Fauna *(Gastropoda, Limacidae).* Arch. Molluskenkde. 99, 171–174, 1969.
KNUTSON, L. V.: Biology and immature stages of snail-killing flies of Europe (*Diptera: Sciomyzidae).* Diss. Abstr. 24, 1–2. Ann Arbor. Mich. 1963.
– Biology and immature stages of malacophagous flies: *Antichaeta analis, A. brevipennis* and *A. obliviosa (Diptera: Sciomyzidae).* Trans Amer. Ent. Soc. 92, 67–191, 1966.
– Biology of snail-killing flies in Sweden (*Dipt., Sciomyzidae*). Ent. Scand. 1, 307–314, 1970.
– Biology and immature stages of *Coremacera marginata* F. a predator of terrestrial snails (*Dipt. Sciomyzidae).* Ent. scand. 4, 123–133, 1973.
– Another approach to biological control of snail-borne disease. Insect. Wld. Digest 3, 12–18, 1976.
– Sciomyzid Flies, another Approach to Biological Control of Snail-Borne Disease. Insect Wld. Digest 3, 13–18, 1976.
– und BERG, C. O.: Biology and immature stages of a snail-killing fly; *Hydromya dorsalis* (Fabricius) *(Diptera: Sciomyzidae).* Proc. Roy. Ent. Soc. London 38 (9), 45–58, 1963.
– und BERG, C. O.: Biology and immature stages of snail-killing flies: The genus *Elgiva (Diptera: Sciomyzidae*). Ann. Ent. Soc. Amer. 57, 173–192, 1964.
– und BERG, C. O.: The Malacophagous Flies of Norway *(Diptera, Sciomyzidae).* Norsk. ent. Tidskr. 18, 119–134, 1971.
– STEPHENSON, J. W., und BERG, C. O.: Biology of a slug-killing fly, *Tetanocera elata (Diptera:*

Sciomyzidae). Proc. Malacol. Soc. London 36, 213–220, 1965.

– und Berg, C. O.: Biology and immature stages of malacophagous Diptera of the genus *Knutsonia* Verbeke *(Sciomyzidae).* Bull. Inst. Roy. Sci. Nat. Belgique 43, 1–60, 1967.

– Neff, S. E., und Berg, C. O.: Biology and immature stages of snail-killing flies from Africa and southern Spain *(Sciomyzidae: Sepedon).* Parasitology 57, 487–505, 1967.

– Stephenson, J. W., und Berg, C. O.: Biosystematic studies of *Salticella fasciata* (Meigen), a snail-killing fly (*Diptera: Sciomyzidae*). Trans. Roy. Ent. Soc. London 122, 81–100, 1970.

– Rozkošný, R., und Berg, C. O.: Biology and immature stages of *Pherbina* and *Psacadina (Diptera, Sciomyzidae).* Acta Sci. nat. Acad. Scient. Bohemoslov. Brno IX (1) (Prag), Nov. Series. 1–38, 1975.

Kocan, A. A.: Some Common North American Aquatic Snails as Experimental Hosts of *Angiostrongylus cantonensis* – with Special Reference to *Lymnaea palustris.* J. Parasit. 58, 186–187, 1972.

Kohn, A., Nybakken, J. W., und Van Mol, J.-J.: Radual tooth structure of the gastropod *Conus imperialis* elucidated by scanning electron microscopy. Science (Wash.) 176, 49–51, 1972.

Koll, L.: Mollusca fennica. 3. Painos. Moniste 5, Helsinki 1961.

Koltermann, A.: Gegen die Schneckenplage. Hannov. Land- u. Forstwirtschaftl. Z. 108 (49), 1659–1660, 1955.

Komiya, Y.: Study and Application of Molluscicides in Japan. Bull. Wld. Hlth. Org. 25, 573–679, 1961.

– Schiao, L., und Hosaka, Y.: Review of studies on the resistance of *Oncomelania* snails to molluscicides in Japan. Bull. Wld. Hlth. Org. 25, 725, 1961.

Koopmans, J.: Cellulases in molluscs: I. The nature of the cellulases in *Helix pomatia* and *Cardium edule.* Netherl. J. Zool. 20, 445–463, 1970.

Kopfler, F. C.: The Accumulation of Organic and Inorganic Mercury Compounds by the Eastern Oyster *(Crassostrea virginica).* Bull. Environm. Contam. Toxicol. 11, 280, 1974.

Korringa, P.: Farming Cupped Oysters of the Genus *Crassostrea.* A Multidisciplinary Treatise. Develop. Aquaculture Fisheries Sci. 2 (XVI). 1–224, 1976.

– Farming the Flat Oysters of the Genus *Ostrea.* Developm. Aquaculture Fisherie Sci. 3 (XIV), 1–238, 1976.

Kosoglazov, A. A.: Bait from Metaldehyde Against Slugs in Greenhouses (Russ.). Khim. Sel. Khos. 12 (10), 769, 1974.

Kothbauer, H., und Schnitzler, St.: Haemagglutinine aus *Cepaea hortensis* (Müller), *C. nemoralis* (L.) und *C. vindobonensis* (Pfeiff.) *(Helicidae, Gastropoda).* Ihre Verwertbarkeit für die Artsystematik. Z. Zool. Syst. Evolutionsforschg. 10, 133–137, 1972.

Kotthoff, P.: Schneckenfraß an Pappeln. Gesunde Pflanzen 14, 212, 1962.

Kozloff, E. N.: The morphology and systematic position of a ciliate parasiting terrestrial pulmonate molluscs. Unveröffentl. Diss. Univ. Calif. Berkeley, USA 1946.

– *Tetrahymena limacis* from the terrestrial pulmonate gastropods *Monadenia fidelis* and *Prophysaon andersoni.* J. Protozool. 3, 204–208, 1956.

– A species of *Tetrahymena* parasitic in the renal organ of the slug *Deroceras reticulatum.* J. Protozool. 4, 75–79, 1957.

Krasts, I. V., Rezanow, A. N., und Veprintsev, B. N.: The effect of sodium, calcium and magnesium on the overshoot amplitude of the action potential from giant neurons of *Planorbarius corneus.* Comp. Biochem. Physiol. A. Comp. Physiol. 42, 487–491, 1972.

Kremzner, L., und Rosenberg, Ph.: Relationship of acetylcholinesterase activity to axonal conduction. Biochem. Pharmacol. 20, 2953–2958, 1971.

Krieg, K.: Experimentelle Kanzerogenese bei Mollusken: 3. Weitere Untersuchungen zur Geschwulstbildung bei der La-Plata-Apfelschnecke *Ampullarius australis* d'Orbigny *(Gastropoda, Prosobranchia)* unter besonderer Berücksichtigung des Methylcholanthrens. Archiv Geschwulstforschg. 33, 255–267, 1969.

– *Ampullarius australis* d'Orbigny *(Mollusca, Gastropoda)* as experimental animal in ecological research: A contribution to the study of cancerogenesis in invertebrates. Neoplasma (Bratisl.) 19, 41–49, 1972.

KUHLMANN, D.: Neurosekretion bei Heliciden *(Gastropoda)*. Z. Zellf. mikr. Anat. 60, 909–932, 1963.

– Der Dorsalkörper der Stylommatophoren *(Gastropoda)*. Z. Wiss. Zool. 173, 218–231, 1966.

– und NOLTE, A.: Spermiogenese, Eireifung und Neurosekretion. Untersuchungen an der Weinbergschnecke *Helix pomatia* L. (*Gastropoda*). Z. Wiss. Zool. 176, 271–286, 1967.

KUHN, D., und LASNIK, G. L.: Bacteriolysis by crystalline styles of marine mollusks. Bact. Proc. 1968, 49.

KUHNA, W., und SCHNELL, P.: *Milax gagates,* eine neue Nacktschnecke in Deutschland *(Gastropoda, Limacidae)*. Arch. Molluskenkde. 92, 137–140, 1963.

KUHR, R. J.: Metabolismus of carbamate insecticide chemicals in plants and insects. J. Agr. Food. Chem. 18, 1023–1030, 1970.

– DAVIS, A. C., und BOURKE, J. B.: DDT Residues in Soil, Water, and Fauna from New York Apple Orchards. Pesticid. Monit. J. 7, 200–204, 1974.

KUIJPER, W. J.: De zoetwatermollusken van het duingebied ten norden van Den Haag. Basteria 37. 1–20, 1973.

KULIEVA, A. M., ROZENGART, V. J., und SHMELEVA, V. G.: [Reactivation of cholinesterase of the squid optic ganglion suppressed by various organophosphorus inhibitors.] russ. Zdravookhr. Turkm. 15 (12), 3–5, 1971.

KULKARNI, A. B.: Studies on the embryology and early development of the land slug *Laevicaulis alte.* Marathwada Univ. J. Sci. Sect. B. Biol. Sci. 10, 165–172, 1971.

KUMAR, V.: Studies on Snail Hosts of *Orientobilharzia turkestanicum* (Skrjabin, 1913) Dutt and Srivastava, 1955 *(Schistomatidae: Trematoda)* in India. Ann. Soc. Belge Méd. trop. 53, 17–23, 1973.

KUNTZ, R. E., und WELLS, W. H.: Laboratory and field evaluations of two dinitrophenols as molluscicides for control of schistosome vectors in Egypt with emphasis on importance of temperature. Amer. J. Trop. Med. Hyg. 31, 784, 1951.

KUPFERMANN, I.: Studies on the neurosecretory control of egg laying in *Aplysia.* Amer. Zool. 12, 513–519, 1972.

KURIS, A. M.: Biological Control: Implications of the Analogy between the Tropic Interactions of Insect Pest-Parasitoid and Snail-Trematode Systems. Experim. Parasitol. 33, 365–379, 1973.

KURSTAK, E., und MARAMOROSCH, K.: Invertebrate Tissue Culture. 1–416. Academic Press, New York 1976.

LÄHDESMÄKI, L.: [On the occurence of gastropods in different vegetation zones of sea coast in Orikari, Oulu (Northern Ostrobothnia, Finland).] finn. Aquilo Ser. Zool 10, 47–52, 1970.

LANCIANI, C., und HARMAN, W. N.: Snail shells as oviposition sites of water mites. Nautilus 82, 34–35, 1968.

LANE, N. J.: Neurosecretory cells in the optic tantacles of certain pulmonate. Quart. J. micr. Sc. 103, 211–226, 1962.

– Thyamine pyrophosphatase, acid phosphatase and alkaline phosphatase in the neurones of *Helix aspersa.* Quart. J. micr. Sc. 104, 401–412, 1963.

– The fine structure of certain secretory cells in the optic tentacles of the snail *Helix aspersa.* Quart. J. micr. Sc. 105, 35–47, 1964.

– Further observations on the secretory cells in the optic tentacles of *Helix* with special reference to the results of vital staining. Quart. J. micr. Sc. 105, 61–71, 1964.

– The fine-structural localization of phosphatase in neurosecretory cells within the ganglia of certain gastropod snails. Amer. Zoologist 6, 139–157, 1966.

LANGE, W. H., und MACLEOD, G. F.: Metaldehyde and Calcium Arsenate in Slug and Snail Baits. J. econ Ent. 34, 321–322, 1941.

– und SCIARONI, R. H.: Metaldehyde dusts for Control of slugs affecting Brussels Sprouts in

Central California. J. econ. Ent. 45, 896–897, 1952.

LANGLOIS, TH.: The conjugal behaviour of the introduced European Giant Garden Slug, *Limax maximus* L., as observed on South Bass Island, Lake Erie. Ohio J. Science 65, 298–304, 1965.

LANKESTER, M. W., und ANDERSON, R.: Gastropods as intermediate hosts of *Pneumostrongylus tenuis* Dougherty of white-tailed deer. Can. J. Zool. 46, 373–383, 1968.

LAROCCA, P. T., CARLSON, G. P., FULLER, G. C., und KINGSTON, R. I.: Relationship Between Sulfhydril Reactivity and Toxicity of Vinyl Sulfone Molluscicidal Agents. Toxicol. Appl. Pharmacol. 31, 222–232, 1975.

LAROCQUE, A.: Catalogue of the recent *Mollusca* of Canada. Can. Dept. Res. Develop. Nat. Mus. Can. Bull. 129, Ottawa, Canada 1953.

LAVAL, J., und PAOLETTI, CL.: Purification and properties of an acid deoxyribonuclease from the snail *Helix aspersa* Müll. Biochemistry 11, 3596–3603, 1972.

LAVIOLETTE, P.: L'évolution de la glande hermaphrodite d'*Arion rufus* L. et ses rapports avec la croissance. C. R. Acad. Sc. Paris 231, 1567–1569, 1950.

– Rôle de la gonade dans la morphogenèse du tractus génital, chez quelques mollusques *Limacidae* et *Arionidae*. Compt. rend. Acad. Sci. 231, 1950.

– Régénérats germinaux multiples chez *Arion rufus* L. C. R. Acad. Sc. Paris, 232, 1139–1141, 1951.

– Etude cytologique et expérimentale de la régénération germinale après castration chez *Arion rufus* L. (Gastéropode, pulmoné). Ann. Sc. Nat., Zool. 16 (11ᵉ S.). 431–535, 1954.

– Rôle de la gonade dans le déterminisme humoral de la maturité glandulaire du tractus génital chez quelques gastéropodes *Arionidae* et *Limacidae*. Bull. Biol. France et Belgique 88, 310–332, 1954.

LE BRETON, J.: Nature endocrine des substances responsables de l'organogénèse et du cycle des tractus génitaux chez les gonochoriques et les hermaphrodites. Haliotis 1, 215–228, 1971.

– Etude experimentale de l'influence du jeûne sur le cycle de la gonade mâle et du pénis chez *Littorina littorea* L. Haliotis 1, 25–26, 1971.

LECOMTE, V.: Essai d'élevage d'*Euparypha (Theba) pisana* Müll. Haliotis 1, 37–38, 1971.

– Le Marché des Mollusques en France. Haliotis 3, 173–175, 1973.

LEE, F. O., und CHENG, TH. C.: Increased Heart Rate in *Biomphalaria glabrata* parasitized by *Schistosoma mansoni*. J. Invertebr. Pathol. 16, 148–149, 1970.

– und CHENG, TH. C.: *Schistosoma mansoni* Infection in *Biomphalaria glabrata:* Alterations in Heart Rate and Thermal Tolerance in the Host. J. Invertebr. Pathol. 18, 412–418, 1971.

– und CHENG, TH. C.: *Schistosoma mansoni:* Respirometric and Partial Pressure Studies in Infected *Biomphalaria glabrata*. Exp. Parasit. 30, 393–399, 1971.

– und CHENG, TH. C.: Incorporation of ^{59}Fe in the snail *Biomphalaria glabrata* parasitized by *Schistosoma mansoni*. J. Parasitol. 58, 481–488, 1972.

– und CHENG, TH. C.: *Schistosoma mansoni:* Alterations in total protein and hemoglobin in the hemolymph of infected *Biomphalaria glabrata*. Exp. Parasit. 31, 203–216, 1972.

LEENTVAAR, P.: Das Vorkommen von *Dreissena polymorpha* in den Niederlanden. R.I.N.-Bericht 37, 1–6, 1971. Vortrag auf der IAWR-Tagung, Rotterdam 1971.

LE GALL, J. G.: Synchronisme entre l'activité sécrétaire des neurons de la jonction cérébropleurale et l'activité dédifférenciatrice de cette zone chez *Crepidula fornicata* Phil. C. R. Acad. Sc. Paris, 278-b, 773–776, 1974.

– und STREIFF, W.: Protandric Hermaphroditisme in Prosobranch gastropods. In: REINBOTH, R.: Intersexuality in the Animal Kingdom, 170–178. Springer, 1975.

LEHOUX, J.-G., und SANDOR, T.: The occurence of steroids and steroid metabolizing enzyme systems in invertebrates. A review. Steroids 16, 141–171, 1970.

– und WILLIAMS, E. E.: Metabolism of progesterone by gonadal tissue *Littorina littorea (Prosobranchia, Gastropoda)*. J. Endocrinol. 51, 411–412, 1971.

LEIBENGUTH, F.: Mikrospektrophotometrische Unterschungen zur Kernhypertrophie nach Infektion mit *Klossia helicina*. Z. Parasitenkde. 39, 211–220, 1972.

LEIMBACHER, F.: Etude de l'effet de quelques facteurs climatiques sur l'évolution épizootique de *Fasciola hepatica,* leur utilisation pour la mise au point d'une méthode de la prévision de l'incidence de la Fasciolose en France. Haliotis 3, 43–55, 1973.

LEMEE, G.: Nouvelle localité du *Limax cinereo-niger* Wolf aux environs d'Allençon. Bull. Soc. 1. Norm. 8e, 7–8, s. 1. 1928.

LEMMA, A.: Laboratory and Field Evaluation of the Molluscicidal Properties of *Phytolacca dodecandra.* Bull. Wld. Hlth. Org. 42, 597–612, 1970.

– BRODY, G., NEWELL, G. W., PARKHURST, R. M., und SKINNER, W. A. S.: Studies on the molluscicidal properties of Endod *(Phytolacca dodecandra):* I. Increased potency with Butanol extraction. J. Parasit. 58, 104–107, 1972.

LE PENNEC, M., und PRIEUR, D.: Développement larvaire de *Mytilus edulis* (L.) en présence d'antibiotique. 1ère partie: Détermination des concentrations actives non toxiques de quatre antibiotiques: Auréomycine, Erythromycine, Chloramphénicol et Sulfamérazine. Rev. intern. Océanogr. méd. 28, 167–180, 1972.

– PRIEUR, D., und CHARDY, P.: Développement larvaire de *Mytilus edulis* (L.) en présence d'antibiotiques. 2. partie: Action sur la croissance de quatre antibiotiques: Auréomycine, Erythromycine, Chloramphénicol et Sulfamérazine. Rev. intern. Océanogr. méd. 30, 115–137, 1973.

LEVER, J.: On the relation between the medio-dorsal bodies and the cerebral ganglia in some pulmonate. Arch. Néerl. Zool. 13, 194–201, 1958.

– JANSEN, J., und VLIEGER, T. A.: Pleural ganglia and water balance in the fresh water pulmonate *Limnaea stagnalis.* Proc. Koninkl. Nederl. Akad. Wetensch. Ser. C 64, 531–542, 1961.

– und JOOSSE, J.: On the influence of the salt content of the medium on some special neurosecretory cells in the lateral lobes of the cerebral ganglia of *Lymnaea stagnalis.* Proc. Koningl. Nederl. Akad. Wetensch. Ser. C. 64. 630–639, 1961.

– KOK, M., MEULEMAN, E. A., und JOOSSE, J.: On the location of gomoripositive neurosecretory cells in the central ganglia of Lymnaea stagnalis. Proc. Koninkl. Nederl. Wetensch. Ser. C. 64, 640–647, 1961.

LEVINE, N. D.: Integrated Control of Snails. Amer. Zoologist 10, 579–582, 1970.

LEVINSON, H. Z.: Possibilities of Using Insectistatics and Pheromones in Pest Control. Die Naturwissenschaften 62, 272–283, 1975.

– The Defensive Role of Alkaloids in Insects and Plants. Experientia 32, 408–411, 1976.

– Ernährungs- und Stoffwechselphysiologie der Insekten und deren Anwendungsmöglichkeiten zur Schädlingsbekämpfung. Z. ang. Ent. 81, 113–132, 1976.

LEVY, M. G., TUNIS, M., und ISSEROFF, H.: Population on Control in Snails by Natural Inhibitors. Nature 241, 65–66, 1973.

LEWIS, H. C., und LA FOLLETTE, J. R.: Control of Brown Snail in Citrus Orchards. J. econ. Ent. 35, 359–362, 1942.

LEWIS, R. D.: Studies on the locomotor activity of the slug *Arion ater* (Linnaeus). I. Humidity, temperature and light reactions. Malacologia 7, 295–306, 1969.

– Studies on the locomotor activity of the slug *Arion ater* (Linnaeus). II. Locomotor activity rhythmus. Malacologia 7, 307–312, 1969.

LIE, K. J.: Antagonism of *Paryphostomum segregatum* rediae to *Schistosoma mansoni* sporocysts in the snail *Biomphalaria glabrata.* J. Parasit. 53, 969–976, 1967.

– Studies on *Echinostomidae (Trematoda)* in Malaya. XV. The Life History of *Echinostoma murinum* (Tubangui, 1931). Proc. Helminth. Soc. 34, 139–143, 1967.

– Larval Trematode Antagonism: Principles and Possible Application as a Control Method. Exp. Parasit. 33, 343–349, 1973.

– BASCH, P. F., und UMATHEVY, T.: Antagonism between two species of larval trematodes in the same snail. Nature (Lond.) 206, 422–423, 1965.

– und BASCH, P. F.: Occurence of *Perezia helminthorum* Canning and Basch, 1968 *(Microsporida: Nosematidae)* in trematode larvae of fresh-water snails. Southeast Asian J. Trop. Med. Publ. Health 1, 419–420, 1970.

– LIM, H.-K., und OW-YANG, C. K.: Synergism and Antagonism between two trematode Species in the Snail *Lymnaea rubiginosa.* Intern. J. Parasit. 3, 729–733, 1973.

LIEPOLT, R., und WEBER, E.: Die Giftwirkung von Cu SO_4 auf Wasserorganismen. Wasser und Abwasser 1958.

LIKHAREV, I. M., und RAMML'MEIER, E. S.: Terrestrial Mollusk of the Fauna of the U.S.S.R. Acad. Scienc. U.S.S.R. Moskva, Leningrad 1–574, 1952 (Engl. transl. Jerusalem 1965).

– und WIKTOR, A.: The fauna of slugs of the USSR and adjacent countries (Gastropoda Terrestria Nuda) (russ.), Fauna SSSR, Molljuski, III, 5, Akademya "Nauck" SSR, Leningrad, 437 pp. 1980.

LIM, H.-K., und HEYNEMAN, D.: Intramolluscan inter-trematode antagonism: a review of factors influencing the host-parasite system and its possible role in biological control. Advances in Parasitology 10, 192–268, 1972.

LIND, H.: The functional significance of the spermatophore and the fate of spermatozoa in the genital tract of *Helix pomatia (Gastropoda: Stylommatophora).* J. Zool. Proc. Zool. Soc. Lond. 169, 39–64, 1973.

LINDNER, G.: Muscheln und Schnecken der Weltmeere. 1–255. BLV Bestimmungsbuch, München-Bern-Wien 1975.

LINDQUIST, R. K., KRUEGER, H. R., SPADAFORA, R. R., und MASON, J. F.: Application of Aldicarb to Greenhouse Tomatoes: Plant Growth, Fruit Yields, Greenhouse Witefly Control, and Residues in Fruits. J. econ. Ent. 65, 862–864, 1972.

LLOYD, R.: The toxicity of mixtures of Zinc and coppersulphates to rainbow trout *(Salmo gairdnerii* Richardson). Ann. Appl. Biol. 49, 535–538, 1961.

LO, C.-T.: Compatibility and host-parasite relationships between species of the genus *Bulinus (Basommatophora: Planorbidae)* and an Egyptian strain of *Schistosoma haematobium (Trematoda: Digenea).* Malacologia 11, 225–280, 1972.

–, BURCH, J. B., und SCHUTTE, C. H. J.: Infection of diploid *Bulinus* s. s. with *Schistosoma haematobium.* Malacol. Rev. 3, 121–126, 1970.

LOMBARD, C.: Cancer et cancerisation des invertebrés. Rev. Med. Vet. (Toulouse) 123, 305–315, 1972.

LOMNICKI, A.: Some results of experimental introduction of new individuals into a natural population of the Roman Snail *Helix pomatia* L. Bull. Acad. Polon. Sci. Cl. II 12, 301–304, 1964.

– Individual differences among adult members of a snail population. Nature (London) 223, 1073–1074, 1969.

LOWELL, L. G.: An attempt to infect mollusks with „*Acanthamoeba*" sp. J. Parasitology 47, 842, 1961.

LU, P. Y.: Model Aquatic Ecosystem Studies of the Environmental Fate and Biodegradability of Industrial Compounds. Diss. Abstr. Intern. B 35, Nr. 1, 1–303; Urbana-Champaign. III. USA 1974.

LUBET, P., und STREIFF, W.: Etude expérimentale de l'action des ganglions nerveux sur la morphogenèse du pénis et l'activité génitale de *Crepidula fornicata* Phil. (Mollusque Gastéropodes). Cours et Documents de Biologie, GORDON et BREACH (édit.), 141–159, Paris 1969.

– und CHOQUET, C.: Cycles et rhythmes sexuels chez les mollusques bivalves et gastéropodes, influence du milieu et étude expérimentale. Haliotis 1. 129–149, 1971.

– STREIFF, W., SILBERZAHN, N., und DROSDOWSKY, M.: Endocrinologie de la différenciation sexuelle chez les Mollusques Prosobranches. Bol. Zool., Biol. Mar. N. S. São Paulo 30, 821–841, 1973.

– und DARDIGNAC, M. J.: Technologie de la Mytiliculture. Haliotis 5, 154–172, 1975.

LUCAS, A.: Les Mollusques des dunes de Bretagne. Penn ar Bed. (Brest) 7, 109–113, 1969.

– Les Gametes des Mollusques. Haliotis 1, 185–214, 1971.

LUCAS, M.: Les pigments des Mollusques. Teil 1, 2. Bruxelles 1974.

LUCHTEL, D.: Gonadal development and sex determination in pulmonate molluscs: I. *Arion circumscriptus.* Z. Zellforsch. Mikrosk. Anat. 130, 279–301, 1972.

– Gonadal development and sex determination in pulmonate molluscs: II. *Arion ater rufus* and *Deroceras reticulatum.* Z. Zellforsch. Mikrosk. Anat. 130, 302–311, 1972.

Lunetta, J. E., und Vernberg, W. B.: Fatty Acid Composition of Parasitized and Nonparasitized Tissue of the Mud-Flat Snail, *Nassarius obsoleta* (Say). Exp. Parasit. 30, 244–248, 1971.

Lupu, D.: Données nouvelles concernant la position systématique de quelques espèces de *Milax* Gray 1855, décrites antérieurement. Trav, Mus. Hist. Natur. „Grigore Antipa" 9, 31–37, 1969.

– Contribution à l'étude des Limacides de Roumanie. Trav. Mus. Hist. Nat. „Grigore Antipa" 10, 60–69, 1970.

– Le conspectus de la famille des Limacides de Roumanie, avec quelques reconsidérations systématiques basées sur des caractères d'anatomie comparée. Trav. Mus. Hist. Nat. „G. Antipa" 11, 121–137, 1971.

– Etude anatomique comparée de quelques espèces du genre *Deroceras* Rafinesque 1820. Trav. Mus. Hist. Nat. „G. Antipa" 12, 85–100, 1972.

– La révision des représentants de la famille des *Arionidae (Gastropoda-Pulmonata)* de Roumanie. Trav. Mus. Hist. Nat. „G. Antipa" 15, 31–44, 1974.

– Le polymorphisme chez quelques espèces appartenant aux familles *Limacidae* et *Arionidae* de Roumanie. Malacologia 16, 21–33, 1977.

– Observations écologiques et éthologiques effectuées sur le terrain et en laboratoire de Roumanie. Trav. Mus. Hist. Nat. „G. Antipa" 15, 45–55, 1974.

Lusis, O.: Postembryonic changes in the reproductive system of the slug *Arion ater rufus* L. Proc. Zool. Soc. London 137, 433–468, 1961.

– Changes induced in the reproductive system of *Arion rufus ater* L. by varying environmental conditions. Proc. malac. Soc. London 37, 19–26, 1966.

Luther, T.: Erfahrungen mit Nacktschnecken. Würt. Wochenbl. Landwirtschaft 122, 1602–1603, 1955.

Lutz, H.: Invertebrate Organ Cultures (Documents on Biology, 2). 1–266. Verlag Gordon und Breach, London 1970.

Lynch, J. I.: The ecology of *Lymnaea tomentosa* (Pfeiffer, 1855) in South Australia. Australian J. Zool. Melbourne 13, 461–473, 1965.

Macan, T. T.: Ecology of fresh-water molluscs in the English Lake district. J. Animal Ecol. 19, 124–146, 1950.

Mac Innis, W. M., Bethel, W., und Cornford, E. M.: Identification of chemicals of snail origin that attract *Schistosoma mansoni* miracidia. Nature 248, 361–363, 1974.

Mac Kay, A., und Gelperin, A.: Pharmacology and reflex responsiveness of the heart in the giant garden slug, *Limax maximus.* Comp. Biochem. Physiol. A. Comp. Physiol 43, 877–896, 1972.

Mac Kenzie, D. L.: Algae Control with GS-14260 (terbutryn). Res. Rept. Can. Weed comm. East Sect. 19 (Meet.), 360–361, 1974.

Mácha, S.: [Kultureinflüsse auf die Molluskenfauna]. tschech. Casop. Acta Musci Silesiae, Ser. A. Sci. nat. 20, 121–134, 1971.

Magendantz, M.: Versuche zur Schneckenbekämpfung und Empfehlungen zur Reduktion der Schistosomiasis-Übertragung in einem Siedlungsgebiet am Viktoria-See. Pflanzenschutz-Nachr. „Bayer" 27, 46–56, 1974.

Mahdi, M. A., und Amin, M. A.: An attempt to control bilharziasis by fish. Hydrobiologia (Den Haag) 26, 66–72, 1966.

Makings, P.: *Agriolimax caruanae* Pollonera new to Ireland. J. Conch. 24, 354–356, 1959.

Malek, E. A., und Cheng, T. C.: Medical and Economic Malacology. 1–408. Academic Press, New York-London 1974.

Mallet, C.: Les limaces, ennemies des jardins mais aussi des grandes cultures. Phytoma 25 (250), 10–12, 1973.

– Methode d'essai d'efficacité practique de molluscicides destinés à combattre les limaces. Soc. Franç. Phytiatrie Phytopharmacie, Méthode 48, 1–8, Mai 1974.

– und Bougaran, H.: Détermination de la valeur attractive des appâts sur les limaces. 20 ème

Symposium Phyth. Phyt. Gent 1968.

– und BOUGARAN, H.: Dégâts de limaces dans les colzas. 245–248. Journ. Int. Colza, Paris, C.E.T.I.O.M. 1970.

– und BOUGARAN, H.: Action molluscicide de divers carbamates. Meded. Landbouw. 207–215. Wetensch. Gent 1971.

MALLET, M.: Expérimentation sur les appâts dans la lutte contre les limaces. La Défense des Végétaux 22 (133), 253–264, 1968.

MANDAHL-BARTH, G.: The Species of the Genus *Bulinus,* Intermediate Hosts of *Schistosoma.* Bull. Wld. Hlth. Org. 33, 33–44, 1965.

MANOHAR, L., VENKATESWARA RAO, P., und SWAMI, K. S.: Variations in aminotransferase activity and total free amino acid level in the body fluid of the snail *Lymnaea luteola* during different larval trematode infections. J. Invert. Pathol. 19, 36–41, 1972.

MARADUFU, A. und OUMA, J. H.: A new chalcone as a natural molluscicide from *Polygonum.* Phytochemistry 17, 823–824, 1978.

MARENBACH, L.: Über den Farbwechsel von *Arion (Lochea) empiricorum* (Férussac, 1819). Z. wiss. Zool. (A) 152, 473–506, 1940.

MARGANIAN, V. M., und WALL, W. J. JR.: Dursban and Diazinon Residues in Biota Following Treatment of Intertidal Plots on Cape Cod – 1967–69. Pesticid. 160–165. Monitor. J. 6, 1972.

MÄRKEL, K.: Bau und Funktion der Pulmonaten-Radula. Z. wiss. Zool. 160, 213–289, 1957.

MARTIN, A., und HARRISON, FL.: Excretion. In: WILBUR, K. und YONGE, C.: Physiology of Mollusca. 353–386. Acad. Press, New York 1966, II.

MARTIN, G. P., COLLINS, H. L., und SPENCE, J. H.: Residues of the Insecticide Mirex Following Aerial Treatment of Cat Island. Bull. Environm. Contam. Toxicol. 12, 233–240, 1974.

MARTIN, T. J. und FORREST, J. D.: Zur Entwicklung von Schneckenkorn Mesurol in Großbritannien. Pflanzenschutz-Nachr. „Bayer" 22, 212–253, 1969.

– DAVIS, M. E. und MORRIS, D. B.: Development work with Methiocarb in Great Britain. Proc. 5 th Brit. Insectic. Fungic. Conf. 2, 434–441, 1969.

MARTOJA, M.: Neurosécrétion et glandes endocrines des Mollusques autres que les Céphalopodes. In: GRASSÉ, P. P.: Traité de Zoologie, (Masson Paris), t. 5, fasc. 3, 927–986, 1968.

– Le rein de *Pomatia* (=*Cyclostoma*) *elegans* (Gastéropode, Prosobranche): Données structurales et analytiques. Ann. Soc. Nat. Zool., Biol. Anim. 12, 535–538 1975.

MARTTI, MARKKULA: Abundance of pests in Greenhouses. Ann. Agricult. Fenniae 8, 320–322, 1969.

MASON, C. F.: Food, feeding rates and assimilation in woodland snails. Oecologia 4, 358–375, 1970.

MASSOUD, J., CHU, K. Y. und ARFAA, F.: Field trials of Bayluscide, sodium pentachlorophenate and copper sulphate in standing waters in Iran. Ann. trop. Med. Parasit. 63, 189–194, 1969.

MATSUDA, H.: Studies on experimental schistosomiasis: II. The fecundity and successive breeding of various species of *Oncomelania* in the laboratory. Jap. J. Parasitol. 20, 481–489, 1972.

MATYSIAK, K.: Zaba trawna sprzymierzencem rolnika (Der Grasfrosch als Helfer des Landwirts). Ochr. Rosl. 14, 23–24, 1970.

MAYER, J.: Digestion of Oysters for the Determination of Mercury. Bull. Environm. Contam. Toxicol. 5, 383–388, 1970.

MAYER, K.: Die Schneckenbekämpfung mit Metaldehydpräparaten. Nachrichtenbl. Dt. Pflanzenschutzd. (Braunschweig) 9, 36–41, 1957.

MAYES, M., und HERMANS, C. O.: Fine structure of the eye of the prosobranch mollusk *Littorina scutulata.* Veliger 16, 166–168, 1973.

MAZEK-FIALLA, K.: Über den Zusammenhang zwischen der Lebensweise einiger Landpulmonaten und deren subepithelialen Drüsen. Z. Morphol. Ökol. Tiere 27, 451–475, 1933.

MCANNALLY, R. D., und MOORE, D. V. Predation by the leech *Helobdella punctata – lineata* upon *Australorbis glabratus* under laboratory conditions. J. Parasitol. 52, 196–197, 1966.

McCaman, R. E. und Dewhurst, S. A.: Metabolism of putative transmitters in individual neurons of *Aplysia californica*: Acetylcholinesterase and catechol-o-methyl transferase. J. Neurochem. 18, 1329–1335, 1971.
McClelland, W. F. J.: Exposure of laboratory bred snails in natural situations. In: East African Inst. for Medical Research. 18-19 Ann. Rep. 1963–64 Mwanza, Tanzania, 1965.
McCoy, L. E. und Joy, J. E.: Tolerance of *Sepedon fuscipennis* and *Dictya* sp. larvae *(Diptera: Sciomyzidae)* to the molluscicides Bayer 73 and Sodium Pentachlorophenate, Environ. Entomol. 6, 198–202, 1977.
McCraw, B. M.: Relaxation of Snails before Fixation. Nature (Lond.) 181, 575, 1958.
– Aspects of the growth of the snail *Lymnaea palustris* (Müller). Malacologia 10, 399–413, 1970.
McCullough, F. S.: The Distribution of *Schistosoma mansoni* and *S. haematobium* in East Africa. Trop. geogr. Med. 24, 199–207, 1972.
McMahon, J. J.: Temperature effects on the developmental rate of squid (*Loligo pealei*) embryos. Biol. Bull. (Woods Hole) 14, 561–567, 1971.
Mead, A. R.: Disease in the Giant African snail *Achatina fulica* Bowdich. Science 123, 1130–1131, 1956.
– Disease transmission in terrestrial mollusks. Amer. Malacol. 26. Union Ann. Rep. 1957.
– The Giant African snail: A problem in economic malacology. 1–257 Univ. Chicago Press, Chicago, Illinois, USA 1961.
– A prognosis in the spread of the Giant African snail to continental United States. Malacologia 14, 427, 1973.
– The Giant African Snail and Economic Malacology. Malacologia 16, 157, 1977.
Medenwald, H. und Shiff, C. J.: Labor- und Feldversuche zur Prüfung des Einflusses von Bayluscid auf Citrusbäume. Pflanzenschutz-Nachr. „Bayer" 18, 146–152, 1965.
Meier-Brook, C.: An improved relaxing technique for mollusks using pentobarbital. Malacological Review 9, 115–117, 1976.
– und Kuang Yung Tan Tjhen: The susceptibility of *Marisa cornuarietis,* a predator of schistosome-bearing snails, to N-Tritylmorpholine. Ann. Trop. Med. Parasitol. 71, 95–100, 1977.
Meincke, K.-F.: Osmotischer Druck und ionale Zusammensetzung der Haemolymphe winterschlafender *Helix pomatia* bei konstanter und sich zyklisch ändernder Temperatur. Z. Vergl. Physiol. 76, 226–232, 1972.
Mekhraliev, A. A.: *Acroloxus lacustris (Ancylidae)* as a new intermediate host of trematodes in the USSR. Parazitologiya (Leningr.) 12, 121–125, 1978.
Mellanby, K.: Slugs at low temperatures. Nature 189, 944, 1961.
Meng, K.: Untersuchungen zur Steuerung der Herztätigkeit bei *Helix pomatia* L. Zool. Jb., Abt. Physiol. Tiere 68, 537–566, 1960.
– Die Beeinflussung der Tätigkeit des Helix-Herzens durch das extrakardiale Nervensystem. Zool. Jb. Physiol. 68, 567–576, 1960.
– Zur Verknüpfung von Atmung und Kreislauf bei *Helix pomatia* L. Zool. Jb. Physiol. 69, 599–608, 1962.
Mengert, H.: Nematoden und Schnecken. Z. Morphol. Oekol. Tiere 41, 311–349, 1953.
Merck, A. G.: Wirkungsvolle Schneckenbekämpfung. Chem. Techn. Landwirtsch. 19, 188–189, 1968.
Mermod, G.: Gastéropodes. In: Catalogue des Invertébrés de la Suisse. George & Cie, Genf 18 (1-12), 1–583, 1930.
Metcalf, R. L.: Carbamate insecticides. Agri. Chemicals 16, 20–21, 104, 106, 1961.
Metri, N. I.: Structure and Activity in Molluscicides. IV. New substitued Tetralins with Potential Molluscidal Effect. J. Pharm. Sci. 62, 323–325, 1973.
Meulemann, E.: Host-Parasite interrelationships between the freshwater Pulmonate *Biomphalaria pfeifferi* and the Trematode *Schistosoma mansoni*. Netherlands J. Zool. 22, 355–427, 1972.

MEUSE, A. D. J. und HUBERT, B.: The mollusc fauna of glasshouses in the Netherlands. Basteria 13, 1–3, 1949.
MEYLING, A. H. und PITCHFORD, R. J.: Physico-chemical properties of substances used as molluscicides. Precipitation and adsorption. Bull. Wld. Hlth. Org. 34, 141–146, 1966.
MICHELSON, E. H.: Studies on the biological control of schistosome-bearing snails. Predators and parasites of freshwater *mollusca*: A review of the literature. Parasitology 47, 413–426, 1957.
– A method for the relaxation and immobilization of pulmonate snails. Trans Amer. Microsc. Soc. 77, 316–319, 1958.
– Development and Specificity of Miracidal Immobilizing. Substances in Extracts of the Snail *Australorbis glabratus* Exposed to Various Agents. Ann. New York Acad. Sci. 113, 486–491, 1963.
– *Plistiphora husseyi*, sp.n., a microsporidian parasite of aquatic pulmonate snails. J. Insect. Path. 5, 28–38, 1963.
– Miracidia-immobilizing substances in extracts prepared from snails infected with *Schistosoma mansoni*. Amer. J. trop. Med. Hyg. 13, 36, 1964.
– The protective action of *Chaetogaster limnaei* on snails exposed to *Schistosoma mansoni*. J. Parasit. 50, 441–444, 1964.
– A neoplasm in the giant African snail *Achatina fulica*. J. Invertebr. Pathol. 20, 264–267, 1973.
– und AUGUSTINE, D. H.: Studies on the biological control of Schistosome-bearing snails. V. The control of *Biomphalaria pfeifferi* populations by the snail *Marisa cornuarietis* under laboratory conditions. J. Parasitol. 43, 135, 1957.
– und DUBOIS, L.: Increased alkaline phosphatase in the tissues and hemolymph of the snail *Biomphalaria glabrata* infected with *Schistosoma mansoni*. Comp. Biochem. Physiol. B. 44, 763–767, 1973.
MIENIS, H. K.: *Cernuella virgata* (Da Costa, 1778) in Nederland. Basteria 33, 31–49, 1969.
– Enkele notities betreffende de land- en zoetwatermollusken van Israel: *Theba pisana* (Müller, 1774) in de Citrusplantages van Netzer Sereni. „De Kreukel" Uitgave van de Malacologische Contactgroep A'dam et Omst., 6 (3), 34–35, 1970.
MOCZAR, M., und MOCZAR E.: Phylogenetic evolution of the cornea: macromolecular composition of the corneal stroma of the squid (*Sepia officinalis*). Comp. Biochem. Physiol. B. Comp. Biochem. 45, 213–223, 1973.
MODIN, J. C.: Residues in fish, wildlife and estuaries. Pesticide Monit. J. 3, 1–7, 1969.
MOENS, R.: *Milax budapestensis* Hazay, een gevaarlijke vijand van onze tuinbouwgewassen. Meded. Landbouwhogesch. Opz. Gent 24, 662–670, 1959.
– Progrès accomplis dans la lutte contre les limaces. Rev. de l'Agriculture 13 (1), 1–9, 1960.
– Recherches sur la limace souterraine *Milax budapestensis* Hazay. C.R. XVI. Int. Hortic. Congr. Brüssel 1962, 372–376.
– Nouvelles possibilités de lutte contre les Lymnaeidae (Mollusques) dans les cressonières. Parasitica 21, 79–97, 1965.
– New possibility for *Limnaeidae (Mollusca)* control in watercress ditches with the molluscicide 5, 2-Prime dichloro-4-prime nitro salicylanalides. Parasitica 21, 79–97, 1965.
– *Hydrobia jenkinsi* Smith (Gastéropode prosobranche) et autres organismes nuisibles au cresson de fontaine. Meded. Rijksfac. Landbouw-Wetensch. Gent 31, 565–571, 1967.
– Techniques chimiques de déstruction des gîtes à Limnées dans les prairies. Colloque d'information scient. „L'Assainissement des prairies par les Traitements molluscicides" Gembloux 1970. Centre Rech. Agron. l'Etat 32–55, Gembloux 1970.
– Test d'efficacité de quelques nouvelles préparations molluscicides pour la lutte contre les limaces. Meded. Facult. Landbouw-Wetensch. Gent 36, 216–223, 1971.
– Controle de l'efficacité de granules helicides en laboratoire. Meded. Fac. Landbouw-Wetensch. Gent 37, 816–823, 1972.
– Denombrement des limnées. Haliotis 3, 97–102, 1973.

– und VAN DEN BRUEL, W. E.: Proeven op bestrijding van slakken (*Agriolimax reticulatus* Müller). Meded. Landbouwhogesch. Opz. Gent 21, 401–410, 1956.

– und VAN DEN BRUEL, W. E.: Praktische bestrijding van *Agriolimax reticulatus* Müll. op rogge. Meded. Landbouw. Opz. Gent 23, 695–703, 1958.

– und VAN DEN BRUEL, E. W.: Verglijkende biologie en schade bij drie Arionidae: *Arion rufus* L., *A. hortensis*, Fér., *A. circumscriptus* Johnston. Meded. Landbouwhogesch. Opz. Gent. 25, 1399–1411, 1960.

– und PECHEUR, M.: Essai de destruction de *Limnaea truncatula* O.-F. Müller au moyens de zirame (Diméthyldithiocarbamate de zinc). Ann. Med. Veterin. 6, 266–274, 1964.

– FRANÇOIS, E., RIGA, A., und VAN DEN BRUEL, W. E.: Les radioisotopes en écologie animale. Premières informations sur le comportement de *Agriolimax reticulatus* Müller. Meded. Landbouwhogesch. Opz. Gent 30, 1810–1823, 1965.

– FRANÇOIS, E., und RIGA, A.: Les radioisotopes en écologie animale: Déplacements dans un milieu naturel de *Agriolimax reticulatus* O.F. Müller. Meded. Rijksfac. Landbouw-Wetensch. Gent 31, 1032–1042, 1966.

– FRANÇOIS, E., RIGA, A., und VAN DEN BRUEL, W. E.: A mechanical barrier against terrestrial Gastropods. Parasitica 23, 22–27, 1967.

MOKRZECKI, Z.: [Das massenhafte Auftreten an Wintersaaten im Jahre 1927 und 1928, im Bezirk Skierniewice, der Feldschnecke (*Agriolimax agrestis* L.) und ihre Bekämpfung.] poln. „Maladies des Plantes“ 1 (3/4), 1–8, 1931.

MOOIJ-VOGELAAR, J. W., und VAN DER STEEN, W. J.: Effects of density on feeding and growth in the pond snail *Lymnaea stagnalis* (L.). Proc. K. Ned. Akad.. (ser.C) 76, 61–68, 1973.

– JAGER, J. C., und VAN DER STEEN, W. J.: Effects of density levels, and changes in density levels on reproduction, feeding and growth in the pond snail *Lymnaea stagnalis* (L.). Proc. K. Med. Akad. (ser.C) 76, 245–256, 1973.

MOORE, M. N., und HALTON, D. W.: Histochemical Changes in the Digestiv Gland of *Lymnaea truncatula* Infected with *Fasciola hepatica*. Z. Parasitenkde. 43, 1–16, 1973.

MORALES, T., und CHALAZONITIS, N.: Ondes lentes du potentiel initiées par le cyclohexamide au niveau du neurone géant d'*Aplysia*. C.R. Séances Soc. Biol. Filiales 164, 1792–1797, 1970, 1971.

MORETON, R. B., und GARDNER, D. R.: Frescon: Neurophysiological Action of a Molluscicide. Experientia 32, 611–612, 1976.

MORETON, B. D.: Preliminary experiments with metaldehyde suspensions against slugs. Plant Pathology 2, 49–51, 1953.

MORPHY, M. J., ROSS, J. G., und TAYLOR, S. M.: Problems of *Lymnaea truncatula* ecology in investigations of fascioliasis. Malacologia 9,127–134, 1969.

– und ROSS, J. G.: An evaluation of techniques for sampling and estimating snail populations. In: The control of fascioliasis, Sect. C., 23–25, London, Febr. 1970.

MORRILL, J., BLAIR, Cl., und LARSEN, W.: Regulative development in the pulmonate gastropod, *Lymnaea palustris*, as determined by blastomere deletion experiments. J. exp. Zool. 183, 47–56, 1973.

MORRISON, J. P. E.: Notes on Hawaiian *Lymnaeidae*. Malacol. Rev. 1. 31–39, 1968.

MORTON, J.E.: Molluscs. 1–244. Hutchinson, London 1967, 4. Aufl.

– und YONGE, C. M.: Classification and Structure of the *Mollusca*. In: Physiology of Mollusca ed. WILBUR, K. M. und YONGE, C. M., Acad. Press, New York-London, 1 (Chapt. 1). 1–58, 1964.

MÖRZER BRUYNS, M. F., VAN REGTEREN-ALTENA, C. O.,und BUTOT,L. J. M.: The Netherlands as an environment for land *Mollusca*. Basteria 23 (suppl.). 132–162, 1959.

MUDRA, A.: Statistische Methoden für landwirtschaftliche Versuche. 1–336 Parey, Berlin-Hamburg 1958.

MÜHLE, E.: Die Gehäuseschnecke *Vallonia pulchella* O.F. Müller als Schädling des Grassamenbaues. Nachrichtenbl. Dt. Pflanzenschutzd. 21, 13–14, 1941.

MUKHERIJ, S. K.: Use of Pentachlorophenol as an Algicide in Paddy Fields in West Bengal.

Weed Res. (Europe) 12, 389–390, 1972.

MÜLLER, F., und BIERINGER, H.: Markierung von kleinen Bodentieren durch Fütterung mit ^{14}C-markierter Glukose. Int. J. Appl. Radiation Isotopen 22, 687–689, 1971.

MÜLLER, R. L.: Molluscicides. Wld.Rev. Pest Control 1, 11–18, 1962.

MUMA, M. H.: Observations on the biology of the citrus tree snail. Citrus Industry 36, 6–9, 1955.

MUSICK, G. J.: Efficacy of Phorate for Control of Slugs in Field Corn. J. econ. Ent. 65, 220–222, 1972.

MVOGO, I., und BARD, J.: Seconde note d'information sur l'*Astatoreochromia alluaudi*, poisson malacophage utilisable dans la prophylaxie de la bilharziose. Bull. Soc. Pathol. Exot. 57, 21–23, 1964.

NAGABHUSHANAM, R., und MANTALE, B. M.: Studies on adenine nucleotides and adenosine activity during hibernation of the Indian snail, *Cryptozoa semirugata*. Marathwada Univ. J. Sci. Sect. B. Biol. Sci. 10, 117–130, 1971.

– und KULKARNI, A. B.: Histochemical observations on the egg capsules of the land slug, *Laevicaulis alte*, Marathwada Univ. J. Sci. Sect. B. Biol. Sci. 10, 121–122, 1971.

– und KULKARNI, A. B.: Reproductive biology of the land slug, *Laevicaulis alte*. Riv. Biol. (Perugia) 64, 15–44, 1971.

– und MANTALE, B. M.: Physiology of hibernation in the land snail, *Cryptozoa semirugata*. Riv. Biol. (Perugia) 65, 53–97, 1972.

NAGUIB, A.: Bilharziose-Bekämpfung in Bewässerungsgebieten. Pflanzenschutz-Nachr. „Bayer" 18, 93–109, 1965.

NAJARIAN, H. H.: Biological Studies on the Snail, *Bulinus truncatus*, in Central Iraq. Bull. Wld. Hlth. Org. 25, 435–446, 1961.

NASIR, P., und SILVA, J.: Freshwater Larval Trematodes. XXIX: Life Cycle of *Guaicaipuria parapseudoconcilia* sp.n. Proc. Helminth. Soc. Wash. 38, 142–146, 1972.

NEFF, S. E.: Natural history and immature stages of certain species of *Sepedon, Hoplodictya* and *Protodictya (Diptera: Sciomyzidae)*. 1–201. Ph. D. Thesis Cornell Univ. (L.C. Card Nr. Mic. 50-6491). Diss. Abstr. 21 (Univ. Microfilms) Ann Arbor, Mich. 1960.

– Snail-killing sciomyzid flies: Application in biological control. Verh. Intern. Verein Limnol. 15, 933–939, 1964.

– und BERG, C. O.: Observations on the immature stages of *Protodictya hondurana (Diptera: Sciomyzidae)*. Bull. Brooklyn Entom. Soc. 56, 46–56, 1961.

– und BERG, C.O.: Biology and immatures stages of *Hoplodictya spinicornis* and *H. setosa (Diptera: Sciomyzidae)*. Trans Amer. Entom. Soc. 88, 77–93, 1962.

NELSON, T. C.: Water filtration by the oyster and a new hormon effect upon the rate of flow. Proc. Soc. Exptl. Biol. Med. 34, 189–190, 1936.

– und ALLISON, J. B.: On the nature and action of dianthin; a new hormon like substance carried by the spermatozoa of the oyster. J. Exptl. Zool. 85. 299–338, 1940.

NEMENZ, H. E.: Der Einfluß des Ionenmilieus auf das Überleben von *Planorbis planorbis* und *Planorbarius corneus*. Verh. Dt. Zool. Ges. 65, 172–176, 1971.

NEUHAUS, W.: Parasitäre Kastration bei *Bithynia tentaculata*. Z. Parasitenkde. 12, 65–77, 1941.

NEUMANN, P.: Gehäuse-Schnecken als Schädlinge im Obstbau. Pflanzenschutz (München) 6, 119–121, 1954.

– Zur Schneckenplage im Garchinger Gemeindeobstgarten. Pflanzenschutz (München) 7, 156–158, 1955.

NEUMÜLLER, O.-A.: Römpps Chemie-Lexikon. Franckh. Stuttgart, 1975, 7. Aufl. 5.

NEWELL, P. F.: Assessing slug populations. Rep. Rothamsted Exp. Sta. 188–189, 1964.

– The behaviour and distribution of slugs. Ph. D. Thesis, Univ. London 1965.

– The nocturnal behaviour of slugs. Medic. Biol. III, 16, 146–159, 1966.

– The measurement of light and temperature as factors controlling the surface activity of the slug *Agriolimax reticulatus* (Müller). Symp. Brit. ecol. Soc. 8, 141–146, 1968.

– Mollusks. In: PHILLIPSON, J. (Ed.): Methods of study in quantitative soil ecology: population, production and energy flow. Oxford 1971.
– The structure and enzyme-histochemistry of slug skin. Malacologia 16, 183–195, 1977.
– und SKELDING, J. M.: Studies on the permeability of the septate junction in the kidney of *Helix pomatia* L. Malacologia 14, 81–91, 1973.
– Etude de l'ultrastructure de l'épithelium dorsal et pédieux des limaces *Arion hortensis* Férussac et *Agriolimax reticulatus* (Müller). Haliotis 3, 131–135, 1973.
NEWTON, W. L.: The development in the laboratory of resistence to sodium pentachlorophenate in *Australorbis glabratus*. Bull. Wld. Hlth. Org. 29, 539–544, 1963.
NICE, G. N., und WILSON, R. A.: A study of the effect of temperature on the growth of *Fasciola hepatica* in *Lymnaea truncatula*. Parasitology 68, 47–56, 1974.
NIETZKE, G.: Die Weinbergschnecke. 1–115, Eugen Ulmer, Stuttgart 1963.
NILSSON, A.: Om Källarsnigeln, *Limax flavus* L., en för svensk fauna ny anthropochorart, jämte några andra i Källare påträffade mollusker. Zool. Revy 24, 88–94, 1962.
NOBRE, A.: Molluscos terrestres fluviais e das aguas solobras de Portugal. 1–259. Ministerio da Agricultura, Direcçáo Geral dos Serviços Florestais e Aquicolas 1930.
NOLAND, L. E., und CARRIKER, M. R.: Observations on the biology of the snail *Lymnaea stagnalis appressa* during twenty generations in laboratory culture. Amer. Midland Naturalist 36, 467–493, 1946.
NOLF, L. O., und CORT, W. W.: On immunity reactions of snails to the penetration of the cercariae of the strigeid trematode, *Cotylurus flabelliformis* (Faust). J. Parasit 20, 38, 1933.
NOLTE, A.: Ultrastruktur des „Neurosekretmantels" des Nervus labialis medius von *Planorbarius corneus* L. (*Basommatophora*). Naturwiss. 6, 148, 1964.
– Neurohämal-Organe bei Pulmonaten. Zool. Jb. Abt. Ontog. 82, 365–380, 1965.
NOPP, R.: Zur Wasseraufnahme und Ionenregulation von *Helix aspersa* O. F. Müll. Verhandl. Zool.-Botan. Ges. Wien 112, 64–66, 1972.
NORDEN, D.: The inhibition of trypsin and some other proteases by ovorubin, a protein from the eggs of *Pomacea canaliculata*. Compl. Biochem. Physiol. B. Comp. Biochem. 42, 569–576, 1972.
NORRIS, R. E.: Tree snails in citrus pest control. Fla Grower 60, 18, 41, 47, 1952.
O'DONALD, P.: Natural selection by glow-worms in a population of *Cepaea nemoralis*. Nature 217, 194, 1968.
OKABE, K., und SHIRAISHI, S.: Experimental infection of *Oncomelania hupensis nosophora* with *Rhabditis oncomelaniae* Yokoo et Okabe. Kurume Med. J. 18, 195–199, 1971 (1972).
ØKLAND, F.: Die Verbreitung der Landgastropoden Norwegens. Skr. Norske Vidensk. Akad. I. Mat.-Naturv. 8. Oslo 1925.
– Quantitative Untersuchungen der Landschneckenfauna Norwegens. Z. wiss. Biol. Abt. A. Z. Morph. Ökol. Tiere 16, 748–804, 1930.
ØKLAND, J.: Distribution and ecology of the fresh-water snails (*Gastropoda*) of Norway. Malacologia 9, 143–151, 1969.
OLDCORN, R. A.: Meta-bran and birds. Gardner's Chron. 103, 32, 1938.
OLIVIER-GONZALEZ, J., BAUMANN, P. M., und BENENSON, A. S.: Effect of the snail *Marisa cornuarietis* on *Australorbis glabratus* in natural bodies of water in Puerto Rico. Amer. J. Trop. Med. Hyg. 5, 290, 1956.
OLIVIER L., und HASKINS, W. T.: The Effects of low Concentrations of Sodium Pentachlorphenate on the Fecundity and Egg Viability of *Australorbis glabratus*. Amer. Trop. Med. Hyg. 9, 199–205, 1960.
OLLERENSHAW, C. B.: Some observations on the epidemiology and control of Fascioliasis in Wales. In: II. Int. Liverfluke Colloquium, Wageningen 2.-6., 34–35,Oct. 1967.
– Some aspects of the epidemiology of fascioliasis in Britain. Proc. malacol. Soc. London 39, 238–243, 1970.
OLSEN, O. W.:Bionomics of the lymnaeid snail, *Stagnicola bulimoides techella*, the intermediate host of the liver fluke in southern Texas. J. Agr. Res. (U.S.) 69, 389–403, 1944.

O'NEIL, L. C.: Occurence of Arion ater (L.) *(Gastropoda: Pulmonata: Arionidae)* in Sherbrooke, Québec. Can. J. Zool. 42,1161–1163, 1964.

OPEL, H., und WOLFGANG, H.: Der Magensaft der Weinbergschnecke *Helix pomatia* L. – eine einfache Quelle kohlenhydratspaltender Enzyme zur Mazeration pflanzlicher Gewebe und zur Spaltung von Oligosacchariden und Glycosiden. Monatsber. Deutsch. Akad. Wissensch. Berlin, Biologie 5, 533–534, 1963.

ORTH, R. E., MOORE, I., FISHER, T. W. and LEGNER, E. F.: A rove beetle, *Ocypus olens,* with potential for biological control of the Brown Garden Snail, *Helix aspersa,* in California, including a key to the nearctic species of *Ocypus.* Can. Entomologist 107, 1111–1116, 1975.

ORTIZ-TORRES, E.: Damage caused by the snail, *Marisa cornuarietis*, to young rice seedlings in Puerto Rico. J. Agricult. Univ. Puerto Rico 46. 241, 1962.

OŠANOVA, H.: [Über die Biologie und Ökologie von zwei für die Fauna Bulgariens neuen Arten der Familie *Arionidae.*] Bull. Inst. Mus. Zool. 15,203–214, 1964.

OSBORNE, N.: The effect of optic tentacle removal on the transmitter content of the giant serotonin cells of *Helix aspersa*. J. Neurochem. 19, 2363–2368, 1972.

– Serotonin, free amino-acids and related substances in the blood and nervous tissue of *Helix aspersa (Gastropoda Mollusca)*. Comp. Gen. Pharmacol. 3, 171–177, 1972.

– The in vivo synthesis of serotonin in an identified serotonin-containing neuron of *Helix pomatia*. Int. J. Neurosci. 3, 215–228, 1972.

– Micro-biochemical and physiological studies on an identified serotonergic neuron in the snail *Helix pomatia*. Malacologia 14, 97–106, 1973.

– POWELL, B., und COTTRELL, G.: The effect of electrical stimulation on the levels of free amino acids and related compounds in the snail brain. Brain Res. 4, 379–386, 1972.

– ALTHAUS, H., und NEUHOFF, V.: Phospholipids in the nervous system of the gastropod mollusc *Helix pomatia* and the in vivo incorporation of ^{32}P into the phospholipids of identified neurons. Comp. Biochem. Physiol. B. Comp. Biochem. 43, 671–679, 1972.

– und NEUHOFF, V.: Neurochemical studies on characterized neurons. Naturwissensch. 60, 78–87, 1973.

OVER, H. J.: Ecological biogeography of *Lymnaea truncatula* in the Netherlands. Doct. Thesis Univ. Utrecht 1967.

OVERBECK, J. Van, HUGHES, W. J., und BLONDEAU, R.: Acrolein for the Control of Water Weeds and Disease-Carrying Water Snails. Science 129, 335–336, 1959.

OWEN, G.: 1. Feeding, 2. Digestion. In: WILBUR, K. und YONGE, C.: Physiologie of *Mollusca* II (Chapt. 1/2). 1–96, Akad. Press New York 1966.

OWEN, D. F.: Ecological Genetics and Natural Selection in Mollusks. Science 145, 376–377, 1974.

– und BENGTSON, S.-A.: Polymorphism in the land snail *Cepaea hortensis* in Iceland. Oikos 23, 218–225, 1972.

PACE, G. L.: The hold-fast function of the preputial organ in *Helisoma*. Malacol. Rev. 1971, 21–24.

PAG, H.: Schnecken als Schädlinge in Orchideenhäusern. Mitt. Berl. Malakol. 11, 188–189, 1957.

PAINE, R. T.: Trophic relationships of 8 sympatric gastropods. Ecology 44, 63–73, 1963.

PALFI, D.: Die Schnecke *Cecilioides petiiana* als neuer Schädling an Kopfsalat in Gewächshäusern. Növenyvédelem, Budapest, 12, 560, 1973.

PAMPIGLIONE, S., und TOFFOLETTO, F.: Molluschi di interesse parasitologico-medico in italia. Rivista di Parasitol. 32, 113–134, 1971.

PAN, C.-T.: Studies on the biological control of schistosome-bearing snails: A preliminary report on pathogenic microorganisms found in *Australorbis glabratus.* J. Parasit. 42 (4 suppl.), 33, 1956.

PAPE, H.: Krankheiten und Schädlinge der Zierpflanzen und ihre Bekämpfung. 1–625, Parey, Berlin, (5 A) 1964.

PAPPAS, J. L., und CARMAN, G. E.: Field screening tests with various materials against the euro-

pean brown snail on citrus in California. J. econ. Ent. 48, 698–700, 1955.
– und CARMAN, G. E.: Control of European brown snail in citrus groves in southern California with Guthion and metaldehyde sprays. J. econ. Ent. 54, 152–156, 1961.
PARDO, L.: Les caracoles. Ministerio de Agricultura, Madrid 1943.
PARHURST, R. M., THOMAS, D. W., und SKINNER, W. A.: Molluscicidal saponins of *Phytolacca dodecandra*: Oleanoglycotoxin-A. Phytochemistry 12, 1437–1442, 1973.
PARTSVANIYA, V. B.: [Temperature effect on the spontaneous activity of *Helix pomatia* neurons.] russ. Soobshch. Akad. Nauk. Gruz. SSR 67, 445–448, 1972.
PASCOE, D.: An electrophoretic examination of the haemocoelic fluid of parasitized and healthy adults in a population of *Littorina saxatilis tenebrosa (Mollusca)*. Acta Parasitol. Polonica 18, 359–368, 1970.
PASCUAL, E.: Contributión al conocimiento de la reproductión del ostión, *Crassostrea angulata* (Lmk.) de la desembocadura del Guadalquivir. Invest. Pesq. 34, 477–498, 1970.
PATTERSON, C. M.: Chromosomes of Molluscs. Proc. Symp. on mollusco, 2, 635–686, 1972.
PAUCK, P.: Die Bekämpfung der Schnecken. Zentralbl. Deutsch. Erwerbsgartenbau 7 (21), 3, 1955.
PAULEY, G. B.: An invasive epithelial neoplasm in a Pacific oyster, *Crassostrea gigas*. J. Nat. Cancer Inst. 49, 897–902, 1972.
PAULINI, E. und DE SOUZA, C. P.: Laboratory evaluation of the molluscicidae „Yurimin-99" [3,5-dibromo, 4-hydroxy, 4'-nitroazobenzene.] Rev. Brasil. Malariol Doencas Trop. 20, 147–154, 1968.
PAULINYI, H. M., und PAULINI, E.: Laboratory observations on the biological control of *Biomphalaria glabrata* by a species of *Pomacea (Ampullariidae)*. Bull WHO 46, 243–247, 1972.
PAYNE, D. W., THORPE, N. A., und DONALDSON, E. M.: Cellulolytic activity and a study of the bacterial population in the digestive tract of *Scrobicularia plana* (Da Costa). Proc. Malacol. Soc. Lond. 40, 147–160, 1972.
PELLUET, D.: On the hormonal control of cell differentiation in the ovotestis of slugs *(Gasteropoda: Pulmonata)*. Canad. J. Zool. 42, 195–199, 1963.
– und HENDERSON, N. E.: The effects of visible radiation on the ovotestis of the slug *Deroceras reticulatum*. Proc. Nova Scotian Inst. Sci. 23, 420–421, 1954.
– und LANE, N. J.: The relation between neurosecretion and cell differentiation in the ovotestis of slugs *(Gasteropoda, Pulmonata)*. Can. J. Zool. 39,789–805, 1961.
PELSENEER, P.: Les parasites des mollusques et les mollusques parasites. Bull. Soc. Zool. France 53, 158–189, 1928.
PERLOWAGORA-SZUMLEWICZ, A.: Survival, growth and fecundity of *Australorbis glabratus* Snails development from eggs exposed to ionizing radiation. Exp. Parasit. 15, 232–241, 1964.
– und BERRY, E. G.: Effects of Ionizing Radiation on *Australorbis glabratus* Eggs. Exp. Parasit. 15, 226–231, 1964.
PERRET, P., EGGER, M., und DEGRÉMONT, A. A.: Essai de lutte anti-mollusque par argumentation de la biomasse planctonique et traitement molluscicide: Association Urée-N-tritylmorpholine. Acta Trop. 29, 175–181, 1972.
PETERS, W.: Occurence of chitin in *mollusca*. Comp. Biochem. Physiol. B. Comp. Biochem. 41, 541–550, 1972.
PETERSEN, G. H.: *Arianta arbustorum* (Linné) *(Mollusca Helicidae)* in Denmark. Vidensk. Meddel. Dansk Naturhist. Foren Kobenhavn 127, 295–302, 1964.
PETERSON, G. D., Jr.: *Lamprophorus tenebrosus* introduced into Guam to combat the giant African snail. J. econ. Ent. 50, 114, 1957.
– Studies on Control of the Giant African Snail on Guam. Hilgardia 26, 643–658, 1957.
PETROCELLI, S. R.: The Interactions of a Chlorinated Hydrocarbon Insecticide Among the Water, Sediments and Biota in an Estuarine System. Diss. Abstr. Intern. B. 34, 7, 3109–3110, 1974.
– HANKS, A. R., und ANDERSON, J.: Uptake and Accumulation of an Organochlorine Insecti-

cide (Dieldrin) by an Estuarine Mollusc, *Rangia cuneata*. Bull. Environ. Contam. Toxicol. 10, 315–320, 1973.

Pflanzenschutz 1974/75: Anleitungen für die Praxis mit Erläuterungen zu den Warnnachrichten und Wochenhinweisen. 1–125. Herausg.: Pflanzenschutzamt Landwirtschaftskammer Weser-Ems, Oldenburg 1975, 8. neu bearb. Aufl.

Picer, M. und Ahel, M.: Separation of polychlorinated biphenyls from DDT and its analogues on a miniature Silica Gel Column. J. Chromatogr. 150, 119–127, 1978.

Piekarski, G.: Lehrbuch der Parasitologie. Springer, 1–760, 1954.

Pielou, D. P., Williams, K., und Brinton, F. A.: Differences in the deposition and persistence of pesticides on the upper and lower surfaces of leaves. Nature 195, 256–257, 1962.

Pietrzak, J. E., Bates, J. M., und Scott, R. M.: Constituents of unionid extrapallial fluid: I. Electrophoretic and immunological studies of protein components. Biol. Bull. (Woods Hole) 144, 391–399, 1973.

Pilsbry, H. A.: Land Mollusca of North America (North of Mexico). I. Part 1 u. 2, Acad. Nat. Sci. Philadelphia Monogr. 1939/40, Philadelphia.

– Land Mollusca of North America (North of Mexico). II. Part 1 u. 2, Acad. Nat. Sci. Philadelphia Monogr. 3. 1946/1948, Philadelphia.

Pilz, I., Glatter, O., und Kratky, O.: Röntgenkleinwinkelstudien über die Struktur von *Helix pomatia* Haemocyanin. Z. Naturforsch. B. 27, 518–524, 1972.

Pinder, L. C. V.: The biology and behaviour of some slugs of economic importance, *Agriolimax reticulatus, Arion hortensis* and *Milax budapestensis*.. Ph. D. Thesis, Univ. Newcastle upon-Tyne 1969.

Pinter, L.: Neue Mollusken aus Bulgarien (*Gastropoda: Helicidae*). Acta Zool. Acad. Sci. Hungar. 15, 91–96, 1969.

– Recent *Zonitidae* in Hungary. J. Conchol. 27, 163–169, 1970.

– Katalog der rezenten Mollusken Ungarns. Fol. Hist.-nat. Mus. Matr. 2, 123–148, 1974.

Pinter, I., und Pinter, L.: Mollusken aus Bulgarien. Malakol. Abh., Mus. Tierkde. 81–98. Dresden 3, 1970.

Pinto de Oliveira, M., und de Oliveira, M. H.: [Malacological communication. 6. On *Bradybaena similaris* Férussac (*Gastropoda, Pulmonata, Stylommatophora, Fructicicolidae*): Copulation and function of the oviduct during the period of fertilization and formation of the egg]. Comun. soc. Malacol. Urug. 3, 155–161, 1971.

Pitois, M.: La Fasciolose ovine et bovine en Grande Bretagne et en Irlande. Haliotis 3, 57–58, 1973.

Plant Quarantine Treatment Manual. 2. Ed. Washington. US Dep. Agric., Agric. Res. Service, Plant Quarantine Division 390, 624 (T-402-403), 1971.

Plate, H.-P.: Die ökologischen Beziehungen zwischen Arthropoden und Mollusken. Z. ang. Ent. 32, 406–432, 1951.

– Schnecken als Vorratsschädlinge. 53–57. Beiträge zur Vorratsschutzforschung, herg.: Ges. Vorratsschutz e.V., Berlin-Steglitz, zum 75. Geb. von F. Zacher. 1959.

– Frucht- und Rindenfraß der Hainschnirkelschnecke an Apfelbäumen. Z. ang. Zool. 51, 107–109, 1964.

– Die Nacktschnecke *Boettgerilla vermiformis* Wiktor 1959 in einem Berliner Gartenbaubetrieb. Z. ang. Zool. 52, 507–511, 1965.

– und Frömming, E.: Untersuchungen über das Verhalten von Landlungenschnecken gegenüber einigen *Liliaceae, Amaryllidaceae* und *Iridaceae*. Z. hyg. Zool. 39, 145–156, 1951.

– und Frömming, E.: Quantitative Untersuchungen über die Wirksamkeit einiger Metaldehydpräparate gegenüber Landlungenschnecken. Anz. Schädlingskde. 25, 133–136, 1952.

– und Frömming, E.: Die Gastropoden der Berliner Gewächshäuser, ihre Biologie und Schadenswirkung. Mitt. Berl. Mal. 1, 5–36, 1953.

– und Frömming, E.: Die tierischen Schädlinge unserer Gewächshauspflanzen. 1–288. Dunker & Humblot, Berlin 1953.

– und Frömming, E.: Über das Verhalten von Landlungenschnecken gegenüber verschiedenen Sorten der Zierpflanze *Chrysanthemum indicum*. Z. ang. Zool. 41, 187–195, 1954.

Plesch, B., De Jong-Brink, M., und Boer, H. H.: Histological and histochemical observations on the reproductive tract of the hermaphroditic pond snail *Lymnaea stagnalis* (L.). Neth. J. Zool. 21, 180–201, 1971.

Poley, D., und Cheng, Th.: Interaction of molluscs and foreign substances: The morphology and behavior of haemolymph cells of the American oyster *Crassostrea virginica*, in vitro. J. Invertebr. Pathol. 19, 383–394, 1972.

Pollard, E.: Growth classes in the adult Roman snail *(Helix pomatia)*. Oecologia (Berl.) 12, 209–212, 1973.

Pomeroy, D. E.: The influence of environment on two species of landsnails in South Australia. Trans Roy Soc. S. Aust. 91, 181–186, 1967.

– Some aspects of the ecology of the land snail *Helicella virgata* in South Australia. Aust. J. Zool. 17, 495–514, 1969.

Poulin, G., und O'Neil, L.-C.: Observations sur les prédateurs de la Limace noire, *Arion ater* (L.) (Gastéropodes, Pulmonés, Arionidés). Phytoprotection 50, 1–6, 1969.

Pozzi, G., und Merlini, M.: The accumulation, distribution, and loss of Zinc (^{65}Zn) in the gastropod *Viviparus ater* (Cristofori & Jan). Malacologia 16, 227–230, 1977.

Price, R. J., und Lee, J. S.: Interaction between paralytic shellfish poison and melanin from butter clam *(Saxidomus giganteus)* and synthetic melanin. J. Fish. Res. Board. Can. 28, 1789–1792, 1971.

Prieur, D.: Les bactéries associées aux élevages de larves de Bivalves marins. 1–133. Thèse, Univ. Bretagne occi, 20. 4. 1974.

Pringle, G., Otieno, L. H., und Chimtawi, M. B.: Notes on the morphology, susceptibility to *Schistosoma haematobium* and genetic relationships of *Bulinus (Physopsis) globosus globosus* and *B. (P.) nasutus* from north-eastern Tanzania. Ann. trop. Med. Parasit. 65, 211–219, 1971.

Prinsloo, J. F., und Van Eeden, J. A.: Temperature and its bearing on the distribution and chemical control of freshwater snails. S. Afr. med. J. 43, 1363–1365, 1969.

– und Van Eeden, J. A.: The influence of temperature on the growth rate of *Bulinus (Bulinus) tropicus* (Krauss) and *Lymnaea natalensis* (Krauss) *(Mollusca: Basommatophora)*. Malacologia 14, 81–88, 1973.

Prokop, O., Uhlenbruck, G., und Kohler, W.: A new source of antibody-like substance having anti-blood group specificity. A discussion on the specificity of *Helix* agglutinin. Vox sang. 14, 321, 1968.

Prota, R.: *Milax gagates* Drap. (Mollusco, Gasteropode) – Brevi osservazioni etologiche e prove di lotta. Studi Sassaresi III. Agr. 8, 160–173, 1960.

Pullin, R. S. V.: Composition of the haemolymph of *Lymnaea truncatula*, the snail host of *Fasciola hepatica*. Comp. Biochem. Physiol. A. 40, 617–626, 1971.

Purchon, R. D.: The Biology of Mollusca (2. Edition), 397 p., Pergamon, Oxford und New York, 1977.

Pussard, R.: Un mollusque endogé, *Cecilioides acicula* Müller, ravageur méconnu *d'Asparagus plumosus* Bak. C.R. Acad. Agric. France 37, 82–84, 1951.

Pusswald, W.: Beiträge zum Wasserhaushalt der Pulmonaten. Z. vergl. Physiol. 31, 227–248, 1948.

Quattrini, D.: Strukture and ultrastructure of the molluscan prostata 3. Observations on *Milax gagates* (Draparnaud) (*Gastropoda, Pulmonata, Stylommatophora*). Monit. Zool. Ital. 1 109–128, 1967.

– und Sacchi, T. B.: [The podocyst of *Milax gagates* (Draparnaud): *Mollusca Gastropoda Pulmonata:* Studies with the light and electron microscope.] Arch. Ital. Anat. Embriol. 76, 39–52, 1971.

Quick, H. E.: *Arion ater* (L.) and *A. rufus* (L.) in Britain and their specific differences. J. Conch. 22, 249–261, 1947.

– Slugs *(Mollusca)*. Synops. Brit. Fauna 8, London 1949.

– Helicellids introduced into Australia. Proc. Malac. Soc. Lond. 30. 1953, 74–79.

– British slugs *(Pulmonata; Testacellidae, Arionidae, Limacidae)*. Bull. Br. Mus. nat.-hist. zool. 6, 105–226, 1961.

Quiot, J. M., Vago, C., und Luciani, J.: Culture in vitro des cellules de gastéropodes. Haliotis 3, 149–154, 1973.

Radke, M. G., Ritchie, L. S., und Ferguson, F. F.: Demonstrated control of *Australorbis glabratus* by *Marisa cornuarietis* under field conditions in Puerto Rico. Amer. J. Trop. Med. Hyg. 10, 370, 1961.

Raehle, W.: Die Knoblauchschnecke *Oxychilus alliarius* (Miller) am Killberg bei Nagold (Wttb.). Jahresb. ges. Naturkd. Württemb. 125, 291–293, 1970.

Raff, R. A.: Polar lobe formation by embryos of *Ilyanassa obsoleta:* Effects of inhibitors of microtubule and microfilament function. Exp. Cell. Res. 71, 455–459, 1972.

Raffy, A.: Effets du métaldéhyde sur les gastéropodes dulçaquicoles. J. Conchyl. 133, 254–258, 1939.

Rainer, E.: Ackerschnecken fressen Kartoffelfelder kahl. Gesunde Pflanzen 14, 74–75, 1962.

Rajala, P.: Über die nördliche Verbreitungsgrenze von *Limax cinereoniger* Wolf in Finnland. Arch. Soc. Vanamo 12, 51–53, 1957.

Ramachandran, N. K.: Two Sarcophagid parasites of phytophagous terrestrial snails in Mysore State, India. Tech. Bull. Commonw. Inst. Biol. Control 10, 113–127, 1968.

Ramadan, M. A.: Sequestration of Parasitic Excretory Products in *Schistosoma mansoni* Infection. Z. Tropenmed. Parasit. 22, 397–405, 1971.

Rammos, G.: Ein Beitrag zur Wirksamkeit und Anwendung fraßmolluscider Substanzen am Beispiel der *Mollusca Pulmonata* 1–83. Diss. Math.-Naturwiss. Fakultät, F.U. Berlin 1962.

Ranaivosoa, H.: Quelques acquisitions préliminaires sur le probleme des escargots nuisibles aux cultures de la côte est de Madagascar et sur l'utilisation en lutte biologique de „*Euglandina rosea*" Fér. contre les gastéropodes. 857–859. Congr. Protect. Cult. Trop., Marseille 1965.

– Lutte Biologique Contre les Excargots Phytophages à Madagascar et aux Comores. Agron. Trop. Agron. Gen. Etudes Sci. 26, 341–347, 1971.

Rands, R. D.: Snails as predisposing agents of sugar cane „root-disease" in Louisiana. J. Agric. Res. Washington 28, 969–970, 1924.

Raven, C. P.: Development. In: Physiology of Mollusca. Ed.: Wilbur, K. M. und Yonge, C. M. 1 (Chapt. 5) 165–195. Akad. Press. New York 1964.

Ravera, O.: Effects of heavy metals (cadmium, copper, chromium and lead) on a freshwater snail: *Biomphalaria glabrata* Say (*Gastropoda, Prosobranchia*). Malacologia 16, 231–236, 1977.

Rawls, H.C, Bernardi, P.J., und White, J. F.: Fluorescence caused by *Pseudomonas* in the mucus of *Helix aspersa* Müller. Sterkiana 49, 36–37, 1973.

Read,. K. R.: Molluscan Haemoglobin and Myoglobin. In: Physiology of Mollusca, ed. Wilbur, K. M. und Yonge, C. M. 2 (Chapt. 6) 209–248, Acad. Press., New York-London 1966.

Reade, P. und Reade, E.: Phagocytosis in invertebrates: II. The clearance of carbon particles by the clam, *Tridacna maxima.* Res. J. Reticuloendothel. Soc. 12, 349–360, 1972.

Reader, T. A. J.: Histochemical observations on carbohydrates, lipids and enzymes in digenean parasites and host tissues of *Bithynia tentaculata*. Parasit. 63, 125–136, 1971.

Real, G.: Répartition en France de *Potamopyrgus jenkinsi* (E.A. Smith, 1889). Haliotis 3, 199–202, 1973.

Reddy, S. R., und Karumuri S. Swami: The occurence of helicorubin in the indian apple-snail *Pila globosa.* Arch. Int. Physiol. Biochim. 76, 20–36, 1968.

Rees, W. J.: The aerial dispersal of mollusca. Malac Soc. London Proc. 36, 269–282, 1965.

Reger, J. F.: A fine structure study on haemocyanin formation in the slug *Limax sp.* J. Ultrastruct. Res. 43, 377–387, 1973.

Rehm, S. und Espig, G.: Die Kulturpflanzen der Tropen und Subtropen. 1–496. Ulmer, Stuttgart 1976.

Reichmuth, W. und Frömming, E.: Pigmentstudien an Gastropoden. I. Farbstoffe als Widerstandsregulativ bei Vergiftungen und Krankheiten. Z. Photographie u. Forschung 8, 97–103, 1960.

– und Frömming, E.: Pigmentstudien an Gastropoden. II. Abhängigkeit der Konstitution von Körperfärbung und Vorzugstemperatur. Biolog. Zentralbl. 80, 67–78, 1961.

Reinboth, R.: Intersexuality in the Animal Kingdom. 1–449. Springer, Berlin-Heidelberg-New York 1975.

Reinmuth, E.: Schnecken als Baumschädlinge. Z. Pflanzenkrankh. 48, 208–210, 1938.

Reygrobellet, D. F.: Histogenèse régénératrice des téguments de l'extrémité postérieure de quelques Limacides. Bull. Soc. Zool. Fr. 100, 349–359, 1975.

Rich, R., und Rouse, W.: Mass producing a tropical snail for biological control. Proc. South Weed Sci. Soc. 23, 288–298, 1970.

Rich, S., und Horsfall, G.: Fungitoxicity of carbamic and thiocarbamic acid esters. The Connecticut Agric. Exp. Sta. New Haven Bull. 639, 1–95, 1961.

Richardson, H. H., und Roth, H.: Ethylene Oxide Fumigants to Eliminate Quarantinable Snails *Cochlicella* or *Theba* in Cargo. J. econ. Ent. 56, 836–839,1963.

– und Roth , H.: Methyl bromide, sulfuryl fluoride and other fumigants against quarantinable *Cochlicella* and *Theba* snails. J. econ. Ent. 58, 690–693, 1965.

Ricou, G.: Les dégâts de la limace grise en grande culture. Phytoma 10 (97), 9–11, 1958.

– Contribution à l'étude des Mollusques nuisibles en Normandie. Rev. Soc. Sav. Hte. Normandie 21 (6), 30–44, 1961.

– Relations entre l'activité des limaces grises et la temperature. Meded. Landbouwhogesch. Gent 29, 1071–1080, 1964.

– Observations sur les peuplements animaux nuisibles aux Cresson. Coll. Anim. Nuis. dans les cressonnières, Gembloux 13–29, 1966.

– Cresson et Distomatose: Les mesures prises en France. 68–74. Coll. Anim. Nuis. dans les cressonnières, Gembloux 1966.

– und Mallet, C.: Données récentes sur la biologie und la lutte contre les limaces. In: Lutte contre les maladies et les ravageurs des Céréales, S. 295–304, Itcf (Institut technique des Céréales et des Fourrages, Paris, 26. Jan. 1977.

– und Marre, R.: Essais de traitements contre les limnées des cressonnières. Phytoma 12 (118), 15–20,1960.

– und Picard, R.: Expériences de désinfection des fonds de bassins en cressiculture. Phytoma 24, 24–25, 1972.

– und Ferret, R.: Méthodologie expérimentale du controle d'efficacité des molluscicides à homologuer. Haliotis 3, 113–123, 1973.

– Ferret, R., und Jouis, M.: Etude sur la dispersion et les conditions d'élevage de *Lymnaea truncatula*. Haliotis 3, 167–172, 1973.

Riedel, A.: Zur Kenntnis der *Zonitidae (Gastropoda)* Spaniens. Ann. Zool. Warszawa 29, 115–145, 1972.

– und Wiktor, A.: *Arionacea (Gastropoda, Stylommatophora)*. 1–139. Polska Akad. Nauk. Inst. Zoologii (Warszawa) 1974.

Riemschneider, R. und Frömming, E.: Über die Wirkung von Insektiziden auf Gastropoden. I. Anz. Schädlingskde. 24, 106–108, 1951.

Rietschel, G.: Das Auftreten von „Fühlermaden" (*Leucochloridium*) in Hessen. Natur u. Museum 100, 57–64, 1970.

Rigby, J. E.: Alimentary and reproductive systems of *Oxychilus cellarius* (Müller) *(Stylommatophora)*. Proc. Zool. Soc. London 141, 311–359, 1963.

Rising, T. L., und Armitage, K. B.: Acclimation to temperature by the terrestrial gastropods *Limax maximus* and *Philomycus carolianus:* Oxygen consumption and temperature preference. Comp. Biochem. Physiol. 30, 1091–1114, 1969.

Ritchie, L. S., Hernandez, A., und Rosa-Amador, R.: Biological Potentials of *Australorbis glabratus:* Life Span and Reproduction. Amer. J. Trop. Med. Hyg. 15, 614–617, 1966.

– und Fox, I.: Comprehensive laboratory evaluations of the molluscicide 2', 5-dichloro-4' nitro-3-phenyl-salicylanilid (PSA): Time concentration relationships and susceptibilities of stage-size arrays of snails. Bull. Wld. Hlth. Org. 39, 312–315, 1968.

– Berrios-Duran, L. A., und Sierra, R.: A Field Screening Test on a Slow-Release Formulation of Sodium Pentachlorophenate for Molluscicidal Use. Bull. Wld. Hlth. Org. 40, 474–476, 1969.

Ritter, M.: Les limaces et les escargots. Importance économique et moyens de lutte. Rev. Zool. Agr. et Appl. 4–6, 1–10, 1955.

Robbs, C. F.: Contribuição as estudo das bacterias que atacam plantas no Brazil. I. Contribuição as estudo da „podridao negra" das Cruciferas no Distrito Federal. Bol. Soc. Brazileira Agron. 9, 195–213, 1946.

Robinson, W. H., und Foote, B. A.: Biology and Immature Stages of *Megaselia aequalis,* a Phorid Predator of Slug Eggs. Ann. Ent. Soc. Amer. 61, 1587–1594, 1968.

Robson, E. M., und Williams, I. C.: Relationships of some species of Digenea with the marine prosobranch *Littorina littorea* (L.). II. The effect of larval *Digenea* on the reproductive biology of *L. littorea.* J. Helminthol. 45, 145–159, 1971.

Rogers, D. C.: The fine structure of sensory neurons and their processes in the optic tentacles of *Helix aspersa*. L. Mikros. Anat. Forschg. 84, 52–64,1971.

Rogers-Lewis, D. S.: Slug damage in potatoes and winter wheat on silt soils. Ann. Appl. Biol. 87, 532–535, 1977.

Rohlack, S.: Über das Vorkommen von Sexualhormonen bei der Meeresschnecke *Littorina littorea* L. Z. vergl. Physiol. 42, 164–180, 1959.

Röhlich, P. und Török, L. J.: Die Feinstruktur des Auges der Weinbergschnecke (*Helix pomatia*). Z. Zellforschg. 60, 348–368, 1963.

– und Bierbauer, J.: Electron microscopic observations on the special cells of the optic tentacle of *Helicella obvia (Pulmonata).* Acta Biol. Acad. Sc. hung. 17, 359–373, 1966.

Rondelaud, D.: La prédation de *Lymnaea (Galba) truncatula* Müller par *Zonitoides nitidus*, moyen de lutte biologique. Ann. Parasit. Hum. Comp. 50, 55–61, 1975.

Rorke, M. A., Gardner, D. R., und Greenhalgh, R.: Lethality and Behavioural Symptoms Produced by Some Organophosphorous Compounds in the Snail *(Helix aspersa)*. Bull. Environm. Contam. Toxicol. 11, 417–424, 1974.

Rose, J. H.: Observations on the Larval Stages of *Muellerius capillaris* within the Intermediate Hosts *Agriolimax agrestis* and *A. reticulatus*. J. Helminth. 31, 1–16, 1957.

– The field-slug *Agriolimax reticulatus* as a vector of the lungworm *Cystocaulus ocreatus.* Nature 185, 180, 1960.

Ross, J. G.,und O'Hagan, J.: *Limnaea truncatula* population studies: Sampling techniques. J. Med. Lab. Technol. 25, 112–116, 1968.

Roth, H.: Fumigation with Ethylene Oxide-Carbon Dioxide Mixture for Quarantine Control of the Giant African Snail. J. econ. Ent. 64, 341–342, 1971.

– und Kennedy, J. W.: *Helicella* Snails Infesting Rosemary Seeds: Methyl Bromide and other Fumigants for Quarantine Control. J. econ. Ent. 66, 935–936, 1973.

Rotschild, L.: „Fertilization", 1–170. Methuen, London 1956.

Rousset-Galangau, V.: Présence de deux catégories de spermies chez *Milax gagates* et *Agriolimax agrestis (Moll. Gast. Pulm. Lim.)*: Etude comparée des ultrastructures au cours de la spermiogenèse. Ann. Sci. Nat. Zool. Biol. Anim. 14, 319–331, 1972.

Rowan, W. B.: Mass cultivation of *Australorbis glabratus*, intermediate host of *Schistosoma mansoni* in Puerto Rico. J. Parasit. 44, 247, 1958.

Rowe, D. R., Canter, L. W., Snyder, P. J., und Mason, J. W.: Dieldrin and Endrin Concentrations in a Louisiana Estuary. Pestic. Monit. J. 4, 177–183, 1971.

Roy, A.: Etude de l'acclimation thermique chez la limace *Arion circumscriptus*. Can. J. Zool. 41, 671–698, 1963.

Roy, S., und Roy W. Jr.: Metabolic effects of commercial ß-glucuronidase preparations on estrogens. Steroids 16, 69–78, 1970.

ROZKOŠNÝ, R.: Neue Metamorphosestadien mancher *Tetanocera-Arten (Diptera, Sciomyzidae)*. Zool. Listy 14, 367–371, 1965.

ROZSA KATALIN S.: Studies on the mechanism of the neural integration. Ann. Inst. Biol. (Tihany) Hung. Acad. Sci. 44, 99–122, 1977.

RUIZ-TIBEN, E., PALMER, J.R., und FERGUSON, F.: Biological control of *Biomphalaria glabrata* by *Marisa cornuarietis* in irrigation ponds in Puerto Rico. Bull. Wld. Hlth. Org. 41, 329–333,1969.

RUNHAM, N. W.:The use of the scanning electron microscope in the study of gastropod radulae: The radula of *Agriolimax reticulatus* and *Nucella lapilus*. Malacologia 9, 179–185, 1969.

– ISARANKURA, K., und SMITH, B. J.: Methods for narcotizing and anaesthetizing gastropods. Malacologia 2, 231–238, 1965.

– und LARYEA, A. A.: Studies on the maturation of the reproductive system of *Agriolimax reticulatus (Pulmonata: Limacidae)*. Malacologia 7, 93–108,1968.

– und HUNTER, P. J.:Terrestrial slugs. 1-184. Hutchinson (Publishers) LTD. London 1970.

– BAILEY, T. G., und LARYEA, A. A.:Studies of the endocrine control of the reproductive tract of the Grey Field Slug *Agriolimax reticulatus*. Malacologia *14*, 135–172, 1973.

RUPPEL, R.: Effectiveness of Sevin against the Grey Garden Slug. J. econ. Ent. 52, 360, 1959.

RYDER, T. A. und BOWEN, I. O.: The slug foot as a size of uptake of copper molluscicide. J. Invertebr. Pathol. 30, 381–386, 1977.

SACCHI, C. F.: Lineamenti biogeografii della Spagna mediterranea su basi malacofaunistiche. Public. Inst. Biol. Aplicada 25, Madrid, 1957.

– Ecological and historical bases for a study of the Iberian terrestrial mollusca. Proc. I. Europ. Malac. Congr. (London 1962). 243–257, 1965.

– Ecologie comparée des Gastéropodes Pulmonés des dunes méditerranéennes et atlantiques. Natura (Milano) 62, 277–358, 1971.

– Points de vue d'un écologiste sur la physiologie de l'estivation chez l'Helicidé dunicole *Euparypha pisana* (Müller). Natura (Milano) 65, 117–133, 1974.

– Le polychromatisme des Littorines (Gastéropodes, Prosobranches) d'Europe; points de vue d'un écologiste. Mém. Soc. Zool. France 37,61–101, 1974.

– und NOS, R.: Quelques distributions intéressantes des Mollusques terrestres ibériques. Inst. Biol. Aplicata 27, Madrid 1958.

SAINT PIE, P. und BEAUCHARD, J.: Les ennemis des cactées. Phytoma 15 (146), 49–59,1963.

SANBORN, J. R.,und CHING-CHIEH, Yu: The Fate of Dieldrin in a Model Ecosystem. Bull. Environm. Contam. Toxicol. 10, 340–346, 1973.

SANCHEZ, S.: Histophysiologie neurohormonale chez quelques mollusques gastéropodes. I. Complexes neuroendocriniens. Bull. Soc. Zool. France 87. 309–318, 1962.

– und SABLIER, H.: Histophysiologie neuro-hormonale chez les mollusques gastéropodes II. Corrélation humorale. Bull Soc. Zool. France 87, 319–330, 1962.

SANCHEZ, J. A.: Estudios sobre morfologia, sistematica y faunistica de invertebrados. I. Sabre la existencia de *Lymnaea (Leptolymnaea) glabra* (Müll.) en Espana. Bol. Real. Soc. Espan Hist. Natur 63, 9–14, 1965.

SANDI, E.: Radioactivity in Snail Shells due to Fall-out. Nature 193, 290, 1962.

SCHARRER, B., und SZABÓ-LOCHHEAD, M.: Tumors in the Invertebrates. A. Review. Cancer Res. 10, 403–419, 1950.

SCHERF, H.: Der Haarbesatz des Gehäuses von *Trichia hispida*. Naturw. Rundschau 20 (3), 120–121, 1967.

SCHILDER, F. A.: Zum Polymorphismus bei *Cepaea*. Zool. Anz., Suppl. 17. 487–492, 1953.

SCHLOOT, W.: Postembryogenese des Schlundrings von *Helix pomatia* L. (Gastropoda) unter Berücksichtigung der Sekretion der Nervenzellen. Z. Zellforsch. 67, 406–426, 1965.

SCHMID, G.: *Boettgerilla vermiformis* Wiktor 1959, eine neue Nacktschnecke in Deutschland *(Gastropoda, Parmacellidae)*. Arch. Moll. 91, 105–108, 1962.

– Zur Verbreitung und Anatomie der Gattung *Boettgerilla*. Arch. Moll. 92, 215–225, 1963.

– *Arion lusitanicus* in Deutschland. Arch. Moll. 100, 95–102, 1971.

SCHMIDT, H. A.: Zur Abhängigkeit der Entwicklung von Gehäuseschnecken vom Kalkgehalt des Bodens. Arch. Moll. 84, 167–177, 1955.

– Die Besiedlung einer Kulturfläche durch Landschnecken. Arch. Freunde Naturgesch. Mecklenburg 1, 193–205, 1955.

– Zur Fortpflanzungsbiologie von *Oxychilus draparnaldi* Beck. Archiv Freunde Naturgesch. Mecklenburg 2, 373–405, 1955/1956.

– Zur Biologie, insbesondere der Fortpflanzung von *Opeas mauritianum* (Pfeiffer). Arch. Moll. 88, 55–67, 1959.

– Zur Verträglichkeit von *Oxychilus draparnaldi* Beck mit *Oxychilus cellarius* Müll. Arch. Naturgesch. Mecklenburg 6. 71–76,1960.

SCHNETTER, M.: Veränderungen der genetischen Konstitution in natürlichen Populationen der polymorphen Bänderschnecken. Verh. Deutsch. Zool. Ges. 44, 192–206, 1950.

– und SEDLMAIR, H.: Verhaltens- und Resistenzunterschiede bei den polymorphen Bänderschnecken. Naturwissensch. 40, 515–516, 1953.

SCHORER, G.: Qualitative und quantitative Untersuchung der Landgastropoden des Siebengebirges und des Rodderberges in ausgewählten Biotopen. Decheniana 126, 69–90, 1973.

SCHOUTEN, A. R.: *Milax (Tandonia) sowerbii* (Fér.) in Nederland. Basteria 13, 45-54, 1949.

SCHREIER, O.: Rübenschädlinge – kurz und bündig –. „Der Pflanzenarzt“ 22, 36–37, 1969.

SCHUTTE, C. H. J., und FRANK, G. H.: Observations on the Distribution of Freshwater *Mollusca* and Chemistry of the Natural Waters in the South-eastern Transvaal and Adjacent Northern Swaziland. Bull. Wld. Hlth. Org. 30, 389–400, 1964.

SCHWALB, H. H.: Beiträge zur Biologie der einheimischen Lampyriden *Lampyris noctiluca* (Geoffr.) und *Phausis splendidula* (Lec.) und experimentelle Analyse ihres Beutefang- und Sexualverhaltens. Zool. Jahrb. Syst., Ökol., Geogr., 88, 399–550, 1961.

SEAMAN, D. E., und PORTERFIELD, W. A.: Control of Aquatic Weeds by the Snail *Marisa cornuarietis*. Weeds 12, 87–92, 1964.

SEDLMAIR, H.: Verhaltens-, Resistenz- und Gehäuseunterschiede bei den polymorphen Bänderschnecken *Cepaea hortensis* Müll. und *Cepaea nemoralis* L. Biol. Zentralbl. 75, 281–313, 1956.

SEGAL, E.: Respiration and temperature acclimation in slugs. Animal Record 134, 636, 1959.

SELIM, S. I.: The Control of Snail and Slug Pests in Agriculture. Part I.: Analysis for Dissipation of Metaldehyde in Vegetable Crops. Part II.: Beer as a Slug Attractant. Diss. Abstr. Intern. B 35, 1(4), Davis, Calif. USA 1974.

SENECA, H., und BERGENDAHL, E.: Toxicity of Antibiotics to snails. Antibiotics Chemotherapy 5, 737–741, 1955.

SERRA, J. A. und KOSHMANN, R. W.: Synchronous divisions in spermatogonial cells of a snail. Canad. J. Genet. Cytol. 9, 38–43, 1967.

SEVO, S.: Note au sujet de l'identification de *Lymnaea truncatula* Müller, hôte intermédiaire de *Fasciola hepatica* Linné. Parasitica 27, 70–83, 1971.

SHAPIRO, Y. S.: Crop pests (russ.) Zashch. Rast. Nr. 12, 59–60, 1975.

– Slug Pests (russ.) Zashch. Rast. Nr. 9, 28–29, 1976.

– Molluscs pests of orchards and market gardens (russ.) Zashch. Rast. Nr. 8, 61, 1976.

SHAPIRO, Y. S.: Control of molluscs (russ.). Zashch. Rast. Nr. 9, 48, 1977.

SHARAF, El Din, H. und WILSON-JOHNES, K.: Copper sulphate as an Aquatic Herbicide. Nature 174, 187, 1954.

SHARAF, El Din, H. und El NAGAR, H.: Control of snails by copper sulphate in the canals of the Gezira irrigated area of the Sudan. J. Trop. Med. 58, 260–263, 1955.

SHARAF, N. S. und GHALEB, A. H.: A laboratory immersion test for assessing the contact toxicity of bait molluscicides. Z. Pfl. krankh. 85, 275–279, 1978.

SHEANON, M. J., und TRAMA, F. B.: Influence of phenol and temperature on the respiration of a freshwater snail: *Helisoma trivolvis*. Hydrobiol. 40, 321–328, 1972.

SHIFF, C. J.:Trials with N-Tritylmorpholine (Shell WL 8008) as a Molluscicide in Southern Rhodesia. Bull. Wld. Hlth. Org. 35, 203–212, 1966.

– Location of *Bulinus (Physopsis) globosus* by miracidia of *Schistosoma haematobium*. J. Parasit. (Lawrence, Can.) 54, 1133–1140, 1968.
– und GARNETT, B.: The short-term effects of three molluscicides on the microflora and microfauna of small, biological stable ponds in Southern Rhodesia. Bull. Wld. Hlth. Org. 25, 543, 1961.
– und WARD, D.: Variation in susceptibility to three molluscicides shown by three species of aquatic snails. Bull. Wld. Hlth. Org. 34, 147–150, 1966.
– und GARNETT, B.: The influence of temperature on the intrinsic rate of natural increase of the freshwater snail *Biomphalaria pfeifferi* (Krauss) *(Pulmonata: Planorbidae)*. Arch. Hydrobiol. 62, 429–438, 1967.
– CLARKE, V. de V., EVANS, A. C., und BARNISH, G.: Molluscicide for the control of Schistosomiasis in irrigation shemes. Bull. Wld. Hlth. Org. 48, 299–307, 1973.

SHILEIKO, A. A.: [Some remarks on the system of the genus *Helicella* Fér. *(Pulmonata, Helicidae)* with a discription of a new species from Talish]. russ. Vestn. Mosk. Univ. Ser. 6, Biol. Pochvoved. 22 (4), 27–34, 1967.

SHIMIZU, M., KAJIHARA, T., SUYAMA, I., und HIYAMA, Y.: Uptake of ^{58}Co by mussel, *Mytilus edulis*. J.Radiat. Res. 12, 17–28, 1971.

SHOREY, H. H., REYNOLDS, H. T., und ANDERSON, L. D.: Effect of Zectran, Sevin and other new carbamate insecticides upon insect populations found on vegetable and field crop in Southern California. J. econ. Ent. 55, 5–11,1962.

SHUKLA, S. S., und ROY, R. M.: Studies on the Comparative Molluscicide Effects of N-Tritylmorpholine and Copper Sulphate on the Common Aquatic Snail in Andhra Pradesh (India). Trop. Anim. Hlth. Prod. 7, 45–50, 1975.

SIEWERT, W.: Schalenbau und Stammesgeschichte von Austern. Stuttgarter Beitr. Naturkde. Ser. B. 1, 1–57,1972.

SIEZEN, R., und VAN DRIEL, R.: Structure and properties of hemocyanins: VIII. Microheterogeneity of α-hemocyanin of *Helix pomatia*. Biochim. Biophys.Acta 295, 131–139, 1973.

SIGNORET, P.: Mise en évidence, chez certains gastropodes marins et terrestres de systemes inhibiteurs agissant sur le virus de la mosaique du tabac. C.R. He. Bd. Séances Acad. Sci. Sér. D. Sci. Natur (Paris) 265 (15), 1093–1095, 1968.

SIMMONDS, F. J.: Possibilities of Biological Control of Liver Flukes. PANS 15, 41–43, 1969.
– und HUGHES, I. W.: Biological control of snails exerted by *Euglandina rosea* (Férussac) in Bermuda. Entomophaga (Paris) 8, 219–222, 1963.

SINGH, O. und AGARWAL, R. A.: Pharmacological studies on the molluskidal activity of phorate on *Pila globosa (Gastropoda)*. Acta Pharmacol. Toxicol. 42, 365–370, 1978.

SKELDING, J. M.:Studies on the renal physiology of *Achatina achatina* (L.). Malacologia 14, 93–96, 1973.

SMINIA, T.: Structure and function of blood and connective tissue cells of the freshwater pulmonate *Lymnaea stagnalis* studied by electron microscopy and enzyme histochemistry. Z. Zellforsch. Mikrosk. Anat. 130, 497–526, 1972.
– Structure and function of blood and connective tissue cells of the pond snail *Lymnaea stagnalis*. Malacologia 16, 255–256, 1977.
– BOER, H., und NIEMANTSVERDRIET, A.: Haemoglobin producing cells in freshwater snails. Z. Zellforsch. Mikrosk. Anat. 135, 563–568, 1972.

SMITH, B. J.: Maturation of the reproductive tract of *Arion ater* (*Pulmonata: Arionidae*). Malacologia 4, 325–349, 1966.

SMITH, F., und BOSWELL, A. L.: New Baits and Attractants for Slugs. J. econ. Ent. 63, 1919–1922, 1970.

SOEDIGDO, R., LIE SIEN NIO, ADIWIKARTA, S., und BARNETT, R. C.: Cellulase from the snail *Achatina fulica* (Fér.). Physiol. Zool. 43, 139–144, 1970.

SOLEM, A. Microarmature and barriers in the apertures of land snails. Veliger 15, 81-87, 1972.

SOLITYS, A., und ZIOMKO, I.: [Investigation on intermediate hosts of *Paramphistomum*]. Wiad Parazytol. 18, 85–88, 1972.

Somasundarao, G.: A preliminary note on the biological control of fresh water snails by aquatic insects. Ind. Vet. J. 40, 50, 1963.

South, A.: Estimation of slug population. Ann. appl. Biol. 53, 251-258, 1964.

– Biology and ecology of *Agriolimax reticulatus* (Müll.) and other slugs: Spatial distribution. J. Anim. Ecol. 34, 403–417, 1965.

– Dégâts causés par les limaces en Grand-Bretagne. Haliotis 3, 19–25, 1973.

– Estimation des populations de limaces. Haliotis 3, 89–95, 1973.

Southgate, V. R.: Observations on the effect of rediae of *Fasciola hepatica* on the lipid composition of the hepatopankreas of *Lymnaea truncatula*. Parasitol. 61, 293–300, 1970.

Speyer, W.: Versuche mit Metaldehyd zur Nacktschneckenbekämpfung. Nachrichtenbl. Dt. Pflanzenschutzd. (Braunschweig) 6, 134–137,1954.

Spiekermann, H.: Die Gastropodenfauna des Siegmündungsgebietes. Decheniana 129, 16–37, 1976.

Sprenger, I., Uhlenbruck, G., und Hermann, G.: Snail albumin gland: A new source of proteinase inhibitors. Enzymologia 43, 83–88, 1972.

Srivastava, P. D., und Srivastava, J. N.: *Orthomorpha sp.* – a new predatory millipede on *Achatina fulica* Bowdich in Andamans. Experientia 23, 776, 1967.

Sundman, R.: *Hyalina lucida* Draparnaud, ny för Finland. Memoranda Soc. Fauna-Flora Fennica 24, 4. 1947–1948, (1948–1949).

Symonds, B. V.: Evaluation of Potential Molluscicides for the Control of the Field Slug *Agriolimax reticulatus* (Müll.). Plant Pathol. 24, 1–9, 1975.

Szabó, I. und Szabó, M.: Altersveränderungen und pathologische Erscheinungen in der Körperwand von *Limax flavus* L. Biol. Zentralbl. 51, 695–701, 1931.

– und Szabó, M.: Epitheliale Geschwulstbildung bei einem wirbellosen Tier *Limax flavus* L. Z. Krebsforschung 40, 540–545,1934.

Szidat, L.: Bemerkungen zur sogenannten parasitären Kastration von Mollusken. Z. Parasitenkde. 12, 251–258, 1941.

Standen, O. D.: Experimental Schistosomiasis. I. The culture of the snail vectors *Planorbis boissyi* and *Bulinus truncatus*. Ann. Trop. Med. Parasit. 43, 13–22, 1949.

Starke, F.-J.: Elektronenmikroskopische Untersuchung der Zwittergonadenacini von *Planorbarius corneus* L. *(Basommatophora)*. Z. Zellforsch. Mikrosk. Anat. 119, 483–514, 1971.

Stauber: Starker Schneckenfraß an Grünmais. Pflanzenschutz (München) 6, 137–138, 1954.

Steiniger, F.: Schnecken und Typhus-Paratyphus-Bakterien. Z. ang. Zool. 44, 93–95, 1957.

Stephens, J. G., und Stephens, G. C.: Photoperiodic Stimulation of Egg Laying in the Land Snail *Helix aspersa*. Nature 212, 1582, 1966.

Stephenson, J. W.: Aldrin controlling slug and wireworm damage to potatoes. Plant Pathol. 8, 53–54, 1959.

– Slug investigations. Rothamsted Exp. Stat. Harpenden, Rep. 1960, 1961, 159.

– The molluscidal properties of three fly repellents. Plant Pathol. 11, 25–27, 1962.

– Slug parasites and predators. Rothamsted Exp. Stat. Rep. 1964, 1965, 188.

– The effect of irrigation on damage to potato tubers by slugs. Europ. Potato J. 8, 145–149, 1965.

– Notes on the rearing and behaviour in soil of *Milax budapestensis* (Hazay). J. Conchol. (London) 26, 141–145, 1966.

– The distribution of slugs in a potato crop. J. appl. Ecol. 4, 129–135, 1967.

– A review of the biology and ecology of slugs of agricultural importance. Proc. Malacol. Soc. London 38, 169–178, 1968.

– Gelatin as a Carrier for S^2-Cyanoethyl N-[(methylcarbamoyl) oxy] thioacetimidate, an Experimental Molluscicide. Pestic. Sci. 3, 81–87, 1972.

– Laboratory Observations on the Effect of Soil Compaction on Slug Damage to Winter Wheat. Plant Pathol. 24, 9–11, 1975.

– Laboratory Observations on the Distribution of *Agriolimax reticulatus* (Müll.) in Different

Aggregate Fractions of Garden Loam. Plant Pathol. 24, 12–15, 1975.
– und KNUTSON, L. V.: A Résumé of Recent Studies of Invertebrates Associated with Slugs. J. econ. Entom. 59, 356–360, 1966.
STERN, G.: Effet de la température sur la production et la consommation chez *Agriolimax reticulatus* (Müll.) en période de croissance. Bull. Ecol. 6, 501–509, 1975.
STIGLINGH, I. und VAN EEDEN, J. A.: Aspects of trematode infection in *Bulinus (Bulinus) tropicus*. Zool. Anz. 187, 61–71, 1971.
STIVA, A.: *Cernuella neglecta* (Drap.) een nieuwe soort voor Belgie. Basteria 37, 29–30, 1973.
STOREY, R.: The importance of mineral particles in the diet of *Limnaea pereger* (Müll.) J. Conchol. 27, 191–195, 1970.
STREIFF, W.: Autodifférenciation ovarienne chez un Mollusque Prosobranche hermaphrodite protandre *Calyptraea sinensis* L. C.R. Acad. Sc. (Paris) Sér. D, 263, 539–542, 1966.
– Etude endocrinologique du déterminisme du cycle sexuel chez un mollusque hermaphrodite protandre *Calyptraea sinensis* L. Ann. d'Endocrin., Paris, 27, 385–400, 1966.
– Recherches cytologiques et endocrinologiques sur le cycle sexuel de *Calyptraea sinensis* L. (Mollusque Prosobranche hermaphrodite protandre). Thèse Doct. Sc. Nat., Univ. Toulouse, 294, 1–261, 1967.
– Etude endocrinologique du déterminisme du cycle sexuel chez un mollusque hermaphrodite protandre *Calyptraea sinensis* L. II. Mise en évidence par culture in vitro de facteurs hormonaux conditionnant l'évolution du tractus femelle. Ann. Endocrin. (Paris) 28, 461–472, 1967.
– Etude endocrinologique du déterminisme du cycle sexuel chez un mollusque hermaphrodite protandre *Calyptraea sinensis* (L.). III. Mise en évidence par culture in vitro de facteurs hormonaux conditionnant l'évolution de la gonade. Ann. Endocrin. (Paris) 28, 641–656, 1967.
– Analyse expérimentale du déterminisme de la morphogenèse et du cycle du tractus génital des Prosobranches. Bull. Soc. Zool. France 95, 451–460, 1970.
– Analyse expérimentale de la différenciation sexuelle chez les Mollusques Gastéropodes. Conf. au Collège de France 1–43, 1970.
– Apports récents de la culture organotypique dans l'étude du déterminisme de la morphogenèse et du cycle du tractus génitale chez les mollusques gastéropodes hermaphrodites et gonochoriques. Ann. Biol. 9, 7–8, 9–10 und 417–426, 1970.
– und PEYRE, A.: Survie, en culture in vitro d'organes de *Calyptraea sinensis* L. (Mollusque, Prosobranche). C.R. Acad. Sci. (Paris) 256, 292–294, 1963.
– und LE BRETON, J.: Etude endocrinologique des facteurs régissant la morphogènese et la règression du pénis chez un Mollusque Prosobranche gonochorique, *Littorina littorea*. L. C.R. Acad. Sci. (Paris) 270, 547–549, 1970.
– und LE BRETON, J.: Etude comparée en culture in vitro des facteurs responsables de la morphogénèse et de la régression du tractus génital mâle externe chez deux Mollusques Prosobranches: *Crepidula fornicata* (Phil.) (espèce protandre) et *Littorina littorea* L. (espèce gonochorique). C.R. Acad. (Paris) 270, 632–634, 1970.
– LE BRETON, J., und SILBERZAHN, N.: Non spécificité des facteurs hormonaux responsables de la morphogénèse et du cycle du tractus génital mâle chez les Mollusques Prosobranches. Ann. d'Endocrin. (Paris) 31, 548–556, 1970.
STRICKLAND, A. H.: Pest control and productivity in British agriculture. 62–81. R. Soc. Arts 1965.
STRINGER, A.: A note on the action of metaldehyde on slugs. 87–88, Ann. Rep. Agric. Hortic. Res. Stat. Long Ashton 1946.
– Snails in black currant plantations. Rep. Long Ashton Res. Stat. 1967, 68.
– Snails in black currant plantations. Rep. Long Ashton Res. Stat. 1969, 68.
– LYONS, C. H., und PICKARD, J. A.: Persistence and degradation of methiocarb. 99–100. Rep. Long Ashton Res. Stat. 1969, 1970.
– und PICKARD, J. A.: The DDT content of herbage and slugs in an apple orchard at Long Ash-

ton. 172–175, Ann. Rep. Agric. Hortic, Res. Stat. Univ. Bristol, 1964.
– und MORGAN, N. G.: Snails in black currant plantations. 83-84. Rep. Long Ashton Res. Stat. 1968. 1969.
– und MORGAN, N. G.: Population and Control of Snails in Black currant Plantations. Proc. 5th Brit. Insectic. Fungic. Conf. 2, 453–457, 1969.
– und MORGAN, N. G.: Control and culture of snails. 97–98. Rep. Long Ashton Res. Stat. 1969. 1970.
– und MORGAN, N.G: Control of snails in black currant plantations. 128. Rep. Long Ashton Res. Stat. 1971, 1972.

STRUFE, R.: Bedeutung und Anwendung von Nachweismethoden für Molluskizide im Freiland. Pflanzenschutz-Nachr. „Bayer" 18. 123–129, 1965.
– Problems and results of residue studies after application of molluscicides. Residue Rev. 24, 79–168, 1968.
–, DAZO, B. C., und DAWOOD, I. K.: Feld- und Laborversuche mit Bayluscid im Bilharziasis Control Pilot Project Egypt 49, Pflanzenschutz-Nachr. „Bayer" 18, 110–122, 1965.

STURROCK, B. M. und STURROCK, R. F.: Laboratory studies of the host-parasite relationship of *Schistosoma mansoni* and *Biomphalaria glabrata* from St. Lucia, West Indies. Ann. Trop. Med. Parasit. 64, 357–363, 1970.

STURROCK, R. F.: An investigation of some factors influencing the survival of St. Lucian *Biomphalaria glabrata* deprived of water. Ann. Trop. Med. Parasit. 64, 365–371, 1970.
– Field Studies on the Transmission of *Schistosoma mansoni* and on the Bionomics of its intermediate Host, *Biomphalaria glabrata*, on St. Lucia, West Indies. Int. J. Parasit. 3, 175–194, 1973.
– Control of *Schistosoma mansoni* Transmission: Strategy for using molluscicides on St. Lucia. Int. J. Parasit. 3, 795–801, 1973.
– Persistence of the molluscicide Bayluscide (clonitralide) emulsifiable concentrate on mud surfaces in the tropics. Ann. Trop. Med. Parasit. 68, 427–434, 1974.
– und WEBBE, G.: The application of catalytic models to schistosomiasis in snails. J. Helminth. (London) 45, 189–200, 1971.
– und STURROCK, B. M.: Shell abnormalities in *Biomphalaria glabrata* infected with *Schistosoma mansoni* and their significance in field transmission studies. J. Helminth. 45, 201–210, 1971.
– und BARNISH, G.: The aerial application of molluscicides with special reference to Schistosomiasis control. Bull. Wld. Hlth. Org. 47, 283–285, 1973.
– und UPATHAM, E. S.: An investigation of the interactions of some factors influencing the infectivity of *Schistosoma mansoni* miracidia to *Biomphalaria glabrata*. Int. J. Parasit. 3, 35–41, 1973.
– BARNISH, G., und UPATHAM, E. S.: Snail findings from an experimental mollusciciding programme to control *Schistosoma mansoni* transmission on St. Lucia. Int. J. Parasit. 4, 231–240, 1974.
– BARNISH, G., und SEEYAVE, J.: Freilandversuche mit drei molluskiziden Substanzen im Bananenbau. Pflanzenschutz-Nachr. „Bayer" 27, 57–62, 1974.

STYCZYNSKA-JUREWICZ, E.: Osmotic properties of capsular fluid in eggs of some snails from fresh and brakish waters. Pol. Arch. Hydrobiol. 19, 53–64, 1972.

TARCZYŃSKI, St. HEIGELMAN, J., KOŚCIAŃSKI, M., SZEPELSKI, L., und WADOWSKI, St.: Studies on the control of *Galba truncatula* in large pastures. I. Attempt to apply planes for spraying pastures with molluscicides. Acta Parasit. Pol. 13, 367–372, 1965.

TARDY, J. Régénération de la gonade après castration chirurgicale chez quelques *Aeolidiidae* (Mollusques, Nudibranches). C.R. Séanc. Soc. Biol. 161, 2013–2016, 1967.
– Contribution à l'étude des Nudibranches. Thèse Doct. d'Etat Sci. Nat. Arch. Orig. centre document (Poitiers), CNRS 3287, 1–196, 1969.
– Organogénèse de l'appareil génital chez les Mollusques. Bull. Soc. Zool. France 95, 407–428, 1970.

– Embryologie et organogénèse sexuelle. Haliotis 1. 151–166, 1971.

– Etude expérimentale de la régénération germinale après castration chez les *Aeolidiidae*. Ann. Sc. Nat. Zool. Biol. Anim. Paris 1971, Sér. 12, XIII (1), 91–147, 1971.

– Effets de la castration chirurgicale sur le tractus génital et la ponte chez les *Aeolidiidae*; Application à la compréhension des mécanismes du controle endocrine de la sexualité. Malacologia 14, 129–133, 1973.

TAUC, L.: Physiology of the Nervous System. In: WILBUR, K. M. und YONGE, C. M.: Physiology of *Mollusca*, 2 (Chapt. 12) 387–454, Acad. Press. New York 1966.

TECHNAU, G.: Bilharziose – ein lösbares Problem. Naturwissensch. 61, 111–116, 1974.

TEESDALE, C., HADMAN, D. F., und NGURIATHI, J. N.: The Use of Continnous Low-Dosage Copper Sulfate as a Molluscicide on an Irrigation Scheme in Kenya. Bull. Wld. Hlth. Org. 25, 563–571, 1961.

TEITELMAN, G.: Protein synthesis during *Ilyanassa* development: Effect of the polar lobe. J. Embryol. Exp. Morphol. 29, 267–281, 1973.

TESHIMA, S.-I., und KANAZAWA, A.: Metabolism of progesterone in the hepatopancreas of the whelk, *Buccinum undatum*. Bull. Jap. Soc. Sci. Fish 27, 529–533, 1971.

– und KANAZAWA, A.: Biosynthesis of 7-cholesterol in the chiton, *Liolophura japonica*. Comp. Biochem. Physiol. B. Comp. Biochem. 44, 881–887, 1973.

THEEDE, H.: Vergleichende ökologisch-physiologische Untersuchungen zur zellulären Kälteresistenz mariner Evertebraten. Mar. Biol. (Berlin) 15, 160–191, 1972.

THIELE, J.: Handbuch der systematischen Weichtierkunde. 2 Bde., 4 Teile (1929–1935), Neudruck (Asher, Amsterdam, 1963).

THIELE, O. W., SCHRÖDER, H. von, und HERLYN, U.: Die Lipiden der Weinbergschnecke *(Helix pomatia* L.). Versuche zum Nachweis von androgenen Steroiden. Naturwissensch. 51, 40–41, 1964.

THOMAS, D. C.: Discussion on slugs. II.Field sampling for slugs. Ann. appl. Biol. 31, 163–164, 1944.

– Some observations on damage to potatoes by slugs. Ann. appl. Biol. 34, 246–251, 1947.

– The use of metaldehyde against slugs. Ann. appl. Biol. 35, 207–227, 1948.

THOMAS, J. D.: Schistosomiasis and the control of molluscan hosts of human schistosomes. Parasitology 11, 307–394, 1973.

THOMPSON, T. E.: Euthyneuran and other molluscan spermatozoa. Malacologia 14, 167–206, 1973.

– und BEBBINGTON, A.: Scanning electron microscope studies of gastropod radulae. Malacologia 14, 147–165, 1973.

TILGNER, P. A.: Jahreszeitliche und klimatische Schwankungen im Calcium- und Phosphat-Gehalt des Blutes von *Helix pomatia* L. Zool. Jahrb. Abt. Allg. Zool., Physiol. 67, 365–372, 1958.

TIMMERMANS, L.: Studies on shell formation in mollusks. Neth. J. Zool. 19,417–523, 1969.

– Mantle activity following shell injury in the pond snail *Lymnaea stagnalis* L. Malacologia 14, 53–61, 1973.

TOMASZEWSKI, W.: Mollusca. In: Sorauer, P.: Handbuch der Pflanzenkrankheiten, 5.A. 5 (1), 100–105. Berlin 1949.

TRAMELL, P., und CAMPBELL, J.: Arginine and urea metabolism in the South American land snail, *Strophocheilus oblongus*. Comp. Biochem. Physiol. B. Comp. Biochem. 42, 439–449, 1972.

TRELKA, D. G.: The behavior of predatory larvae of *Tetanocera plebeia (Diptera: Sciomyzidae)*, and toxicological and neurological aspects of a toxic salivary secretion used to immobilize slugs. Ph. D. Thesis, Cornell Univ., 1–125, 1972.

– und FOOTE, B. A.: Biology of slug-killing *Tetanocera (Diptera: Sciomyzidae)*. Ann. Entom. Soc. Amer. 63, 877–895, 1970.

TSCHESCHE, H., und DIETL, Th.: Identifizierung der Proteaseinhibitoren synthetisierender Organe der Weinbergschnecke *(Helix pomatia)*. Hoppe-Seyler's Z. Physiol. Chem. 353,

1189–1193, 1972.

TURK, F. A., und PHILLIPS, ST. M.: A monograph of the Slug Mite *Riccardoella limacum* (Schrank). Proc. Zool. Soc. London 115, 448–472, 1945.

TURNER, G. I.: Snail transmission of the species of *Phytophtora* with special reference to foot rot of *Piper nigrum*. Transact. Brit. Mycol. Soc. 50, 251–258, 1967.

TUZET, O., SANCHEZ, S., und PAVANS de CECCATTY, M.: Données histologiques sur l'organisation neuroendocrine de quelques mollusques gastéropodes. C.R. Acad. Sci. (Paris) 244, 2962–2964, 1957.

UBELAKER, J., und OLSEN, W.: Effect of Age of the Pelecypod Host on the Development of the Larvae of the Trematode *Phyllodistomum bufonis*. J. Invert. Pathol. 16, 367–370, 1970.

UHLENBRUCK, G., und PROKOP, D.: An agglutinin from *Helix pomatia* with reacts with terminal N-acetyl-D-galactosamine. Vox Sang. 11, 519, 1966.

– SPRENGER, J., und HERMANN, G.: Heterogeneity of polyvalent proteinase-inhibitors from certain snails demonstrated by fibrin-agar-electrophoresis. Z. Klin. Chem. Klin. Biochem. 9, 494–496, 1971.

– VAITH, P., und SCHUMACHER, K.: A novel antimitogenic principle occuring in the albumin gland of snails. Naturwiss. 59. 220–221, 1972.

UNDERWOOD, A. J.: Observations on the reproductive cycles of *Monodonta lineata, Gibbula umbilicalis* and *G. cineraria*. Mar. Biol. (Berl.) 17, 333–340, 1972.

– Spawning, larval development and settlement behaviour of *Gibbula cineraria (Gastropoda: Prosobranchia)* with a reappraisal of torsion in gastropods. Mar. Biol. (Berl.) 17, 341–349, 1972.

UNTERSTENHÖFER, G.: Mesurol – ein polyvalentes Insektizid und Akarizid. Pflanzenschutz-Nachr. „Bayer" 15, 181–194, 1962.

UPATHAM, E. S.: Interference by Unsusceptible Aquatic Animals with the Capacity of the Miracidia of *Schistosoma mansoni* Sambon under Field-Simulated Conditions in St. Lucia, West Indies. J. Helminth. 46, 277–283, 1972.

– Exposure of caged *Biomphalaria glabrata* (Say) to investigate dispersion of miracidia of *Schistosoma mansoni* Sambon in field-simulated and field habitats in St. Lucia, West Indies. J. Helminth. 46, 297–306, 1972.

– Effects of Some Physico-chemical Factors on the Infection of *Biomphalaria glabrata* (Say) by Miracidia of *Schistosoma mansoni* Sambon in St. Lucia, West Indies. J. Helminth 46, 307–315, 1972.

– und STURROK, R. F.: Preliminary trials against *Biomphalaria glabrata* of a new molluscicide formulation: gelatin granules containing Bayluscide wettable powder. Ann. Trop. Med. Parasitol. 71, 85–93, 1977.

URBAŃSKI, J.: Krajowe Slimaki i maze. 1–276. Warszawa 1957.

– und WIKTOR, A.: Beiträge zur Kenntnis bulgarischer Nacktschnecken (*Moll., Pulm.*). Systematische Studien über die Mollusken der Balkan-Halbinsel. Bull. Soc. amis. sc. lett. Poznań D 8, 47–97, 1967.

VALA, J. C.: Etude écologique du parasitsme des mollusques, hôtes intermédiaires de Trématodes. 1–174, Thèse, Univ. Sc. et. Techn. du Languedoc (Montpellier) 1973.

VALOVIRTA, I.: List of Finnish land gastropods and their distribution. Ann. Zool. Fennici 4, 29–32, 1967.

– First record of *Milax gagates* (Drap.) *(Gastropoda, Milacidae)* in Finland. Ann. Zool. Fennici 6, 345–347, 1969.

VAN BRUGGEN, A. C.: *Lymnaea columella* Say und *Zonitoides arboreus* (Say) in den Gewächshäusern des Basler Botanischen Gartens *(Moll., Gastr., Pulm.)*. Verh. Naturf. Ges. Basel 66, 1–4, 1955.

– New data on recent Dutch *Mollusca,* a critical compilation 2. Basteria 21, 74–82, 1957.

VAN DINTHER, J.: Control of *Pomacea (Ampullaria)* Snails in Rice Field's. Bull. 68, Agr. Exp. Stat. Surinam. 1–20, 1956.

– Levenswijze en bestrijding van slakken in de rijstvelden van Suriname. Meded. Landbouw-

hogesch. Opz. Stat. Gent 22, 670–676, 1957.
– Molluscs in agriculture and their control. Meded. Lab. Entom. Wageningen 232, 281–286. World Crops 11/12 1973.
– and Stubbs, R.: Summary of research on the control of rice snails in Surinam. Agric. Exp. Stat. Paramaribo, Surinam, Bull. 82, 415–420, 1963.
Van den Biggelaar, J. A. M.: RNA synthesis during cleavage of the *Lymnaea* egg. Exp. Cell. Res. 67, 207–210, 1971.
Van den Bruel, W. E.: Le problème de la Grande Douve du foie *Fasciola hepatica* L. (Trématode) pourrait-il être résolu par la lutte contre *limnaea truncatula* O. F. Müller (Mollusque, Gastéropode) dans les prairies et dans les cressonnières? Bull. Inst. Agronom. Stat. Rech. Gembloux 31, 418–422, 1963.
– Examen critique des possibilités actuelles de destruction par le cyanide calcique de *Limnaea truncatula* O.F. Müller, hôte intermédiaire de *Fasciola hapatica.* Parasitica 23, 1–21, 1967.
– Groupe de Travail pour l'Etude de la lutte contre les Limnées. Rapport d'activité 1968. 1–35, Minist. Agricult. Stat. Zool., Centre Rech. Agronom. Gembloux 1968.
– Exposé introductif „L'Assainissement des Prairies par les Traitements molluscicides". 1–66. Coll. d'Inform. Scientif. Gembloux 1970.
– und Moens, R.: Une méthode de lutte efficace, utilisable en plein champ contre les limaces. Parasitica 12, 8–15, 1956.
– und Moens, R.: A propos des propriétés hélicides de la cyanamide calcique. Comp. Rend. Séanc. Soc. Biol. C.L. 12, 2281–2282, 1956.
– und Moens, R.: Essais de destruction d'un mollusque pulmoné aquatique: *Limnaea (Radix) peregra* O.F. Müller dans les cressonnières. Parasitica 13, 97–116, 1957.
– und Moens, R.: Nouvelles observations sur les propriétés des hélicides. Bull. Inst. Agron. Stat. Rech. Gembloux 26, 281–301, 1958.
– und Moens, R.: Nouvelles observations sur la destruction des *Limnaeidae* dans les cressonnières. Parasitica 14, 89–106, 1958.
– und Moens, R.: Remarques sur les facteurs écologiques influençant l'efficacité de la lutte contre les limaces. Parasitica 14, 135–147, 1958.
– und Moens, R.: Les propriétés des hélicides et la protection des cultures. Proc. IVe Cong. Int. Crop. Prot. Hamburg 1957 (Braunschweig 1960) 2, 1255–1275, 1960.
– und Moens, R.: Des possibilités d'utilisation de la cyanamide calcique pour la lutte contre la distomatose bovine par la destruction de *Limnaea truncatula* O. F. Müller, hôte intermédiaire de *Fasciola hepatica* L. Parasitica 20, 41–70, 1964.
Van der Borght, S. C. K., und Van Puymbroeck, S.: Kinetics of the direct uptake of Ca and Sr ions from water by freshwater gastropods: Influence of temperature. Environ. Physiol. 1, 83–95, 1971.
Van der Horst, J.: Investigation of the synthesis and distribution of fatty acids in the lipids of the snail *Cepaea nemoralis* (L.). Netherl. J. Zool. 20, 1970.
– und Voogt, P. A.: Biosynthesis and composition of sterols and sterol esters in the land snail *Cepaea nemoralis* (L.). *(Gastropoda, Pulmonata, Stylommatophora)*. Comp. Biochem. Physiol. B. Comp. Biochem. 42, 1–6, 1972.
Van der Meer-Mohr, J. C.: Über Ameisen als Schneckenfeinde. Miscellanea Zoologica Sumatra 52, 1–3, 1931.
– Über Ameisen als Schneckenfeinde. Tropische Natuur 20, 119–120, 1931.
Van der Schalie, H. N.: Nembutal as a relaxing agent for mollusks. Amer. Midl. Natur. 50, 511–512, 1953.
– und Davis, G. M.: Culturing *Oncomelania* snails *(Prosobranchia: Hydrobiidae)* for studies of oriental Schistosomiasis. Malacologia 6, 321–367, 1968.
Van der Steen, W. J.: The influence of environmental factors on the oviposition of *Lymnaea stagnalis* (L.) under laboratory conditions. Arch. néerl. Zool. 17, 403–468, 1967.
– Jager, J. C., und Tiemersma, D.: The influence of food quantity on feeding, reproduction and growth in the pond snail *Lymnaea stagnalis* (L.) with some methodological comments.

Proc. K. Med. Akad. Wet. (Ser.C) 76, 47–60, 1973.

VAN MOL, J. J.: Neurosecretory phenomena in the slug *Arion rufus* (Gasteropoda: *Pulmonata*). Gen. comp. Endocrinol. 2. 616–617, 1962.

– Anatomie et physiologie de la limace rouge (*Arion rufus*). Les Naturalistes belges 43, 1–17, 1962.

– Etude morphologique et phylogénétique du ganglion cérébroide des Gastéropodes Pulmonés (Mollusques). Acad. Roy. Belg. Cl. Sc. Mém. 37 (5), 1–168, 1967.

VAN REGTEREN-ALTENA, C. O.: Notes sur les limaces. 3. Sur la présence en France d'*Arion lusitanicus* Mabille (1). J. Conch. (Paris) 95, 89–99, 1956.

– Notes on land slugs. 11. *Arionidae, Milacidae* and *Limacidae* from South Africa *(Mollusca, Gastropoda, Pulmonata)*. Zool. Meded. Rijksmus. Nat. Hist. Leiden 41, 269–298, 1966.

– Notes sur les limaces. 17. Sur deux espèces de *Deroceras*. Basteria 34, 1970, 2.

– Mitteilungen über Nacktschnecken. 18. Neue Fundorte von *Arion lusitanicus* Mabille. Arch. Moll. 101, 183–185, 1972.

VASCONCELOS, G. J., und LEE, J. S.: Microbial flora of pacific oysters *(Crassostrea gigas)* subjected to ultraviolet-irradiated seawater. Appl. Microbiol. 23, 11–16, 1972.

VASIAGINA, R. D., CHAMTL, B. D., und DOBRIANSKII, V. M.: Chromatographic determination of the molluscicide. Metaldehyde residues in water and in cabbage. Khim. Sel. SK Khoz. 14, 17, 1976.

VASIL'EV V. P.: [Schädliche Nematoden, Mollusken, Gliederfüßler]. In: Schädlinge Land und Forstwirtschaftlicher Kulturpflanzen 1. Kiew, (russ.) 1973.

VAUGHN, Ch. M.: Effects of Temperature on hatching and growth of *Lymnaea stagnalis appressa* Say. Amer. Middl. Naturalist 49, 214–228, 1953.

VERBEKE, J.: Contribution à l'étude des diptères malacophages. II. Données nouvelles sur la taxonomie et la répartition géographique des Sciomyzidae paléarctiques. Bull. Inst. roy. Sci. nat. Belg. 36, 1–15, 1964.

VERDCOURT, B.: The Zoogeography of the non-marine Mollusca of East-Africa. J. Conchol. 27, 291–348, 1972.

VERSCHUUREN, H. G., KROES, R., TONKELAAR, E. M. DEN, BERKWENS, J. M., HELLEMAN, P. W., und ESCH, G. J. VAN: Long-Term Toxicity and Reproduction Studies with Metaldehyde. Toxicology 4, 97–115, 1975.

VIANEY-LIAUD, M.: Etude du contrôle de la maturité des tractus génitaux de la ponte par castration chirurgicale chez le Planorbe *Australorbis glabratus* (Say) (Pulmoné, Basommatophore). Bull. Soc. Zool. Fr. 97, 675–690, 1972.

– Influence du jeûne sur la production d'oeufs et la ponte chez le Planorbe *Australorbis glabratus* (Gastéropode, Pulmoné). Bull. Soc. Zool. Fr. 98, 349–361, 1973.

– La variation pondérale au cours de la croissance de la limace *Milax gagates* Drap. (Limacidés, Milacinés). Arch. Zool. exp. gén. 116, 5–25, 1975.

VICENTE-PICCHIOCHI, M.-M.: Contribution à l'étude corrélative des infrastructures et des phosphatases au cours de la spermiogenèse chez *Rana* et *Helix*. 1–80. Thèse Doct. Biol. Anim., Spéc. Physiol. Cell. Paris 1967.

VIETS, K., und PLATE, H.-P.: Die ökologischen (parasitologischen) Beziehungen zwischen Wassermilben *(Hydrachnellae, Acari)* und Süßwasser-Mollusken. Z. ang. Ent. 35, 459–494, 1954.

VIK, R.: Vinbergsneglen overvintrer og formerer seg i det sørøstlige Norge. Fauna (Oslo) 24 (1), 1968.

VILLIERS, J. P., und MACKENZIE, J. G.: Structure and activity in Molluscicides: The Phenacyl Halides, a group of potentially useful molluscicides. Bull. Wld. Hlth. Org. 29, 424–427, 1963.

– und ROSSOUW, M. M.: Structure and activity in molluscicides. Nature (Lond.) 213, 1208–1209, 1967.

VILPO, J. A., ISO-HEIKKILA, P., und TEIR, H.: Elimination of India Ink phagocytizing cells in the fresh water clam. Exp. Pathol. (Jena) 6. (1/2) 12–26, 1972.

Visser, G. J. M.: Malacologische mededelingen van het eiland Texel: 2. Basteria 30, 49–52, 1966.

Voelker, J.: Der chemische Einfluß von Kalziumkarbonat auf Wachstum, Entwicklung und Gehäusebau von *Achatina fulica* Bowdich *(Pulmonata)*. Mitt. Hamburg. Zool. Mus. Inst. 57, 37–78, 1959.

– Wasserwanzen als obligatorische Schneckenfresser im Nildelta *(Limnogeton fieberi* Mayr: *Belostomatidae, Hemiptera)*. Z. Tropenmed. Parasitol. 17, 155–165, 1966.

– Untersuchungen zur Ernährung, Fortpflanzungsbiologie und Entwicklung von *Limnogeton fieberi* Mayr *(Belostomatidae, Hemiptera)* als Beitrag zur Kenntnis von natürlichen Feinden tropischer Süßwasserschnecken. Entom. Mitt. Zool. Staatsinst., Zool. Mus. Hamburg 3 (60), 1–24, 1968.

– Schnecken als Zwischenwirte von Helminthen. In: Nauck, E. G.: Lehrbuch der Tropenkrankheiten. Georg Thieme, Stuttgart 1975, 4. Aufl.

Vogel, H.: Über eine Dauerzucht von *Oncomelania hupensis* und Infektionsversuche mit *Bilharzia japonica*. Z. Parasitenk. 14, 70–91, 1948.

Voogt, P. A.: Biosynthesis of 3ß-sterols in a snail *Arion rufus* L. from 1-^{14}C-Acetate. Arch. Int. Physiol. Biochim. 75, 492–500, 1967.

– Sterols of some prosobranchs. 391–400. Arch. Int. Physiol. Biochim. 1971.

– Analysis of the sterols of the pulmonates *Arion rufus, Arianta arbustorum* and *Succinea putris*. Neth. J. Zool. 22, 489–496, 1972.

– Investigations of the capacity of synthesizing 3ß-sterols in *Mollusca*: VI. The biosynthesis and composition of 3ß-sterols in the neogastropods *Purpura lapillus* and *Murex brandaris*. Comp. Biochem. Physiol. B. Com. Biochem. 41, 831–841, 1972.

– A gaschromatographic analysis of the sterols of *„Lymnaea stagnalis“* and *„Planorbarius corneus“*. Arch. Int. Physiol. Biochim. 80, 697–704, 1972.

– und Van der Horst, D. J.: Investigations of the capacity of synthesizing 3ß-sterols in *Mollusca*: VIII. The biosynthesis of 3ß-sterols in the pulmonates *Succinea putris* and *Arianta arbustorum*. Arch. Int. Physiol. Biochim. 80, 293–297, 1972.

Vulfius, E. A., und Yurchenko, O. P.: [Effect of anticholinesterase substances on the effect of acetylcholine and its analogues in neurons of gastropods]. Dok. Akad. Nauk SSSR, Ser. Biol. 205, 1254–1257. (russ.), 1972.

Waerden, B. L., und Nievergelt, E.: Tafeln zum Vergleich zweier Stichproben mittels X-Test und Zeichentest. Berlin 1956.

Wagge, L. E.: Quantitative studies of calcium metabolism in *Helix aspersa*. J. Exp. Zool. 120, 311–342, 1952.

– und Mittler, T.: Shell Regeneration in some British Mollusca. Nature (London) 171, 528, 1953.

Wagner, E. D. und Chi, L. W.: Methods on the rearing of the snail, *Oncomelania* spp. Trans Am. Microsc. Soc. 78, 421–423, 1959.

Wain, R. L.: 3,5-dihalogeno-4-hydroxy-benzonitriles: New herbicides with molluscicidal activity. Nature 200, 28, 1963.

– Ioxynil – a new Herbicide and Molluscicide. Meded. Landbouwhogesch. Opz. Gent 29, 628–632, 1964.

Waldén, H. W.: The land gastropoda of the vicinity of Stockholm. Arch. för Zoologi 7, 391–448, 1954.

– Om ett par för Sverige nya, anthropochora landmollusker *Limax valentianus* Férussac och *Deroceras caruanae* (Pollonera), jämte några andra, kulturbundna arter. [On two anthropochorous terrestrial gastropods, *Limax valentianus* Férussac and *Deroceras caruanae* (Pollonera) found in Sweden, with notes on some further species, living under synanthropous conditions]. Göteborgs Kungl. Vetensk. Vitterhets Samhalles Handl. Sjatte Foljden Ser. B. 8 (8), 5–48, 1960.

– On two anthropochorous terrestrial Gastropods, *Limax valentianus* Férussac and *Deroceras caruanae* (Pollonera), found in Sweden. Göteborgs Kungl. Vet. Vitterh. Samh. Handl. F. 6,

Ser. B. 8., 42–47, 1960.
– On the variation, nomenclature, distribution and taxonomical position of *Limax (Lehmannia) valentianus* Férussac *(Gastropoda, Pulmonata)*. Arch. Zool. Ser. 2., 15, 1961.
– Terrestrial faunistic studies in Sweden. Proc. Ith Europ. Malac. Congr. (1962), 95–109, 1965.
– Einige Bemerkungen zum Ergänzungsband zu Ehrmann's „Mollusca" in „Die Tierwelt Mitteleuropas". Arch. Moll. 95, 49–68, 1966.
– Recent advances in land mollusc research in Sweden. Malacologia 9, 297–299, 1969.
– A nomenclatural list of the land *mollusca* of the british isles. J. Conch. 29, 21–25, 1976.

WALKER, G.: The digestive system of the slug, *Agriolimax reticulatus* (Müller): Experiments on phagocytosis and nutrient absorption. Proc. Malacol. Soc. (London) 40, 33–43, 1972.

WALKER, J. O.: A trial to assess the consequences of feeding chickens with methiocarb-poisoned slugs and methiocarb pellets. Baywood Internal Report No. T.R. 541 vom 5.4.1968.

WALKER, R., RALPH, K., WOODRUFF, G., und KERKUT, G.: Evidence for acetylcholine and 5-hydroxytryptamine as excitatory transmitters on to a single neuron in the brain of the snail, *Helix aspersa.* Comp. Gen. Pharmacol. 3, 52–60, 1972.

WALLACE, G. D., und ROSEN, L.: Experimental Infection of Pacific Island Mollusks with *Angiostrongylus cantonensis.* Am. J. Trop. Med. Hyg. 18, 13–19, 1969.
– und ROSEN, L.: Studies on eosinophilic meningitis. V. Molluscan hosts of *Angiostrongylus cantonensis* on Pacific islands. Am. J. Trop. Med. Hyg. 18, 1969.

WALTER, H. J.: Evolution, Taxonomy revolution and Zoogeography of the *Lymnaeidae*. Amer. Malacol. Union Ann. Rep. 35, 18–20, 1968.

WALTON, B. C., WINN, M. M., und WILLIAMS, J. E.: Development of resistance to molluscicides in *Oncomelania nosophora*. Amer. J. Trop. Med. Hyg. 7, 618, 1958.

WAREBORN, I.: Environmental factors influencing the distribution of land molluscs of an oligotrophic area in southern Sweden. Oikos 21, 285–291, 1970.

WARLEY, A. P.: Some aspects of the biology, ecology and control of slugs in S.E. Scotland with particular reference to the potato crop. Ph. D. Thesis, University of Edinburgh 1970.
– and WRIGHT, A. N.: A large-scale snail control trial with Trifenmorph in the Gezira irrigation scheme, Sudan. Bull. WHO. 54, 573–585, 1977.

WARREN, K. S., und WEISBERGER, A. S.: The treatment of molluscan schistosomiasis mansoni with Chloramphenicol. Amer. J. Trop. Med. 15, 342–350, 1966.
– und WEISBERGER, A. S.: Molluscan schistosomiasis mansoni: Effect of 2 analogues of Chloramphenicol on both parasite and host. Proc. Soc. exp. Biol. Med. 124, 789–791, 1967.

WATTEZ, Ch.: Influence des tentacules oculaires dans la différenciation génitale chez *Arion subfuscus* Draparnaud (Mollusque Gastéropode Pulmoné). C.R. HEBD Séances Acad. Sci. Sér. D. Sci. Nat. 274, 2328–2331, 1972.

WATTS, A. H. L.: Spermatogenesis in the slug *Arion subfuscus*. J. Morph. 91,53–77, 1952.

WATTS, St. D. M.: Effects of larval digenea on the free amino acid pool of *Littorina littorea* (L.). Parasitol. 62, 361–366, 1971.

WEBBE, G., und JAMES, C.: Host-parasite Relationships of *Bulinus globosus* and *B. truncatus* with Strains of *Schistosoma haematobium.* J. Helminth. 46, 185–199, 1972.

WEBLEY, D.: Experiments with slug baits in S. Wales. Ann. appl. Biol. 50, 129–136, 1962.
– Experiments with slug baits in 1959. Plant Pathol. 12, 19, 1963.
– Slug activity in relation to weather. Ann. appl. Biol. 53, 407–414, 1964.
– Aspects of trapping slugs with metaldehyde and bran. Ann. appl. Biol. 56, 37–54, 1965.
– Waterproofing of Metaldehyde on bran baits for slug control. Nature (London) 212, 320–321, 1966.
– A comparison of methiocarb and metaldehyde baits for the control of four species of slugs. Proc. 5. Brit. Insecticide-Fungicide Conf. 2, 442–444, 1969.
– Observations on the effects of distribution and numbers of slug pellets on the catch of slugs. Ann. appl. Biol. 66, 347–352, 1970.

WEINBACH, E. C., und NOLAN, M. O.: The effect of pentachlorophenol on the metabolism of the snail *Australorbis glabratus*. Expt. Parasitol. (New York) 5, 276–284, 1956.

WEINMANN, W. D.: Über die Kontamination der Umwelt mit persistenten Insektiziden. Nachrichtenbl. Dt. Pflanzenschutzd. (Braunschweig) 21, 86–92, 1968.

WEISS, J. C.: La lutte anti-limaces historique et évolution. Haliotis 3, 107–111, 1973.

WEISS, M.: Zur embryonalen und postembryonalen Entwicklung des Mitteldarms bei Limaciden und Arioniden *(Gastropoda, Pulmonata)*. Rev. Suisse Zool. 75, 157–225, 1968.

WELLS, G. P.: The water relations of snails and slugs. III. Factors determining activity in *Helix pomatia* L. J. exp. Biol. 20, 79–87, 1944.

WELLS, M. J., und WELLS, J.: Optic glands and the state of the testis in *Octopus. Mar. Behav. Physiol. 1, 71–83, 1972.*

WELSH, J. H.:Neurohormons of mollusca. Amer. Zool. 1, 267–272, 1961.

WENDELAAR BONGA, S. E.: Neuroendocrine involvement in osmoregulation in a freshwater mollusc, *Lymnaea stagnalis*. Gen. Comp. Endocrin. (Suppl.) 3, 308–316, 1972.

WESTER, R. E., GOTH, R. W., und WEBB, R. E.: Transmission of downy mildew *(Phytophtora phaseoli)* of lima beans by slugs. Phytopathol. 54, 749, 1964. Abstracts.

WHITE, A. R.: Observation on slug activity in a Northumberland Garden. Plant Pathol. 8, 62–68, 1959.

WHITE, J. C.: On the ecology of an adult digenetic trematode *Proctoeces subtenuis* from a lamellibranch host *Scrobicularia plana*. J. Mar. Biol. Assoc. UK 52, 457–467, 1972.

WHITTEN, M. J., und FOSTER, G. G.: Genetical Methods of Pest Control. Ann. Rev. Entomol. 20, 461–476, 1975.

WHO, Molluscicides – Second report of the expert committee on bilharziasis. Wld. Hlth. Org. 214, 1–50, 1961.

WIKTOR, A.: Description of *Boettgerilla vermiformis* Wiktor (*Mollusca, Pulmonata*). Bull. Soc. Amis Sci. Lettr. Poznań D 1, 151–156, 1960.

– Eine neue Nacktschneckenart *(Gastropoda, Limacidae)* aus Polen. Ann. Zool. 23, 449–457, 1966.

– [Die für Polen neuen Nacktschnecken von den Familien *Limacidae* und *Milacidae* (*Gastropoda*)]. (poln.) Fragm. Faunist. 13, 323–335, 1967.

– Die Nacktschnecken Polens *Arionidae, Milacidae, Limacidae (Gastropoda, Stylommatophora)*. Monogr. Fauny Polski 1, 1–289, Polska Akad. Nauk, Krakow 1973.

– Verbreitung der Nacktschnecken *(Arionidae, Milacidae, Pulmonata, Gastropoda)* in Bulgarien. Malacologia 16, 291–294, 1977.

– und LIKHAREV, I. M.: Phylogenetische Probleme bei Nacktschnecken aus den Familien *Limacidae* und *Milacidae (Gastropoda, Pulmonata)* Malacologia 18, 123–131, 1979.

WILBUR, K. M.: Shell formation and regeneration. In: Wilbur, K. M. und Yonge, C. M.: Physiology of *Mollusca* 1 (Chapt. 8) 242–282, Acad. Press., New York 1964.

– und OWEN, G.: Growth. In: WILBUR, K. M. und YONGE, C. M.: Physiology of *Mollusca* 1 (Chapt. 7) 211–242, Acad. Press. New York 1964.

– und YONGE, C. M.: Physiology of *Mollusca*. Bd. 1 1–473, 1964. Bd. 2, 1–645, 1966. Akad. Press. New York.

WILKINS, L. A., und GREENBERG, M. J.: Effects of acetylcholine and 5-hydroxytryptamine and their ionic mechanisms of action on the electrical and mechanical activity of molluscan heart smooth muscle. – Comp. Biochem. Physiol. A., Comp. Physiol. 45, 637–651, 1973.

WILKINSON, A. T. S.: Control of Slugs. Canada Departm. Agric. 1213, 1–4, 1964.

WILSON, G. L., und SAMSON, K. S.: The incidence of Fascioliasis of Sheep and Cattle in the Southwest with Observations on the Snail Vectors. Proc. Helminth. Soc. Washington 38, 52–56, 1971.

WINDOM, H., SMITH, R., und BOUSQUET, F.: Distribution of K, Mg, Sr, Fe, Mn and Zn in *Crassostrea virginica* shells. Southeastern Geol. 13, 275–281, 1971.

WINFIELD, A. L., WARDLOW, L. R., und SMITH, B. F.: Further observations on the susceptibility of maincrop potato cultivars to slug damage. Plant Pathol. 16, 136–138, 1967.

WINKLER, L. R., und CHI, L. W.: Defensive mechanisms of the Schistosome snail host *Oncome-*

lania formosana against copper sulfat in adult and egg stages. Amer. J. Trop. Med. Hyg. 13, 897–907, 1964.

WIRTH, W.: *Platygymnopa*, a new genus of *Ephydridae* reared from decaying snails in north America (*Diptera*). Canad. Entomologist 103, 266–270,1971.

WIUM-ANDERSEN, G.: Electrophoresis as a support for the identification of various african *Biomphalaria*. Malacologia 14, 287–289, 1973.

WOLDA, H.: Some preliminary observations on the distribution of the various morphs within natural populations of the polymorphic landsnail *Cepaea nemoralis* (L.). Arch. Néer. Zool. 16, 280–292, 1965.

– The effect of drought on egg production in *Cepaea nemoralis* (L.). Arch. Néer. Zool. 16, 387–399, 1965.

– The effect of temperature on reproduction in some morphs of the landsnail *Cepaea nemoralis* (L.). Evolution 21, 117–129, 1967.

– Variation in growth rate in the landsnail *Cepaea nemoralis*. Res. Pop. Ecol. 12, 185–204, 1970.

– Ecological variation and its implications for the dynamics of population of the landsnail *Cepaea nemoralis*. Proc. Adv. Study Inst. Dynamics Numbers Popul. (Oosterbeek), 98–108, 1970.

– Ecology of some experimental populations of the landsnail *Cepaea nemoralis* (L.). I. Adult numbers and adult mortality. Netherl. J. Zool. 22, 428–455, 1972.

– Changes in shell size in some experimental populations of the landsnail *Cepaea nemoralis* (Linnaeus). Argamon Israel J. Malac. (Haifa) 3, 63–71, 1972.

– ZWEEP, A., und SCHUITEMA, K. A.: The role of food in the dynamics of populations of the landsnail *Cepaea nemoralis*. Oecologia (Berlin) 7, 361–381, 1971.

– und KREULEN, D. A.: Ecology of some experimental populations of the landsnail *Cepaea nemoralis* (L.). II. Production and survival of eggs and juveniles. Netherl. J. Zool. 23, 168–188, 1973.

WOLFF, ET., und HAPPEN, K.: Sur une méthode de culture d'organes embryonnaires in vitro. Texas Rep. Biol. and Med. 10, 463–472, 1962.

WOLFF, H. G.: Schwerkraft-Orientierung von Schnecken und Tintenfischen. Umschau 77. 22–24, 1977.

WOUTERS, L.: Schneckenbekämpfung in Oost-Flevoland. Pflanzenschutz-Nachr. „Bayer" 23, 173–177, 1970.

WRIGHT, C. A.: Some views on biological control of trematode diseases. Trans. Roy. Soc. trop. Med. Hyg. 62, 320–324, 1968.

– Ecology of life cycles. In: Flukes and Snails. Allen and Urwin Ltd. London 1971.

WÜTHRICH, M.: *Oxychilus alliarius* neben *O. helveticus* in der Schweiz. Arch. Moll. 92, 131–135, 1963.

YARINSKY, A., und FREELE, H.: A comparison of molluscicidal and mollusc inhibitory activity of hycanthone and lucanthone and the effect of the drugs on the development of *Schistosoma mansoni* in the snail intermediate host *Australorbis glabratus*. J. Trop. Med. Hyg. 73, 23–27, 1970.

YASURAOKA, K., HOSAKA, Y., und KOMIYA, Y.: Some laboratory investigations on Yurimin P-99 as an experimental molluscicide. Jap. J. Parasit. 17, 376–381, 1968.

– und HOSAKA, Y.: The problem of resistance of *Oncomelania* snail to sodium pentachlorophenate. Jap. J. Med. Sci. Biol. 24, 393–394, 1971.

YATES, R. L., und RAWLS, H. C.: The anatomical distribution of fluorescence caused by *Pseudomonas* in *Anguispira kochi* (Pfeiffer). Sterkiana 45, 14–20, 1972.

YEO, D.: Non-Ovicidal Molluscicides in the Control of Bilharziasis. Bull. Wld. Hlth. Org. 33, 144–146, 1965.

YEO, R. R., und FISHER, T. W.: Progress and potential for biological weed control with fish, pathogens, competitive plants and snails. 450–463. FAO. Int. Conf. on Weed Control, Davis, Calif. 1970.

YOM-TOV, Y.: The effect of predation on population densities of some desert snails. Ecology 51, 907–911, 1970.

YORK, B., und TWAROG, B. M.: Evidence for the release of serotonin by relaxing nerves in molluscan muscle. Comp. Biochem. Physiol. A. Comp. Physiol. 44, 423–430, 1973.

YOSHIOKA, S., MAKINO, N., und YAMASHITA, E.: The study on sex-phenomenon in Pulmonata. II. Sexual vicissitudes in hermaphroditic gland of the Slug, *Limax flavus* L. Zool. Mag. Japan, 72, 177–182, 1963.

ZAJÍČEK, D.: Snails as the second intermediate hosts of the trematode *Plagiorchis muris* Tanabe, 1922, *(Trematoda: Plagiorchidae)*. Věstn. čsl. Společ. zool. 35, 75–78, 1971.

ZAKARIA, H.: *Heteropneustes fossilis* (Bloch), a possible agent for the biological control of the snail hosts of schistosomes. Ann. Trop. Med. Parasitol. 57, 157, 1963.

ZATRAVKIN, M. N.:Principles of russian names of mollusks (superfamily) and lower (russ.). Vestn. Zool. 6, 88, 1976.

ZHADIN, V. I.: Mollusks of fresh and brackish waters of the USSR. Acad. Sciences USSR, Moskva-Leningrad 1952; (Engl. transl. Jerusalem 1965.).

ZILCH, A., und JAECKEL, S. G. A.: Die Weichtiere (Mollusca) Mitteleuropas. 1. Ergänzungen und Berichtigungen zur Nomenklatur und Systematik in P. Ehrmann's Bearbeitung. 2. Ergänzungen und Berichtigungen zum rezenten und quartären Vorkommen der mitteleuropäischen Mollusken. In: (Herausg.) BROHMER, EHRMANN, und ULMER: Die Tierwelt Mitteleuropas 2 (1), 1–294, Quelle und Meyer, 1962.

ZSOLNAI, T.: Die molluskizide Wirkung von Thiol-Reagentien. Zbl. Bakteriol. Parasitenk. Infektionskr. Hyg. (1.Abt.-Orig.) 213, 270–283, 1970.

ZYLSTRA, U.: Histochemistry and ultrastructure of the epidermis and the subepidermal gland cells of the freshwater snails *Lymnaea stagnalis* and *Biomphalaria pfeifferi*. Z. Zellforsch. Mikrosk. Anal. 130, 93–134, 1972.

– Distribution and ultrastructure of epidermal sensory cells in the freshwater snails *Lymnaea stagnalis* and *Biomphalaria pfeifferi*. Neth. J. Zool. 22, 283–298, 1972.

Additional References for the English Edition

BURCH, j. B.: Genera and subgenera of recent freshwater gastropods of North America (north of Mexico). Malacological Review 12 (1–2), 97–100, 1979.

CHEYMOL, J. P., CHARRIER, P., THUONG, N. T., THIZY, A., DEMOZAY, D., and PILLON, D.: Nouveaux organophosphores molluscicides. – Comp. Rend. Sér. D. 281, Nr. 21, 1633–1635, 1975.

FITZPATRICK, G.: Impact of Temefos and other mosquito larvicides on the Salt Marsh Snail *Melampus bidentatus* Say (Basommatophora: Ellobiidae). – Diss. Abstr. Intern. B. 36, Nr. 5, 2045–2046, 1975.

KERNEY, M. P., and CAMERON, R. A.: A Field Guide to the Land Snails of Britain and North-west Europe. Colling, London 1–287, 1979.

MEAD, A. R.: Economic Malacology with particular reference to *Achatina fulica*. In: Pulmonates, Bd. 2 B, ed. Fretter, V. und Peake, J. Academic Press, 150 S., 1979.

MIENIS, H. K.: *Theba pisana* in droppings of the Stone Curlew in Israel. Argamon, Israel J. Malac. 6, 3–4, 61–63, 1978.

MIENIS, H. K.: A Further Note on *Zonitoides arboreus* and Orchids. The Conchologist's Newsletter, 73, 230, 1980.

MIENIS, H. K.: The Ground Beetle *Carabus impressus* Feeding on Snails in Israel. The Conchologist's Newsletter, 75, 279–281, 1980.

MIENIS, H. K.: New records of predation on snails by the Hooded Crow in Israel. Levantina No. 28, 331–332, 1980.

MIENIS, H. K.: Two additional cases of predation on land snails by the Song Thrush in Israel. Levantina No. 29, 341–343, 1980.

MIENIS, H. K.: Land snails in faecal pellets of the Golden Sking. Basteria 43 (5–6), 106, 1979

MOENS, R.: Au sujet de la lutte biologique contre *Lymnaea truncatula* Müller (Gastropoda), hôte intermédiare de *Fasciola hepatica* L. (Trematoda): Bull. Rech. agron. Gembloux, 15 (3), 227–238, 1980.

MOENS, R.: Lutte contre les gastéropodes en serre. In: Gembloux Rapport d'Activité 1977, 166 pp., 1977.

MOENS, R., and FRASELLE, J.: La lutte contre les limaces en cultures d'orchidées. Med. Fac. Landbouw. Rijksuniv. Gent. 45 (3), 613–625, 1980.

NASSI, H.: Rédies à localisation cérébrale chez *Biomphalaria glabrata* parasite par *Ribeiroia marini guadeloupensis*. Coincidence entre cette localisation et l'arrêt de la ponte du mollusque dans la phase initiale de l'invasion parasitaire. Haliotis, 10 (2), 107, 1980.

REICHHOLF, J.: Gefleckte Schnirkelschnecken *Arianta arbustorum*, als Nahrung der Singdrossel, *Turdus philomelos*, im Auwald: Größen- und Typenwahl. Bonn. zool. Beitr. 30 (3/4), 404–409, 1979.

PLATES

1

2

3

4

5

6

Plate I. *1 Limax flavus* L., length 80–100 mm. *2 Limax maximus* L., light and dark animal, length 100–150 mm. *3 Lehmannia valentiana* (Fér.), length 60–70 mm. *4 Deroceras reticulatum* (Müller), length about 50 mm. *5 Arion rufus* (L.), red animal, length 120–150 mm. *6 Arion rufus* (L.), black animal, length 120–150 mm.

Plate II. *1 Arion lusitanicus* Mabille, length 70–100 mm. *2 Milax sowerbyi* (Fér.), length 60–70 mm. *3 Milax gagates* (Drap.), length 45–55 mm. *4 Arion hortensis* Fér., length 25–30 mm. *5 Milax budapestensis* (Hazay), length 60–70 mm. *6 Boettgerilla pallens* Simroth, length 40–45 mm

Plate III. *1 Helix pomatia* L., shell 38–45 mm high and broad. *2 Cepaea nemoralis* (L.), shell 17–18 mm high, 22–23 mm broad. *3* Difference between *Cepaea hortensis* (Müller), shell 15–16 mm high, 19–21 mm broad (*left*) and *Cepaea nemoralis* (L.), shell 17–18 mm high, 22–23 mm broad (*right*), dorsal view. *4* Difference between *Cepaea hortensis* (Müller) (*left*) and *Cepaea nemoralis* (L.), (*right*), ventral view (umbilical area), shell dimensions as in *3*. *5 Helicigona arbustorum* (L.), shell 20–21 mm high, 18–25 mm broad, dorsal view. *6 Helicigona arbustorum* (L.), shell 20–21 mm high, 18–25 mm broad, ventral view (umbilical area)

Plate IV. *1 Helix aperta* Born, shell about 20 mm high, about 28 mm broad. *2 Pomatias elegans* (Müller), shell about 15 mm high, about 10 mm broad. *3 Theba pisana* (Müller), shell about 15 mm high, 18 mm broad, ventral view (umbilical area). *4 Theba pisana* (Müller), lateral view, shell dimensions as in *3*. *5 Achatina fulica* Bowdich, shell 150–200 mm high, 40–50 mm broad, length of animal 10–12 cm. *6 Achatina fulica* Bowdich, mouth of shell

Plate V. *1 Cernuella virgata* (DaCosta), shell very variable, about 11 mm high, 15 mm broad. *2 Trichia hispida* (L.), shell 4–5 mm high, about 8 mm broad, dorsal view. *3 Helicella obvia* (Hartmann), shell about 8 mm high, about 16 mm broad, ventral and lateral views. *4 Helicella obvia* (Hartmann), shell about 8 mm high, about 16 mm broad, dorsal view. *5 Helicella itala* (L.), shell about 7 mm high, about 15 mm broad, ventral and lateral views. *6 Helicella itala* (L.), shell about 7 mm high, about 15 mm broad, dorsal view. *7 Discus rotundatus* (Müller), shell 2.5–3 mm high, 6–7 mm broad, ventral view (umbilical area). *8 Discus rotundatus* (Müller), shell 2.5–3 mm high, 6–7 mm broad, dorsal view

Plate VI. *1 Zonitoides arboreus* (Say), shell 2–2.5 mm high, 3.5–5 mm broad, dorsal view. *2 Zonitoides arboreus* (Say), shell 2–2.5 mm high, 3.5–5 mm broad, ventral view (umbilical area). *3 Succinea putris* (L.), shell 16–22 mm high, 8–11 mm broad. *4 Planorbarius corneus* (L.), shell 11–13 mm high,27–33 mm broad. *5 Biomphalaria glabrata* (Say), shell in one plane, at least 12–15 mm, adults more than 20 mm broad, ventral view (umbilical area). *6 Bulinus truncatus* (Audouin), shell 8–10 mm high, 6–8 mm broad, lateral view. *7 Lymnaea stagnalis* (L.), shell 48–60 mm high, 22–30 mm broad. *8 Lymnaea stagnalis* (L.), shell 48–60 mm high, 22–30 mm broad, shell mouth

Plate VII. *1* Feeding damage by *Radix peregra* (Müller) to watercress, damaged sucking roots. *2* Slug feeding damage to strawberries (fruit). *3* Feeding damage by *Limax maximus* L. to tomatoes (fruit). *4* Field Slugs on green bean pods. *5* Field Slug damage to bean seedlings. *6* Damage by *Deroceras reticulatum* (Müller) to wheat grains. *7* Deformed seedlings from wheat seeds damaged by *Deroceras reticulatum* (Müller)

Index of Scientific Names of Molluscs, Their Pathogens, Parasites, and Predators

Subject Index

E. Bresler, B. L. McNeal, D. L. Carter

Saline and Sodic Soils

Principles - Dynamics - Modeling

1982. 78 figures. X, 236 pages (Advanced Series in Agricultural Sciences, Volume 10). ISBN 3-540-11120-4

Saline and Sodic Soils is a comprehensive exposition of the principles and processes involved in the genesis, formation and reclamation of saline and sodic soils. The coverage includes critical interpretations of models characterizing the physical and chemical behavior of salt concentration and composition and their effects on soils and plants, as well as practical suggestions for the control of soil salinity to improve economic potential.
Each topic in this volume is clearly explained and readily accessible for all students and professionals in the agricultural and environmental sciences with a general background in mathematics, physics, chemistry and biology.

H. C. Coppel, J. W. Mertins

Biological Insect Pest Suppression

1977. 46 figures, 1 table. XIII, 314 pages (Advanced Series in Agricultural Sciences, Volume 4). ISBN 3-540-07931-9

"...*Another* book on biological control? Yes and no. The title of this book gives the clue that the authors have a somewhat different viewpoint regarding "biological control." The authors restricted their discussion to those cases in which insect pests are suppressed through human intervention... it appears to be most suitable for a general course in insect pest suppression that excludes the use of pesticides; ...As an introductory level textbook it has its virtues - among them its compactness. The book has **1199 references,** of which a number are as recent as 1976. The book is **attractively printed** and should be a useful introduction for students of the various alternatives to chemical insect pest suppression."

The Quarterly Review of Biology

J. J. Hanan, W. D. Holley, K. L. Goldsberry

Greenhouse Management

1978. 283 figures, 117 tables. XIV, 530 pages (Advanced Series in Agricultural Sciences, Volume 5). ISBN 3-540-08478-9

"...*Greenhouse Management* strongly emphasizes the mechanical and engineering information backgrounding our greenhouse construction and management practices. It is especially strong in its descriptions and discussions of the equipment needed for greenhouse environmental control. ...The Table of Contents is unusually detailed and will be invaluable to the reader trying to locate specific information. ...The abundant illusttrative material used includes **many photographs of excellent quality.** ...We truly believe that most readers will appreciate the monumental task that has been accomplished in gathering together and organizing such a large mass of data and information not previously available to the general greenhouse manager, floriculture student, or research specialist. It will surely serve as **an excellent sourcebook for a long time to come..."**

Journal of Hort. Science

J. Palti

Cultural Practices and Infectious Crop Diseases

1981. 43 figures. XVI, 243 pages (Advanced Series in Agricultural Sciences, Volume 9). ISBN 3-540-11047-X

The prevention of plant disease is generally not a major consideration in crop planning, sowing, fertilization and harvesting. Although these and other agricultural practices have long been recognized as a major component in integrated disease control, far more research has been invested in resistance breeding and the application of chemical agents, leaving growers to rely on experience and tradition rather than on sound scientific principles in the conduct of successful antipathogenic agriculture.
Cultural Practices and Infectious Crop Diseases is the first monograph to deal comprehensively with the effects of agricultural practices on plant health. Following the effects of background factors (climate, soil, stress, and crop age), the author describes the impact of the major farming operations on the development of crop diseases, among them crop planning and alternation, multiple cropping, tillage, fertilization, moisture management, sowing, harvesting, and sanitation. The author concludes with a consideration of integrated disease control combining agricultural practices, resistance breeding and the use of chemical agents.
This volume will prove an invaluable aid to agricultural advisors, growers, researchers, students and teachers in appreciating the importance of appropriate practices management in the prevention of crop disease.

Springer-Verlag
Berlin
Heidelberg
New York